Steel-Rolling Technology

MANUFACTURING ENGINEERING AND MATERIALS PROCESSING

A Series of Reference Books and Textbooks

SERIES EDITORS

Geoffrey Boothroyd

*Chairman, Department of Industrial
and Manufacturing Engineering
University of Rhode Island
Kingston, Rhode Island*

George E. Dieter

*Dean, College of Engineering
University of Maryland
College Park, Maryland*

1. Computers in Manufacturing, *U. Rembold, M. Seth and J. S. Weinstein*
2. Cold Rolling of Steel, *William L. Roberts*
3. Strengthening of Ceramics: Treatments, Tests and Design Applications, *Henry P. Kirchner*
4. Metal Forming: The Application of Limit Analysis, *Betzalel Avitzur*
5. Improving Productivity by Classification, Coding, and Data Base Standardization: The Key to Maximizing CAD/CAM and Group Technology, *William F. Hyde*
6. Automatic Assembly, *Geoffrey Boothroyd, Corrado Poli, and Laurence E. Murch*
7. Manufacturing Engineering Processes, *Leo Alting*
8. Modern Ceramic Engineering: Properties, Processing, and Use in Design, *David W. Richerson*
9. Interface Technology for Computer-Controlled Manufacturing Processes, *Ulrich Rembold, Karl Armbruster, and Wolfgang Ülzmann*
10. Hot Rolling of Steel, *William L. Roberts*
11. Adhesives in Manufacturing, *edited by Gerald L. Schneberger*
12. Understanding the Manufacturing Process: Key to Successful CAD/CAM Implementation, *Joseph Harrington, Jr.*
13. Industrial Materials Science and Engineering, *edited by Lawrence E. Murr*
14. Lubricants and Lubrication in Metalworking Operations, *Elliot S. Nachtman and Serope Kalpakjian*
15. Manufacturing Engineering: An Introduction to the Basic Functions, *John P. Tanner*
16. Computer-Integrated Manufacturing Technology and Systems, *Ulrich Rembold, Christian Blume, and Ruediger Dillmann*
17. Connections in Electronic Assemblies, *Anthony J. Bilotta*

18. Automation for Press Feed Operations: Applications and Economics, *Edward Walker*
19. Nontraditional Manufacturing Processes, *Gary F. Benedict*
20. Programmable Controllers for Factory Automation, *David G. Johnson*
21. Printed Circuit Assembly Manufacturing, *Fred W. Kear*
22. Manufacturing High Technology Handbook, *edited by Donatas Tijunelis and Keith E. McKee*
23. Factory Information Systems: Design and Implementation for CIM Management and Control, *John Gaylord*
24. Flat Processing of Steel, *William L. Roberts*
25. Soldering for Electronic Assemblies, *Leo P. Lambert*
26. Flexible Manufacturing Systems in Practice: Applications, Design, and Simulation, *Joseph Talavage and Roger G. Hannam*
27. Flexible Manufacturing Systems: Benefits for the Low Inventory Factory, *John E. Lenz*
28. Fundamentals of Machining and Machine Tools, Second Edition, *Geoffrey Boothroyd and Winston A. Knight*
29. Computer-Automated Process Planning for World-Class Manufacturing, *James Nolen*
30. Steel-Rolling Technology: Theory and Practice, *Vladimir B. Ginzburg*

OTHER VOLUMES IN PREPARATION

Computer Integrated Electronics Manufacturing and Testing, *Jack Arabian*
Robot Technology and Applications, *edited by Ulrich Rembold*

Steel-Rolling Technology

Theory and Practice

Vladimir B. Ginzburg

International Rolling Mill Consultants, Inc.
Pittsburgh, Pennsylvania

Marcel Dekker, Inc.   •   New York and Basel

Library of Congress Cataloging-in-Publication Data

Ginzburg, Vladimir B.
    Steel-rolling technology : theory and practice / Vladimir B.
Ginzburg
        p. cm.-- (Manufacturing engineering and materials processing
; 30)
    Includes index.
    ISBN 0-8247-8124-4 (alk. paper)
    1. Rolling (Metal-work) 2. Steelwork. I. Title. II. Series.
TS340.G534 1989
672.8'2--dc 20
                                                89-33531
                                                    CIP

This book is printed on acid-free paper.

MARCEL DEKKER, INC.
270 Madison Avenue, New York, New York   10016

Current printing (last digit):
10 9 8 7 6 5 4 3 2

PRINTED IN THE UNITED STATES OF AMERICA

Preface

The steel industry has had a long history of development, yet, despite all the time that has passed, it still demonstrates all the signs of longevity. New ideas continue to revolutionize the steel-producing process today as much as they did a hundred years ago. The latest advances – making of 'clean' steel, development of the continuous casting process for thin slabs and strip, introduction of the ingenious strip profile and shape control technologies in rolling mills – are only few examples that illustrate the great potential for further innovations and discoveries. It is no wonder that many specialists, engineers and scientists from different countries still find the steel industry an exciting field for implementation of their creativity.

The new developments in the steel industry go along three major paths. The first path is determined by the multi-stage character of the steel production process (steelmaking, casting, rolling, etc.), with developments concentrated within each of these stages. The second path reflects the multi-discipline character of the science describing the steel production process (physics, metallurgy, computer science, etc.), with each of these branches of the science being independently developed.

The third path of the development, which is attracting more and more attention, is **along the interfaces** among the different stages of the steel-producing process as well as among the different branches of science related to steel production.

Today, the quality control engineer working in a hot strip mill realizes that to make an objective determination of the origin of some steel defect he has to understand the technology of steelmaking and casting processes. A metallurgist feels a need to comprehend the basic concept of the computer control that maintains the strip temperature in order to define an acceptable compromise between the desired tolerances for strip temperature and further sophistication of the process control.

These desires can be generalized by the idea of a **system approach**, where the problem solving is based on an intelligent analysis of the problems that arise at the interfaces among different disciplines.

Creation of the experts who can utilize the system approach is not the easiest task. Courses on system approach are taught in few schools. Nevertheless, the author firmly believes that this area of expertise will attract many talented engineers and scientists. He also hopes that this book will be helpful to them in their exciting endeavor.

Vladimir B. Ginzburg

Contents

PREFACE   iii

PART I   MAIN PROPERTIES AND CLASSIFICATIONS OF STEELS AND ALLOYS

1. The Crystalline Structure of Metals   3
2. Physical Properties of Metals   13
3. Classification of Steels and Alloys   45

PART II   PRINCIPLES OF METALLURGICAL DESIGN OF STEELS

4. Phase Transformation in Steel   73
5. Alloying Elements and Impurities in Steel   88
6. Metallurgical Factors Controlling Properties of Steels   107
7. Heat Treatment of Steel   124

PART III   MAKING AND CASTING OF STEEL FOR FLAT PRODUCTS

8. Primary Steelmaking Processes   137
9. Secondary Steelmaking Processes   151
10. Casting of Steel for Flat Products   165
11. Defects in Ingots and Slabs   187

PART IV   THEORY OF PLASTIC DEFORMATION

12. Principles of Microscopic Plasticity   199
13. Principles of Macroscopic Plasticity   209
14. Slab Analysis of Plastic Deformation   225
15. Upper-Bound Analysis of Plastic Deformation   247
16. Slip-Line Field Analysis of Plastic Deformation   255

PART V   CALCULATION OF ROLLING PARAMETERS

17. Resistance to Deformation in Hot Rolling   269
18. Roll Force, Torque and Power in Hot Rolling   298
19. Roll Force, Torque and Power in Cold Rolling   324

PART VI   TRIBOLOGY IN THE ROLLING PROCESS

20. Basic Concept of Friction   343
21. Basic Principles of Lubrication and Wear   353
22. Friction, Lubrication and Wear in Rolling   366

PART VII   HEAT TRANSFER IN ROLLING MILLS

23. Steel Heating for Hot Rolling   397
24. Heat Transfer During the Rolling Process   419

PART VIII   METALLURGICAL ASPECTS OF THE ROLLING PROCESS

25. Structural Changes in Steel During Hot Rolling   439
26. Thermomechanical Treatments Combined with Rolling   454
27. Scaling of Steel in Hot Strip Mill   483

PART IX   ROLLING MILLS FOR FLAT PRODUCTS

28. Classification of Rolling Mills   511
29. High Reduction Rolling Mills   529
30. Optimization and Modernization of Hot Strip Mills   550

PART X   GEOMETRY OF FLAT PRODUCTS

31. Geometrical Characteristics of Flat Products   571
32. Measurement of Geometrical Parameters of Flat Products   589

PART XI   GAUGE AND WIDTH CONTROL

33. Principle of Gauge Control   615
34. Modeling of Dynamic Characteristics of HAGC   648
35. Principles of Width and Plan View Control   669
36. Width Change and Control in Rolling Mills   685

PART XII   STRIP PROFILE AND FLATNESS CONTROL

37. Strip Profile and Flatness Actuators   711
38. Roll Deformation Models   730
39. Roll Contour and Strip Flatness Models   749
40. Selection of Strip Profile and Flatness Actuators   762

INDEX

                                                                    777

Steel-Rolling Technology

Part I

Main Properties and Classifications of Steels and Alloys

1

The Crystalline Structure of Metals

1.1 SPACE LATTICES

Substances can exist in either amorphous or crystalline state. In the amorphous state the elementary particles are intermixed in a disorderly manner; their positions are not fixed relative to those of their neighbors. In the crystalline state the substance consists of atoms, or more properly, ions which are arranged according to some regular geometric pattern. This pattern varies from one substance to another.

All metals are crystalline in nature. Crystallization of the metals takes place during solidification when atoms of the liquid metal group themselves in an orderly arrangement, forming a definite space pattern. This pattern is known as a 'space lattice'.

There are several types of the lattices in which metallic atoms can arrange themselves upon solidification, but the three most common are shown in Fig. 1.1 and are known as the body-centered cubic (b.c.c.), face-centered cubic (f.c.c.), and close-packed hexagonal (c.p.h.) crystal lattices [1].

1.2 LATTICE CONSTANT

The side of the elementary cube or hexagon is known as the lattice constant (Fig. 1.2). These lengths are usually expressed in the Angstrom units (designated also by symbols Å or A.U.). One Angstrom unit

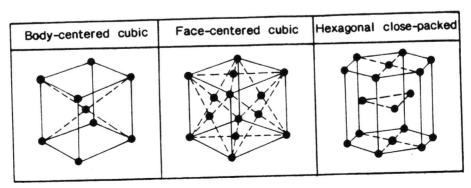

Body-centered cubic	Face-centered cubic	Hexagonal close-packed

Fig. 1.1 Three principal types of crystal lattices. Adapted from Higgins (1983).

3

is equal to 10^{-8} centimeters. In metals, the length of the cube edge varies between 2.878 Å in chromium to 4.941 Å in lead [2].

1.3 METALLIC BOND

The term 'metallic bond' is used to explain the forces which hold the ions within the crystalline structure. During solidification of the metals the atoms relieve their outer (valence) electrons. As a result, the atoms transform into ions which carry like positive charges and so tend to repel each other. However, the relieved valence electrons are donated to a common negatively charged 'cloud' which is shared by all atoms present (Fig. 1.3). Thus, while the positively charged ions repel each other, they are held in the equilibrium positions by their mutual attraction to the negatively-charged electron cloud.

1.4 ALLOTROPIC CHANGES

When a metal undergoes a transformation from one crystal pattern to another, it is known as an allotropic change. The allotropic forms of iron are temperature dependent as illustrated in Table 1.1. In case of an allotropic change from a body-centered to a face-centered cubic lattice, this is accompanied by a marked quantitative change in the characteristics and properties of the metal involved. The allotropic change in density of iron is shown in Fig. 1.4 [1, 3].

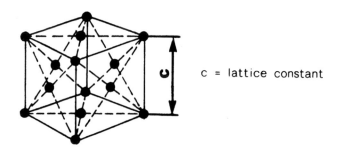

c = lattice constant

Fig. 1.2 Lattice constant.

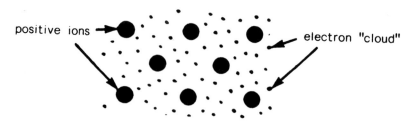

positive ions →

electron "cloud"

Fig. 1.3 Diagrammatic representation of the metallic bond. Adapted from Higgins (1983).

Table 1.1 Allotropic forms of iron. From The Making, Shaping and Treating of Steel (1985).

Allotropic forms	Crystallographic forms	Temperature range
Alpha	Body-centered cubic (b.c.c.)	Up to 910°C (1670°F)
Gamma	Face-centered cubic (f.c.c.)	910–1403°C (1670–2557°F)
Delta	Body-centered cubic (b.c.c.)	1403–1535°C (2557–2795°F)

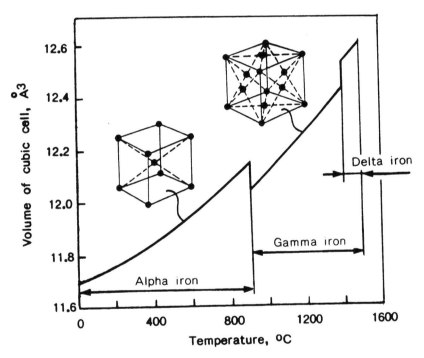

Fig. 1.4 Allotropic changes in density of iron. Adapted from Higgins (1983).

1.5 ATOMIC PLANES

It is possible to draw several sets of parallel planes through a crystal so that all atoms constituting it would be located in any one set. They are known as the atomic planes.

Figure 1.5 illustrates the atomic planes in a cubic cell which may be considered as made up of eight atoms, one at each corner. The atomic planes are described by the 3-digit numbers. Each digit corresponds to a unit coordinate in relation to the crystallographic axes X, Y, and Z. For example, the set of planes designated by the symbol (001) would cut neither X nor the Y axis [2].

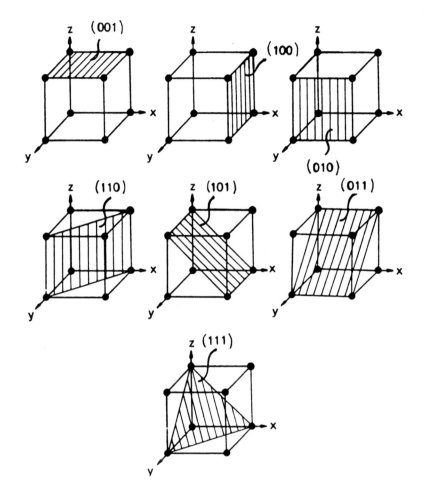

Fig. 1.5 Atomic planes in a cubic cell. Adapted from Sauveur (1935).

1.6 CRYSTALLOGRAPHIC ANISOTROPY

Physical properties of a crystal depend on the direction of their measurement in respect to their crystallographic axis. In the body-centered cubic crystalline structure of iron, the greatest strength lies in the direction of the atomic plane (111) and the weakest along the atomic plane (100) as indicated in Fig. 1.6a.

Any deviation from a perfectly random orientation of the many crystals in a metallic material will therefore result in anisotropy of the mechanical properties of the material [4].

The same can be said in regard to the anisotropic character of the magnetic properties of some alloys such as a silicon steel. Magnetic permeability of a silicon steel crystal is maximum when it is measured along the atomic plane (100) and it is minimum when measured along the atomic plane (111) as shown in Fig. 1.6b [5].

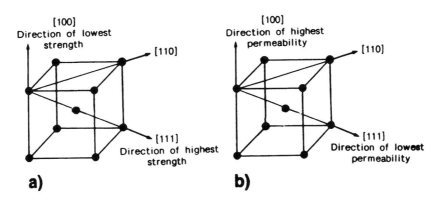

Fig. 1.6 Anisotropy of physical properties in a cubic crystal:
a) mechanical strength, b) magnetic permeability. Adapted from Bozorth (1951) and Roberts (1978).

1.7 COOLING CURVE

A pure liquid metal solidifies into a crystalline solid at a fixed temperature called the freezing point (Fig. 1.7). This is due to the fact that the amount of internal heat relieved during the crystallization process is equal to the amount of heat given up by the material. The latter is known as the latent heat of solidification [1].

1.8 METALLIC DENDRITE

When a pure metal solidifies, each crystal begins to grow independently from a nucleus, or 'center of crystallization'. The crystal develops by the addition of atoms according to the lattice pattern evolving in what is called a 'dendrite' as shown in Fig. 1.8. A metallic crystal grows in this way because heat is dissipated more quickly from a point, so it will be there that the temperature decreases at the highest rate leading to a formation of a rather elongated skeleton [1].

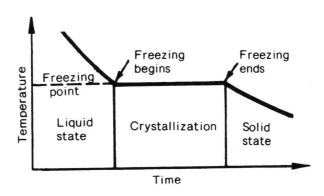

Fig. 1.7 Typical cooling curve of a pure metal. Adapted from Higgins (1983).

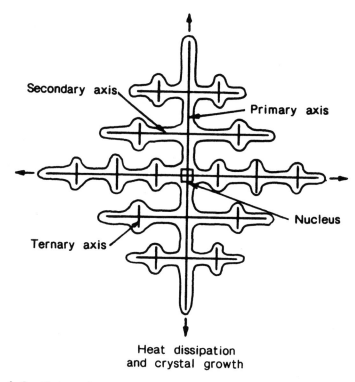

Fig. 1.8 The early stages in the growth of a metallic dendrite.

1.9 DENDRITE GROWTH

The dendrite arms continue to grow and thicken (Fig. 1.9a,b) until ultimately the space between them will become completely solid. Meanwhile, outer arms begin to make contact with those of neighboring dendrites (Fig. 1.9c). At this time the dendrite growth ceases and solidification will be complete. The

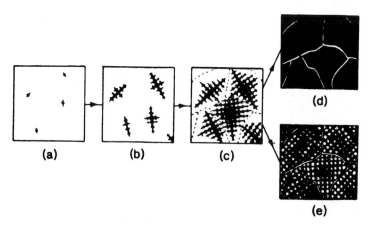

Fig. 1.9 Dendrite growth of metallic crystals from liquid state.
(From ENGINEERING METALLURGY, Part 1, by R.A. Higgins, 1983. Copyright Hodder & Stoughton Ltd., England. Reprinted with permission).

remaining liquid is used in thickening the existing dendrite arms leading to the irregular overall shape of crystals (Fig. 1.9d). An impure metal (Fig. 1.9e) carries the impurities between the dendrite arms, thus revealing the initial skeleton.

1.10 CRYSTAL BOUNDARY

A solidified metal consists of a mass of separate crystals irregular in shape but interlocking with each other rather like a three-dimensional jigsaw puzzle. It is now widely held that at the crystal boundaries there exists a film of metal, some three atoms thick, in which the atoms do not conform to any pattern as illustrated in Fig. 1.10 [1, 6].

1.11 GRAIN SHAPE AND SIZE

The shape and size of crystals depend on the cooling rate of a molten metal when it reaches its freezing point. A slow decrease in temperature promotes the formation of relatively few nuclei, so as a result the crystal will be large. Rapid cooling triggers formation of a large number of nuclei producing a large number of small crystals, or grains (throughout this book the terms 'grain' and 'crystal' are used synonymously).

In a large ingot the grain size may vary considerably from the outside surface to the center as shown in Fig. 1.11. This is due to variation in the temperature gradient as the ingot solidifies and the heat is transferred from the metal to a mold. It is possible to distinguish three forms in the grain

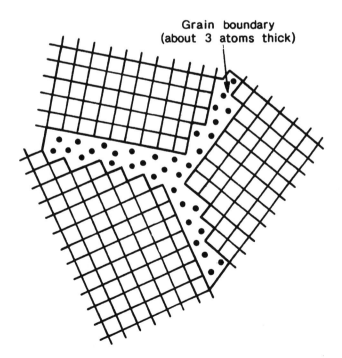

Fig. 1.10 Diagrammatic representation of a grain boundary. Adapted from Wusatowski (1969) and Higgins (1983).

structure: a) small chill grains located at the surface of an ingot, b) columnar grains located in the intermediate zone, and c) large equi-axed grains located at the center of an ingot [1].

1.12 ASTM GRAIN SIZE

The grain size is commonly given as the ASTM grain number. This is an arbitrary exponential number (n) that refers to the mean number of grains per square inch (N) at a magnification of 100X according to the following equation [2, 7] as shown in Fig. 1.12:

$$N = 2^{n-1}$$

(1-1)

1.13 PHASES IN METALS

A phase is a distinct and physically, chemically or crystallographically homogeneous portion of an alloy. Only three solid phases shall be considered, namely: a) pure metals, b) intermetallic compounds, and c) solid solutions. When an alloy contains more than one of these phases, it is generally referred to as an **aggregate** [2].

1.14 INTERMETALLIC COMPOUNDS

Chemical compounds between metals and metalloids are known as intermetallic compounds. A large portion of the known intermetallic compounds contains one of the following metalloids: carbon, phosphorus, silicon, sulfur, arsenic or the metal aluminum.

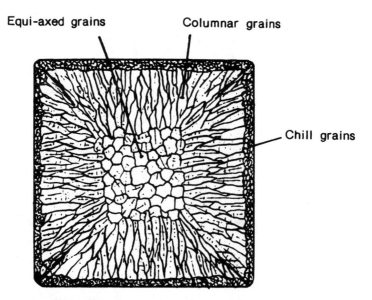

Fig. 1.11 The grain structure in a section of an ingot.
(From ENGINEERING METALLURGY, Part 1, by R.A. Higgins, 1983. Copyright Hodder & Stoughton Ltd., England. Reprinted with permission).

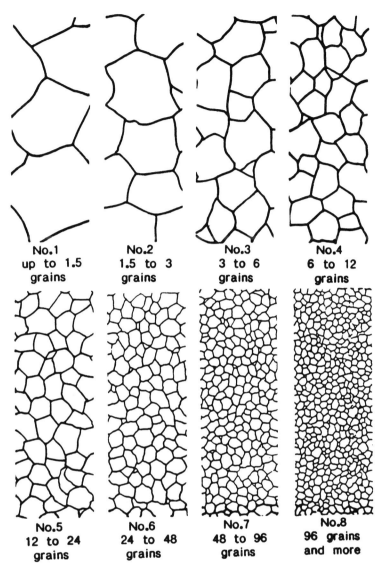

No.1   No.2   No.3   No.4
up to 1.5   1.5 to 3   3 to 6   6 to 12
grains   grains   grains   grains

No.5   No.6   No.7   No.8
12 to 24   24 to 48   48 to 96   96 grains
grains   grains   grains   and more

Fig. 1.12 ASTM grain size chart. (From METALLURGY by C.G. Johnson
and W.R. Weeks, 1977. Copyright ASTM. Reprinted with permission).

The important compound present in alloys of iron and carbon is the carbide Fe_3C or cementite. It has an orthorhombic space lattice.

1.15 SOLID SOLUTIONS

A complete merging in the solid state of the two phases, pure metals and intermetallic compounds, are known as the solid solutions. There can be solid solutions of two metals, of a metal and an intermetallic compound, or of two compounds.

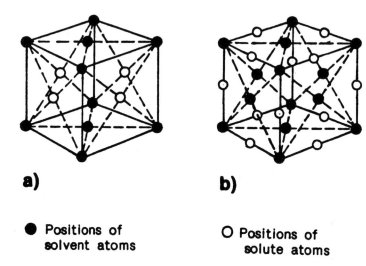

Fig. 1.13 Space lattices of (a) substitutional solid solution and (b) interstitial solid solution.

When metal A forms a solid solution with another metal B having a different space lattice, the space lattice of the solid solution may be either similar to the lattice of one of the metals, or different from both. In case the lattice of the solution is identical to one of the metals, A, for instance, that metal may logically be considered the solvent, and B the solute. If the space lattice of the solution is unlike the lattice of either metal, the metal present in larger proportions can be considered the solvent.

There are two following types of solid solutions [2]:

1. Substitutional solid solution - This is a solid alloy in which the solute atoms replace some atoms in the space lattice of the solvent as shown in Fig. 1.13a.

2. Interstitial solid solution - This is a solid alloy in which the solute atoms are located at random except at lattice points within the space lattice as shown in Fig. 1.13b.

REFERENCES

1. R.A. Higgins, Engineering Metallurgy, Part 1, Applied Physical Metallurgy, Robert E. Krieger Publishing Company, Melbourne, Florida, pp. 1-77 (1983).

2. A. Sauveur, The Metallography and Heat Treatment of Iron and Steel, Fourth Edition, McGraw-Hill Book Company Inc., New York and London, pp. 1-11 (1935).

3. The Making, Shaping and Treating of Steel, 10th Edition, eds. W.T. Lankford, Jr., et al, Association of Iron and Steel Engineers, Pittsburgh, Pennsylvania, pp. 1231-1240 (1985).

4. W.L. Roberts, Cold Rolling of Steel, Marcel Dekker, Inc., New York and Basel, p. 736 (1978).

5. R.M. Bozorth, Ferromagnetism, D.Van Nostrand Company, Inc., Toronto, New York and London, pp. 90-91 (1951).

6. Z. Wusatowski, Fundamentals of Rolling, Pergamon Press, Oxford, pp. 4-8 (1969).

7. C.G. Johnson and W.R. Weeks, Metallurgy, Fifth Edition, American Technical Society, Chicago, p.164 (1977).

2

Physical Properties of Metals

2.1 THREE KINDS OF STRESS

When a body is subjected to some external forces, the internal reactive forces between the particles within the body are produced. The internal reactive forces are in equilibrium with the external forces. The internal force per unit of an area is called **stress** [1].

No matter how produced, there are only three kinds of stress:

1. **Tensile stress** - It occurs in a body subjected to a tensile force P (Fig. 2.1a).

2. **Compressive stress** - It occurs in a body subjected to a compressive force P (Fig. 2.1b).

3. **Shear stress** - It occurs in a body subjected to a shearing force P_s (Fig 2.1c) resulting in one part of the body to slide in respect to the adjacent part.

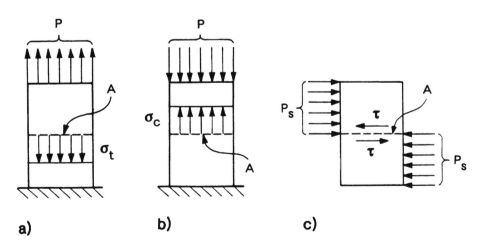

Fig. 2.1 Three kinds of stress: a) tensile, b) compressive, and c) shear.

13

2.2 ENGINEERING STRESS AND TRUE STRESS

Engineering stress or nominal stress, is defined as a ratio of the applied load P to the original cross-sectional area A_0 [2]. In application to tension or compression it is equal to

$$\sigma = P/A_0 \qquad\qquad (2\text{-}1)$$

In application to shear it is expressed as

$$\tau = P_s/A_0 \qquad\qquad (2\text{-}2)$$

True stress is defined as the ratio of the applied force P to the instantaneous cross-sectional area A_i which results from this force. In application to tension or compression it is equal to

$$\sigma = P/A_i \qquad\qquad (2\text{-}3)$$

In case of shear force it is expressed as

$$\tau = P_s/A_i \qquad\qquad (2\text{-}4)$$

2.3 ENGINEERING STRAIN AND TRUE STRAIN

Generally, strain is defined as deformation per unit of length. Figure 2.2 illustrates three basic types of strain, namely, tension, compression, and shear.

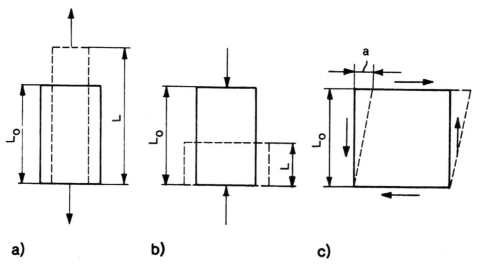

a)   **b)**   **c)**

Fig. 2.2 Three types of strain: a) tension, b) compression, and c) shear.

Engineering strain is defined as a relative change in the length of a body subjected to deformation. In application to tension or compression (Fig. 2.2a,b), it is equal to

$$e = (L - L_0)/L_0 \qquad\qquad (2\text{-}5)$$

When applied to shear (Fig. 2.2c), it is equal to

$$\gamma = a/L_0 \qquad\qquad (2\text{-}6)$$

True strain is defined as a sum of small increments in length of a body subjected to deformation, i.e.

$$\epsilon = \int_{L_0}^{L} dL/L = \ln(L/L_0) \qquad\qquad (2\text{-}7)$$

Once necking begins, the true strain at any point in a metal specimen can be calculated from the change in either cross-sectional area A or diameter D. Since the volume of the specimen in the plastic region remains constant, the true strain within the uniform elongation can be expressed as

$$\epsilon = \ln(A_0/A) = 2\ln(D_0/D) \qquad\qquad (2\text{-}8)$$

2.4 ENGINEERING STRESS-STRAIN DIAGRAMS

Figure 2.3 illustrates typical engineering stress-strain diagram for both ductile and brittle materials obtained from tension tests. In these diagrams, the engineering stress is defined by the Eq. (2-1) and the engineering strain is defined by the Eq. (2-5).

The engineering stress-strain diagrams are used to determine the following strength-deformation characteristics of materials [1-3]:

Proportional limit - This is the maximum stress at which strain remains directly proportional to stress. It corresponds to the point P in Fig. 2.3.

Elastic limit - This is the maximum stress to which a material may be subjected without any permanent strain remaining upon complete release of the stress. For most structural materials the elastic limit has nearly the same numerical value as the proportional limit.

Yield point - It is defined as the first stress in a material, usually less than the maximum attainable stress, at which an increase in strain occurs without an increase in stress (point Y', Fig. 2.3a.). This phenomenon occurs only in certain ductile materials. If there is a decrease in stress after yielding, a distinction is made between upper and lower yield points.

Yield strength - This is the stress at which a material exhibits a specified limiting deviation from the proportionality of stress to strain. It is defined by the 'offset method' as shown in Fig. 2.3b. The specified offset OX (usually 0.2%) is laid off along the strain axis. Then the line XW is drawn parallel to OP, and thus point Y', the intersection of the line XW with the stress-strain diagram, is located.

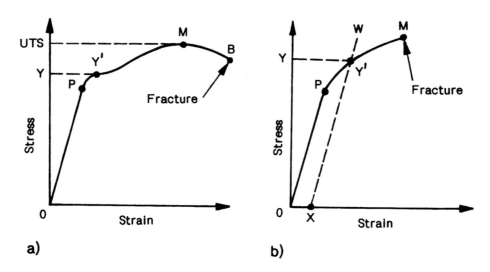

Fig. 2.3 Typical engineering stress-strain diagram for a) ductile, b) brittle material.

Ultimate strength - It is equal to the maximum stress that a material can withstand. In case of a ductile material (Fig. 2.3a) it corresponds to point M representing the limit of uniform elongation (descending of the curve MB is due to necking of a specimen). In case of a brittle material (Fig. 2.3b), it corresponds to point M at which the material breaks. In tensile testing it is called the **ultimate tensile strength** (UTS) and is equal to

$$UTS = P_m/A_o \, , \tag{2-9}$$

where P_m = maximum tensile load which material can withstand.

Modulus of elasticity -It is a measure of the rigidity of metal. It is expressed as a ratio of stress, within proportional limit, to corresponding strain

$$E = \sigma/e \tag{2-10}$$

Modulus of resilience - This parameter indicates the specific energy that a material can store elastically. This specific energy is equal to the area under the stress-strain curve up to the yield point of a material

$$E_r = (Y^2)/(2E) \, , \tag{2-11}$$

where Y = yield strength.

Poisson's ratio -It is the absolute ratio of a lateral strain to a longitudinal strain.

2.5 TRUE STRESS-STRAIN DIAGRAMS

In the true stress-strain diagrams obtained from a tension test, true stress is defined by the Eq. (2-3) and true strain is defined by the Eq. (2-7).

Unlike the slope in an engineering stress-strain diagram, the slope of a true stress-strain diagram is always positive.

The true stress-strain curve can be approximated by the equation

$$\sigma = K\epsilon^n , \qquad\qquad (2\text{-}12)$$

where K = strength coefficient

n = strain hardening exponent of the material.

Figure 2.4 illustrates three kinds of true stress-strain diagrams [2]:

1. When n = 1, the stress-strain relationship is linear with slope equal to K. It corresponds to a perfectly elastic material.

2. When n = 0, the stress does not change with strain ($\sigma = K$). It corresponds to a rigid, perfectly plastic material.

3. The intermediate curve (0<n<1) corresponds to a typical engineering material.

The values for K and n for some steels at room temperature are listed in Table 2.1.

True stress-true strain curve allows evaluation of the **toughness** of a metal. Toughness is defined as the energy per unit volume that has been dissipated up to fracture and is equal to the area under the true stress-true strain curve.

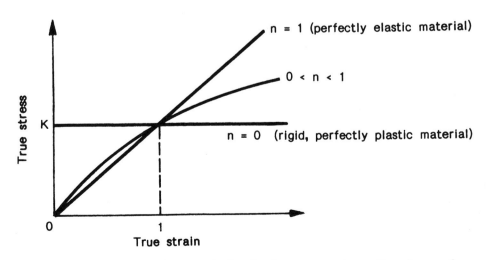

Fig. 2.4 The effect of strain-hardening exponent on the shape of true-strain curves. Adapted from Kalpakjian (1985).

Table 2.1 Typical values for K and n in Eq. (2-12).
Data from Kalpakjian (1985).

Material	K psi x 10^3	K MPa	n
Steel			
Low-carbon, annealed	77	530	0.26
1045 hot-rolled	140	965	0.14
1112 annealed	110	760	0.19
1112 cold-rolled	110	760	0.08
4135 annealed	147	1015	0.17
4135 cold-rolled	160	1100	0.14
4340 annealed	93	640	0.15
17-4 P-H annealed	175	1200	0.05
52100 annealed	210	1450	0.07
302 stainless, annealed	190	1300	0.30
304 stainless, annealed	185	1275	0.45
410 stainless, annealed	140	960	0.10

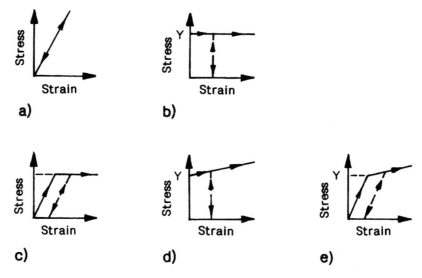

Fig. 2.5 Idealized stress-strain diagrams of: a) perfectly elastic
material, b) rigid, perfectly plastic material, c) combination of
perfectly elastic and plastic materials, d) rigid, linearly strain-
hardening material, and e) elastic, linearly strain-hardening material. (From Serope Kalpakjian, MANUFACTURING PROCESSES FOR
ENGINEERING MATERIALS, Copyright 1985 Addison-Wesley Publishing Co.,
Reading, Massachusetts. Reprinted with permission).

2.6 IDEALIZED STRESS-STRAIN DIAGRAMS

Some of the major types of idealized stress-strain diagrams are shown in Fig. 2.5. They represent the following kinds of materials [2]:

Perfectly elastic material (Fig. 2.5a) - It behaves like a spring, i.e., when the load is released the material undergoes a complete elastic recovery.

Rigid, perfectly plastic material (Fig. 2.5b) - It undergoes deformation at the same stress level. In this material, there is no elastic recovery after the load is released.

Combination of a perfectly elastic and plastic material (Fig. 2.5c) - It undergoes elastic recovery when the load is released.

Rigid, linearly strain-hardening material (Fig. 2.5d) - It produces an increasing stress level with increase in strain. There is no elastic recovery upon unloading.

Elastic, linearly strain-hardening material (Fig. 2.5e) - It is an approximate representation of most engineering materials.

2.7 DERIVATIVE TYPES OF STRESS

Other names are given to the three kinds of stress described above when it is appropriate to indicate either the manner in which the stress acts or the manner by which it is produced. Below are the names which are commonly used [1, 2]:

Axial stress - This is the tensile or compressive stress acting in the direction of the axis of an axially loaded prism (Fig. 2.6a).

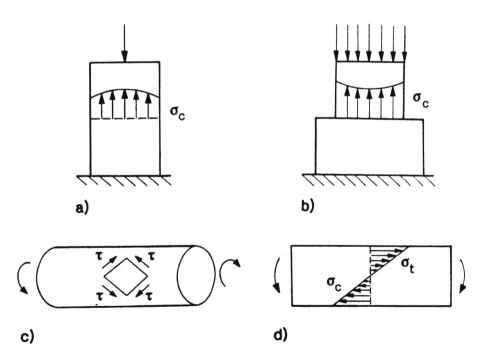

Fig. 2.6 Derivative types of stress: a) axial, b) contact, c) torsional, and d) bending.

Normal stress - This is another name for tensile or compressive stresses.

Contact, or **bearing stress** - This is the compressive stress between two bodies at their surface contact (Fig. 2.6b).

Torsional stress - This is the shearing stress in a shaft subjected to twisting (Fig. 2.6c).

Bending stress - This is the tensile and compressive stress in a beam subjected to bending (Fig. 2.6d).

2.8 DUCTILITY

Ductility is the ability of a material to deform plastically without fracturing.

The following two quantities are commonly used to define ductility in a tension test [2]:

1. **Percent elongation.** It is defined as

$$\% \text{ elongation} = (L_f/L_o - 1) \times 100\%, \tag{2-13}$$

where L_f = specimen length at fracture.

2. **Percent reduction of area.** It is defined as

$$\% \text{ reduction of area} = (1 - A_f/A_o) \times 100\%, \tag{2-14}$$

where A_o = original cross-section area of a specimen

A_f = cross-section area of a specimen at fracture.

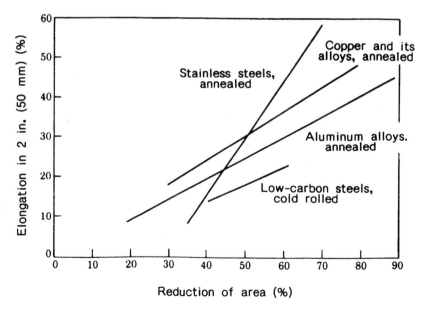

Fig. 2.7 Approximate relationship between elongation and reduction of area for different groups of metals. (From Serope Kalpakjian, MANUFACTURING PROCESSES FOR ENGINEERING MATERIALS, Copyright 1985 Addison-Wesley Publishing Co., Reading, Massachusetts. Reprinted with permission).

Relationship between percent elongation and percent reduction of area for some materials is shown in Fig. 2.7.

2.9 STRAIN RATE

The strain rate is the time rate for straining. Since a strain is dimensionless, the units of strain rate are reciprocal time. Typical values of strain rates for different metalworking processes are shown in Table 2.2.

For tension and compression, the **engineering strain rate** is defined as

$$\dot{\epsilon} = d\epsilon/dt = V/L_0 , \qquad (2\text{-}15)$$

where V = speed of deformation.

The **true strain rate** is defined as [2]

$$\dot{\epsilon} = d\epsilon/dt = v/L \qquad (2\text{-}16)$$

The effect of strain rate on the strength of materials is generally expressed by

$$\sigma = C\dot{\epsilon}^m , \qquad (2\text{-}17)$$

Table 2.2 Typical ranges of true strain and strain rates in metalworking processes. Data from Kalpakjian (1985).

Process	True strain	Strain rate, s^{-1}
Cold working		
Forging, rolling	0.1–0.5	$1.0\text{–}10^3$
Wire and tube drawing	0.05–0.5	$1.0\text{–}10^4$
Explosive forming	0.05–0.2	$10\text{–}10^5$
Hot working and warm working		
Forging, rolling	0.1–0.5	$1.0\text{–}10^3$
Extrusion	2.0–5.0	$10^{-1}\text{–}10^2$
Machining	1.0–10	$10^3\text{–}10^6$
Sheet-metal forming	0.1–0.5	$1.0\text{–}10^2$
Superplastic forming	0.2–3.0	$10^{-4}\text{–}10^{-2}$

where C = strength coefficient

     m = strain-rate sensitivity exponent of the material.

2.10 COMPRESSION TEST

The compression test is usually carried out by compressing a solid cylindrical specimen between flat dies (Fig. 2.8). Compression of the specimen is accompanied by its barreling. It is due to the friction between the specimen and the dies resulting in restricted expansion of the top and the bottom surfaces of the specimen.

During the compression test, friction dissipates energy through continuously increasing contact surfaces. In order to establish true properties of the material, the effect of friction has to be reduced.

It has been demonstrated that compressive stress reduces when the surface roughness of the dies and specimen is reduced (Fig. 2.9). Similar effect can be achieved by providing more effective lubrication between surfaces [3].

Unlike the tension test, where a necking of the specimen starts after relatively little elongation, the compression test can be carried out uniformly for ductile materials if effective means are provided to minimize the detrimental effect of friction.

It has also been shown that the compressive stress reduces with increase in the initial height of the specimen (Fig. 2.10) and also with decrease of its initial contact surface area (Fig. 2.11). In more general terms, it can be said that the compressive stress increases with an increase in an **aspect ratio** which in case of cylindrical specimen with initial diameter D_0 and an initial length L_0 is equal to

$$Z_c = D_0/L_0 \tag{2-18}$$

One of the purposes of the compression test is to determine the **maximum compressive stress** that a material is capable of developing, based on an original cross-section area [2].

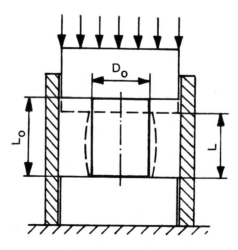

Fig. 2.8 Compression test.

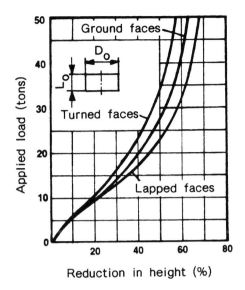

Fig. 2.9 Effect of surface finish on compressive load. (From THE ROLLING OF STRIP, SHEET, AND PLATE, by E.C. Larke. Copyright Chapman and Hall Ltd., England. Reprinted with permission).

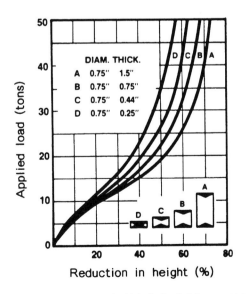

Fig. 2.10 Effect of the specimen initial height on compressive load. (From THE ROLLING OF STRIP, SHEET, AND PLATE, by E.C. Larke. Copyright Chapman and Hall Ltd., England. Reprinted with permission).

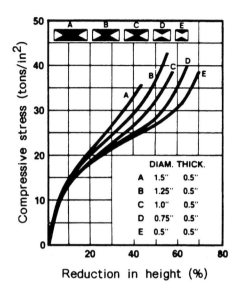

Fig. 2.11 Effect of the specimen initial contact surface area on compressive load. (From THE ROLLING OF STRIP, SHEET, AND PLATE, by E.C. Larke, 1963. Copyright Chapman and Hall Ltd., England. Reprinted with permission).

For the material with low ductility, the maximum compressive stress corresponds to the stress at which the material fails by a shattering fracture. For ductile materials, the maximum compressive stress may be related to an arbitrary selected degree of distortion that is regarded as an indication of complete failure of the material.

2.11 TORSION TEST

The torsion test allows one to determine the material properties while avoiding the disturbing effect of necking (as in a tensile test) or friction (as in a compression test). The test is generally carried out on a tubular specimen with reduced wall thickness in its middle [2].

The shear stress can be determined from the equation

$$\tau = M/(2\pi r^2 h) \, , \tag{2-19}$$

where M = torque

    r = mean radius of the reduced section

    h = wall thickness of the reduced section.

The shear strain is equal to

$$\gamma = (r\theta)/L, \tag{2-20}$$

where θ = angle of twist, radian

    L = length of the reduced section.

In regard to shear deformation, the modulus of elasticity is known as the **shear modulus,** or the modulus of rigidity. Similar to modulus of elasticity it is also applied to the elastic range and is equal to

$$G = \tau/\gamma \qquad\qquad (2\text{-}21)$$

The relationship between the modulus of elasticity E and the shear modulus is expressed by the equation

$$G = 0.5E/(1 + \nu) , \qquad\qquad (2\text{-}22)$$

where ν = Poisson's ratio.
For plain carbon steel ν = 0.33. Thus, E is 2.66 times G.

2.12 COMBINED DEFORMATION TESTS

In conducting the combined deformation tests, the specimen is simultaneously subjected to more than one type of deformation. Two following combined deformation tests are considered below [2]:

1. Torsion-compression test.
2. Torsion-tension test.

These tests are conducted on round bars that are simultaneously twisted and also either axially compressed or stretched.

It was found from the torsion-compression tests, that the shear strain at fracture increases substantially as the axial compressive stress increases (Fig. 2.12). It indicates that the compressive stress improves ductility of materials. On the other hand, it was found from the torsion-tension tests that the axial tensile stress results in reduction of the shear strain at fracture, which means that ductility of materials is being reduced.

The axial compressive stress has been found to have no effect on the magnitude of shear stress required to cause yielding or to continue the deformation.

2.13 HARDNESS TESTS

Generally, hardness is defined as resistance of metal to plastic deformation, usually by indentation. Hardness is not a fundamental property of a material but is related to its elastic and plastic properties.

Below is a brief description of the most common standardized hardness tests [2].

Brinell hardness test - This is a test for determining the hardness of material by forcing a hard steel or a carbide ball of specified load (Table 2.3). The Brinell hardness number (HB) is defined as the ratio of the load to the curved area of an indentation.

Tungsten carbide balls are generally recommended for a Brinell hardness number higher than 500. This test is generally applicable to a material of low to medium hardness [4].

Rockwell hardness test - This is a test for determining the hardness of a material based upon the depth of penetration of a specified penetrator into a specimen under certain arbitrary fixed conditions of

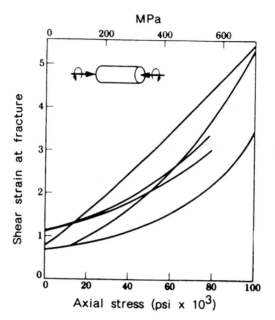

Fig. 2.12 The effect of axial compressive stress on the shear strain at fracture for various steels. (From Serope Kalpakjian, MANUFACTURING PROCESSES FOR ENGINEERING MATERIALS, Copyright 1985 Addison-Wesley Publishing Co., Reading, Massachusetts. Reprinted with permission).

the test. The most commonly used penetrators and the values for loads are shown in Table 2.3. This test is used for a wide range of hardnesses.

Vickers hardness test- It is also known as the diamond pyramid hardness test employing a diamond pyramid indenter and variable loads (ranging from 1 to 120 kg). One hardness scale is used for all ranges of hardness from very soft lead to a tungsten carbide. The Vickers test gives essentially the same hardness number (HV) regardless of the load (Table 2.3).

Scleroscope test - This hardness test measures the loss in kinetic energy of a falling metal 'tup', absorbed by an indentation upon impact of the tup on the metal being tested. The hardness is indicated by the height of the rebound. Since the scleroscope is portable, it is useful for measuring the hardness of large objects.

Mohs hardness test - This test is based on the capability of one material to scratch another. The Mohs hardness is based on a scale of 10 with 1 for talc and 10 for a diamond.

Knoop hardness test - This test uses a pyramidal diamond indenter, making a rhombohedral impression with one long and one short diagonal. The loads range from 25 g to 50 kg. This test is suitable for very small and brittle materials.

2.14 HARDNESS VERSUS STRENGTH

Performing a hardness test is similar to performing a compression test on a small portion of the surface of the material. However, there is also a substantial difference between these two tests. Indeed, during

Table 2.3 Approximate hardness relations for steel. Data from ASM Metals Reference Book (1981).

Brinell hardness (HB) 3000-kg load, 10-mm ball				Rockwell hardness Brale indenter				
Indent. dia- meter mm	Standard ball	Tungsten carbide ball	Vickers hardness	A 60-kg load	C 150-kg load	D 100-kg load	Super- ficial 30-kg load	Tensile strength ksi
2.25	...	(745)	840	84.1	65.3	74.8	82.2	...
2.30	...	(712)	783	83.1	63.4	73.4	80.5	...
2.35	...	(682)	737	82.2	61.7	72.0	79.0	...
2.40	...	(653)	697	81.2	60.0	70.7	77.5	...
2.45	...	627	667	80.5	58.7	69.7	76.3	347
2.50	...	601	640	79.8	57.3	68.7	75.1	328
2.55	...	578	615	79.1	56.0	67.7	73.9	313
2.60	...	555	591	78.4	54.7	66.7	72.7	298
2.65	...	534	569	77.8	53.5	65.8	71.6	288
2.70	...	514	547	76.9	52.1	64.7	70.3	273
2.75	(495)	...	539	76.7	51.6	64.3	69.9	269
	...	495	528	76.3	51.0	63.8	69.4	263
2.80	(477)	...	516	75.9	50.3	63.2	68.7	257
	...	477	508	75.6	49.6	62.7	68.2	252
2.85	(461)	...	495	75.1	48.8	61.9	67.4	244
	...	461	491	74.9	48.5	61.7	67.2	242
2.90	444	...	474	74.3	47.2	61.0	66.0	231
	...	444	472	74.2	47.1	60.8	65.8	229
2.95	429	429	455	73.4	45.7	59.7	64.6	220
3.00	415	415	440	72.8	44.5	58.8	63.5	212
3.05	401	401	425	72.0	43.1	57.8	62.3	202
3.10	388	388	410	71.4	41.8	56.8	61.1	193
3.15	375	375	396	70.6	40.4	55.7	59.9	184
3.20	363	363	383	70.0	39.1	54.6	58.7	177
3.25	352	352	372	69.3	37.9	53.8	57.6	172
3.30	341	341	360	68.7	36.6	52.8	56.4	164
3.35	331	331	350	68.1	35.5	51.9	55.4	159
3.40	321	321	339	67.5	34.3	51.0	54.3	154
3.45	311	311	328	66.9	33.1	50.0	53.3	149
3.50	302	302	319	66.3	32.1	49.3	52.2	146
3.55	293	293	309	65.7	30.9	48.3	51.2	142
3.60	285	285	301	65.3	29.9	47.6	50.3	138
3.65	277	277	292	64.6	28.8	46.7	49.3	134
3.70	269	269	284	64.1	27.6	45.9	48.3	131
3.75	262	262	276	63.6	26.6	45.0	47.3	127
3.80	255	255	269	63.0	25.4	44.2	46.2	123
3.85	248	248	261	62.5	24.2	43.2	45.1	120
3.90	241	241	253	61.8	22.8	42.0	43.9	116
3.95	235	235	247	61.4	21.7	41.4	42.9	114
4.00	229	229	241	60.8	20.5	40.5	41.9	111

the compression test, the specimen has freedom for lateral expansion, although it is restricted near contact surfaces. During the hardness test, lateral expansion is prevented by the material which surrounds a penetrator. It results in higher compressive yield stress.

It was found that hardness values are approximately three times that of the uniaxial yield stress of the metal when both values are expressed in the same units (Fig. 2.13)

When ultimate tensile strength (UTS) is expressed in psi and Brinell hardness number is in kg/mm^2, the following equation may be used [2]:

$$UTS = 500(HB) \qquad [psi], \qquad\qquad\qquad (2-23)$$

where HB = Brinell hardness number with a load of 3000 kg, kg/mm^2.

2.15 HARDENABILITY TEST

Hardenability is the relative ability of a ferrous alloy to form martensite after being quenched from a temperature above the upper critical temperature Ac_3.

It is usually measured as the distance below a quenched surface where the metal hardness is equal to a prerequisite value. In some cases it is determined as a specific percentage of martensite in the microstructure.

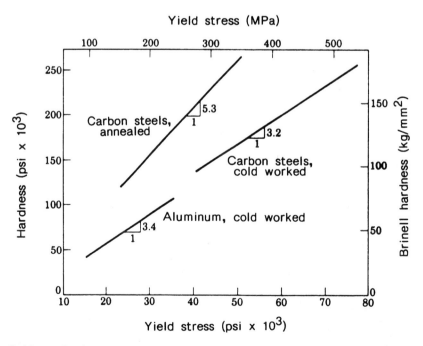

Fig. 2.13 Relationship between Brinell hardness number and yield stress for aluminum and steels. (From Serope Kalpakjian, MANUFACTURING PROCESSES FOR ENGINEERING MATERIALS, Copyright 1985 Addison-Wesley Publishing Co., Reading, Massachusetts. Reprinted with permission).

One of the most common tests used for evaluation of hardenability is the end-quench test, also known as the **Jominy test** [5]. During this test, a cylindrical specimen of 1 inch in diameter is cooled at one end by a column of water; thus the entire specimen experiences a range of cooling rates between those associated with water and air cooling. After quenching opposite ends of the specimen are grounded to be parallel to each other, and the hardness readings are taken every 1/16 inch from the quenched end as shown in Fig. 2.14.

2.16 DYNAMIC IMPACT TEST

The dynamic impact tests are conducted to determine the behavior of materials when subjected to high rates of loading, usually in bending, tension or torsion. The quantity measured is the energy absorbed in breaking a specimen by a single blow.

The following two standard tests are commonly used [6]:

1. Charpy test.

2. Izod test.

Both of them are the pendulum type of a single impact tests in which a specimen is usually notched.

In the Charpy test, the specimen is supported at both ends as a simple beam (Fig. 2.15a) and is broken by a falling pendulum. In the Izod test, the specimen is fixed at one end (Fig. 2.15b) and is also

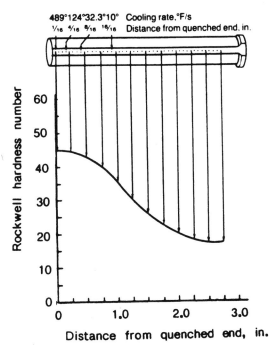

Fig. 2.14 Method of plotting hardness data from an end-quenched Jominy specimen. (From PRINCIPLES OF HEAT TREATMENT OF STEEL, by G. Krauss, 1980, Copyright ASM International. With permission).

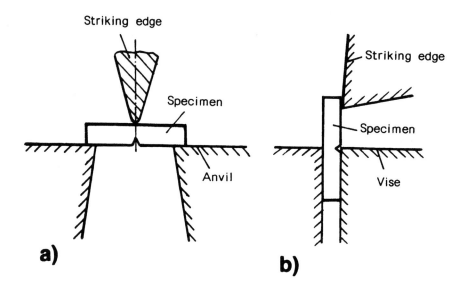

Fig. 2.15 Schematic diagrams of dynamic impact tests: a) Charpy test, b) Izod test.

broken by a falling pendulum. In both tests, the rise of the pendulum after the specimen has been broken is used as a measure of impact strength or notch toughness. These parameters are also frequently evaluated by the percentage of brittle fracture on the fracture faces.

To determine the ductile-to-brittle transition, the dynamic impact tests are conducted for different temperatures of the test samples. The purpose of this test is to establish the **impact-transition temperature (ITT)** that corresponds to a transition for a material from a brittle to ductile state. This temperature is usually assumed to be equal to the temperature at which the test specimen breaks with a certain, relatively low, level of energy absorbtion, frequently 15 foot-pounds (20 J) force as shown in Fig. 2.16 [7].

2.17 TOUGHNESS

Toughness is the ability of a metal to absorb energy and deform plastically before fracturing. Toughness can be defined as a combination of two factors: ductility and strength [8].

Under any load the metal will flow if the maximum shear stress exceeds a certain value, and it will fracture or fail when the maximum normal stress exceeds a certain value.

In torsion testing, the ratio of the maximum normal stress causing fracture to the maximum shear stress is 1 to 1. In testing by compression, tension or bending, this ratio is 2 to 1. In the dynamic impact test, such as the Charpy or Izod tests, the ratio of the maximum normal stress to the maximum shear stress may be considerably higher than 2 to 1. The greater this ratio the more likely that the fracture occurs before the flow [2].

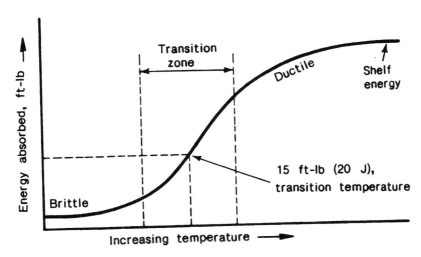

Fig. 2.16 Impact-energy absorbtion as a function of test temperature, obtained with Charpy V-notch impact tests. From Fletcher (1979). Reprinted with permission.

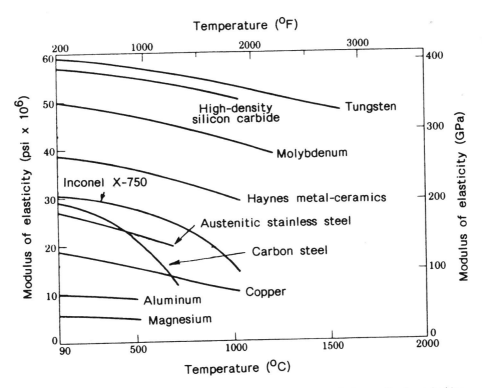

Fig. 2.17 The effect of temperature on the modulus of elasticity for various materials. Adapted from Kalpakjian (1985).

2.18 HEAT RESISTANCE

Heat resistance is generally known as a capability of a metal to maintain its mechanical properties at elevated temperatures. This property of the metal is usually presented by the following parameters expressed as a function of temperature as shown in Figs. 2.17 and 2.18 [9]:

1. Modulus of elasticity
2. Tensile strength
3. 0.2% offset yield strength
4. Elongation
5. Reduction in area

There are two following parameters which are also frequently used to evaluate the heat resistance of metals as shown in Fig. 2.19 [9]:

1. Creep strength
2. Rupture strength

The creep strength is the constant nominal stress that will cause a specified quantity of creep in a given time or a specified creep rate at constant temperature.

The rupture strength is the constant nominal stress that will cause rupture in a specified time at constant temperature.

2.19 THERMAL CONDUCTIVITY

Thermal conductivity is a parameter describing a capability of a material to transfer heat by conduction. The quantity of heat passing through a cross-section area per unit of time is determined by a Fourier's law [10]

$$q = kA(dT/dL) \qquad [Btu/h] , \qquad (2-24)$$

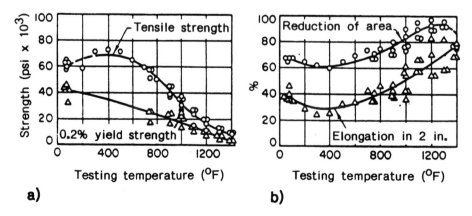

Fig. 2.18 Mechanical properties of killed carbon steel, 0.15% C, annealed at elevated temperature: a) strength, b) ductility. From METALS HANDBOOK (1961). Reprinted with permission.

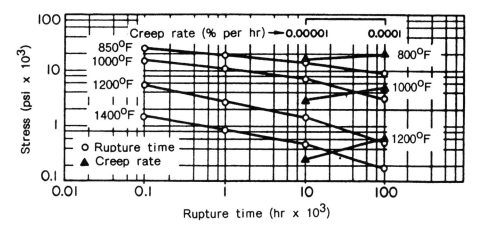

Fig. 2.19 Stress versus rupture time and creep rate curves based on an average data for killed carbon steel, 0.15% C, annealed. From METALS HANDBOOK (1961). Reprinted with permission.

where (dT/dL) = temperature gradient, °F/in.

A = cross-section area, ft^2

k = thermal conductivity, Btu in./ft^2/h/°F.

Thus, the thermal conductivity is equal to a heat transfer rate by conduction through a unit cross-section area of a material having a unit temperature gradient in the direction of the heat transfer.

As can be seen from Fig. 2.20, the thermal conductivity of austenitic stainless steel gradually increases with temperature. The same can be said of thermal conductivity of low-carbon steel in austenitic conditions. However, when low-carbon steel is in ferritic conditions, thermal conductivity decreases with increase in temperature.

2.20 SPECIFIC HEAT

Specific heat is a parameter describing the capability of a material to absorb heat. The relationship between the amount of heat absorbed by a body and the corresponding temperature rise is expressed by the equation [10]

$$Q = cm(dT) \qquad [Btu], \qquad (2-25)$$

where dT = temperature rise, °F

m = body mass, lb

c = specific heat, Btu/lb/°F.

Thus, the specific heat is equal to the amount of heat that is necessary to transfer to a unit mass of a body in order to increase its temperature by a unit degree. Specific heat is temperature dependent and also is sensitive to allotropic transformations in a metal as shown in Fig. 2.21.

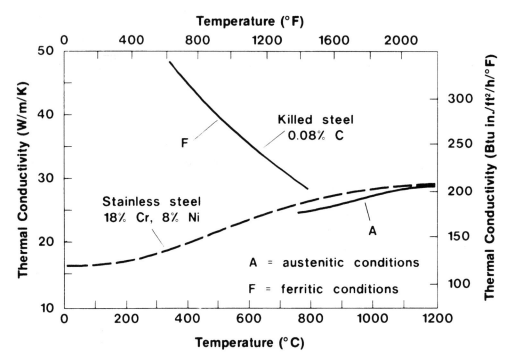

Fig. 2.20 Thermal conductivity of two types of steels at elevated temperatures. Based on data from BISRA (1953).

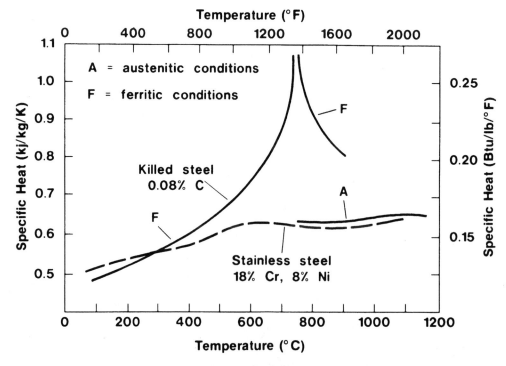

Fig. 2.21 Specific heat of two types of steels at elevated temperatures. Based on data from BISRA (1953).

2.21 DENSITY

Density is usually defined as a mass of a matter in a unit of its volume as it follows from the equation

$$\rho = m/V \qquad [lb/in^3] ,$$

(2-26)

where m = mass of the matter, lb

V = volume of the matter, in.^{3}.

Density usually decreases with increase of temperature (Fig. 2.22). There is a pronounced change in density at the temperature corresponding to an allotropic change in a metal.

2.22 THERMAL DIFFUSIVITY

Thermal diffusivity is a parameter describing a capability of a material to transfer heat when the temperature field of the body changes with time.

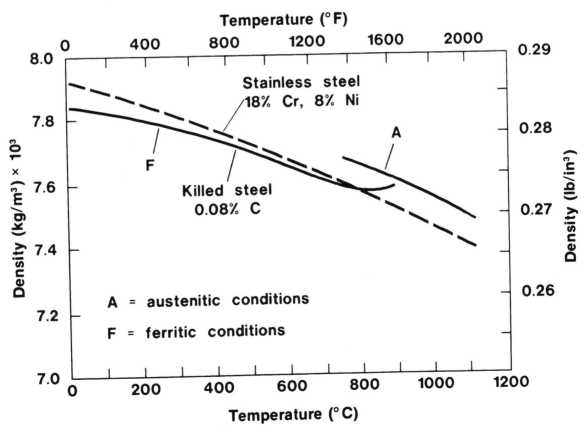

Fig. 2.22 Density of two types of steels at elevated temperatures. Based on data from BISRA (1953).

Temperature distribution in a wall of a body, with respect to time and position and with heat flow along the x-axis only (Fig. 2.23), is given by the following partial differential equation of the second order, based on Fourier's general law of heat conduction [10]

$$dT/dt = a(d^2T/dx^2) \qquad [^{\circ}F\ in^2/h/ft^2], \qquad (2-27)$$

where dT/dt = temperature rate, $^{\circ}F/h$

$\quad dT/dx$ = temperature gradient, $^{\circ}F/in$.

$\quad\quad a$ = thermal diffusivity, $in^4/ft^2/h$.

Thermal diffusivity may be expressed as a function of thermal conductivity k, specific heat c, and density ρ by the formula

$$a = (k/c)/\rho \qquad (2-28)$$

Equation (2-28) shows that the temperature increase propagates faster in bodies exhibiting higher thermal conductivity and lower specific heat and density.

2.23 THERMAL EMISSIVITY

Thermal emissivity describes a capability of a solid body to radiate heat. The heat radiation is known to be electro-magnetic oscillatory phenomena. A solid body emits radiation over a wide range of wavelengths [10].

In order to compare the radiation capabilities of different materials, the intensity of the radiation provided by these materials is compared with that of a so-called **black body.** The black body is the one that absorbs all radiation which it receives, irrespective of wavelength, and reflects none.

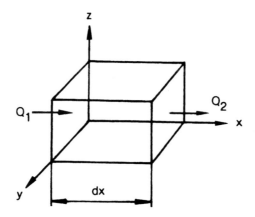

Fig. 2.23 Variable heat flow in a body element.

The radiation process is described by the Steffan-Boltzmann and Kirchhoff's laws of radiation which are expressed as following

$$Q = S\xi A_R [(T_1 + 460)^4 - (T_2 + 460)^4] ,$$

(2-29)

where S = emissivity of a black body, BTU/in.^{2}/h/$^\circ$F^4

A_R = area of radiating or reradiating surface, in.^{2}

T_1 = temperature of radiating surface, $^\circ$F

T_2 = temperature of reradiating surface, $^\circ$F

ξ = coefficient of thermal emissivity.

The coeffecient of thermal emissivity of a given body is the ratio of the amount of heat radiated from a unit area of its surface to the amount of heat radiated from a unit area of a black body under the same conditions. The coefficient of thermal emissivity is affected by the surface finish and temperature as shown in Table 2.4.

2.24 THERMAL EXPANSION

Linear expansion of a solid body with temperature can be defined by the equation

Table 2.4 Coefficient of thermal emissivity for some metals. Data from Schack (1965).

Material	Surface finish	Temperature ($^\circ$F)	Coefficient of thermal emissivity
Mild steel			
	Oxidized, smooth	797	0.83
	" "	1112	0.96
	" "	1292	0.95
	" "	1472	0.92
Austenitic stainless steel type 310			
	As delivered	824	0.22
	Sand blasted	824	0.62
	As delivered	1427	0.58
	Sand blasted	1427	0.78
	Oxidized, smooth	824	0.85
	Sand blasted, oxidized	824	0.91
	Oxidized, smooth	1427	0.90
	Sand blasted, oxidized	1427	0.93
	Oxidized, smooth	2012	0.96
	Sand blasted, oxidized	2012	0.99

$$L_T = L_0[1 + \alpha (T-T_0)] \ ,$$

(2-30)

where L_0, L_T = body lengths at the temperatures T_0 and T correspondingly, in.

α = coefficient of linear thermal expansion, in./in./$^\circ$F.

Thus, the coefficient of linear thermal expansion is equal to relative linear expansion of the body per unit of the temperature change. This coefficient is temperature dependent as shown in Fig. 2.24.

2.25 CORROSION RESISTANCE

Corrosion is the deterioration of a metal by chemical or electrochemical reaction with its environment.

Two types of corrosion processes are known [3, 9]:

1. Electrochemical corrosion - It occurs when a current flows between cathodic and anodic areas on metallic surfaces.

2. Galvanic corrosion - It is associated with electric current produced by a galvanic cell consisting of two different conductors in an electrolyte or two similar conductors in different electrolytes. In case of two different steels (bimetal), the resulting reaction, or 'couple action', is regulated by: a) the metal

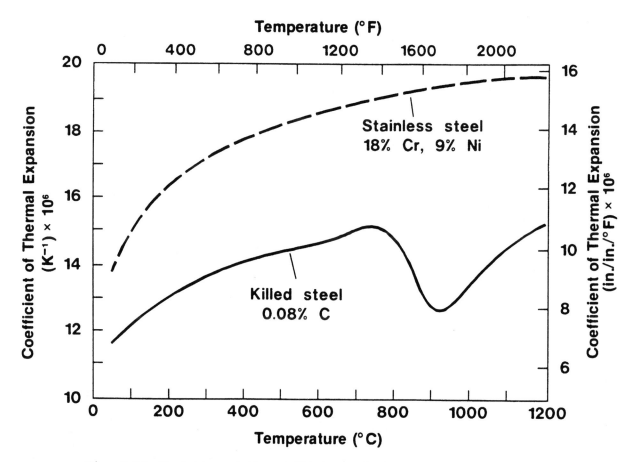

Fig. 2.24 Variation of the coefficient of linear expansion with temperature for two types of steels. Based on data from BISRA (1953).

to which the steel is coupled, b) the conductivity of the solution in which it is in service, c) the contact area between metals, and d) presence or absence of oxygen or other depolarizing agents.

Steel is affected adversely by galvanic corrosion only when in contact with a metal below it in the galvanic series such as shown in Table 2.5 for a specific environment.

The corrosion rate is usually measured as a loss of weight per unit of surface area per unit of time.

2.26 ELECTRICAL RESISTIVITY

Electrical resistance of a solid body can be defined by the equation

$$R = (\gamma L)/A \quad [\text{ohm}], \tag{2-31}$$

where L = body length, in.

A = cross-section area, in.^{2}

γ = electrical resistivity, ohm in.

Thus, the electrical resistivity is equal to the electrical resistance between opposite faces of a cube of unit dimensions. Generally, electrical resistivity of metals increases with temperature as shown in Fig. 2.25.

2.27 MAGNETIC PROPERTIES

Magnetic properties are the most important characteristics of electrical sheet steels. They are defined as follows (Fig. 2.26):

Permeability - A measure of the relative ease with which a metal can be magnetized. It is equal to

$$\mu = B/H, \tag{2-32}$$

Table 2.5 Solution potentials of metals or alloys in a solution containing 53 g/1 NaCl plus 3 g/1 H_2O_2. Data from Metals Handbook (1961).

Metal or alloy	Potential (volts)*
Magnesium	-1.73
Aluminum	-0.84
Iron	-0.63
Stainless steel	-0.15

* Measured against a 0.1 calomel half-cell.

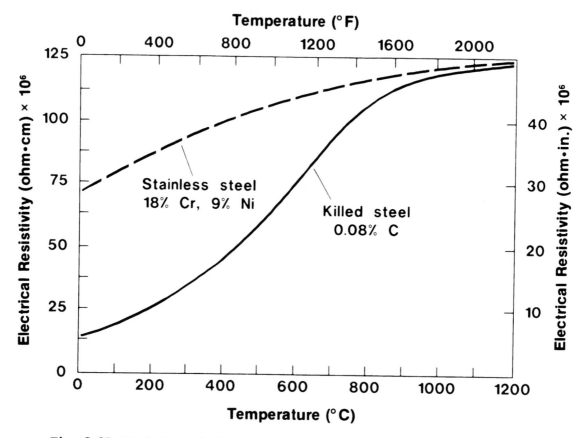

Fig. 2.25 Variation of electrical resistivity with temperature for two types of steels. Based on data from BISRA (1953).

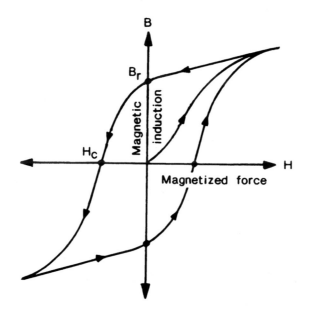

Fig. 2.26 Magnetic hysteresis loop.

where B = magnetic induction, teslas

H = magnetized force, amp-turns/in.

Saturation induction B_s - The maximum intrinsic induction possible in a magnetic material.

Residual induction B_r - The induction remaining in a material when the magnetizing force has been reduced to zero.

Coercive force H_c - The direct-current magnetizing force that must be applied in a direction opposite to the residual induction to reduce the magnetic induction to zero.

Total core loss P_c - The energy dissipated as heat within magnetic core in the presence of a cyclically alternating induced current. It consists of two components: a) magnetic hysteresis loss P_h, and b) eddy-current loss P_e [6].

2.28 MALLEABILITY AND MACHINABILITY

Malleability is the capability of metals to undergo plastic deformation in compression without rupturing [3, 8]. Machinability is the capability of a material to be machined with relative ease with regard to tool life, surface finish and power consumption.

The physical condition of steel, principally the microstructure and the hardness in the annealed state, are two major factors affecting the machining characteristics of tool steel.

In order to provide the ratings of the machinability of various steels, the machinability index is utilized which is a relative measure of the machinability of an engineering material under specified standard conditions.

2.29 WEAR RESISTANCE

The wear resistance or abrasion resistance is an important characteristic of tool steels. The wear tests are usually conducted by measuring the loss of weight per unit surface area of a sample after it has been in moving contact with a standard hardened and finished ground surface for a definite number of strokes and with a definite pressure [8].

When the test is conducted under actual service conditions, it should give valuable information for that particular application. However, because of the diversity of the testing procedures, comparison and correlation of the results from one application to another are virtually impossible.

REFERENCES

1. F.E. Miller and H.A. Doeringsfield, Mechanics of Materials, International Textbook Company, Scranton, Pennsylvania, pp. 3-55 (1955).

2. S. Kalpakjian, Manufacturing Processes for Engineering Materials, Addison-Wesley Publishing Company, Inc., Reading, Massachusetts, pp. 25-65 (1984).

3. E.C. Larke, The Rolling of Strip, Sheet, and Plate, Second Edition, Chapman and Hall, London, pp. 186-214 (1963).

4. ASM Metals Reference Book, American Society for Metals, Metals Park, Ohio, pp. 1-80, 92, 93 (1981).

5. G. Krauss, Principles of Heat Treatment of Steel, American Society for Metals, Metals Park, Ohio, pp. 134-159 (1980).

6. The Making, Shaping and Treating of Steel, 10th Edition, eds. W.T. Lankford, Jr., et al, Association of Iron and Steel Engineers, Pittsburgh, Pennsylvania, pp. 1321-1446 (1985).

7. E.E. Fletcher, A Review of the Status, Selection, and Physical Metallurgy of High-Strength, Low-Alloy Steels, Metals and Ceramics Information Center Columbus, Ohio, Report 79-39, March 1979.

8. G.A. Roberts and R.A. Cary, Tool Steels, Fourth Edition, American Society for Metals, Metals Park, Ohio, pp. 51-70 (1980).

9. Metals Handbook, Eight Edition, Volume 1 Properties and Selection of Metals, American Society for Metals, Metals Park, Ohio, pp. 491-523, 986-988 (1961).

10. A. Schack, Industrial Heat Transfer, Chapman and Hall, London, pp. 4-60, 205-216, 429, (1965).

11. Physical Constants of Some Commercial Steels at Elevated Temperatures, eds. British Iron and Steel Research Association (BISRA), Butterworths Scientific Publications, London (1953).

3

Classification of Steels and Alloys

3.1 MAJOR CLASSIFICATIONS AND SPECIFICATIONS

Classification is the systematic arrangement or division of metals into groups on the basis of (a) composition, (b) finishing methods, (c) product forms, etc. [1-4].

The most commonly known classification systems for metals are listed below:

1. Unified numbering system (UNS)
2. American Iron and Steel Institute (AISI) designation system
3. Society of Automotive Engineers (SAE) designation system
4. AISI-SAE designation system.

Specification is a published document that describes a product acceptable for a wide range of applications and that can be produced by many manufacturers of such items. It describes the requirements, both technical and commercial, that a product must meet.

The most comprehensive and widely used standard specifications are listed below:

1. American Society for Testing and Materials (ASTM) specification
2. Aerospace Materials Specifications (AMS)
3. American Society of Mechanical Engineers (ASME) specification.

3.2 UNIFIED NUMBERING SYSTEM

The Unified Numbering System (UNS) established 15 series of numbers for metals and alloys as shown in Table 3.1. Each number consists of a single-letter prefix followed by five digits. In most cases the letter is suggestive of the family of metals identified, for example, A = aluminum, P = precious metals, H = H-steels, etc.

Table 3.2 shows the secondary division of some primary series of numbers.

3.3 AISI-SAE DESIGNATION SYSTEM

This system classifies the carbon and constructional alloy steels [1]. According to this system, a steel is identified by a numerical index that is partially descriptive of the composition. The first digit identifies the type of steel. In the case of a simple alloy steel, the second number usually indicates the percentage of the predominant alloying element. The last two digits indicate the carbon content in hundredths of a percent. Thus, '2517' indicates a nickel steel of approximately 5% Ni (4.75 to 5.25) and 0.17% C (0.15 to 0.20).

3.4 CARBON STEELS

Carbon steels, or plain carbon steels, represent the largest group of engineering materials which are processed by both hot and cold rolling.

Plain carbon steels can be classified by:

1. Chemical composition

2. Method of manufacturing

According to American Iron and Steel Institute definition, plain carbon steels may have the following maximum content of some chemical elements [1, 2]:

Carbon (C)	up to 1.04%
Manganese (Mn)	up to 1.65%
Silicon (Si)	up to 0.60%
Copper (Cu)	up to 0.60%

On the basis of the carbon content, the plain carbon steel is customarily divided into the following groups [1, 5]:

Hypoeutectoid steels	carbon not over 0.80%
Hypereutectoid steels	carbon over 0.80%
Very low carbon steels	carbon not over 0.10%
Low carbon steels	carbon not over 0.25%
Medium soft carbon steels	carbon 0.26 to 0.40%
Medium high carbon steels	carbon 0.41 to 0.60%
High carbon steels	carbon over 0.60%.

On the basis of the manufacturing method, the plain carbon steels can be specified by:

A. Type of steelmaking process: basic open-hearth, acid open-hearth, basic oxygen, or basic electric-furnace steels

B. Type of deoxidation practice: rimmed, capped, semi-killed, aluminum killed, fully killed, etc.

C. Type of cast product: steels produced by teeming into ingots, as opposed to continuous cast steels.

According to AISI-SAE designation system, the carbon steels are classified as shown in Table 3.3 [2]. In addition to this, the AISI-SAE system identifies the steels which meet hardenability specification and are known as H-steels (Table 3.4).

Table 3.5 shows the composition ranges and limits for carbon steel plate according to ASTM specification.

Some mechanical properties of selected carbon steels are shown in Table 3.6 and Fig. 3.1.

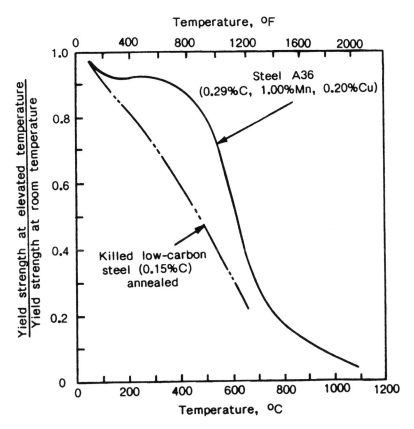

Fig. 3.1 Effect of temperature on yield strength of some low-carbon and alloy steels. Based on data from Metals Handbook (1961) and USS Design Manual (1974).

3.5 ALLOY STEELS

Alloy steels are usually described as those containing specified quantities of alloying elements (other than carbon and the commonly accepted amounts of manganese, copper, silicon, sulfur and phosphorus) within the limits recognized for constructional alloy steels. These alloying elements are added to affect changes in mechanical or physical properties of the steels[2].

Table 3.7 shows the AISI-SAE system of designations for alloy steels.

One of the important groups of alloy steels are so-called 'high-strength low-alloy steels' - HSLA (Table 3.8) that have specified minimum yield points above 275 MPa (40,000 psi) and for some steels may be as high as 1035 MPa (150,000 psi). These steels typically contain small amount of alloying elements (Table 3.9) in order to achieve their strength in hot-rolled or heat-treated conditions.

Some mechanical properties of selected alloy steels are shown in Table 3.10.

3.6 TOOL STEELS

The tool steels may be defined as a class of carbon and alloy steels commonly used to make tools [8]. Tool steels are characterized by high hardness and resistance to abrasion, often accompanied by high toughness and resistance to softening at elevated temperature. These properties are generally attained with high carbon and alloy contents.

According to AISI classification, the tool steels are divided into twelve major categories as shown in Table 3.11. This classification is based on the following criteria:

A. Common end use (e.g. high-speed steels)

B. Common properties (e.g. shock-resisting steels)

C. Common method of heat treatment (e.g. oil-hardening steels)

D. Common composition.

Some mechanical properties of selected tool steels are shown in Table 3.12.

3.7 STAINLESS STEELS

'Stainless steel' is a generic term covering a large group of alloys commonly known for their corrosion resistance [9].

The term 'stainless' implies a resisting to staining, rusting, and pitting in gaseous and aqueous environments. Stainless steels contain in excess of 11% but less than 30% chromium as a principal alloying element.

The stainless steels can be classified into the following four major classes as shown in Tables 3.13 and 3.14:

1. **Ferritic stainless steels** - They are so named because the crystal structure of the steels is the same as that of iron at room temperature.

2. **Martensitic stainless steels** - These are heat treatable steels. With a properly adjusted composition of iron, chromium, and carbon (and other elements) they can be quenched for maximum hardness and subsequently tempered in order to improve ductility.

3. **Precipitation-hardening stainless steels** - They are so designed that their composition is amenable to precipitation hardening.

4. **Austenitic stainless steels** - They are so called because of their austenitic microstructure that is retained at room temperature. It is accomplished by adding up to 37% nickel which is known as a strong austenite former. In some types of steels nickel is partially substituted by manganese and nitrogen.

Some mechanical properties of selected stainless steels are shown in Tables 3.15-3.18 and Fig. 3.2.

3.8 ELECTRICAL SHEET STEELS

Electrical sheet steels are divided into two general groups [1]:

1. **Oriented steels** - They are designed to yield exceptionally good magnetic properties in the rolling, or lengthwise, direction of the steel.

2. **Non-oriented steels** - They are made with a mill treatment that yields a grain structure, or texture, of a random nature. As a result, the magnetic properties in the rolling direction of the strip are not significantly different from those in the transverse direction.

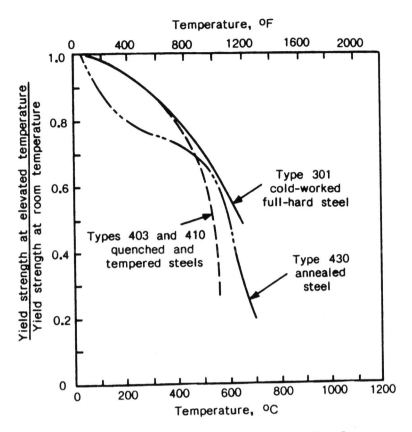

Fig. 3.2 Effect of temperature on yield strength of some stainless steels. Based on data from Metals Handbook (1961).

Main alloying elements in the electrical sheet steels are silicon and aluminum (Table 3.19) with predominant content of silicon.

3.9 HEAT-RESISTING ALLOYS

Heat-resisting alloys are developed for a very high-temperature service where relatively high tensile, thermal, vibratory or shock stresses are encountered and where an oxidation resistance is frequently required [2].

Wrought heat-resisting alloys can be divided into following two groups:

1. Iron-based heat-resisting alloys - These are mainly ferritic, martensitic, precipitation-hardening, and austenitic stainless steels with special compositions (Table 3.20) that provide the heat-resisting characteristics.

2. Superalloys - These are iron, cobalt and nickel-base alloys with outstanding heat-resisting characteristics. Their compositions are shown in Table 3.21. Some mechanical properties of selected superalloys are shown in Table 3.22 and Fig. 3.3.

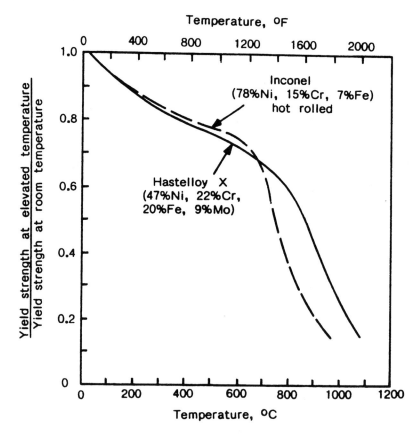

Fig. 3.3 Effect of temperature on yield strength of some superalloys. Based on data from Metals Handbook (1961).

Table 3.1 UNS primary series of numbers for ferrous metals and alloys. Adapted from SAE Handbook (1979).

UNS series	Metal
D00001 - D99999	Specified mechanical properties steels
F00001 - F99999	Cast irons
G00001 - G99999	AISI and SAE carbon and alloy steels (except tool steels)
H00001 - H99999	AISI H-steels
J00001 - J99999	Cast steels (except tool steels)
K00001 - K99999	Miscellaneous steels and ferrous alloys
S00001 - S99999	Heat and corrosion-resistant (stainless) steels
T00001 - T99999	Tool steels

Table 3.2 UNS secondary division of some reactive and refractory metals and alloys. Adapted from SAE Handbook (1979).

UNS series	Metal
R01001 - R01999	Boron
R02001 - R02999	Hafnium
R03001 - R03999	Molybdenum
R04001 - R04999	Rhenium
R05001 - R05999	Tantalum
R06001 - R06999	Thorium
R07001 - R07999	Tungsten
R08001 - R08999	Vanadium
R10001 - R19999	Beryllium
R20001 - R29999	Chromium
R30001 - R39999	Cobalt
R40001 - R49999	Columbium
R50001 - 549999	Titanium
R60001 - R69999	Zirconium

Table 3.3 AISE-SAE system of designations for carbon steels. Adapted from ASM Metals Reference Book (1981).

Numerals and digits	Type of steel and nominal alloy content
10XX (a).......	Plain carbon (Mn 1.00% max)
11XX..........	Resulfurized
12XX..........	Resulfurized and rephosphorized
15XX..........	Plain carbon (1.00-1.65% Mn max)

(a) XX in the last two digits of these designations indicates that the carbon content (in hundredths of a percent) is to be inserted.

Table 3.4 Composition ranges and limits for AISI-SAE standard carbon H-steels. Data from ASM Metals Reference Book (1981).

AISI-SAE designation	UNS designation	Heat composition ranges and limits, % (a)		
		C	Mn	Si
1038H...........H10380		0.34-0.43	0.50-1.00	0.15-0.30
1045H...........H10450		0.42-0.51	0.50-1.00	0.15-0.30
1522H...........H15220		0.17-0.25	1.00-1.50	0.15-0.30
1524H...........H15240		0.18-0.26	1.25-1.75	0.15-0.30
1526H...........H15260		0.21-0.30	1.00-1.50	0.15-0.30
1541H...........H15410		0.35-0.45	1.25-1.75	0.15-0.30
15B21H (b)......H15211		0.17-0.24	0.70-1.20	0.15-0.30
15B35H (b)......H15351		0.31-0.39	0.70-1.20	0.15-0.30
15B37H (b)......H15371		0.30-0.39	1.00-1.50	0.15-0.30
15B41H (b,c)....H15411		0.35-0.45	1.25-1.75	0.15-0.30
15B48H (b,c)....H15481		0.43-0.53	1.00-1.50	0.15-0.30
15B62H (b)......H15621		0.54-0.67	1.00-1.50	0.40-0.60

(a) Typical limits are: 0.040% maximum phosphorus and 0.050% maximum sulfur. (b) Can be expected to contain 0.0005 to 0.003% boron. (c) AISI grade only.

Table 3.5 Composition ranges and limits for carbon steel plate (ASTM specifications). Data from ASM Metals Reference Book (1981).

ASTM specifications	Form, type or grade	UNS designations	Heat composition ranges and limits, % (a)			
			C max	Mn	Si	Cu (b)
A36........Plate		...	0.29	0.80-1.20	...	0.20
A283.......Plate		...	...	...	...	0.20
A284.......Grade A		K01804	0.24	0.90 max	0.10-0.30	...
	Grade B	K02001	0.24	0.90 max	0.15-0.30	...
	Grade C	K02401	0.36	0.90 max	0.15-0.30	...
	Grade D	K02702	0.35	0.90 max	0.15-0.30	...
A529.......Plate		K02703	0.27	1.20 max	...	0.20
A573.......Grade 58		K02301	0.23	0.60-0.90	0.10-0.35	...
	Grade 65	K02404	0.26	0.85-1.20	0.15-0.30	...
	Grade 70	K02701	0.28	0.85-1.20	0.15-0.30	...
A678.......Grade A		K01600	0.16	0.90-1.50	0.15-0.50	0.20
	Grade B	K02002	0.20	0.70-1.60	0.15-0.50	0.20
	Grade C	K02204	0.22	1.00-1.60	0.20-0.50	0.20

(a) Typical limits are: 0.040% maximum phosphorus and 0.050% maximum sulfur. (b) Minimum copper content applicable only if copper-bearing steel is specified.

Table 3.6 Mechanical properties of selected carbon steels in the hot rolled, normalized and annealed condition. Data from ASM Metals Reference Book (1981).

AISI No.(a)	Treat-ment	Austenitizing temperature		Tensile strength		Yield strength		Elongation
		°C	°F	MPa	ksi	MPa	ksi	%
1015...As-rolled				420.6	61.0	313.7	45.4	39.0
	Normalized	925	1700	424.0	61.5	324.1	47.0	37.0
	Annealed	870	1600	386.1	56.0	284.4	41.3	37.0
1020...As-rolled				448.2	65.0	330.9	48.0	36.0
	Normalized	870	1600	441.3	64.0	346.5	50.3	35.8
	Annealed	870	1600	394.7	57.3	294.8	42.8	36.5
1030...As-rolled				551.6	80.0	344.7	50.0	32.0
	Normalized	925	1700	520.6	75.5	344.7	50.0	32.0
	Annealed	845	1550	463.7	67.3	341.3	49.5	31.2
1040...As-rolled				620.5	90.0	413.7	60.0	25.0
	Normalized	900	1650	589.5	85.5	374.0	54.3	28.0
	Annealed	790	1450	518.8	75.3	353.4	51.3	30.2
1050...As-rolled				723.9	105.0	413.7	60.0	20.0
	Normalized	900	1650	748.1	108.5	427.5	62.0	20.0
	Annealed	790	1450	636.0	92.3	365.4	53.0	23.7
1060...As-rolled				813.6	118.0	482.6	70.0	17.0
	Normalized	900	1650	775.7	112.5	420.6	61.0	18.0
	Annealed	790	1450	625.7	90.8	372.3	54.0	22.5
1080...As-rolled				965.3	140.0	586.1	85.0	12.0
	Normalized	900	1650	1010.1	146.5	524.0	76.0	11.0
	Annealed	790	1450	615.4	89.3	375.8	54.5	24.7
1095...As-rolled				965.3	140.0	572.3	83.0	9.0
	Normalized	900	1650	1013.5	147.0	499.9	72.5	9.5
	Annealed	790	1450	656.7	95.3	379.2	55.0	13.0
1117...As-rolled				486.8	70.6	305.4	44.3	33.0
	Normalized	900	1650	467.1	67.8	303.4	44.0	33.5
	Annealed	855	1575	429.5	62.3	279.2	40.5	32.8
1137...As-rolled				627.4	91.0	379.2	55.0	28.0
	Normalized	900	1650	668.8	97.0	396.4	57.5	22.5
	Annealed	790	1450	584.7	84.8	344.7	50.0	26.8
1144...As-rolled				703.3	102.0	420.6	61.0	21.0
	Normalized	900	1650	667.4	96.8	399.9	58.0	21.0
	Annealed	790	1450	584.7	84.8	346.8	50.3	24.8

(a) All grades are fine-grained except for those in the 1100 series which are coarse-grained. Heat treated specimens were oil quenched unless otherwise indicated.

Table 3.7 AISE - SAE system of designations for alloy steels. Adapted from ASM Metals Reference Book (1981).

Numerals and digits	Type of steel and nominal alloy content	Numerals and digits	Type of steel and nominal alloy content
Manganese Steels		**Nickel-Chromium-Molybdenum Steels**	
13XX............Mn 1.75		43XX....Ni 1.82; Cr 0.50 and 0.80 Mo 0.25	
Nickel Steels		43BVXX..Ni 1.82; Cr 0.50; Mo 0.12 and 0.25; V 0.03 min	
23XX............Ni 3.50		47XX....Ni 1.05; Cr 0.45; Mo 0.20 and 0.35	
25XX............Ni 5.00		81XX....Ni 0.30; Cr 0.40; Mo 0.12	
Nickel-Chromium Steels		86XX....Ni 0.55; Cr 0.50; Mo 0.20	
		87XX....Ni 0.55; Cr 0.50; Mo 0.25	
31XX....Ni 1.25; Cr 0.65 and 0.80		88XX....Ni 0.55; Cr 0.50; Mo 0.35	
32XX....Ni 1.75; Cr 1.07		93XX....Ni 3.25; Cr 1.20; Mo 0.12	
33XX....Ni 3.50; Cr 1.50 and 1.57		94XX....Ni 0.45; Cr 0.40; Mo 0.12	
34XX....Ni 3.00; Cr 0.77		97XX....Ni 0.55; Cr 0.20; Mo 0.20	
		98XX....Ni 1.00; Cr 0.80; Mo 0.25	
Molybdenum Steels		**Chromium-Vanadium Steels**	
40XX....Mo 0.20 and 0.25			
44XX....Mo 0.40 and 0.52		61XX....Cr 0.60, 0.80 and 0.95 V 0.10 and 0.15 min	
Chromium-Molybdenum Steels		**Tungsten-Chromium Steels**	
41XX....Cr 0.50, 0.80 and 0.95 Mo 0.12, 0.20, 0.25 and 0.30		72XX....W 1.75; Cr 0.75	
Nickel-Molybdenum Steels		**Silicon-Manganese Steels**	
46XX....Ni 0.85 and 1.82; Mo 0.20 and 0.25		92XX....Si 1.40 and 2.00 Mn 0.65, 0.82 and 0.85 Cr 0.00 and 0.65	
48XX....Ni 3.50; Mo 0.25		**High-Strength Low-Alloy Steels**	
Chromium Steels		9XX.....Various SAE grades	
50XX....Cr 0.27, 0.40, 0.50 and 0.65		**Boron Steels**	
51XX....Cr 0.80, 0.87, 0.92, 0.95 1.00 and 1.05		XXBXX...B denotes boron steel	
50XXX...Cr 0.50; C 1.00 min		**Leaded Steels**	
51XXX...Cr 1.02; C 1.00 min			
52XXX...Cr 1.45; C 1.00 min		XXLXX....L denotes leaded steel	

Table 3.8 SAE system of designations for HSLA steels. Data from ASM Metals Reference Book (1981).

SAE designation(b)	Heat composition limits, %(a)			SAE designation(b)	Heat composition limits, %(a)		
	C max	Mn max	P max		C max	Mn max	P max
942X	0.21	1.35	0.04	950D	0.15	1.00	0.15
945A	0.15	1.00	0.04	950X	0.23	1.35	0.04
945C	0.23	1.40	0.04	955X	0.26	1.35	0.04
945X	0.22	1.35	0.04	960X	0.26	1.45	0.04
950A	0.15	1.30	0.04	965X	0.26	1.45	0.04
950B	0.22	1.30	0.04	970X	0.26	1.65	0.04
950C	0.25	1.60	0.04	980X	0.26	1.65	0.04

(a) Maximum contents of sulfur and silicon for all grades: 0.050% S, 0.90% Si. (b) Second and third digits of designation indicate minimum yield strength in ksi. Suffix "X" indicates that the steel contains niobium, vanadium, nitrogen or other alloying elements. A second suffix "K" indicates that the steel is produced fully killed using fine grain practice. Otherwise, the steel is produced semikilled.

Table 3.9 ASTM classification of HSLA steels. Adapted from Making, Shaping and Treating of Steel (1985).

Types	Composition max, %	Types	Composition max, %
Type 1:		**Type 5:**	
Titanium, min	0.05	Columbium, min (c)	0.03
Silicon, max	0.10	Molybdenum, min (c)	0.20
Type 2:		Silicon, max	0.30
Vanadium, min	0.02	**Type 6:**	
Silicon, max (a)	0.60	Columbium	0.005–0.10
Nitrogen, min (a)	0.005	Silicon, max	0.90
Type 3:		**Type 7:**	
Columbium, min	0.005	Columbium or vanadium or both, min	0.005
Vanadium, max (a)	0.08	Silicon, max	0.60
Silicon, max (a)	0.60	Nitrogen, max	0.02
Nitrogen, max (a)	0.02	**Type 8:**	
Type 4:		Columbium	0.005–0.15
Zirconium, min	0.05	Zirconium, min	0.05
Silicon, max	0.90	**Type 9:**	
Chromium, max (a)	0.80	Columbium, min	0.01
Titanium, max (a)	0.10	Vanadium, min (a)	0.05
Boron, max (a)	0.0025	Silicon, max (a)	0.60
Columbium (b)	0.005–0.06		

(a) Not added to Grades 50 and 60. (b) Might not be added to Grade 50. (c) Available as Grade 80 only.

Table 3.10 Mechanical properties of selected alloy steels in the hot rolled, normalized and annealed condition. Data from ASM Metals Reference Book (1981).

AISI No.(a)	Treatment	Austenitizing temperature		Tensile strength		Yield strength		Elongation
		°C	°F	MPa	ksi	MPA	ksi	%
1340...Normalized		870	1600	836.3	121.3	558.5	81.0	22.0
	Annealed	800	1475	703.3	102.0	436.4	63.3	25.5
3140...Normalized		870	1600	891.5	129.3	599.8	87.0	19.7
	Annealed	815	1500	689.5	100.0	422.6	61.3	24.5
4140...Normalized		870	1600	1020.4	148.0	655.0	95.0	17.7
	Annealed	815	1500	655.0	95.0	417.1	60.5	25.7
4340...Normalized		870	1600	1279.0	185.5	861.8	125.0	12.2
	Annealed	810	1490	744.6	108.0	472.3	68.5	22.0
4620...Normalized		900	1650	574.3	83.3	366.1	53.1	29.0
	Annealed	855	1575	512.3	74.3	372.3	54.0	31.3
4820...Normalized		860	1580	75.0	109.5	484.7	70.3	24.0
	Annealed	815	1500	681.2	98.8	464.0	67.3	22.3
5150...Normalized		870	1600	870.8	126.3	529.5	76.8	20.7
	Annealed	825	1520	675.7	98.0	357.1	51.8	22.0
6150...Normalized		870	1600	939.8	136.3	615.7	89.3	21.8
	Annealed	815	1500	667.4	96.8	412.3	59.8	23.0
8630...Normalized		870	1600	650.2	94.3	429.5	62.3	23.5
	Annealed	845	1550	564.0	81.8	372.3	54.0	29.0
8740...Normalized		870	1600	929.4	134.8	606.7	88.0	16.0
	Annealed	815	1500	695.0	100.8	415.8	60.3	22.2
9255...Normalized		900	1650	932.9	135.3	579.2	84.0	19.7
	Annealed	845	1550	774.3	112.3	486.1	70.5	21.7
9310...Normalized		890	1630	906.7	131.5	570.9	82.8	18.8
	Annealed	845	1550	820.5	119.0	439.9	63.8	17.3

(a) All grades are fine-grained. Heat treated specimens were oil quenched unless otherwise indicated.

Table 3.11 Composition ranges of principal types of tool steels. Data from ASM Metals Reference Book (1981).

AISI designation	Composition (a), %								
	C	Mn	Si	Cr	Ni	Mo	W	V	Co
Molybdenum high-speed steels, M1......M47									
min	0.75	0.10	0.15	3.50	...	3.25	...	0.95	...
max	1.40	0.60	0.65	4.75	0.30	10.0	7.00	4.50	13.0
Tungsten high-speed steels, T1........T15									
min	0.65	0.10	0.15	3.75	...	...	11.75	0.80	...
max	1.60	0.40	0.40	5.00	0.30	1.25	21.0	5.25	13.0
Chromium hot work steels, H10........H19									
min	0.30	0.20	0.20	3.00	...	0.30	...	0.25	...
max	0.45	0.70	1.20	5.50	0.30	3.00	5.25	1.20	4.5
Tungsten hot work steels, H21........H26									
min	0.22	0.15	0.15	1.75	...	...	8.50	0.25	...
max	0.55	0.40	0.60	12.75	0.30	...	19.0	1.25	...
Molybdenum hot work steels, H42									
min	0.55	0.15	...	3.75	...	4.50	5.50	1.75	...
max	0.70	0.40	...	4.50	0.30	5.50	6.75	2.20	...
Air-hardening medium-alloy cold work steels, A2......A10									
min	0.45	0.40	0.50	...	0.30	0.90	...	...	...
max	2.85	2.50	1.50	5.75	2.05	1.80	1.50	5.15	...
High-carbon, high-chromium cold work steels, D2.....D7									
min	1.40	...	...	11.0	...	...	...	1.00	...
max	2.50	0.60	0.60	13.5	0.30	1.20	1.00	4.40	3.5
Oil-hardening cold work steels, O1......O7									
min	0.85	0.30	0.50	0.35	...	...	...	1.00	1.0
max	1.55	1.80	1.50	0.85	0.30	0.30	1.00	4.40	3.5
Shock-resisting steels, S1......S7									
min	0.40	0.10	0.15	0.35	...	0.20	...	0.15	...
max	0.65	1.50	2.50	3.50	0.30	1.80	3.00	0.50	...
Low-alloy special purpose steels, L2......L6									
min	0.45	0.10	...	0.60	...	0.25	...	0.10	...
max	1.00	0.90	0.50	1.20	2.00	0.50	...	0.30	...
Low-carbon mold steels, P2......P21									
min	0.05	0.10	0.10	0.20	...	...	...	...	...
max	0.40	1.00	0.80	5.25	4.25	1.00	...	0.25	1.25
Water-hardening tool steels, W1......W5									
min	0.70	0.10	0.10	0.15	...	...	...	0.10	...
max	1.50	0.40	0.40	0.60	0.20	0.10	0.15	0.35	...

(a) All steels contain 0.25% max Cu, 0.03% max P and 0.03% max S.

Table 3.12 Nominal room-temperature mechanical properties of group L and group S tool steels. Data from ASM Metals Reference Book (1981).

Type	Condition	Tensile strength		0.2% yield strength		Elongation
		MPa	ksi	MPa	ksi	%, (a)
L2	Annealed............................710		103	510	74	25
	Oil quenched from 855 °C (1575 °F) and single tempered at:					
	205 °C (400 °F)............... 2000		290	1790	260	5
	315 °C (600 °F)..................1790		260	1655	240	10
	425 °C (800 °F)..................1550		225	1380	200	12
	540 °C (1000 °F)..................1275		185	1170	170	15
	650 °C (1200 °F)................. 930		135	760	110	25
L6	Annealed............................655		95	380	55	25
	Oil quenched from 845 °C (1550 °F) and single tempered at:					
	315 °C (600 °F)..................2000		290	1790	260	4
	425 °C (800 °F)..................1585		230	1380	200	8
	540 °C (1000 °F)..................1345		195	1100	160	12
	650 °C (1200 °F)................. 965		140	830	120	20
S1	Annealed............................690		100	415	60	24
	Oil quenched from 930 °C (1700 °F) and single tempered at:					
	205 °C (400 °F)..................2070		300	1895	275	..
	315 °C (600 °F)..................2030		294	1860	270	4
	425 °C (800 °F)..................1790		260	1690	245	5
	540 °C (1000 °F)..................1680		244	1525	221	9
	650 °C (1200 °F)..................1345		195	1240	180	12
S5	Annealed............................725		105	440	64	25
	Oil quenched from 870 °C (1600 °F) and single tempered at:					
	205 °C (400 °F)..................2345		340	1930	280	5
	315 °C (600 °F)..................2240		325	1860	270	7
	425 °C (800 °F)..................1895		275	1690	245	9
	540 °C (1000 °F)..................1520		220	1380	200	10
	650 °C (1200 °F)..................1035		150	1170	170	15

a) In 50 mm or 2 in.

Table 3.13 Composition of ferritic, martensitic and precipitation-hardening stainless steels. Data from ASM Metals Reference Book (1981).

Type	Composition, % (a)						
	C	Mn	Si	Cr	P	S	Other
Ferritic types							
405....0.08	1.00	1.00	11.5-14.5	0.04	0.03	0.10-0.30 Al
409....0.08	1.00	1.00	10.5-11.75	0.045	0.045	6x%C min Ti(d)
429....0.12	1.00	1.00	14.0-16.0	0.04	0.03	...
430....0.12	1.00	1.00	16.0-18.0	0.04	0.03	...
434....0.12	1.00	1.00	16.0-18.0	0.04	0.03	0.75-1.25 Mo
442....0.20	1.00	1.00	18.0-23.0	0.04	0.03	...
446....0.20	1.50	1.00	23.0-27.0	0.04	0.03	0.25 N
Martensitic types							
403....0.15	1.00	0.50	11.5-13.0	0.04	0.03	...
410....0.15	1.00	1.00	11.5-13.0	0.04	0.03	...
414....0.15	1.00	1.00	11.5-13.5	0.04	0.03	1.25-2.5 Ni
416....0.15	1.25	1.00	12.0-14.0	0.04	0.03	0.6 Mo (c)
420....0.15(e)	1.00	1.00	12.0-14.0	0.04	0.03	...
431....0.20	1.00	1.00	15.0-17.0	0.04	0.03	1.25-2.5 Ni(b)
440A...0.75	1.00	1.00	16.0-18.0	0.04	0.03	0.75 Mo
440B...0.95	1.00	1.00	16.0-18.0	0.04	0.03	0.75 Mo
440C...1.20	1.00	1.00	16.0-18.0	0.04	0.03	0.75 Mo
502....0.10	1.00	1.00	4.0-6.0	0.04	0.03	0.4-0.65 Mo
503....0.15	1.00	1.00	6.0-8.0	0.04	0.04	0.45-0.65 Mo
504....0.15	1.00	1.00	8.0-10.0	0.04	0.04	0.9-1.1 Mo
Precipitation-hardening types							
15-5PH..0.07	1.00	1.00	14.0-15.5	0.04	0.03	2.5-4.5 Cu; 3.5-5.5 Ni 0.15-0.45 Nb+Ta
17-4PH..0.07	1.00	1.00	15.5-17.5	0.04	0.03	3.0-5.0 Cu 3.0-5.0 Ni(b) 0.15-0.45 Nb+Ta
17-7PH..0.09	1.00	1.00	16.0-18.0	0.04	0.03	0.75-1.5 Al; 6.5-7.75 Ni(b)

(a) Single values are maximum values unless otherwise indicated.
(b) For some tubemaking processes, the nickel content of certain austenitic types must be slightly higher than shown. (c) Optional.
(d) 0.75% maximum. (e) Minimum value.

Table 3.14 Compositions of austenitic stainless steels. Data from ASM Metals Reference Book (1981).

Type	Composition (a), %							
	C	Mn	Si	Cr	Ni(b)	P	S	Others
201	0.15	7.50	1.0	16.0–18.0	3.5–5.5	0.06	0.03	0.25 N
202	0.15	10.00	1.0	17.0–19.0	4.0–6.0	0.06	0.03	0.25 N
205	0.25	15.50	1.0	16.5–18.0	1.0–1.75	0.06	0.03	0.32–0.40 N
301	0.15	2.00	1.0	16.0–18.0	6.0–8.0	0.045	0.03	...
302	0.15	2.00	1.0	17.0–19.0	8.0–10.0	0.045	0.03	...
302B	0.15	2.00	3.0	17.0–19.0	8.0–10.0	0.045	0.03	...
303Se	0.15	2.00	1.0	17.0–19.0	8.0–10.0	0.20	0.06	0.15 min Se
304	0.08	2.00	1.0	18.0–20.0	8.0–10.5	0.045	0.03	...
304L	0.03	2.00	1.0	18.0–20.0	8.0–12.0	0.045	0.03	...
304LN	0.03	2.00	1.0	18.0–20.0	8.0–10.5	0.045	0.03	0.10–0.15 N
S30430	0.08	2.00	1.0	17.0–19.0	8.0–10.0	0.045	0.03	3.0–4.0 Cu
304N	0.08	2.00	1.0	18.0–20.0	8.0–10.5	0.045	0.03	0.10–0.16 N
305	0.12	2.00	1.0	17.0–19.0	10.5–13.0	0.045	0.03	...
308	0.08	2.00	1.0	19.0–21.0	10.0–12.0	0.045	0.03	...
309	0.20	2.00	1.0	22.0–24.0	12.0–15.0	0.045	0.03	...
309S	0.08	2.00	1.0	22.0–24.0	12.0–15.0	0.045	0.03	...
310	0.25	2.00	1.5	24.0–26.0	19.0–22.0	0.045	0.03	...
310S	0.08	2.00	1.5	24.0–26.0	19.0–22.0	0.045	0.03	...
314	0.25	2.00	3.0	23.0–26.0	19.0–22.0	0.045	0.03	...
316	0.08	2.00	1.0	16.0–18.0	10.0–14.0	0.045	0.03	2.0–3.0 Mo
316H	0.10	2.00	1.0	16.0–18.0	10.0–14.0	0.045	0.03	2.0–3.0 Mo
316L	0.03	2.00	1.0	16.0–18.0	10.0–14.0	0.045	0.03	2.0–3.0 Mo
316N	0.08	2.00	1.0	16.0–18.0	10.0–14.0	0.045	0.03	2.0–3.0 Mo 0.10–0.16 N
317	0.08	2.00	1.0	18.0–20.0	11.0–15.0	0.045	0.03	3.0–4.0 Mo
317L	0.08	2.00	1.0	18.0–20.0	11.0–15.0	0.045	0.03	3.0–4.0 Mo
321	0.08	2.00	1.0	17.0–19.0	9.0–12.0	0.045	0.03	5 x %C min Ti
321H	0.10	2.00	1.0	17.0–19.0	9.0–12.0	0.045	0.03	5 x %C min Ti
329	0.10	2.00	1.0	25.0–30.0	3.0–6.0	0.045	0.03	1.0–2.0 Mo
330	0.08	2.00	1.5	17.0–20.0	34.0–37.0	0.040	0.03	...
347	0.08	2.00	1.0	17.0–19.0	9.0–13.0	0.045	0.03	10 x %C min Nb + Ti(c)
348	0.08	2.00	1.0	17.0–19.0	9.0–13.0	0.045	0.03	0.2 Cu; 10 x %C min Nb + Ta(d)
384	0.08	2.00	1.0	15.0–17.0	17.0–19.0	0.045	0.03	...

(a) Single values are maximum values unless otherwise indicated. (b) For some tubemaking processes, the nickel content of certain austenitic types must be slightly higher than shown. (c) Optional. (d) 0.10% Ta max.

Table 3.15 Minimum mechanical properties of ferritic stainless
steels. Data from ASM Metals Reference Book (1981).

Product form (a)	Condition	Tensile strength		0.2% yield strength		Elongation
		MPa	ksi	MPa	ksi	%
Type 405 (UNS S40500)						
P, Sh, St	Annealed........415		60	170	25	20
Type 409 (UNS S40900)						
P, Sh, St	Annealed........415		60	205	30	22(c)
Type 429 (UNS S42900)						
P, Sh, St	Annealed........450		65	205	30	22(c)
Type 430 (UNS S43000)						
P, Sh, St	Annealed........450		65	205	30	22(c)
Type 434 (UNS S43400)						
Sh	Annealed........530(b)		77(b)	365(b)	53(b)	23(b)
Type 436 (UNS S43600)						
Sh, St	Annealed........530(b)		77(b)	365(b)	53(b)	23(b)
Type 442 (UNS S44200)						
P, Sh, St	Annealed........515		75	275	40	20
Type 444 (UNS S44400)						
P, Sh, St	Annealed........415		60	275	40	20
Type 446 (UNS S44600)						
P, Sh, St	Annealed........515		75	275	40	20

(a) P-plate; Sh - sheet; St - strip. (b) Typical values.
(c) 20% reduction for 1.3 mm (0.050 in.) in thickness and under.

Table 3.16 Minimum mechanical properties of martensitic stainless steels. Data from ASM Metals Reference Book (1981).

Product form (a)	Condition	Tensile strength		0.2% yield strength		Elongation
		MPa	ksi	MPa	ksi	%
Type 403 (UNS S40300)						
P, Sh, St	Annealed........485		70	205	30	25(b)
Type 410 (UNS S41000)						
P, Sh, St	Annealed........450		65	205	30	22(b)
Type 410S (UNS S41008)						
P, Sh, St	Annealed........415		60	205	30	22
Type 501 (UNS S50100)						
B, P	Annealed........485(c)		70(c)	205(c)	30(c)	28(c)
B, P	Tempered 540 °C (1000 °F).......1210(c)		175(c)	965(c)	140(c)	15(c)
Type 502 (UNS S50200)						
B, P	Annealed........485(c)		70(c)	205(c)	30(c)	30(c)

(a) B – bar; P – plate; Sh – sheet; St – strip. (b) 20% elongation for 1.3 mm (0.050 in.) and under in thickness. (c) Typical values.

Table 3.17 Minimum mechanical properties of precipitation-hardening stainless steels. Data from ASM Metals Reference Book (1981).

Product form (a)	Condition	Tensile strength		Yield strength		Elongation
		MPa	ksi	MPa	ksi	%
PH 13-8 Mo (UNS S13800)						
B, P, Sh, St	H950........1520		220	1410	205	6-10 (b)
B, P, Sh, St	H1000.......1380		200	1310	190	6-10 (b)
15-5 PH (UNS S15500) and						
17-4 PH (UNS S17400)						
B, P, Sh, St	H900........1310		190	1170	170	10 (c)
B, P, Sh, St	H925........1170		170	1070	155	10 (c)
B, P, Sh, St	H1025.......1070		155	1000	145	12 (c)
B, P, Sh, St	H1075.......1000		145	860	125	13 (c)
B, P, Sh, St	H1100........965		140	795	115	14 (c)
B, P, Sh, St	H1150........930		135	725	105	16 (c)
B, P, Sh, St	H1150M.......795		115	515	75	18 (c)
17-7 PH (UNS S17700)						
P, Sh, St	RH950.......1450		210	1310	190	1-6 (c)
P, Sh, St	TH1050......1240		180	1030	150	3-7 (c)
P, Sh, St	Cold rolled.1380		200	1210	150	1
P, Sh, St	CH900.......1650		240	1590	230	1

(a) B - bar; P - plate; Sh - sheet; St - strip. (b) Where minimum value is also given, maximum value applies only to the flat-rolled products. Both max and min values may vary with thickness for flat-rolled products. (c) Value varies with thickness for flat-rolled products.

Table 3.18 Minimum room-temperature mechanical properties of austenitic stainless steels. Data from ASM Metals Reference Book (1981).

Product form (a)	Condition	Tensile strength		0.2% yield strength		Elongation
		MPa	ksi	MPa	ksi	%
Type 201 (UNS S20100)						
W, P, Sh, St	Annealed......	655	95	310	45	40
Sh, St	1/4 hard.......	860	125	515	75	20
Sh, St	1/2 hard......	1030	150	760	110	10
Sh, St	3/4 hard......	1210	175	930	135	7
Sh, St	Full hard....	1280	185	965	140	5
Type 202 (UNS S20200)						
W, P, Sh, St	Annealed......	655	95	310	45	40
Sh, St	1/4 hard.......	860	125	515	75	12
Sh, St	1/2 hard......	1030	150	760	110	10
Type 301 (UNS S30100)						
B, W, P, Sh, St	Annealed......	515	75	205	30	40
Sh, St	1/4 hard.......	860	125	515	75	25
Sh, St	1/2 hard......	1030	150	760	110	18
Sh, St	3/4 hard......	1210	175	930	135	12
Sh, St	Full hard....	1280	185	965	140	9
Type 302 (UNS S30200)						
P, Sh, St	Annealed......	515	75	205	30	40
P, Sh, St	High tensile, grade B........	585	85	310	45	40
P, Sh, St	High tensile, grade C........	860	125	515	75	..
P, Sh, St	High tensile, grade D.......	1030	150	760	110	..
Type 304 (UNS S30400)						
P, Sh, St	Annealed......	515	75	205	30	40
Sh, St	High tensile, grade B........	550	80	310	45	..
Sh, St	High tensile, grade C........	860	125	515	75	..
Sh, St	High tensile, grade D.......	1030	150	760	110	..
Types: 309 (UNS S30900), 309S (UNS S30908), 310 (UNS S31000), 310S (UNS S31008) and 316 (UNS S31600)						
P, Sh, St	Annealed......	515	75	205	30	40
Type 330 (UNS N08330)						
P, Sh, St	Annealed......	480	70	210	30	30
Type 330HC						
B, W, St	Annealed......	585(b)	85(b)	290(b)	42(b)	45(b)

(a) B - bar; P - plate; Sh - sheet; St - strip. (b) Values given are typical.

Table 3.19 Typical mechanical properties of some electrical sheet grades. Data from Making, Shaping and Treating of Steel (1985).

Common trade designation	Normal thickness		Approximate alloy content (Si + Al)%	Yield point(a)		Tensile strength(a)		Elongation- %,(b)	Electrical resistivity μΩcm
	mm	in.		MPa	ksi	MPa	ksi		
Oriented									
M-6	0.35	0.0138	3.0	345	50	379	35	18	50
Non-oriented fully processed									
M-19	0.36	0.014	3.3	365	53	503	73	21	52
M-27	0.36	0.014	2.8	365	53	503	73	22	49
M-36	0.47	0.0185	2.65	365	53	496	72	24	49
M-43	0.64	0.025	2.35	345	50	476	69	28	39
M-45	0.64	0.025	1.85	345	50	476	69	30	35
M-47	0.64	0.025	1.05	352	51	434	63	27	23
Non-oriented semi-processed									
M-36	0.47	0.0185	2.65	400	58	517	75	24	49
M-43	0.64	0.025	2.35	441	64	545	79	20	39
M-45	0.64	0.025	1.85	462	67	517	75	27	35
Non-oriented carbon									
Type 2S	0.61	0.024	0.15	379	55	434	63	24	16
Type 1	0.61	0.024	0.15	269	39	372	54	33	14

(a) Average of lengthwise and transverse test specimens. (b) In 50 mm or 2 in.

Table 3.20 Nominal composition of wrought iron-base heat-resisting alloys. Data from ASM Metals Reference Book (1981).

Designation	Composition, %					
	C	Cr	Ni	Mo	Nb	Others
Ferritic stainless steels						
405.............0.15*	13.0	...	...	...	0.2 Al
406.............0.15*	13.0	...	...	...	4.0 Al
409.............0.08*	11.0	0.5	...	...	6 x %C min Ti
430.............0.12*	16.0	...	...	...	...
434.............0.12*	17.0	...	1.0	...	...
18SR............0.05	18.0	0.5	...	...	2.0 Al, 0.4 Ti
18Cr-2Mo............	18.0	...	2.0	...	...
446.............0.20*	25.0	...	...	...	0.25 N
E-Brite 26-1....0.01*	26.0	...	1.0	0.10	0.015* N
26-1Ti..........0.04	26.0	...	1.0	...	10 x %C min Ti
29Cr-4Mo........0.01*	29.0	...	4.0	...	0.02* N
Quenched and tempered martensitic stainless steels						
403.............0.15*	12.0	...	...	...	...
410.............0.15*	12.5	...	...	...	...
416.............0.15*	13.0	...	0.6	...	0.15 min C
422.............0.20	12.5	0.75	1.0	...	1.0 W, 0.22 V
H-46............0.12	10.75	0.50	0.85	0.30	0.2 V, 0.07 N
Moly Ascoloy....0.14	12.0	2.4	1.8	...	0.35 V, 0.05 N
Greek Ascoloy...0.15	13.0	2.0	...	...	3.0 W
Jethete M-152...0.12	12.0	2.5	1.7	...	0.30 V
Almar 363.......0.05	11.5	4.5	...	...	10 x %C min Ti
431.............0.20*	16.0	2.0	...	...	...
Precipitation-hardening martensitic stainless steels						
Custom 450......0.05*	15.5	6.0	0.75	8 x %C min	1.5 Cu
Custom 455......0.03	11.75	8.5	...	0.30	2.25 Cu, 1.2 Ti
15-5 PH.........0.07	15.0	4.5	...	0.03	3.5 Cu
17-4 PH.........0.04	16.5	4.25	...	0.25	3.6 Cu
PH 13-8 Mo......0.05	12.5	8.0	2.25	...	1.1 Al

(continued)

* Maximum content.

Table 3.20 Nominal composition of wrought iron-base heat-resisting alloys (continued). Data from ASM Metals Reference Book (1981).

Designation	Composition, %					
	C	Cr	Ni	Mo	Nb	Other
Precipitation-hardening semiaustenitic stainless steels						
AM-350..........0.10	16.5	4.25	2.75	...	0.10 N	
AM-355..........0.13	15.5	4.25	2.75	...	0.10 N	
17-7 PH.........0.07	17.0	7.0	...	...	1.15 Al	
PH 15-7 Mo......0.07	15.0	7.0	2.25	...	1.15 Al	
Austenitic stainless steels						
304.............0.08*	19.0	10.0	...	...	...	
304L............0.03*	19.0	10.0	...	...	...	
304N............0.08*	19.0	9.25	...	...	0.13 N	
309.............0.20*	23.0	13.0	...	...	...	
310.............0.25*	25.0	20.0	...	...	...	
316.............0.08*	17.0	12.0	2.5	...	...	
316L............0.03*	17.0	12.0	2.5	...	...	
316N............0.08*	17.0	12.0	2.5	...	0.13 N	
317.............0.08*	19.0	13.0	3.5	...	...	
321.............0.08*	18.0	10.0	...	...	5 x %C min	
347.............0.08*	18.0	11.0	...	10 x %C min	...	
19-9 DL.........0.30	19.0	9.0	1.25	0.40	1.25 W, 0.30 Ti	
19-9 DX.........0.30	19.2	9.0	1.5	...	1.20 W, 0.55 Ti	
17-14-CuMo......0.12	16.0	14.0	2.5	0.40	3.0 Cu, 0.30 Ti	
202.............0.09	18.0	5.0	...	...	8.0 Mn, 0.10 N	
216.............0.05	20.0	6.0	2.5	...	8.5 Mn, 0.35 N	
21-6-9..........0.04*	20.25	6.5	...	...	9.0 Mn, 0.30 N	
Nitronic 32.....0.10	18.0	1.6	...	...	12.0 Mn, 0.34 N	
Nitronic 33.....0.08*	18.0	3.0	...	...	13.0 Mn, 0.30 N	
Nitronic 50.....0.06*	21.0	12.0	2.0	0.20	5.0 Mn, 0.30 N	
Nitronic 60.....0.10*	17.0	8.5	2.0	...	8.0 Mn, 0.20 V, 4.0 Si	

* Maximum content.

Table 3.21 Nominal composition of wrought superalloys. Data from ASM Metals Reference Book (1981).

Designation	Composition (a), %						
	C	Cr	Ni	Co	Mo	Ti	Other
Iron-base solid-solution alloys							
Incoloy 800......0.05		21.0	32.5	...	...	0.38	0.38 Al
Incoloy 801......0.05		20.5	32.0	...	...	1.13	...
Incoloy 802......0.35		21.0	32.5	...	...	0.75	0.58 Al
Cobalt-base solid-solution alloys							
Haynes 25........0.10		20.0	10.0	50.0	...	...	1.5 Mn, 15.0 W
Haynes 188.......0.10		22.0	22.0	37.0	...	...	0.90 La, 14.5 W
Nickel-base solid-solution alloys							
Hastelloy B......0.05*		1.0*	63.0	2.5*	28.0	...	0.03 V
Hastelloy B-2....0.02*		1.0*	69.0	1.0*	28.0	...	...
Hastelloy C......0.15*		16.5	56.0	...	17.0	...	4.5 W
Hastelloy C-4....0.015*		16.0	63.0	2.0*	15.5	0.7*	...
Hastelloy C-276..0.02*		15.5	59.0	...	16.0	...	3.7 W
Hastelloy N......0.06		7.0	72.0	...	16.0	0.5*	...
Hastelloy S......0.02*		15.5	67.0	...	15.5	...	0.2 Al, 0.02 La
Hastelloy W......0.12*		5.0	61.0	2.5*	24.5	...	0.6 V
Hastelloy X......0.15		22.0	49.0	1.5*	9.0	...	2.0 Al, 0.6 W
Inconel 600......0.08		15.5	76.0	...	...	...	0.25* Cu
Inconel 601......0.05		23.0	60.5	...	...	...	0.5* Cu, 1.35 Al
Inconel 625......0.05		21.5	61.0	...	9.0	0.2	0.2 Al, 3.6 Nb
Iron-base precipitation-hardening alloys							
A-286...........0.04		15.0	26.0	...	1.25	2.0	0.005 B, 0.3 V, 0.2 Al
Discaloy........0.06		14.0	26.0	...	3.0	1.7	0.25 Al
W-545...........0.08		13.5	26.0	...	1.5	2.85	0.05 B, 0.2 Al

(continued)

* Maximum content. (a) Remaining composition is iron.

CLASSIFICATION OF STEELS AND ALLOYS

Table 3.21 Nominal composition of wrought superalloys (continued). Data from ASM Metals Reference Book (1981).

Designation	Composition (a), %						
	C	Cr	Ni	Co	Mo	Ti	Other
Cobalt-base precipitation-hardening alloys							
MP-35N...........		20.0	35.0	35.0	10.0	...	...
MP-159...........		19.0	25.0	36.0	7.0	3.0	0.2 Al, 0.6 Nb
Nickel-base precipitation-hardening alloys							
Astroloy.........0.06	0.06	15.0	56.5	15.0	5.25	3.5	0.03 B, 0.06 Zr, 4.4 Al
Incoloy 901......0.10*	0.10*	12.5	42.5	...	6.0	2.7	...
Inconel 706......0.03	0.03	16.0	41.5	...	...	1.75	2.9 (Nb + Ta), 0.2 Al
Inconel 718......0.08*	0.08*	19.0	52.5	...	3.0	0.9	0.15* Cu, 0.5 Al, 5.1 Nb
Inconel 751......0.05	0.05	15.5	72.5	...	...	2.3	0.25* Cu, 1.2 Al, 1.0 Nb
Inconel X750.....0.04	0.04	15.5	73.0	...	...	2.5	0.25* Cu, 0.7 Al, 1.0 Nb
Nimonic 80A......0.05	0.05	19.5	73.0	1.0	...	2.25	0.1* Cu, 1.4 Al
Nimonic 90.......0.06	0.06	19.5	55.5	18.0	...	2.4	1.4 Al
Nimonic 100......0.30*	0.30*	11.0	56.0	20.0	5.0	1.5	+ B, + Zr, 5.0 Al
Nimonic 263......0.06	0.06	20.0	51.0	20.0	5.9	2.1	0.45 Al
Pyromet 860......0.05	0.05	13.0	44.0	4.0	6.0	3.0	0.01 B, 1.0 Al
Refractory 26....0.03	0.03	18.0	38.0	20.0	3.2	2.6	0.015 B, 0.2 Al
Rene 41..........0.09	0.09	19.0	55.0	11.0	10.0	3.1	0.01 B, 1.5 Al
Rene 95..........0.16	0.16	14.0	61.0	8.0	3.5	2.5	0.01 B, 0.05 Zr, 3.5 Al, 3.5 Nb, 3.5 W
Rene 100.........0.16	0.16	9.5	61.0	15.0	3.0	4.2	0.015 B, 0.06 Zr, 5.5 Al
Udimet 500.......0.08	0.08	19.0	48.0	19.0	4.0	3.0	0.005 B, 3.0 Al
Udimet 630.......0.04	0.04	17.0	50.0	...	3.0	1.0	0.004 B, 0.7 Al, 6.5 Nb 3.0 W
Waspaloy.........0.07	0.07	19.5	57.0	13.5	4.3	3.0	0.006 B, 0.09 Zr, 1.4 Al

* Maximum content. (a) Remaining composition is iron.

Table 3.22 Typical mechanical properties of cobalt-base and nickel-base superalloys. Data from ASM Metals Reference Book (1981).

Temperature		Tensile strength		Yield strength		Elongation
°C	°F	MPa	ksi	MPa	ksi	%
Inconel 601, sheet						
21	70.......740		107	340	49	45
540	1000.......725		105	150	22	38
650	1200.......525		76	180	26	45
760	1400.......290		42	200	29	73
870	1600.......160		23	140	20	92
Inconel 718, sheet						
21	70......1280		185	1050	153	22
540	1000......1140		166	945	137	26
650	1200......1030		150	870	126	15
760	1400...... 675		98	625	91	8
Haynes 25, sheet						
21	70......1010		146	460	67	64
540	1000.......800		116	250	36	59
650	1200.......710		103	240	35	35
760	1400.......455		66	260	38	12
870	1600.......325		47	240	35	30
Haynes 188, sheet						
21	70.......960		139	485	70	56
540	1000.......740		107	305	44	70
650	1200.......710		103	305	44	61
760	1400.......635		92	290	42	43
870	1600.......420		61	260	38	73
Hastelloy X, sheet						
21	70.......785		114	360	52	43
540	1000.......650		94	290	42	45
650	1200.......570		83	275	40	37
760	1400.......435		63	260	38	37
870	1600.......255		37	180	26	50

REFERENCES

1. The Making, Shaping and Treating of Steel, 10th Edition, eds. W.T. Lankford, Jr., et al, Association of Iron and Steel Engineers, Pittsburgh, Pennsylvania, pp. 1277-1328 (1985).

2. ASM Metals Reference Book, American Society for Metals, Metals Park, Ohio, pp. 3, 69-217 (1981).

3. Metals Handbook, Volume 1, Properties and Selection: Irons and Steels, American Society for Metals, Metals Park, Ohio, pp. 117-119 (1985).

4. SAE Handbook 1979, Society of Automotive Engineers, Warrendale, Pennsylvania, pp. 1.14 (1979).

5. J.F. Keller, Lectures on Steel and Its Heat Treatment, Evangelical Press, Cleveland, Ohio, p. 59 (1928).

6. Metals Handbook, Eigth Edition, Volume 1, Properties and Selection of Metals, American Society for Metals, Metals Park, Ohio, pp. 491-515 (1961).

7. R.L. Brockenbrough and B.G. Johnston, USS Steel Design Manual, US Steel Corp., Pittsburgh, Pennsylvania, p. 12 (1974).

8. G.A. Roberts and R.A. Cary, Tool Steels, Fourth Edition, American Society for Metals, Metals Park, Ohio, pp. 227-229 (1980).

9. R.A. Lula, Stainless Steel, American Society for Metals, Metals Park, Ohio, pp. 1-5 (1980).

Part II

Principles of
Metallurgical Design
of Steels

4

Phase Transformation in Steel

4.1 PHASE DIAGRAM

A phase diagram is a graphical representation of the temperature, pressure and composition limits of phase fields in an alloy system as they exist under conditions of complete equilibrium. It is also known as **equilibrium** or **constitutional diagram.**

In a phase diagram, temperature is plotted vertically and composition is plotted horizontally. Any point on the diagram represents a definite composition of a constituent and its temperature, each value being found by projecting to the proper reference axes. For illustration, let us consider the changes that take place during cooling of an alloy containing 50 percent element A and 50 percent element B (Fig. 4.1). The alloy remains homogeneous liquid solution until temperature drops to a value indicated by the intersection of the liquidus line at c^0. The crystals which form from 50-50 liquid consist of a solid solution, the composition of which is found on the solidus line at c^1, 80 percent element B and 20 percent element A.

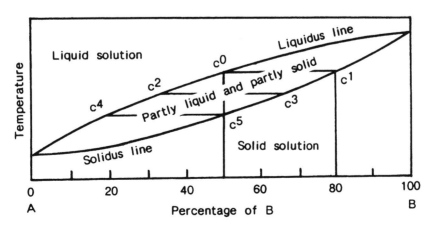

Fig. 4.1 Phase diagram of the A-B alloys. Adapted from Johnson and Weeks (1977).

As the mass cools, the composition of the growing crystals changes along the solidus line from c^1 to c^5, while the remaining liquid alloy varies in composition along the liquidus line from c^0 to c^4 [1].

Figure 4.2 illustrates the iron-cementite phase diagram which is also known as iron-carbon phase diagram.

4.2 CONSTITUENTS IN STEELS

Plain carbon steels are generally defined as the alloys of iron and carbon which contain up to 2.0% carbon. For the present, we will neglect the effects of such elements as manganese which may be present in most ordinary steels and regard steels as being simple iron-carbon alloys.

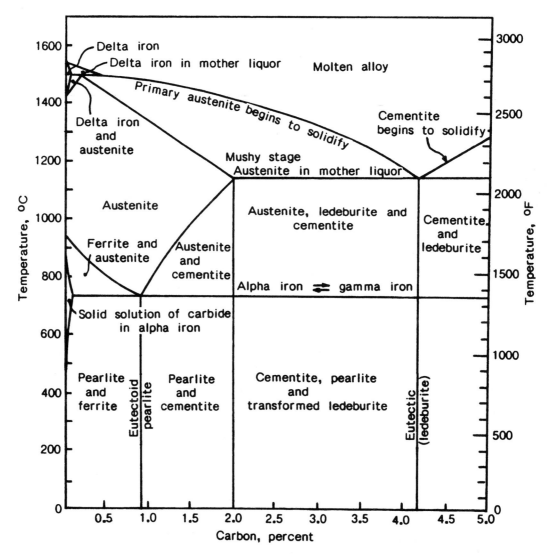

Fig. 4.2 Iron-cementite phase diagram.

Constituents in steels exist mainly as phases. They include molten alloy, delta ferrite, austenite (gamma phase), ferrite (alpha phase), cementite and graphite. Another constituent in steels is pearlite. It is not a phase but an aggregate [2].

4.3 AUSTENITE

In iron-carbon alloys austenite is the solid solution formed when carbon dissolves in face-centered cubic (gamma) iron in amounts up to 2%. Its microstructure is usually large grained as shown in Figure 4.3.

Austenite is a difficult structure to retain at room temperature unless a steel contains a large percentage of alloy, such as manganese or nickel. Austenitic steel is characterized by high tensile strength and unusually great ductility. The tensile strength is often around 125,000 pounds per square inch with elongation in two inches of 35 to 40 percent [2, 3].

4.4 FERRITE

In iron-carbon alloys ferrite is a very dilute solid solution of carbon in a body-centered cubic (alpha) iron and containing at the most only 0.02% carbon. Its microstructure (Fig. 4.4) appears as polyhedral grains.

Ferrite is very ductile and soft and has a low tensile strength but high elongation. Its tensile strength is about 40,000 pounds per square inch and an elongation in 2 inches of about 40% [2, 3].

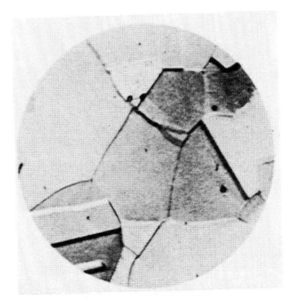

Fig. 4.3 The microstructure of austenite. Magnification: 500X. (From THE MAKING, SHAPING AND TREATING OF STEEL, by W.T. Lankford, et al, Tenth Edition. Copyright AISE, Pittsburgh, Pennsylvania. Reprinted with permission).

4.5 GRAPHITE

Graphite, or graphitic carbon, is a free carbon in steel or cast iron. The carbon is amorphous, having no particular form.

Figure 4.5 illustrates the metallographic appearance of graphite in a low-carbon steel which has been subjected to a prolonged heating at a temperature below that at which austenite is formed [2].

4.6 CEMENTITE

Cementite, or iron carbide, is an interstitial compound of iron and carbon containing 6.69% carbon. Its approximate chemical formula is Fe_3C. When it occurs as a phase in steel, the chemical composition will be altered by the presence of manganese and other carbide-forming elements.

In case of a slow cooled, relatively high-carbon steel, microstructure of cementite appears as a brilliant white network around the pearlite colonies or as some needles interspersed with the pearlite (Fig. 4.6a). Figure 4.6b shows the metallographic appearance of spheroidized cementite in a steel which has been heated to a temperature just below that at which austenite first forms.

Cementite is a very hard compound. Its tensile strength is about 5,000 pounds per square inch and an elongation in 2 inches is equal to zero. Cementite is an unstable phase. Given sufficient time, cementite decomposes into two complete equilibrium constituents, iron and graphite [2, 3].

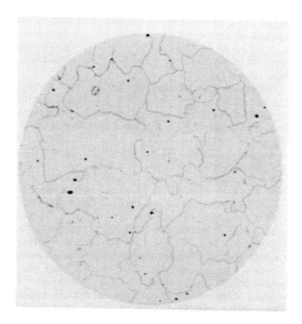

Fig. 4.4 The microstructure of ferrite. Magnification: 100X. (This sample has a coarse grain size.) (From THE MAKING, SHAPING AND TREATING OF STEEL, by W.T. Lankford, et al, Tenth Edition. Copyright AISE, Pittsburgh, Pennsylvania. Reprinted with permission).

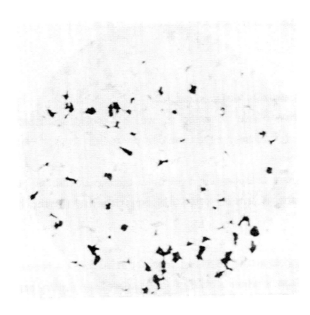

Fig. 4.5 Micrograph showing graphite particles in low-carbon
steel. Magnification: 100X. (From THE MAKING, SHAPING AND
TREATING OF STEEL, by W.T. Lankford, et al, Tenth Edition.
Copyright AISE, Pittsburgh, Pennsylvania. Reprinted with
permission).

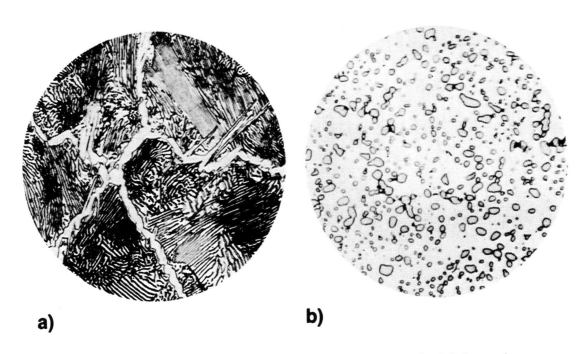

a)

b)

Fig. 4.6 The microstructures of: a) slow-cooled, high-carbon
steel showing pearlite with cementite in the grain boundaries,
b) spheroidized cementite in matrix of ferrite.
Magnification: 1000X. (From THE MAKING, SHAPING AND TREATING
OF STEEL, by W.T. Lankford, et al, Tenth Edition. Copyright
AISE, Pittsburgh, Pennsylvania. Reprinted with permission).

4.7 EUTECTOID

The term 'eutectoid' is usually defined as [4]:

1. An isothermal reversible reaction in which a **solid solution** is converted into two or more intimately mixed solids on cooling, the number of solids formed being the same as the number of components in the system.

2. An alloy having the composition indicated by the eutectoid point on an equilibrium reaction.

3. An alloy structure of intermixed solid constituents formed by a eutectoid.

4.8 PEARLITE

Pearlite is a lamellar aggregate of ferrite and cementite. It is a result of the eutectoid reaction which takes place when a plain carbon steel of approximately 0.08% carbon is cooled slowly from the temperature range at which austenite is stable.

Pearlite has lamellar micrographic structure (Fig. 4.7) known as the **eutectoid structure.** It exerts maximum hardening power of any constituent. It has a tensile strength of around 125,000 pounds per square inch and an elongation in 2 inches of 10 percent [2, 3].

4.9 EUTECTIC

The term 'eutectic' is usually defined as [4]:

1. An isothermal reversible reaction in which a **liquid solution** is converted into two or more intimately mixed solids on cooling, the number of solids formed being the same as the number of components in the system.

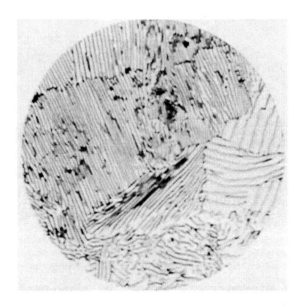

Fig. 4.7 The microstructure of pearlite. Magnification: 1000X. (From THE MAKING, SHAPING AND TREATING OF STEEL, by W.T. Lankford, et al, Tenth Edition. Copyright AISE, Pittsburgh, Pennsylvania. Reprinted with permission).

2. An alloy having the composition indicated by the eutectic point on an equilibrium diagram.

3. An alloy structure of intermixed solid constituents formed by an eutectic reaction.

4.10 LEDEBURITE

Ledeburite is an eutectic of the iron-carbon system, the constituents being an austenite and a cementite.

The eutectic contains 4.3% carbon. This eutectic is a constituent of iron-carbon alloys containing more than 2.0% carbon and for this reason the dividing line between steels and cast iron is set at 2.0% carbon. Figure 4.8 illustrates the microstructure of ledeburite in cast iron [2].

4.11 TRANSFORMATION TEMPERATURE

The temperature at which a change in phase takes place is known as the transformation temperature. It is also known as the **critical point**, or **critical temperature**.

The following symbols are used to define the transformation temperatures in the iron-carbon phase diagrams [3, 4]:

Ar_1 - The temperature at which transformation of austenite to ferrite or to ferrite plus cementite was completed during cooling.

Ac_1 - The temperature at which austenite begins to form during heating.

Ar_3 - The temperature at which austenite begins to transform to ferrite during cooling.

Ac_3 - The temperature at which transformation of ferrite to austenite is completed during heating.

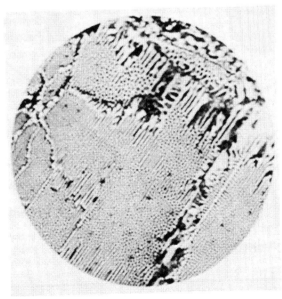

Fig. 4.8 The microstructure of ledeburite. Magnification: 150X. (From THE MAKING, SHAPING AND TREATING OF STEEL, by W.T. Lankford, et al, Tenth Edition. Copyright AISE, Pittsburgh, Pennsylvania. Reprinted with permission).

Ar$_4$ - The temperature at which delta ferrite transforms to austenite during cooling.

Ac$_4$ - The temperature at which delta ferrite begins to form during heating.

Ar$_{cm}$ - In hypereutectoid steel, the temperature at which precipitation of cementite starts during cooling.

Ac$_{cm}$ - In hypereutectoid steel, the temperature at which the solution of cementite starts during heating.

Ae$_{cm}$, Ae$_1$, Ae$_3$, Ae$_4$ - The temperatures of phase changes at equilibrium.

A$_2$ - The previously designated transformation temperature at about 768°C (1414°F). It has since been found that behavior at this temperature differs from those at A$_1$ and A$_3$ in that it does not involve a phase change. In the neighborhood of 768°C (1414°F) and up to about 790°C (1454°F) there is a gradual magnetic change, ferrite being ferromagnetic below this temperature range, and paramagnetic above. The change is also accompanied by a heat effect.

4.12 PHASES IN HYPOEUTECTOID STEEL

Hypoeutectoid steels are those containing less than the eutectoid percentage of carbon which is about 0.80% in plain carbon steel.

At some temperature above Ae$_3$, a steel containing 0.40% carbon is completely austenitic (Fig 4.9a). On slow cooling below Ae$_3$ the austenite first rejects ferrite which concentrates at grain boundaries (Fig. 4.9b). As the temperature falls down to Ae$_1$, the crystals of austenite shrink (Fig. 4.9c) and their

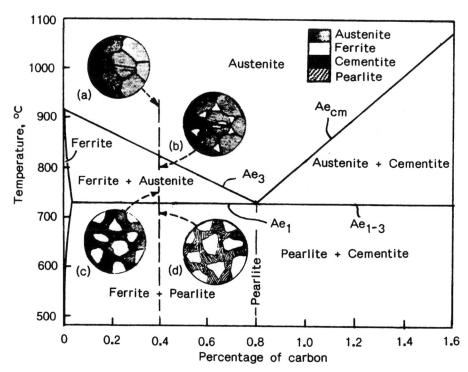

Fig. 4.9 Phase transformation in a steel containing 0.4% carbon. Adapted from Higgins (1983).

carbon content increases to 0.80%. On cooling below Ae₁, the austenite changes to pearlite so that the final constituents in steels below Ae₁ are ferrite and pearlite as illustrated in Fig. 4.9d [5].

4.13 PHASES IN EUTECTOID STEEL

Eutectoid steel is a steel containing the eutectoid percentage of carbon which is about 0.80% in plain carbon steels.

The eutectoid steel will not begin to transform from austenite (Fig. 4.10e) on cooling until the critical temperature Ae₁₋₃ is reached. Then the transformation will begin and end at the same temperature (723°C or 1333°F). The final structure will be entirely pearlite as shown in Fig. 4.10f [5].

4.14 PHASES IN HYPEREUTECTOID STEEL

Hypereutectoid steels are those containing more than the eutectoid percentage of carbon which is about 0.80% in plain carbon steels.

At some temperature above Ae$_{cm}$, a steel containing 1.2% carbon is completely austenitic (Fig. 4.11g). On slow cooling below Ae$_{cm}$ the carbon will precipitate as needle-shaped crystals of cementite around the austenite grain boundaries (Fig. 4.11h,i). As a result the carbon content in an austenite will be gradually reduced down to 0.80% at the temperature Ae₁₋₃. Below this point the remaining austenite will then transform to pearlite as shown in Fig. 4.11j [5].

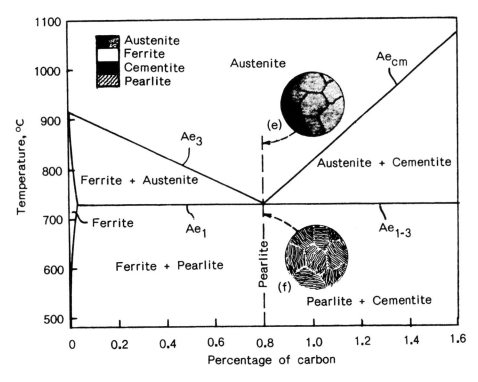

Fig. 4.10 Phase transformation in a steel containing 0.8% carbon. Adapted from Higgins (1983).

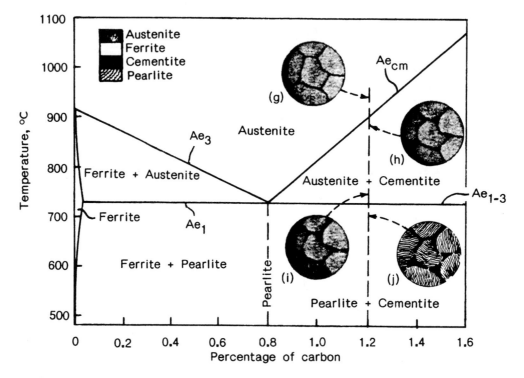

Fig. 4.11 Phase transformation in a steel containing 1.2% carbon. Adapted from Higgins (1983).

4.15 PHASE TRANSFORMATION HYSTERESIS

The phase transformations do not occur at the same temperature in heating as in cooling [4, 6]. The metal is rather reluctant to change its physical state so that on heating, the Ac temperatures are somewhat higher than equilibrium temperatures Ae. Likewise, the Ar temperatures on cooling are lower than equilibrium temperatures Ae (Fig. 4.12). The difference in temperature between the Ac and the Ar varies. In some cases it is as great as 24ºC, or 75ºF.

4.16 SUPERCOOLING OF AUSTENITE

As it has been shown in this chapter that austenite transforms to pearlite when it is cooled slowly below the Ar critical temperature. When more rapidly cooled, however, this transformation is retarded. The faster the cooling rate, the lower the temperature at which transformation occurs resulting in a formation of the micro-constituents shown in Table 4.1 [2].

The following symbols are used to define the transformation temperatures of austenite to martensite [4]:

M_s - The temperature at which transformation of austenite to martensite starts during cooling.

M_f - The temperature at which martensite formation finishes during cooling.

M_n - The temperature at which the microstructure will consist of n% martensite and (1-n)% retained austenite.

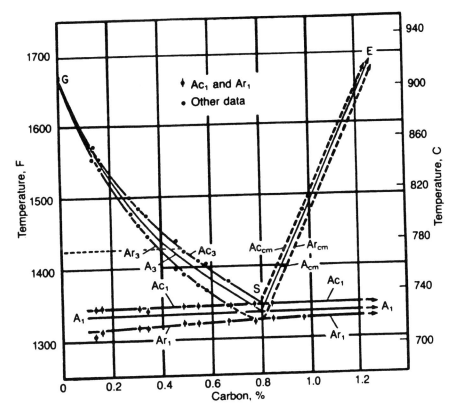

Fig. 4.12 Phase transformation temperature in steels. (From PRINCIPLES OF HEAT TREATMENT OF STEEL, by G. Krauss. Copyright ASM International. Reprinted with permission).

Table 4.1 Constituents formed during supercooling of austenite. Data from The Making, Shaping and Treating of Steel (1985).

Constituents	Temperature range
Pearlite	705°C to 535°C (1300°F to 1000°F)
Bainite	535°C to 230°C (1000°F to 450°F)
Martensite	Below 230°C (450°F)

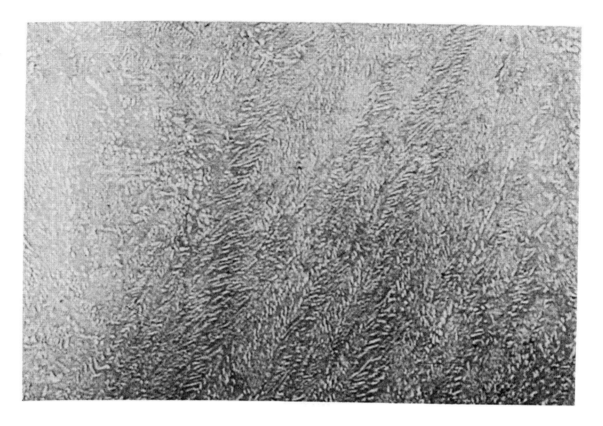

Fig. 4.13 Electron micrograph of bainitic microstructure formed on complete transformation of eutectoid steel at 260°C (500°F). Magnification: 15,000X. (From THE MAKING, SHAPING AND TREATING OF STEEL, by W.T. Lankford, et al, Tenth Edition. Copyright AISE, Pittsburgh, Pennsylvania. Reprinted with permission).

4.17 BAINITE

Bainite is a decomposition of austenite consisting of an aggregate of ferrite and carbide. Its appearance is featherlike if formed in the upper part of the temperature range and acicular if formed in the lower part (Fig 4.13). The hardness increases as the transformation temperature decreases. This is due to a finer distribution of carbide in bainite formed at a lower temperature.

In a given steel, the bainitic microstructures will generally be found both harder and tougher than pearlite, although the hardness will be lower than that of martensite [2].

4.18 MARTENSITE

Martensite is a metastable phase of steel formed by a transformation of austenite below M_s temperature. It is an interstitial supersaturated solid solution of carbon in iron having a body-centered tetragonal lattice.

Transformation to martensite occurs almost instantly during cooling and the percentage of transformation is dependent only on the temperature to which it is cooled. It is the hardest of the

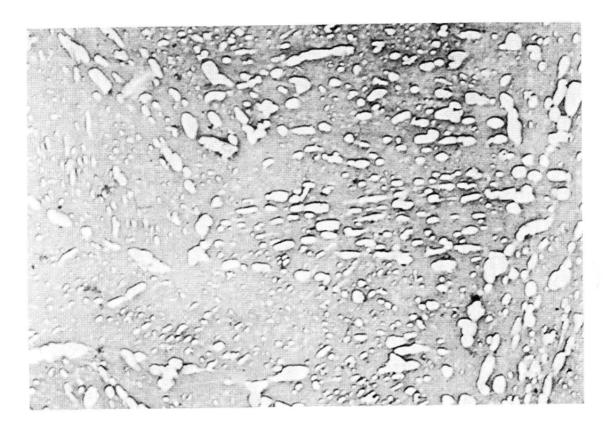

Fig. 4.14 Electron micrograph of the structure of tempered martensite. Tempering temperature 595°C (1100°F).
Magnification: 15,000X. (From THE MAKING, SHAPING AND TREATING OF STEEL, by W.T. Lankford, et al, Tenth Edition. Copyright AISE, Pittsburgh, Pennsylvania. Reprinted with permission).

transformation products of austenite. The microstructure of martensite is acicular, or needlelike.

Figure 4.14 shows a microstructure of so called tempered martensite. This structure is formed when martensite is reheated to a subcritical temperature after quenching [2].

4.19 ISOTHERMAL TRANSFORMATION DIAGRAM

The isothermal transformation diagrams, or **Time-Temperature-Transformation (T.T.T.) curves** (Fig. 4.15), are constructed by completing the following five steps [2, 5]:

1. Heating a number of similar specimens of a steel to just above the Ac_1 temperature.

2. Quenching these specimens to a different temperature in an 'incubation' liquid bath.

3. Holding the specimens in the bath during predetermined time intervals.

4. Final quenching of the specimens in water.

5. Examining the microstructure to reveal the extent to which the phase transformation had taken place at the holding temperature.

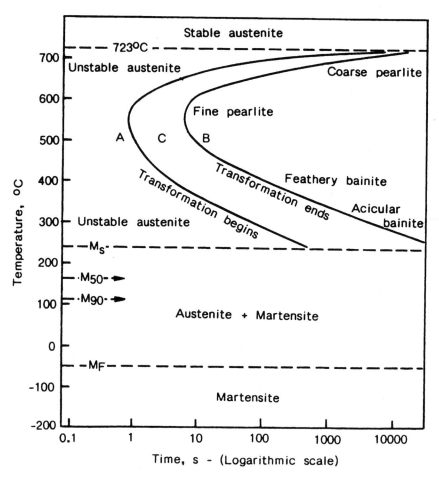

Fig. 4.15 Time-Temperature-Transformation (T.T.T.) curves for plain carbon steel of eutectoid composition. (From ENGINEERING METALLURGY, Part 1, by R.A. Higgins, 1983. Copyright Hodder & Stoughton Ltd., England. Reprinted with permission).

4.20 CONTINUOUS-COOLING TRANSFORMATION DIAGRAM

A continuous-cooling transformation diagram reveals the phase transformations which take place during continuous cooling of selected specimen at predetermined constant cooling rate [5]. The continuous-cooling diagram lies below and to the right of corresponding isothermal diagram as shown in Fig. 4.16 in application to a plain carbon steel of eutectoid composition. Curves A, B, and D represent cooling rates of approximately 5, 400 and 50°C per second correspondingly.

For example, in the case of the curve D, transformation would begin at P with formation of some fine pearlite. Transformation, however, is interrupted at Q and does not begin again until the M_s temperature is reached at R. Curve C represents the **critical cooling rate**. This is a minimal cooling rate that has to be maintained in order to obtain a completely martensite microstructure.

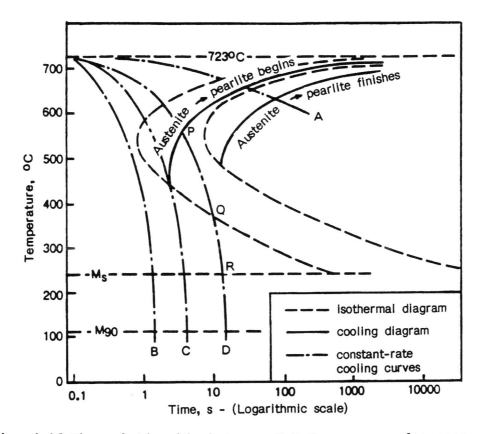

Fig. 4.16 The relationship between T.T.T. curves and curves representing continuous cooling. (From ENGINEERING METALLURGY, Part 1, by R.A. Higgins, 1983. Copyright Hodder & Stoughton Ltd., England. Reprinted with permission).

REFERENCES

1. C.G. Johnson and W.R. Weeks, Metallurgy, Fifth Edition, American Technical Society, Chicago, pp. 131-148 (1977).

2. The Making, Shaping, and Treating of Steel, 10th Edition, W.T. Lankford, Jr., et al, eds, Herbick and Held, Pittsburgh, Pennsylvania, pp. 1231-1249 (1985).

3. J.F. Keller, Lectures on Steel and Its Heat Treatment, Evangelical Press, Cleveland, Ohio, pp. 35-41 (1928).

4. S.H. Avner, Introduction to Physical Metallurgy, Second Edition, McGraw-Hill Book Company, New York, pp. 667-687 (1974).

5. R.A. Higgins, Engineering Metallurgy, Part 1, Applied Physical Metallurgy, Robert E. Krieger Publishing Company, Melbourne, Florida, pp. 1-164 (1983).

6. G. Krauss, Principles of Heat Treatment of Steel, American Society for Metals, Metal Park, Ohio, pp. 13-16 (1980).

5

Alloying Elements and Impurities in Steel

5.1 MAJOR EFFECTS OF ALLOYING ELEMENTS

Steel by definition is an alloy of iron with 0.03 to 1.7% carbon. Many steels contain other alloying elements in addition to carbon in order to obtain other specific properties.

It is customarily regarded to divide steels into two following groups [1]:

1. Low alloy steels containing less than 5% elements other than iron
2. High alloy steels containing more than 5% elements other than iron.

Alloying elements and impurities are incorporated into austenite, ferrite and cementite. If the alloy or impurity atoms are roughly the same size as iron atoms, the incorporation is usually done by replacement of the iron atoms. If the alloy or impurity atoms are significantly smaller than iron atoms, as is nitrogen, the incorporation may occur by filling in the interstitial sites. In the cases when solubility level is exceeded, phases other than those in plain carbon steel may form. A typical example of these phases is a combination of the cementite structure M_3C (M is standing for a combination of chromium and iron atoms), which is formed when small additions of chromium is added to iron-carbon alloys at 890°C (1634°F). Larger additions of chromium cause a formation of the carbides M_7C_3 and $M_{23}C_6$ [2].

Depending on their effect on formation and stabilization of phases in steels, the alloying elements are usually regarded as [3, 4]:

1. Austenite formers
2. Ferrite formers
3. Carbide formers.

5.2 AUSTENITE FORMERS

Austenite formers are the alloying elements which enlarge the austenite field.

The following elements are considered as the austenite formers [4]:

1. Carbon (C)
2. Cobalt (Co)
3. Copper (Cu)
4. Manganese (Mn)
5. Nickel (Ni)
6. Nitrogen (N).

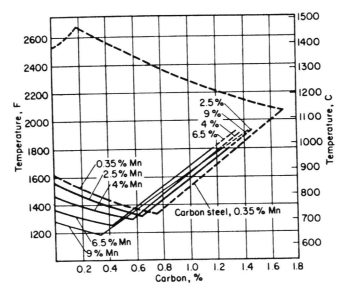

Fig. 5.1 Effect of several uniform manganese contents on the carbon limitations for pure austenite at elevated temperatures. (From ALLOYING ELEMENTS IN STEEL, by E.C. Bain and H.W Paxton, Second Edition. Copyright ASM International. Reprinted with permission).

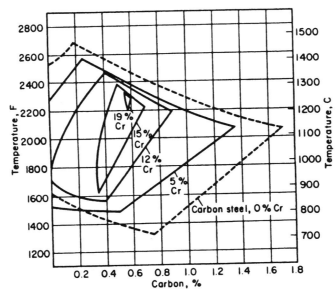

Fig. 5.2 Effect of several uniform chromium contents on the carbon limitations for pure austenite at elevated temperatures. (From ALLOYING ELEMENTS IN STEEL, by E.C. Bain and H.W. Paxton, Second Edition. Copyright ASM International. Reprinted with repmission).

Figure 5.1 illustrates a typical effect of the austenite former on the austenitic field.

5.3 FERRITE FORMERS

Ferrite formers are the alloying elements which enlarge the ferrite field as shown in Fig. 5.2 when alloying element is chromium.

The following elements are considered as ferrite formers [4]:

1. Aluminum (Al)   6. Silicon (Si)
2. Chromium (Cr)   7. Titanium (Ti)
3. Columbium (Nb)   8. Tin (Sn)
4. Molybdenum (Mo)   9. Tungsten (W)
5. Phosphorus (P)   10. Vanadium (V).

5.4 CARBIDE FORMERS

The elements such as chromium, tungsten, vanadium, molybdenum, titanium, and columbium form very stable carbides when added to a steel (Fig. 5.3). This generally increases hardness of steel, particularly when the carbides formed are harder than iron carbide [5].

Other elements, such as manganese, are not strong carbide formers. However, they contribute to the stability of other carbides present.

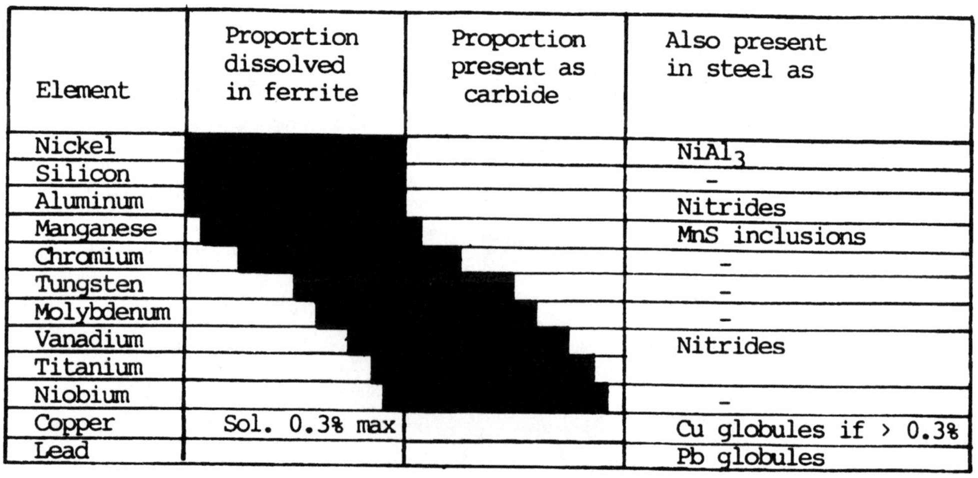

Element	Proportion dissolved in ferrite	Proportion present as carbide	Also present in steel as
Nickel			$NiAl_3$
Silicon			-
Aluminum			Nitrides
Manganese			MnS inclusions
Chromium			-
Tungsten			-
Molybdenum			-
Vanadium			Nitrides
Titanium			
Niobium			-
Copper	Sol. 0.3% max		Cu globules if > 0.3%
Lead			Pb globules

Fig. 5.3 The physical states in which the principal alloying elements exist when in steel. (From ENGINEERING METALLURGY, Part 1, by R.A. Higgins, 1983. Copyright Hodder & Stoughton Ltd., England. Reprinted with permission).

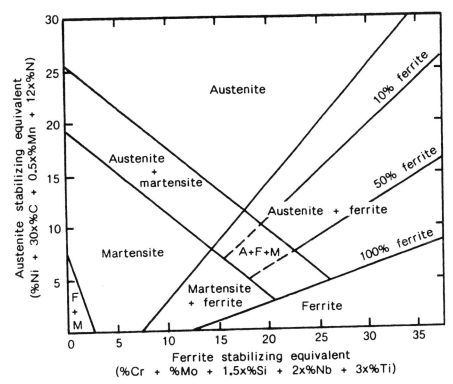

Fig. 5.4 A modified Schaeffler diagram. (From ENGINEERING METALLURGY, Part 1, by R.A. Higgins, 1983. Copyright Hodder & Stoughton Ltd., England. Reprinted with permission).

5.5 EFFECT OF ALLOYING ON MICROSTRUCTURE

When several elements are present in a steel, their combined effect in respect to formation of an austenite may be described by the Austenite Stabilizing Equivalent (ASE) and by the Ferrite Stabilizing Equivalent (FSE) which are equal to [5]:

$$ASE = Ni + 30C + 0.5Mn + 12N \qquad (5\text{-}1)$$

$$FSE = Cr + Mo + 1.5Si + 2Nb + 3Ti \qquad (5\text{-}2)$$

The symbols in Eqs. (5-1, 5-2) represent the weight percent of the element indicated.

By using ASE and FSE as the coordinates in a modified Schaeffler diagram (Fig. 5.4), it is possible to define microstructure of alloyed steel after air-cooling from its austenite conditions.

5.6 EFFECT OF ALLOYING ON CRITICAL TEMPERATURES

Effect of alloying elements on the critical temperatures can be illustrated by the following empirical formulas which have been developed by Andrews [4, 6], based on statistical analysis of an experimental data:

$$Ac_3(^oC) = 910 - 203\sqrt{C} - 15.2Ni + 44.7Si + 104V + 31.5Mo + 13.1W - 30Mn$$
$$-11Cr - 20Cu + 700P + 120As \tag{5-3}$$

$$Ac_1(^oC) = 723 - 10.7Mn - 16.9Ni + 29.1Si + 16.9Cr + 290As + 6.38W \tag{5-4}$$

$$M_s(^oC) = 539 - 423C - 30.4Mn - 17.7Ni - 12.1Cr - 11Si - 7.5Mo \tag{5-5}$$

In these formulas, the symbols represent the weight percent of the element indicated. The formulas are applicable for steel containing less than 0.6% C and less than 5% of each of the other alloying elements.

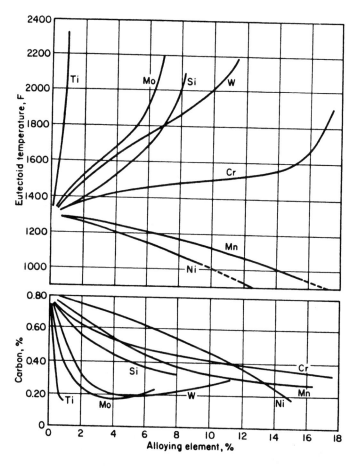

Fig. 5.5 Eutectoid composition and eutectiod temperature, as influenced by several alloying elements. (From ALLOYING ELEMENYS IN STEEL, by E.C. Bain and H.W. Paxton, Second Edition. Copyright ASM International. Reprinted with permission).

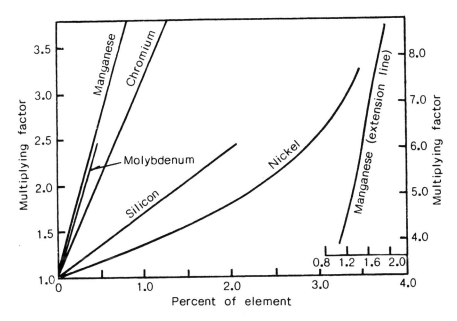

Fig. 5.6 Hardenability multiplying factors for a variety of alloying elements. Adapted from HIGH-STRENGTH SHEET STEEL SOURCE GUIDE SC-603D, American Iron and Steel Institute (1981).

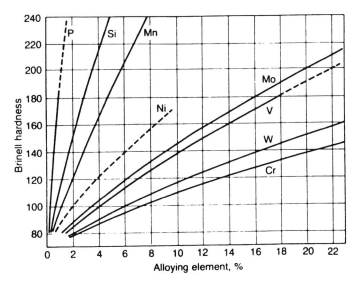

Fig. 5.7 Probable hardening effects of the various elements as dissolved in pure iron. (From ALLOYING ELEMENTS IN STEEL, by E.C. Bain and H.W. Paxton, Second Edition. Copyright ASM International. Reprinted with permission).

5.7 EFFECT OF ALLOYING ON TRANSFORMATION RATE

The addition of all alloying elements, with exception of cobalt, results in retardation of the transformation rates at temperatures between 500°C (932°F) and 700°C (1292°F). Therefore, it reduces the critical cooling rate necessary to obtain a martensite [5].

5.8 EFFECT OF ALLOYING ON EUTECTOID COMPOSITION AND TEMPERATURE

When alloying elements are added to a carbon steel, the solubility of carbon in austenite diminishes [3]. An alloy steel will be completely pearlitic when it contains less than 0.8% carbon. As it follows from Fig. 5.5, the addition of manganese and nickel lowers the eutectoid temperature whereas the other elements raise it [3].

5.9 EFFECT OF ALLOYING ON HARDENABILITY

The effect of alloying elements on hardenability may be evaluated by the hardenability multiplying factor (Fig. 5.6) which is developed by a measurement of hardenability on a series of steels in which a single alloying element is the only variable [4].

It was found that the cumulative hardenability h_C of an alloy steel containing more than one alloying elements can be calculated as follows:

$$h_C = h_B(K_1 \times K_2 \times ... \times K_n), \tag{5-6}$$

where h_B = hardenability of the base iron-carbon alloy

K_1, K_2, K_n = hardenability multiplying factors of alloying elements.

5.10 EFFECT OF ALLOYING ON HARDNESS

Alloying elements contribute to hardness in steel by being as either

(a) carbide formers, or

(b) ferrite strengtheners [4].

The carbide-forming elements, such as chromium, molybdenum and vanadium, greatly affect the tempering behavior. These elements raise the tempering temperature required to obtain a given hardness.

The ferrite strengtheners increase hardness of ferrite by forming solid solutions. This furnishes a method of increasing the hardness of steel in the unhardened state (Fig. 5.7). This hardening effect is, of course, small as compared with that obtainable by changes in the dispersion of the carbide.

5.11 EFFECT OF ALLOYING ON TENSILE STRENGTH

As in the case of hardness, the ferrite strengthening effect on tensile strength is small as compared with that obtained by changes in the dispersion of carbides.

Increase in tensile strength as a function of content of the principal alloying elements can be estimated by using the multiplication factors (Fig. 5.8) derived by Walters [5] in application to the pearlitic steels in the normalized condition. The multiplication factor equal to 1 corresponds to a pure

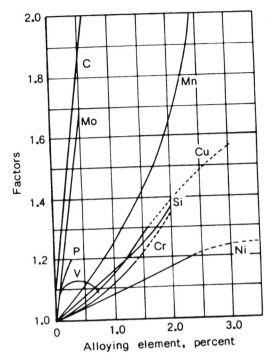

Fig. 5.8 Walters factors for estimating the tensile strength of pearlitic steels in the normalized condition. (From ENGINEERING METALLURGY, Part 1, by R.A. Higgins, 1983. Copyright Hodder & Stoughton Ltd., England. Reprinted with permission).

iron with tensile strength of 250 N/mm² (36,000 psi). The most accurate results are obtained for steels with less than 0.25% carbon and within the intermediate alloy range.

5.12 EFFECT OF ALLOYING ON GRAIN GROWTH

Presence of some elements, notably chromium, results in an increased rate of grain growth in a steel. In case of overheating the steel, it may promote brittleness which is usually associated with coarse grain [5]. The opposite effect, retardation of the grain growth, is produced by vanadium, titanium, columbium, aluminum, and to some degree, nickel.

Vanadium, titanium and columbium inhibit the grain growth by forming finely-dispersed carbides and nitrides which are relatively insoluble at high temperatures and act as barriers to the grain growth. The carbides of tungsten and molybdenum present in high-alloy tool steels reduce grain growth at heat-treatment temperatures. Aluminum present in an aluminum-killed steel makes it inherently fine-grained.

5.13 EFFECT OF ALLOYING ON CORROSION-RESISTANCE

Addition of elements such as aluminum, silicon and chromium results in substantial improvement of the corrosion resistance. It is due to a thin but dense and adherent film of oxide which these elements form on the surface of the steel thereby protecting it from further attack [5].

Chromium is the most useful when corrosion resistance at high temperatures is required.

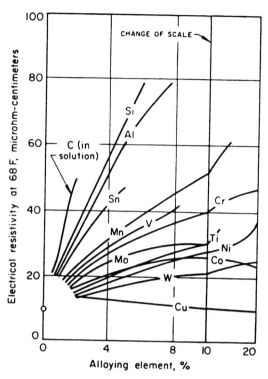

Fig. 5.9 Effect of alloying elements on the electrical resistance of iron. (From METALS HANDBOOK, Eight Edition, Vol.1, 1961. Copyright ASM International. Reprinted with permission).

5.14 EFFECT OF ALLOYING ON ELECTRICAL RESISTANCE

In general, addition of majority of alloying elements with some exceptions such as copper (Cu) increases electrical resistance of steels [7]. As follows from Fig. 5.9, the most effective contributors to electrical resistance are carbon (C), silicon (Si), aluminum (Al), and tin (Sn). An increase in content of molybdenum (Mo) and cobalt (Co) up to approximately 9% produces an increase in electrical resistivity, whereas any further additions of these alloying elements result in decrease of the electrical resistivity.

5.15 RESIDUAL ELEMENTS AND IMPURITIES

Residual elements are the elements which are present in an alloy in small quantities, but not added intentionally [4, 8].

The most common residual elements and their main sources are listed below:

1. Hydrogen, oxygen and nitrogen may be introduced during the steelmaking process.

2. Nickel, copper, molybdenum, chromium and tin may be transferred into steel from scrap.

3. Aluminum, titanium, vanadium and zirconium may be introduced during the deoxidation process.

Impurities are the elements or compounds whose presence in a metal is not desired. Nonmetallic substances in a solid metallic matrix are often referred to as **inclusions** [7]. Impurities can be present in steel in both solid and gaseous forms. They also can be present as pure elements or as oxides.

Nitrogen, hydrogen, carbon monoxide and oxygen are typical gaseous impurities which are present in steel as bubbles or may be dissolved, or in combination of both. Main effect of the gaseous impurities is a decrease in plasticity. In addition to that hydrogen embrittles the steel.

The residual elements such as nickel, copper, molybdenum, chromium and tin increase hardenability of carbon steels. This may change the heat-treating characteristics and for some applications in which the ductility is important, the increased hardness from these residual elements may be harmful.

In electrical sheet steels carbon, sulfur, nitrogen and oxygen are especially harmful because they occupy interstitial spaces in the crystal structure and, even in small amounts, may result in significant deterioration of the magnetic properties of these steels [9].

Iron oxides are another type of impurities. They are formed during steelmaking process and may be trapped in solid steel during its solidification. These impurities are known as **dirt**. Oxides which remain in a steel are mechanically mixed with the steel, breaking up the continuity of the steel structure and disrupting a directional uniformity of physical properties of the steel. These oxides may also serve as points of weaknesses from which a fracture may commence.

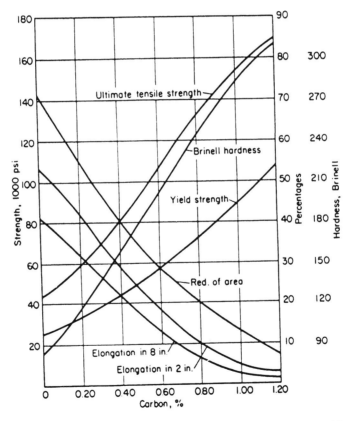

Fig. 5.10 Approximate influence of carbon content on tensile properties of carbon steels in the as-rolled condition. Applicable to sections about 1/2" to 3/4" thick. (From ALLOYING ELEMENTS IN STEEL, by E.C. Bain and H.W. Paxton, Second Edition. Copyright ASM International. Reprinted with permission).

5.16 CARBON (C)

Carbon is often called **the Master** because of its influence on the steel properties. Although by itself it does not posses strength or hardness, in solid solution as an iron carbide (Fe_3C), the carbon is a chief controller of strength and hardness (Fig. 5.10).

Principal functions of carbon are [1]:

1. It increases strength, hardness and hardenability

2. It lowers ductility, malleability, magnetic characteristics and electrical conductivity.

Ranges of carbon content (percent) in some steels are shown below [8]:

Plain carbon steels   0.03 - 1.04

High-speed tool steels   0.75 - 1.60

Hot-work tool steels   0.22 - 0.70

Cold-work tool steels   0.45 - 2.85

5.17 MANGANESE (Mn)

Manganese is widely used in steel production for deoxidation and desulfurization of the molten steel. It remains in steel in the amount less than 1%. When its content exceeds 1% it is regarded as a deliberate alloying element.

Principal functions of manganese are [3, 5]:

1. Manganese increases tensile strength (Fig. 5.11).

2. It moderately increases hardenability and promotes both toughness and machinability.

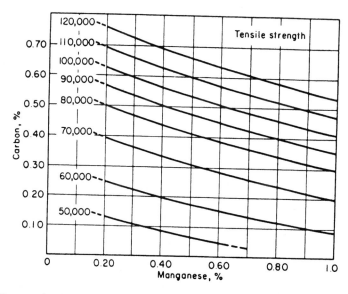

Fig. 5.11 Approximate carbon and manganese compositions required to produce the indicated tensile strengths in as-rolled steel. (From ALLOYING ELEMENTS IN STEEL, by E.C. Bain and H.W. Paxton, Second Edition. Copyright ASM International. Reprinted with permission).

3. In steels containing sulfur, manganese minimizes the resulting hot and cold brittleness ('shortness').

Ranges of manganese content (percent) in some steels are shown below [8]:

Plain carbon steels	0.25 - 1.65
Manganese steels	1.60 - 1.90
Cold-work tool steels	0.30 - 2.50
Chromium-nickel austenitic stainless steels	2.00 - 15.5

5.18 MOLYBDENUM (Mo)

Molybdenum is a strong carbide former. It forms a hard, stable carbide Mo_2C and also double carbides such as Fe_4Mo_2C and $Fe_{21}MoC_6$.

Principal functions of molybdenum are [5]:

1. Small amount of molybdenum effectively reduces the transformation rates displacing the nose of T.T.T. curve to the right.

2. Molybdenum raises high-temperature strength and creep resistance of high-temperature alloys.

3. It enhances the corrosion resistance of stainless alloys particularly in chloride solutions.

4. It replaces much of the tungsten in high-speed tool steels.

5. Additions of about 0.3% molybdenum are usually sufficient to reduce the tendency to produce temper-brittleness in low-nickel, low-chromium steels.

6. Molybdenum is one of the important components of the nickel-chrome-moly steels which posses a combination of high tensile strength with good ductility.

7. In chromium steels, it is added in order to improve machinability and mechanical properties.

8. In nickel-molybdenum steels, it enhances a case-hardening.

Ranges of molybdenum content (percent) in some steels are shown below [8]:

Molybdenum steels	0.15 - 0.60
Tool steels, types:	
high-speed steels	0 - 10.0
hot-work steels	0 - 5.50
cold-work steels	0 - 1.80
Stainless steels, types:	
austenitic chromium-nickel steels	0 - 4.00
martensitic chromium steels	0 - 1.25
ferritic chromium steels	0 - 1.25

5.19 CHROMIUM (Cr)

Chromium is a carbide former. It forms the hard carbides Cr_7C_3 or $Cr_{23}C_6$ or, alternatively, double carbides with iron. All of these carbides are harder than ordinary cementite. As a ferrite former, chromium lowers the A_4 temperature and raises the A_3 temperature thereby stabilizing the alpha-phase at the expense of gamma-phase. If more than 11% chromium is added to a pure iron, the gamma-phase is entirely eliminated.

Principal functions of chromium are [3, 5]:

1. Chromium increases hardness of steels when a sufficient amount of carbon is present. Low-chromium steels containing 1.0% carbon are extremely hard.

2. In low-carbon steels the addition of chromium increases strength with some loss of ductility.

3. It contributes to high-temperature strength.

4. It provides moderate contribution to hardenability.

5. In high-carbon steels, it improves abrasion resistance.

6. When added in large amounts, up to 25%, it improves corrosion resistance due to the protective layer of oxide which forms on the surface.

7. Chromium promotes grain growth resulting in increased brittleness of steel.

Ranges of chromium (percent) in some steels and alloys are shown below [8]:

Chromium steels	0.30 - 1.60
Stainless steels, types:	
austenitic chromium-nickel steels	15.0 - 30.0
martensitic chromium steels	4.0 - 18.0
ferritic chromium steels	10.5 - 27.0
precipitation hardening steels	12.2 - 18.0
Heat-resistant casting alloys, types:	
nickel-base alloys	0 - 21.0
cobalt-base alloys	0 - 27.0

5.20 NICKEL (Ni)

Nickel does not form carbides. Its presence in steel makes iron carbides less stable and therefore it promotes graphitization. Nickel acts as a ferrite strengthener by forming a simple substitutional solid solution. As an austenite former, nickel stabilizes austenite by raising the A_4 temperature and depressing the A_3 temperature. If more than 25% nickel is added to pure iron, the resulting alloy becomes purely austenitic even after slow cooling to ambient temperatures.

Principal functions of nickel are [3, 5]:

1. In alloy steels with nickel content up to 5%, it increases strength and toughness.

2. Nickel makes high-chromium composition austenitic as it is in chromium-nickel austenitic stainless steels.

3. It induces grain-refining effect.

4. In high-nickel steels containing small amount of carbon, it substantially increases thermal hysteresis of the allotropic transformation so that martensite can be retained in steels as it takes place in martensite aging ('maraging') steels after heating up to 600^oC (1112^oF).

5. Nickel reduces the coefficient of thermal expansion.

6. In high-nickel alloys, it increases magnetic permeability.

Ranges of nickel (percent) in some steels and alloys are shown below [8]:

Nickel steels	3.25 - 5.25
Stainless steels, types:	

austenitic chromium-nickel steels	1.00 - 37.0
martensitic chromium steels	0 - 2.50
precipitation hardening steels	3.00 - 8.50
Heat- resistant casting alloys, types:	
nickel-base alloys	50.0 - 75.0
cobalt-base alloys	0 - 27.0

5.21 VANADIUM (V)

Vanadium is a strong carbide former. It forms carbide VC. Vanadium is present in the microstructure as finely dispersed carbides and nitrides which are not dissolved at normal heat-treatment temperatures, their presence produces a barrier to a grain growth. Sufficient amount of carbon and vanadium soluble at elevated temperature causes maximum observed secondary hardness effects.

Principal functions of vanadium are [3, 5]:

1. Vanadium is a very important grain refiner. It restricts austenitic growth. As little as 0.1% vanadium effectively restricts grain growth during hardening processes. The grain growth, however, starts immediately after the steel is being heated to the temperature at which the grain-growth restricting particles of carbide and nitride dissolve.

2. Vanadium promotes hardenability even when it is dissolved in small amounts.

3. It induces resistance to softening at high temperature.

4. Vanadium readily combines with oxygen and nitrogen and is often used as a 'scavenger' or 'cleanser' in the final stage of deoxidation to produce a gas-free ingot.

5. Chromium-vanadium steels show higher yield stress and percentage reduction in area.

Ranges of vanadium (percent) in some steels and alloys are shown below [8]:

Chromium-vanadium steels	0.10 - 0.20
Tool steels, types:	
high-speed steels	0.90 - 5.25
hot-work steels	0 - 2.20
cold-work steels	0 - 5.15.

5.22 TUNGSTEN (W)

Tungsten is a very strong carbide former. It forms extremely hard and stable carbides W_2C, WC, and a double carbide Fe_4W_2C. These carbides dissolve very slow and only at very high temperatures. Therefore tungsten is an important constituent of tool steels and particularly high-speed tool steels. In these steels, a substantial increase in hardness can be achieved after secondary hardening. As a ferrite former, tungsten lowers the A_4 temperature and raises the A_3 temperature.

Principal functions of tungsten are [3, 5]:

1. Tungsten inhibits grain growth and therefore has grain refining effect.

2. It reduces decarburization during hot working and heat treatment processes.

3. Tungsten induces abrasion resistance.

4. It develops high-temperature ('red') hardness in quenched and tempered steels.

5. In some high-temperature alloys, it contributes to creep strength.

6. Its contribution to hardenability is very considerable.

7. Tungsten opposes softening in tempering. Steels containing tungsten can be heated in the range 600-700°C (1112-1292°F) before carbides begin to precipitate resulting in softening the steel.

Ranges of tungsten (percent) in some carbon and tool steels are shown below [8]:

Tungsten-chromium steels   1.75 (nom.)

Tool steels, types:

high-speed steels	1.15 - 21.0
hot-work steels	0 - 19.0
cold-work steels	0 - 2.0
shock-resisting	0 - 3.0.

5.23 COBALT (Co)

Cobalt is a carbide former. Its carbide-forming tendency is about or slightly stronger than that of an iron. Cobalt is an essential constituent of some tool steels and superalloys as well as of some permanent-magnet alloys.

In some alloys containing 18% nickel, 8 to 12% cobalt, 3 to 5% molybdenum and small amounts of titanium and aluminum, cobalt enhances precipitation hardening process by producing more sites for nucleation of $(Ti, Al, Mo)Ni_3$ precipitates. In these alloys the original iron-nickel martensite contributes about half of the strength whereas the other half is due to subsequent precipitation hardening. The latter process is known as 'aging' of martensite and therefore these types of alloys are sometimes referred to as the **maraging steels**.

Principal functions of cobalt are [3, 5]:

1. Cobalt promotes high residual induction and high coercive force in permanent-magnet alloys.

2. Cobalt enhances high strength and considerable toughness by stimulating the precipitation hardening process.

3. When cobalt is dissolved in ferrite or austenite, it resists softening with an elevation in temperature.

4. Cobalt provides negative contribution to hardenability.

Ranges of cobalt (percent) in some steels and superalloys are shown below [8]:

Tool steels, types:

high-speed steels	0 - 13.0
hot-work steels	0 - 4.5

Heat-resistant casting alloys, types:

nickel-base alloys	0 - 18.5
cobalt-base alloys	42.0 - 67.5

5.24 ALUMINUM (Al)

Aluminum has a carbide-forming tendency less than an iron. It promotes a graphitization.

Principal functions of aluminum are [3]:

1. It restricts grain growth by forming an effective fine dispersion with nitrogen or oxygen.

2. It forms an effective surface-hardening layer by relatively low-temperature diffusion of nitrogen (nitriding).

3. It restricts corrosion by forming a strong layer of aluminum oxide on the steel surface.

4. Aluminum is an excellent deoxidizer.

5. Its contribution to hardenability is moderate.

Ranges of aluminum (percent) in some steels and alloys are shown below [8]:

Ferritic stainless steels	0 - 0.30
Precipitation hardening stainless steels	0 - 1.5
Heat-resistant casting alloys, types:	
nickel-base alloys	0 - 6.5
cobalt-base alloys	0 - 4.3

5.25 TITANIUM (Ti)

Titanium is a very strong carbide former. Its principal functions are [3, 5]:

1. In medium-chromium steels, titanium withdraws carbon from solution and reduces martensitic hardness and hardenability.

2. In high-chromium steels, titanium prevents formation of austenite.

3. In austenitic stainless steels, titanium withdraws carbon from solution at elevated temperature and thereby prevents intergranular deterioration as a result of chromium-carbide formation at grain boundaries with accompanied depletion of local chromium.

4. In austenitic high-temperature alloys, titanium promotes precipitation hardening.

5. In killed high-strength low-alloy steels, it contributes to malleability by making improvements in inclusion characteristics, mainly by rounding the sulphides. (The same purpose can be achieved by utilizing zirconium and rare earths.)

6. It is used as a deoxidizer.

Ranges of titanium (percent) in some steels and alloys are shown below [8]:

High-strength low-alloy sheet steels	0 - 0.10
Heat-resistant casting alloys, types:	
nickel-base alloys	0 - 5.0
cobalt-base alloys	0 - 3.8.

5.26 SILICON (Si)

Silicon is a ferrite former. It raises both A_1 and A_3 temperatures. Since silicon has a graphitizing effect, it is usually combined in steels with manganese as a carbide stabilizer.

Principal functions of silicon are [3, 5]:

1. In electrical sheet steels, silicon raises magnetic permeability and electrical resistivity and allows one to obtain very low magnetic hysteresis losses. Silicon content range in these steels is between 0.5 and 4.5%.

2. Silicon contributes to the oxidation resistance in some heat-resistant steels.

3. In combination with manganese, silicon increases hardenability, strength, and impact toughness.

5.27 COPPER (Cu)

Copper is an austenite former. Since it has a graphitizing effect, it is added to low-carbon steels only and in quantities of no more than 1.5%.

Principal functions of copper are [5]:

1. Copper improves corrosion resistance.

2. It produces an alloy with increased tensile strength obtained by a precipitation hardening.

3. Copper produces a rather small increase in yield strength in those steels which are not precipitation hardened.

4. In carbon steels, it increases hardenability that may result in reduced ductility.

5.28 LEAD (Pb)

Lead has a limited solubility in molten steel. Because it has a much greater density than steel, it settles in the molten steel leading to a heavy lead segregation. It is present in steel in an elemental form as small inclusions. These inclusions are soft and act as the internal lubricants. Lead is an important constituent of leaded steels [4, 8] used for making the tools.

5.29 BORON (B)

Boron is a very hard solid which melts at 2300°C (4172°F). It is added to some steels known as **boron steels.**

Principal functions of boron are [5]:

1. Hardenability of fully-deoxidized steels increases when very small amounts (from 0.0005 to 0.005%) of boron are added. It is due to a reduction in the phase transformation rates during cooling. This effect is useful the most in low-carbon steels.

2. Addition of small quantities of boron allows one to reduce the amounts of other more expensive alloying elements by as much as half while maintaining the same transformation rates.

3. Boron improves malleability and machinability in steels.

5.30 NIOBIUM (Nb)

Niobium (also known as columbium) provides strengthening by both precipitation hardening and ferrite grain refinement.

Principal functions of niobium are [4]:

1. Small quantities of niobium, about 0.02%, significantly increase tensile strength and yield strength of carbon steels.

2. Increase in strength is accompanied by a marked deterioration of notch toughness. This can be avoided by special rolling practices involving lower-than-normal temperature for the last rolling passes.

Niobium is an important constituent of some high-strength low-alloy sheet steels.

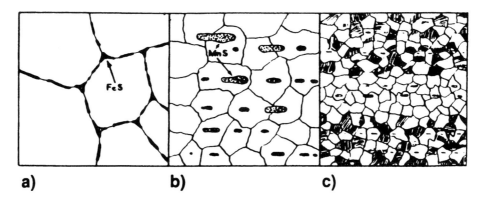

Fig. 5.12 (a) The segregation of iron sulphide (FeS) at the crystal boundaries in steel. Magnification: 750X.
(b) The formation of isolated manganese sulphide (MnS) globules when manganese is present in a steel. Magnification: 200X.
(c) 'Ghost Bands' or areas lacking in pearlite, which indicate the presence of phosphorus. Magnification: 75X. (From ENGINEERING METALLURGY, Part 1, by R.A. Higgins, 1983. Copyright Hodder & Stoughton Ltd., England. Reprinted with permission).

5.31 SULFUR (S)

Sulfur produces the most harmful effect on steel quality by forming a brittle sulfide FeS. When sulfide is present in steel in quantities as low as 0.01%, it may trigger a precipitation of the sulfide at the crystal boundaries (Fig. 5.12a).

Principal functions of the sulfur are [5]:

1. By forming brittle sulfides having a low melting point, sulfur makes a steel to crumble during hot working. The steel also becomes unsuitable for cold-working processes.

2. When a sufficient amount of manganese is added to a steel, the sulfur forms the manganese sulfides MnS which are plastic at a hot-working temperature. These sulfides are distributed throughout the steel as rather large globules (Fig. 5.12b), which are not soluble in steel and thus are not associated with the steel structure.

5.32 PHOSPHORUS (P)

Phosphorus dissolves in solid steel when it is in quantities less than 1%. When this amount is exceeded, phosphorus precipitates in form of brittle phosphide Fe_3P.

Principal functions of phosphorus are [5]:

1. In solution phosphorus produces a significant hardening effect. However, its amount is usually limited to 0.05% due to an increase in brittleness. This is especially true in the presence of the phosphide Fe_3P as a separate constituent in the microstructure.

2. In hot rolled or forged steel phosphorus forms so-called 'ghost bands', which are the areas containing no pearlite but with high concentration of phosphorus (Fig. 5.12c). These areas are becoming the planes of weakness in steel.

5.33 NITROGEN (N)

In steel nitrogen either forms nitrides or remains dissolved interstitially after solidification. In both these forms, nitrogen makes steel brittle and unsuitable for severe cold work [5]. Therefore, in order to produce mild steel with high ductility, the nitrogen has to be kept as low as 0.002%. Reduction in nitrogen content also reduces the quench aging.

5.34 HYDROGEN (H)

Hydrogen contents of more than about 0.0005% will result in reduced ductility of a steel. Hydrogen may be diffused out of steel at room temperature. This diffusion is more effective at slightly elevated temperatures [4].

Hydrogen, in excess of 0.0005%, also contributes to internal cracking which usually occurs during the cooling of a metal after rolling or forging.

REFERENCES

1. A.T. Peters, Ferrous Production Metallurgy, John Wiley & Sons, New York, pp. 77-81 (1982).

2. G. Krauss, Principles of Heat Treatment of Steel, American Society for Metals, Metals Park, Ohio, pp. 10-13 (1980).

3. E.C. Bain and H.W. Paxton, Alloying Elements in Steel, Second Edition, American Society for Metals, Metals Park, Ohio (1966).

4. The Making, Shaping and Treating of Steel, 10th Edition, eds. W.T. Lankford, Jr., et al, Association of Iron and Steel Engineers, Pittsburgh, Pennsylvania, pp. 1231-1288 (1985).

5. R.A. Higgins, Engineering Metallurgy, Part 1, Applied Physical Metallurgy, Robert E. Krieger Publishing Company, Melbourne, Florida, pp. 295-343 (1983).

6. K.W. Andrews, "Empirical Formulae for the Calculation of Some Transformation Temperatures," Journal of the Iron and Steel Institute, Vol. 203, pp. 721-727 (1965).

7. Metals Handbook, Eighth Edition, Vol. 1, Properties and Selection of Metals, American Society for Metals, Metals Park, Ohio, pp. 785-787 (1961).

8. ASM Metals Reference Book, American Society for Metals, Metals Park, Ohio, pp. 1-80, 137-223 (1981).

9. R.M. Bozorth, Ferromagnetism, D. Van Nostrand Company, Inc., Toronto, New York and London, pp. 83-88 (1951).

6

Metallurgical Factors Controlling
Properties of Steels

6.1 GENERAL PROPERTIES OF STEELS

The following three general properties are usually considered in the design of a steel [1]:

A. **Functional properties,** or the properties required by the alloy in order to allow the component manufactured from the alloy to perform the designed function.

B. **Cost properties,** or the properties required by the alloy in order to reduce the cost of production of the alloy and its fabrication into the finished or semi-finished component.

C. **Surface quality properties,** or the properties which affect the appearance of the finished components in terms of surface finish, aesthetical appeal, etc.

Typical examples of the functional properties are strength, formability, toughness, weldability, machinability, etc. The cost properties may include such characteristics as prime material cost, cost of microalloying elements and conversion cost of the prime metal into a designed component.

The main goal of the metallurgical design is to produce the required combination of properties for a particular product such as the design of an alloy for an oil pipe-line in which an optimum combination of strength with toughness and weldability would be required.

6.2 THE CONTROL OF MICROSTRUCTURE

There is a strong correlation between properties of a steel and its microstructure. The control of the microstructure can be achieved by controlling the following three aspects of the steel production [1]:

A. **Composition** - The steel composition controls the phases present in steel and their proportion. It also controls the evolution of the phases during the phase transformation.

B. **Heat treatment** - The heat treatment also controls the proportion, size and distribution of the phases. It also affects the grain size, composition of phases, dislocation structure and defect structure.

C. **Hot and cold deformation** - Most of the features listed previously are affected by the hot and cold deformation. In addition to that, the deformation process results in the crystallographic textures developed by the phases in the structure.

In the process of the metallurgical design it is necessary not only to obtain the required final

107

microstructure but also to ensure that, throughout the whole manufacturing process, the structure at any given stage is optimum for processing at the next stage.

6.3 METHODOLOGY OF THE METALLURGICAL DESIGN

The methodology which is used for the metallurgical design of steels may be described as consisting of the following main steps [1]:

1. Identifying the most important properties of the alloy being designed.

2. Establishing the relationship between those properties and microstructural and compositional parameters, often using the metallurgical models.

3. Defining the design criteria such as a simple ratio of the effects of the various microstructural and compositional parameters on the relevant properties of the designed steel.

4. Optimizing the microstructural and compositional parameters by using the design criteria and taking into account the economics of production of the designed steel.

5. Controlling the processing and heat treatment of the steel so the designed compositional and microstructural parameters are obtained.

The multiple-regression technique is often used for relating the microstructural and compositional parameters to the properties of steel such as yield strength, toughness, etc.

As an example of the design criterion, let us consider the following ratio

$$C = \frac{\Delta (ITT)}{\Delta \sigma_y} , \qquad\qquad (6\text{-}1)$$

where $\Delta(ITT)$ = change in impact transition temperature

$\Delta \sigma_y$ = change in yield strength.

For optimum combination of strength and toughness the microstructural or compositional parameters are implemented for which the ratio C has to have the largest negative value.

6.4 METALLURGICAL FACTORS CONTROLLING STRENGTH

The following metallurgical factors are usually considered as influencing the strength of steel [1, 2]:

1. Grain size

2. Substitutional solid-solution strengthening

3. Interstitial solid-solution strengthening

4. Substitutional-interstitial solute interaction strengthening

5. Precipitation strengthening

6. Dislocation strengthening

7. Second-phase strengthening.

All these factors, except substitutional solid-solution strengthening, are sensitive to processing. They also often interact with one another, so the process of metallurgical design of steel may become very complicated. It is possible, however, to apply some metallurgical models to predict an increase in strength of steels as a function of composition and metallurgical structure as it will be described below.

6.5 EFFECT OF GRAIN SIZE ON STRENGTHENING

The strengthening effect associated with a refinement of the grain size is analytically described by Hall-Petch relationship [1]:

$$\sigma_y = \sigma_i + k_y d^{-1/2}, \qquad\qquad (6\text{-}2)$$

where σ_y = yield stress

σ_i = friction stress opposing the movement of dislocation in the grains

k_y = constant.

A typical plot of Eq. (6-2) is shown in Fig. 6.1.

Grain refinement can be achieved with the addition of aluminum and nitrogen. The addition of other additives, such as columbium, vanadium and titanium are found to be more advantageous because, besides their contribution to the grain refinement, they also contribute to precipitation hardening and higher yield strengths.

6.6 SUBSTITUTIONAL SOLID-SOLUTION STRENGTHENING

The intensity of solid solution strengthening is a function of:

a) the difference in size between the solute and solvent atoms,

b) disturbances to the electronic structure which may be presented in terms of the difference in shear modulus between the solute and solvent.

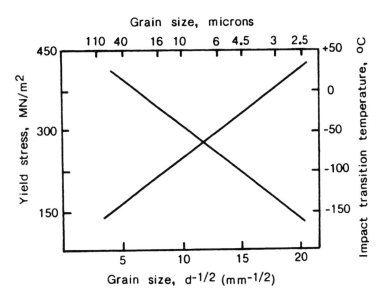

Fig. 6.1 Effect of grain size of ferrite on yield stress and impact transition temperature (0.10% C, 0.50% Mn, 0.2% Si, 0.006% N). (From PHYSICAL METALLURGY AND THE DESIGN OF STEELS, by F.B. Pickering. Copyright Elsevier Applied Science Publishers Ltd., England. Reprinted with permission).

The overall effect of substitutional solutes on the strength is given by [1]:

$$\sigma \sim C^{1/2},\tag{6-3}$$

where σ = flow stress

C = concentration of solutes.

Substitutional solid solutions are formed by various elements. However, formation of these solutions is not accompanied by a substantial strengthening of steel at room temperature.

Since most of steels have diluted solid solutions, Eq. (6-3) can be simplified to a linear dependence of solid solution strengthening upon atomic percentage of solute as shown in Fig. 6.2.

6.7 INTERSTITIAL SOLID SOLUTION STRENGTHENING

Carbon and nitrogen are the two elements which can substantially increase the strength of steels by forming interstitial solid solutions (Fig. 6.2). Since the solubility of these elements in austenite is greater than in ferrite, the interstitial solid solution strengthening is more effective in steels in which austenite is stable at room temperature.

Carbon is practically not present in most hot-rolled or box-annealed coils whereas a substantial amount of nitrogen may be retained in solution depending on hot rolling process. Interstitial solid solution strengthening is particularly effective in strengthening martensite.

6.8 SUBSTITUTIONAL-INTERSTITIAL SOLUTE INTERACTION STRENGTHENING

When the substitutional solutes interact with interstitial solutes, it will produce:

a) formation of soluble compounds such as TiC, TiN, etc. Although it will decrease the overall solid solution strengthening effect, this may be offset by introduction of some degree of precipitation

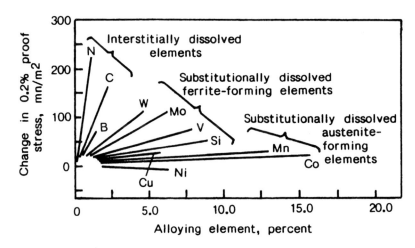

Fig. 6.2 Solid solution strengthening effects in austenitic stainless steels. (From PHYSICAL METALLURGY AND DESIGN OF STEELS, by F.B. Pickering. Copyright Elsevier Applied Science Publishers Ltd., England. Reprinted with permission).

hardening or pinning of the grain boundaries causing the grain refinement.

b) formation of association of substitutional and interstitial atoms without forming separate phases resulting in increase of strength.

6.9 PRECIPITATION STRENGTHENING

The precipitation strengthening mechanism is caused by precipitation of a constituent from a supersaturated solid solution. This effect is especially strong in alloy steels containing strong carbide-forming elements such as vanadium, niobium, columbium, titanium, molybdenum and tungsten [2]. These elements form very fine particles which precipitate during transformation process. The particles are located in the interface between the decomposed austenite and newly formed ferrite. These alloy carbide particles produce high-strength microstructures.

This mechanism is explained by the model of dispersion hardening proposed by Ashby-Orowan that allows one to show the roles of particle size and volume fraction [3]. According to this model, the tensile strength is equal to

$$TS = \frac{5.9\sqrt{f}}{\bar{x}}\ln(4000\bar{x}), \qquad (6\text{-}4)$$

where TS = tensile strength, MPa

f = precipitate fraction

$\bar{x}$ = mean planar-intercept diameter of a precipitate, μm.

Since $1/\bar{x}$ is the predominant function of $\bar{x}$, the stress increment due to fine precipitates increases with the reduction in precipitate size and the increase in fine precipitate fraction as shown in Fig. 6.3.

The main precipitation-hardening systems used in commercial HSLA steels are listed in Table 6.1. The effects of some systems are briefly described below [4].

6.10 STRENGTHENING WITH COLUMBIUM

Columbium (niobium) is a very effective strengthening agent for hot-rolled carbon steels. During conventional rolling, columbium produces marked precipitation strengthening which is complemented with some strengthening due to refining the ferrite grain size [4].

Columbium carbide (Cb_4C_3) and columbium nitride (CbN) are mutually soluble. In combination with carbon and nitrogen, they form the compound known as the carbonitride Cb(C,N).

Figure 6.4 shows the effect of microalloying with columbium on yield strength of hot-rolled strip [5]. The strengthening effect increases with decrease in size of the precipitated columbium carbide particles.

The solution temperature for columbium is very high. For a steel containing 0.20% C, the temperature of about 1300°C (2372°F) is desired for complete solution of 0.07% columbium. Columbium has relatively low affinity for oxygen that allows its use in semikilled steel, thus providing maximum production yield from the ingot.

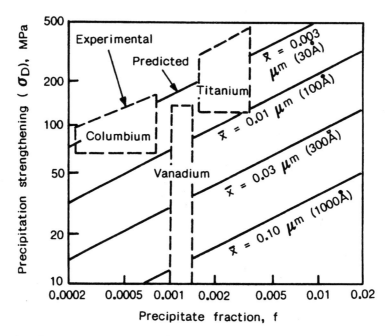

Fig. 6.3 The dependence of precipitation strengthening on precipitate size
($\bar{x}$) and fraction according to the Ashby-Orowan Model, compared with
experimental observations for given microalloying additions. (From
MICROALLOYING '75. Copyright STRATCOR, Pittsburgh, Pennsylvania.
Reprinted with permission).

Table 6.1 Important precipitation-strengthening
mechanisms in HSLA steels. Data from Fletcher
(1979).

Elements	Main precipitates
Columbium..............	$Cb(C,N)$*, Cb_4C_3
Vanadium...............	$V(C,N)$**, V_4C_3
Columbium + vanadium.....	$Cb(C,N)$, $V(C,N)$, Cb_4C_3, V_4C_3
Vanadium + nitrogen......	VN
Copper..................	Cu
Copper + columbium.......	Cu, $Cb(C,N)$
Titanium................	$Ti(C,N)$***, TiC
Aluminum + nitrogen......	AlN

* Columbium carbonitride.
** Vanadium carbonitride.
*** Titanium carbonitride.

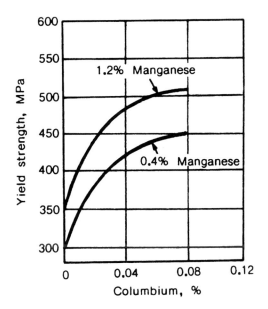

Fig. 6.4 The effect of columbium on the yield strength of hot-rolled 8 mm (0.315 in.) - thick strip with 0.08% C and 0.3% Si. (From MICROALLOYING '75. Copyright STRATCOR, Pittsburgh, Pennsylvania. Reprinted with permission).

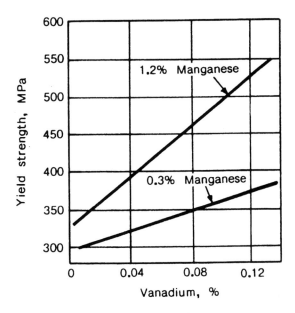

Fig. 6.5 The effect of vanadium on the yield strength of hot-rolled 8 mm (0.315 in.) - thick strip with 0.08% C and 0.3% Si. (From MICROALLOYING '75. Copyright STRATCOR, Pittsburgh, Pennsylvania. Reprinted with permission).

6.11 STRENGTHENING WITH VANADIUM AND NITROGEN

When vanadium is present in carbon-manganese steels, it forms carbides (V_4C_3) which are more stable than an iron or manganese carbide. Vanadium carbide, however, is less stable than columbium carbide. Much lower temperature is required for the vanadium carbide to be taken into solution in comparison with that for columbium carbide [4].

Generally, in order to provide the same increase in yield stress in hot-rolled, high-strength structural steels, the vanadium content has to be from two to four times the amount of columbium that would be added. Because its strengthening effect results mainly from precipitation strengthening, the effect of vanadium in increasing the yield strength of carbon-manganese steels is approximately linear function of the amount of vanadium present [4, 5] as shown in Fig. 6.5.

The strength of conventionally-hot-rolled steels is derived from precipitation of vanadium in ferrite as very fine particles of vanadium carbonitride V(C,N) during cooling. The properties of conventionally rolled steels containing vanadium are less process-sensitive than the properties of conventionally rolled steels containing columbium.

Vanadium is less likely than columbium to produce undesirable nonpolygonal transformed structures (bainite or acicular ferrite). The combination of vanadium with nitrogen results in the yield strength more than the sum of the increases produced by each element alone (Fig.6.6). The principal strengthening mechanism in the hot-rolled vanadium-nitrogen steels is precipitation strengthening, with vanadium nitride VN as the agent [4, 6].

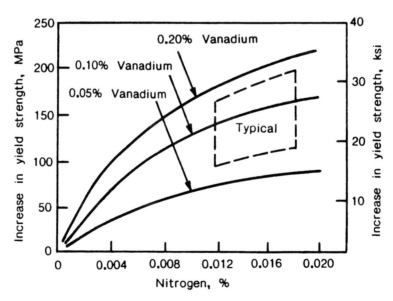

Fig. 6.6 Increase in yield strength from nitrogen and vanadium as a result of the precipitation of vanadium nitride. Data for a controlled-cooled coil product with a coiling temperature of 593°C (1100°F). (From Grozier, MICROALLOYING '75. Copyright STRATCOR, Pittsburgh, Pennsylvania. Reprinted with permission).

6.12 STRENGTHENING WITH TITANIUM

In low-carbon steels, titanium carbide (TiC) and titanium carbonitrides Ti(C,N) increase strength by both the grain refinement and precipitation strengthening [4]. A given addition of titanium produces a larger carbide fraction in steel than does the same addition of columbium. However, it has to be considered that some amount of titanium is spent on formation of titanium nitride (TiN) that is insoluble and does not participate in precipitation hardening.

Increasing amounts of titanium give strong increase in yield strength [4, 6] as shown in Fig. 6.7. However, because of the affinity of titanium for nitrogen, the addition of up to 0.025% titanium does not increase the yield strength.

High affinity of titanium for oxygen promotes formation of undesirable titanium oxides. This problem is usually solved by thorough deoxidizing a molten steel with aluminum prior to addition of titanium.

6.13 DISLOCATION STRENGTHENING

The stress required to maintain deformation at any given strain (flow stress) can be related to the dislocation density by an equation of the type [1]:

$$\sigma_f = \sigma_0 + k\sqrt{\rho} , \qquad\qquad (6\text{-}5)$$

where σ_f = total flow stress

σ_0 = flow stress attributed to other strengthening mechanisms

ρ = dislocation density

k = constant.

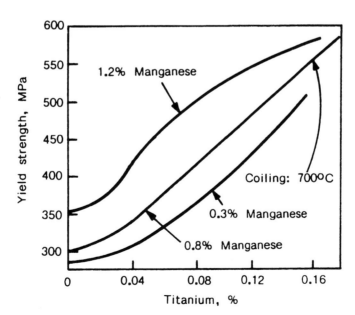

Fig. 6.7 The effect of titanium on the yield strength of hot-rolled 8 mm (0.315 in.) - thick strip with 0.08% C and 0.3% Si. (From Meyer et al, MICROALLOYING '75. Copyright STRATCOR, Pittsburgh, Pennsylvania. Reprinted with permission).

As it follows from Eq. (6-5), the flow stress increases with increase in the density of dislocations.

Since most hot-rolled steels have polygonal ferrite microstructure that is virtually dislocation free, the increase in yield strength due to dislocation strengthening is negligible. The contribution of dislocation strengthening, however, is much greater in some high-strength sheet steels which have acicular ferrite microstructures containing a high dislocation density.

6.14 SECOND-PHASE STRENGTHENING

Second phases such as pearlite and martensite produce strengthening effect in steels.

The law of mixtures which describes the yield stress of ferrite-pearlite structures varying from 0 to 100% pearlite is described in the form [1]:

$$YS = X_f^{1/3}(YS)_f + (1 - X_f^{1/3})(YS)_p,$$

$(6-6)$

where X_f = volume fraction of ferrite

$(YS)_f$ = yield stress of the ferrite

$(YS)_p$ = yield stress of the pearlite.

The following empirical formulae have been proposed for calculation of the mechanical properties of ferritic-pearlitic steels [7]:

Yield stress, MPa:

$$YS = X^{1/3}(35 + 58Mn + 17.4d_f^{-1/2}) + (1 - X^{1/3})(178 + 3.8s^{-1/2})$$
$$+ 63Si + 3535\sqrt{V} + 3535\sqrt{N}$$

$(6-7)$

Tensile strength, MPa:

$$TS = X^{1/3}(246 + 18.2d_f^{-1/2}) + (1 - X^{1/3})(720 + 3.5s^{-1/2}) + 97Si$$
$$+ 1047V + 2294N$$

$(6-8)$

Impact-transition temperature, ^{o}C:

$$ITT = - (19 + 11.5d^{-1/2}) + 44Si + 2.2(\% \text{ pearlite}) + 919\sqrt{V} + 919\sqrt{N},$$

$(6-9)$

where X = transformed volume fraction

d = ferrite grain size, mm

s = mean interlamellar spacing of pearlite, mm.

In this formulas, the symbols represent the weight percent of the element indicated.

Martensite is used as a second-phase strengthening component in dual-phase steels in which martensite is dispersed within a fine-grained ferrite matrix. Both yield strength and tensile strength are found to be a function of the volume fraction of martensite.

6.15 EFFECT OF COMPOSITION AND STRUCTURE ON YIELD STRESS

The ferrite-grain size is the most important structural parameter controlling the yield stress.

A typical equation describing the yield stress of plain low-carbon steel as a function of the compositional and structural factors is [1]:

$$Y = 15.4 \, (3.5 + 2.1Mn + 5.4Si + 23\sqrt{N_f} + 1.13d^{-1/2}), \qquad (6\text{-}10)$$

where   Y = yield stress, MPa

N_f = weight percent of free nitrogen.

It follows from the Eq. (6-10) that:

a) pearlite (i.e. carbon) has little or no effect on the yield stress of low carbon steels

b) free nitrogen substantially increases yield stress

c) grain refinement would increase the yield stress.

6.16 EFFECT OF COMPOSITION AND STRUCTURE ON TENSILE STRENGTH

The tensile strength of low-carbon steels with ferrite-pearlite microstructures is described by [1]:

$$\sigma_t = 15.4[19.1 + 1.8Mn + 5.4Si + 0.25(\% \text{ pearlite}) + 0.5d^{-1/2}], \qquad (6\text{-}11)$$

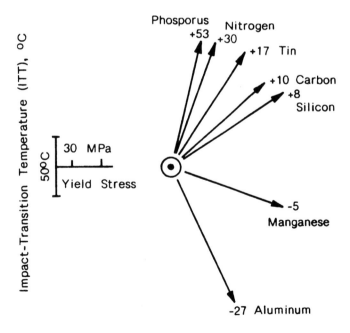

Fig. 6.8 Effect of composition on impact-transition-temperature (ITT). Ratios indicate the change in transition temperature per 15 MPa increase in yield strength. (From Pickering, MICROALLOYING '75. Copyright STRATCOR, Pittsburgh, Pennsylvania. Reprinted with permission).

where σ_t = tensile strength, MPa

As it follows from Eq. (6-11), both pearlite content and grain refinement contribute to the tensile strength. The tensile strength of low-carbon steels with a bainitic microstructure is given by [8]:

$$\sigma_t = 15.4[16 + 125C + 15(Mn + Cr) + 12Mo + 6W$$
$$+ 8Ni + 4Cu + 25(V + Ti)] \tag{6-12}$$

6.17 METALLURGICAL FACTORS CONTROLLING TOUGHNESS

In designing a steel with high toughness, it is assumed that notches or stress concentrations of various types are present in nearly all structures. Good notch-toughness behavior of the steel is promoted by [4]:

   a) low carbon content
   b) high ratio of manganese to carbon
   c) low phosphorus and sulfur content
   d) deoxidation of the steel with aluminum
   e) a fine grain size
   f) normalizing
   g) quenching and tempering.

Toughness is evaluated by the impact-transition temperature (ITT). The lower values of ITT the higher toughness of steel. It follows from Fig. 6.8 showing the effect of composition on ITT, that low-carbon, high-manganese, aluminum grain-refined steels are preferred, because aluminum is particularly beneficial in removing nitrogen as aluminum nitride [8].

The effect of grain size on ITT is shown in Fig. 6.1.

The impact-transition temperature for HSLA steels with ferrite-pearlite microstructure is expressed as a function of compositional and structural factors by an equation [8]:

$$ITT = -19 + 44Si + 700\sqrt{N_f} + 2.2(\% \text{ pearlite}) - 11.5d^{-1/2}, \tag{6-13}$$

where ITT = impact-transition temperature, ^oC.

The impact-transition temperature is affected by different strengthening mechanisms [9] as shown in Fig. 6.9. Both the grain refinement and solid solution of manganese and nickel improve toughness with increase in yield strength. However, precipitation hardening, dislocation strengthening and solid solution of carbon produce deterioration in toughness with increase in yield strength.

6.18 METALLURGICAL FACTORS CONTROLLING DEEP DRAWABILITY

The material property controlling deep drawability can be described by R-value determined in a strip tensile test and can be expressed as [1]:

$$R = \frac{\epsilon_w}{\epsilon_t}, \tag{6-14}$$

METALLURGICAL FACTORS

where ϵ_w = true-width strain

ϵ_t = true-thickness strain.

When the properties in the plane of the sheet are sensitive to direction, the average R-value is used [1]:

$$R = (R_L + 2R_{45} + 2R_T)/4, \qquad (6\text{-}15)$$

where R_L, R_T and R_{45} = R-values measured parallel, transverse to and at 45° to the rolling direction.

It has been shown that the greater the R-value, the more readily is the material capable of deep drawing. The variation in the R-value in the actual plane of the sheet can serve as a measure of planar anisotropy, and can be expressed as [1]:

$$\Delta R = (R_L + R_T - 2R_{45})/2 \qquad (6\text{-}16)$$

A cube-on-corner texture with (111) planes in the plane of the sheet results in high R-values, whereas a cube-on-face texture in which (110) planes are parallel to the plane of the sheet reduces the R-value. Normally processed rimming steels have R-values of 1.0-1.2. The R-values in the aluminum-killed

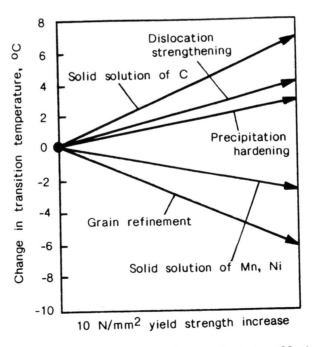

Fig. 6.9 Different strengthening mechanisms and their effect on transition temperature. (From Meyer and Boer, JOURNAL OF METALS, Vol. 29, Jan. 1977. Publication of The Metallurgical Society, Warrendale, Pennsylvania, 15086, USA. Reprinted with permission).

steels can be as high as 1.4-1.8. Rimming steel requires finely dispersed carbides to pin the boundaries to produce large grains of preferable (111) texture and smaller grains of less preferable (110) orientation.

The R-value increases with increase in grain size as described in application for low-carbon steel by a relationship [1, 10]:

$$R = R_o - kN, \tag{6-17}$$

where R_o, k = constants

N = ASTM ferrite grain size.

6.19 METALLURGICAL FACTORS CONTROLLING STRETCH FORMING

The stretch forming is usually controlled by maximum uniform elongation, ϵ_u, prior to necking.

It was shown in Part I, that the work-hardening behavior in tensile testing can be expressed as:

$$\sigma = k\epsilon^n, \tag{6-18}$$

where σ = true flow stress

ϵ = true strain

k = constant

n = work-hardening index (or exponent).

It can be shown [1] that the maximum uniform elongation, ϵ_u, is numerically equal to n.

The maximum uniform elongation, ϵ_u, for low-carbon steel is expressed by [1, 8]:

$$\epsilon_u = 0.28 - 0.2C - 0.25Mn - 0.044Si - 0.039Sn - 1.2N_f \tag{6-19}$$

The stretch-forming capability is also evaluated by work-hardening rate, $d\sigma/d\epsilon$, at constant low strain value. For low-carbon steel the work-hardening rate may be presented by [1, 11]:

$$\frac{d\sigma}{d\epsilon} = 15.4(24 + 7.8C + 1.5Mn + 7.5Si + 36P$$
$$+ 9.3Sn + 98N_f - d^{-1/2}), \tag{6-20}$$

where $d\sigma/d\epsilon$ = work-hardening rate at strain $\epsilon = 0.2$, MPa

For HSLA steels with ferrite-pearlite structures, the maximum uniform elongation and work-hardening rate are given by [8]:

$$\epsilon_u = 0.27 - 0.016(\% \text{ pearlite}) - 0.015Mn - 0.040Sn - 0.04Si - 1.1N_f \tag{6-21}$$

$$\frac{d\sigma}{d\epsilon} = 15.4[25 + 7.2Si + 30P + 9.9Sn + 89N_f + 0.09(\% \text{ pearlite}) + d^{-1/2}] \tag{6-22}$$

6.20 METALLURGICAL FACTORS CONTROLLING BENDING

A number of parameters are used to evaluate bending capabilities of steels. It has been shown that in simple bending increasing maximum uniform elongation, ϵ_u, is beneficial [12]. Bending under tension requires a low yield-stress to ultimate-tensile strength ratio, and high R and n values [12, 13].

Another important parameter is the total strain at fracture, ϵ_T, in a tensile test. The higher the ϵ_T-value, the better capabilities of a steel sheet in bending.

The total strain at fracture for low-carbon steels is described by [1, 11]:

$$\epsilon_T = 1.4 - 2.9C + 0.2Mn + 0.16Si - 2.2S - 3.9P - 0.25Sn + 0.017d^{-1/2} \qquad (6-23)$$

For HSLA steels with ferrite-pearlite structures, the total strain at fracture is given by [8]:

$$\epsilon_T = 1.3 - 0.02(\% \text{ pearlite}) + 0.3Mn + 0.2Si - 3.4S$$
$$- 4.4P + 0.29Sn + 0.015d^{-1/2} \qquad (6-24)$$

6.21 METALLURGICAL FACTORS CONTROLLING WELDABILITY

Welding involves the melting and solidification of steel, and the heat-affected zone of the base metal is subjected to a heat treatment. Major metallurgical factors controlling quality of welding are briefly discussed below [4].

Hydrogen-induced cold cracking is the most important source of weld defects in the welding and hardenable materials [4]. A convenient way to evaluate weldability from standpoint of hydrogen cracking is by using the **carbon equivalent (CE)** formula. One of the best-known carbon-equivalent formulae was developed by Deardon-O'Neill in application for carbon-manganese steels. It is expressed as follows [4]:

$$(CE) = Mn/6 + (Cr + Mo + V)/5 + (Ni + Cu)/15, \qquad (6-25)$$

where the numbers for the elements are the alloy content in weight percent.

The weldability ratings based on carbon equivalent used may vary from one steel producer to another. Below is one of the examples:

CE	Weldability
Up to 0.35	Excellent
0.36 to 0.40 incl.	Very good
0.41 to 0.45 incl.	Good
0.46 to 0.50 incl.	Fair
Over 0.50	Poor

Lamellar tearing may take place in the parent metal during welding as a result of shrinkage forces in the plate-thickness direction. The problem can be minimized by [4]:

a) reduction in the sulfur level,

b) sulfide-inclusion-shape control with addition of rare earths, zirconium or calcium.

Another source for lamellar tearing is the presence of excessive hydrogen.

Spot-weld peeling is mainly due to excessive hardenability of weld buttons. The following tentative carbon equivalent formula has been proposed for spot welding of HSLA steels [4]:

$$CE = C + Mn/30 + (Cr + Mo + Zr)/10 + Ti/2$$
$$+ Cb/3 + V/7 + (UTS)/900 + h/20, \tag{6-26}$$

where UTS = ultimate tensile strength, ksi

h = strip thickness, in.

A carbon-equivalent values of 0.30 or less are considered to be acceptable.

6.22 INCLUSION-SHAPE AND TEXTURE CONTROL

The inclusion-shape control has been found beneficial for decreasing premature ductile fracture of HSLA steels [8]. These fractures are caused by elongated stringers of ribbons of manganese sulfide, especially in more or less co-planar aggregates of discontinuous stringers of alumina. Furthermore, because manganese sulfide becomes more plastic than steel as the temperature decreases, rolling to a low-finishing temperature may result in an undesirable distribution of non-metallic inclusions.

The inclusion-shape control involves additions of zirconium, cerium, or calcium so that elongated stringers of inclusions are not produced. Zirconium and cerium are mainly effective in modifying sulfides whereas calcium also alters the constitution of alumina and prevents it from forming discontinuous stringers of segregated particles.

It was found that low finishing-rolling temperature during controlled rolling may result in a pronounced crystallographic texture into the rolled product. This texture significantly influences properties, especially the impact transition temperature [8].

REFERENCES

1. F.B. Pickering, Physical Metallurgy and the Design of Steels, Applied Science Publishers, London, pp. 1-88 (1978).

2. M.Cohen and W.S. Owen, "Thermo-Mechanical Processing of Microalloyed Steels, " Microalloying '75, Union Carbide Corp., New York, 1977, pp. 2-8.

3. T. Gladman, et al, "Structure-Property Relationships in High-Strength Microalloyed Steels", Microalloying '75, Union Carbide Corp., New York, 1977, pp. 32-58.

4. E.E. Fletcher, "A Review of the Status, Selection, and Physical Metallurgy of High-Strength, Low-Alloy Steels", Metals and Ceramics Information Center Report, No.MCIC-79-39, Batelle Columbus Laboratories, Columbus, Ohio, March 1979.

5. L. Meyer, et al, "Columbium, Titanium, and Vanadium in Normalized, Thermo-Mechanically Treated and Cold-Rolled Steels", Microalloying '75, Union Carbide Corp., New York, 1977, pp. 153-171.

6. J.D. Grozier, "Production of Microalloyed Strip and Plate by Controlled Cooling", Microalloying '75, Union Carbide Corp., New York, 1977, pp. 241-250.

7. S. Licka, et al, "Mathematical Model to Calculate Structure Development and Mechanical Properties of Hot-Rolled Plates and Strips", Thermomechanical Processing of Microalloyed Austenite, Metallurgical Society of AIME, New York, 1981, pp. 521-528.

8. F.B. Pickering, "High-Strength, Low-Alloy Steels - A Decade of Progress", Microalloying '75, Union Carbide Corp., New York, 1977, pp. 9-31.

9. L. Meyer and H.de Boer, "HSLA Plate Metallurgy: Alloying, Normalizing, Controlled Rolling", Journal of Metals, Vol. 29, No. 1, Jan. 1977, pp. 17-23.

10. D.J. Blickwede, Transaction of the American Society for Metals, Vol. 61, 1968, p. 653.

11. T. Gladman, et al, Journal of the Iron and Steel Institute, Vol. 208, 1970, p. 172.

12. R.D. Butler, Sheet Metal Industries, Vol. 39, 1964, p. 705.

13. R.D. Butler and J.F. Wallace, Iron and Steel Institute Special Report, No. 79, 1963, p. 131.

7
Heat Treatment of Steel

7.1 TYPES OF HEAT TREATMENT

Heat treatment is a process of heating a solid metal or alloy to a certain temperature (Figs. 7.1 and 7.2) with consequent cooling at a certain rate in order to obtain desired conditions or properties. Table 7.1 shows major types of heat treatment processes and the resulting microstructures [1, 2].

7.2 FULL ANNEALING

The main purposes of annealing are to relieve cooling stresses induced during hot and cold working, and to soften a metal in order to improve its machinability or formability. The process produces a uniform microstructure of ferrite and pearlite [2].

Full annealing process allows one to develop a coarse pearlite microstructure in low or medium-carbon steels. The process consists of heating at relatively high temperature in austenitic region so that full carbide solution is obtained, followed by a slow cooling which produces a complete transformation in the high-temperature end of the pearlite range as shown in Fig. 7.3. The process is considered to be simple and reliable for most steels but rather slow. A great deal of time is required to cool from an austenizing temperature to a room temperature.

7.3 ISOTHERMAL ANNEALING

Isothermal annealing allows one to obtain a coarse pearlite microstructure with a very considerable time saving in comparison with the time required by the full annealing [2]. It is achieved by rather rapid cooling from austenizing temperature to the transformation temperature, holding at this temperature until the transformation is complete and subsequent rapid cooling from the transformation temperature to room temperature as shown in Fig. 7.4. If the extreme softness of the coarse pearlite is not necessary, the transformation may be carried out at a 'nose' of the curve. This allows for further speedup of the annealing process.

Isothermal annealing is utilized in a continuous heat treatment process commonly known as 'cycle annealing'. The cycle annealing furnace is equipped with an air-blast chamber for rapid cooling from high austenizing temperature down to holding temperature at which isothermal transformation to pearlite takes place.

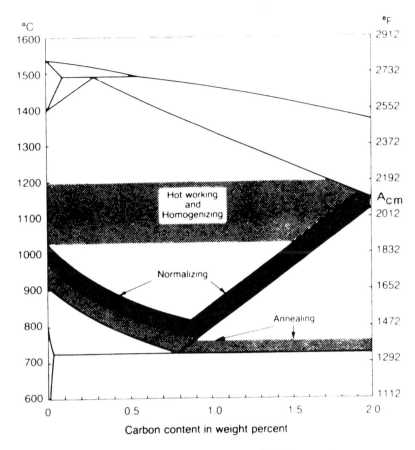

Fig. 7.1 Portion of the Fe-C diagram with temperature ranges for full annealing, normalizing, hot working and homogenizing indicated. From PRINCIPLES OF HEAT TREATMENT OF STEEL, by G. Krauss, ASM International (formerly American Society for Metals), 1980. With permission.

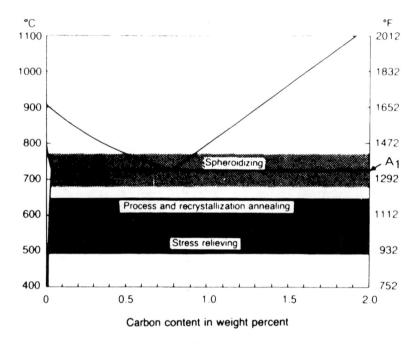

Fig. 7.2 Portion of the Fe–C diagram with temperature ranges for process annealing, recrystallization annealing, stress relieving and spheroidizing indicated. From PRINCIPLES OF HEAT TREATMENT OF STEEL, by G. Krauss, ASM International (formerly American Society for Metals), 1980. With permission.

Table 7.1 Microstructures produced by major heat treatment processes. Adapted from Krauss (1980) and The Making, Shaping and Treating of Steel (1985).

Heat treatment process	Microstructure
Full annealing	Ferrite and pearlite
Isothermal annealing	Ferrite and pearlite
Normalizing	Ferrite and pearlite
Spheroidizing	Ferrite and carbide
Quenching and tempering	Tempered martensite
Martempering	Tempered martensite
Austempering	Bainite
Dual-phase	Ferrite and martensite

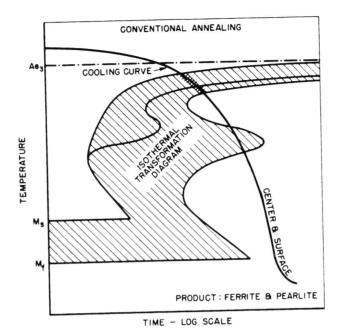

Fig. 7.3 Schematic transformation diagram for full annealing. (From THE MAKING, SHAPING AND TREATING OF STEEL, by W.T. Lankford, et al, Tenth Edition. Copyright AISE, Pittsburgh, Pennsylvania. Reprinted with permission).

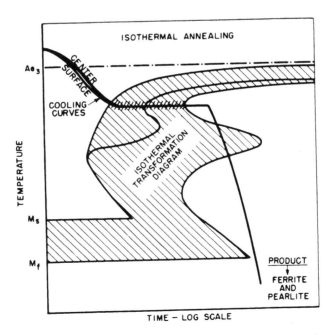

Fig. 7.4 Schematic transformation diagram for isothermal annealing. (From THE MAKING, SHAPING AND TREATING OF STEEL, by W.T. Lankford, et al, Tenth Edition. Copyright AISE, Pittsburgh, Pennsylvania. Reprinted with permission).

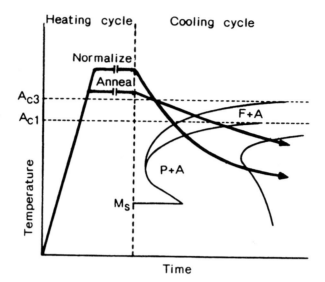

Fig. 7.5 Schematic time-temperature cycles for normalizing and full annealing. From PRINCIPLES OF HEAT TREATMENT OF STEEL, by G. Krauss, ASM International (formerly American Society for Metals), 1980. With permission.

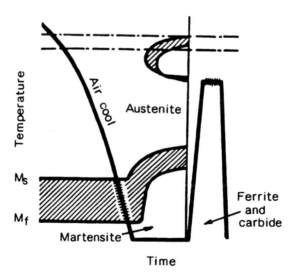

Fig. 7.6 Schematic heat treatment cycle for spheroidizing an air-hardening steel. Martensite forms first and then is tempered close to the Ac_1 to produce a spheroidized structure. From PRINCIPLES OF HEAT TREATMENT OF STEEL, by G. Krauss, ASM International (formerly American Society for Metals), 1980. With permission.

7.4 NORMALIZING

Normalizing is a heat treatment similar to full annealing. Its main objective is to produce a uniform microstructure of ferrite and pearlite. Another objective is to refine the grain size that frequently becomes very coarse during hot working or that is present in steel casting [1, 2].

Normalizing in hypoeutectoid steels is performed at temperatures higher than those used for annealing. Hypereutectoid steels are heated above the temperature A_{cm} as shown in Fig. 7.1. For both steels, heating is followed by an air cooling which provides a much faster cooling rate in comparison with that of full annealing (Fig 7.5). The faster cooling rate the lower the temperature range over which ferrite and pearlite are being formed resulting in more refine microstructure and therefore higher strength, hardness and lower ductility.

7.5 SPHEROIDIZING

Spheroidized microstructures in steels consist of spherical carbide particles uniformly dispersed in a ferrite matrix (Fig. 4.6). This microstructure is the most stable microstructure in steels. It allows one to improve ductility which is important for low and medium-carbon steels and also to obtain lower hardness which is important for high-carbon steels that undergo machining prior to final hardening [1, 2].

Spheroidizing can be accompanied by either complete or partial austenizing with subsequent holding just below critical temperature Ac_1 and then cooling very slow through the Ac_1 (Fig. 7.6) or cycling above and below Ac_1. Spheroidizing of martensite microstructures is frequently performed on highly alloyed tool steels that form martensite on air cooling.

7.6 HOMOGENIZING

Homogenizing is a type of annealing treatment which is usually performed in earlier stages of steel processing prior to hot rolling or forging [1]. Main purpose of homogenizing is to obtain uniformity or homogeneity of austenite which improves hot workability and contributes to uniformity of microstructure obtained during subsequent annealing or hardening operations.

Homogenizing is usually performed at high temperatures in the austenitic phase field (Fig. 7.1) in order to speed up the reduction of segregations or chemical concentrations that are produced during a solidification of an ingot or a slab. Also, at higher temperatures second phases such as carbides are most fully dissolved.

7.7 STRESS RELIEVING

The primary purpose of stress relieving is to relieve stresses that have been introduced to a workpiece from hot and cold working processes or welding. The stress relief is accomplished by recovery mechanisms that precede recrystallization. For steels the stress relieving is performed by heating the workpiece to some temperature below the lower critical temperature A_1 (Fig. 7.2). Holding time at this temperature has to be sufficient enough to reduce the residual stresses to an acceptable level. This is followed by cooling at a relatively slow rate in order to avoid creation of new stresses [1].

Some steels exhibit loss of toughness on slow cooling from temperature of about 535°C (1000°F) and above. This phenomenon is known as 'temper brittleness'. A rapid cooling is desirable in this case.

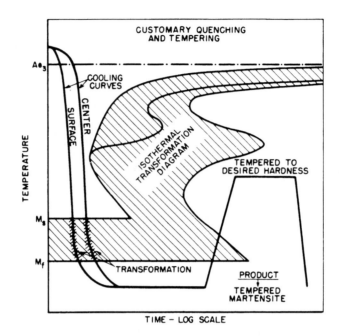

Fig. 7.7 Schematic transformation diagram for quenching and tempering. (From THE MAKING, SHAPING AND TREATING OF STEEL, by W.T. Lankford, et al, Tenth Edition. Copyright AISE, Pittsburgh, Pennsylvania. Reprinted with permission).

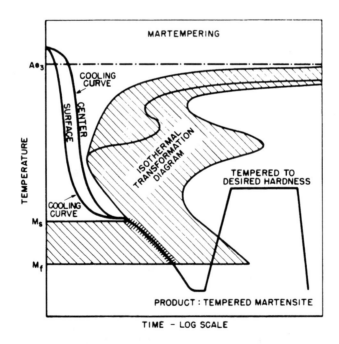

Fig. 7.8 Schematic transformation diagram for martempering. (From THE MAKING, SHAPING AND TREATING OF STEEL, by W.T. Lankford, et al, Tenth Edition. Copyright AISE, Pittsburgh, Pennsylvania. Reprinted with permission).

7.8   QUENCHING AND TEMPERING

Quenching and tempering is the heat treatment commonly used in order to obtain tempered martensite microstructure which is characterized by relatively high toughness and good ductility [2]. During quenching, a workpiece is heated to a temperature at which an austenite is formed and then it is cooled rapidly enough that no transformation occurs at the temperature above the martensite range. Water, oils or brine is most commonly used as the quenching medium.

Formation of a hard and brittle martensite is accompanied by creation of a high residual stress. Tempering allows one to relieve these stresses and also to improve ductility. This heat treatment is similar to stress relieving. Tempering operation should immediately follow the quenching (Fig. 7.7) in order to minimize cracking.

7.9   MARTEMPERING

Martempering is a modified quenching procedure used to reduce high residual stresses which are usually created during rapid cooling through the martensite temperature range [2]. This heat treatment is carried out by quenching a workpiece into a molten-salt bath at a temperature just above the martensite transformation temperature M_S, holding in this bath in order to establish the same temperature throughout the workpiece, and then air cooling to a room temperature. As a result, transformation to martensite takes place during a relatively slow air cooling. After martempering, the workpiece may be tempered to a desired hardness as shown in Fig. 7.8.

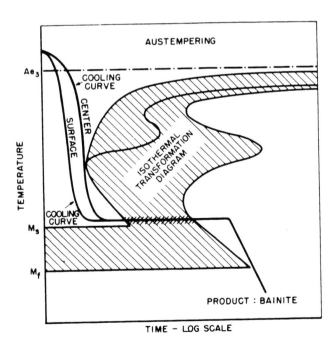

Fig. 7.9 Schematic transformation diagram for austempering.
(From THE MAKING, SHAPING AND TREATING OF STEEL, by W.T. Lankford, et al, Tenth Edition. Copyright AISE, Pittsburgh, Pennsylvania. Reprinted with permission).

Tools, bearings, and dies are the examples of the workpieces which are usually subjected to martempering.

7.10 AUSTEMPERING

Austempering is the heat treatment used in order to obtain lower bainite microstructure which is similar in strength and superior in ductility to those of tempered martensite [2]. This heat treatment includes quenching to a desired temperature in lower bainite region, usually in molten-salt bath, and holding at this temperature until transformation to bainite is completed as shown in Fig. 7.9. After austempering, the workpiece may be quenched or air cooled to room temperature. It may also be tempered to a lower hardness level if required.

A major advantage of this process over conventional quenching and tempering is minimum distortion and absence of quench cracking. Plain high-carbon steels in small section sizes, such as sheet, strip, and wire products are typical subjects of austempering.

7.11 DUAL-PHASE HEAT TREATMENT

The main purpose of dual-phase heat treatment is to obtain a microstructure representing a mixture of ferrite and martensite. It is achieved by heating the steels into the intercritical temperature region (between the A_1 and A_3 temperature). In this region, the microstructure consists of ferrite and austenite. Because austenite areas of a steel in this region are high in carbon, it has a higher ability to form martensite during quenching than a steel which was heated above the A_3 temperature.

The amount of martensite formed during dual-phase heat treatment depends on the cooling rate. Intercritical continuous annealing allows low-carbon hot or cold-rolled sheet steels with dual-phase microstructure to be obtained. The microstructure consists of ferrite with 15 to 20 percent martensite. Sheets with this dual-phase microstructure have increased strength and superior ductility.

7.12 SURFACE HARDENING

Surface hardening is the heat treatment process allowing to obtain a hard, wear-resistant surface and a strong, fracture-resistant core.

The following two processes are known [2]:

1. Decremental hardening.
2. Case hardening.

Decremental hardening is the process of increasing surface hardness without a change in the chemical composition of the surface. It is achieved by heating the workpiece at a high rate in order to produce a steep temperature gradient. This heating is continued until the A_3 temperature is reached at the desired depth below the surface at which time the piece is then quenched.

Decremental hardening can be accomplished by one of the following processes:

A. Heating in the furnace maintained at high temperature.
B. Flame hardening, or direct impingement of a high temperature.
C. Induction hardening, the process which utilizes induction heating.

Case hardening process involves a change in chemical composition of the surface portion of a workpiece. The following modifications of this process are utilized:

A. **Carburizing** - This process produces a high-carbon layer at the surface of low-carbon steel by heating it in contact with either carbonaceous solid material or in a carbon-rich liquid (liquid carburizing) with subsequent quenching. Depending on the process, the heating temperature varies from 870°C to 1050°C (1600°F to 1920°F).

B. **Nitriding** - This process introduces nitrogen into the surface layer of a solid ferrous alloy by holding it in contact with nitrogen material. For ferritic steel, the holding temperature is below Ac$_1$. Quenching is usually not required. The gas-phase nitronization treatment is known as **bright annealing.**

C. **Carbonitriding** - This process combines both carbonizing and nitriding processes.

7.13 QUENCH AGING

In application to carbon steel, the quench aging may be described as a process of gradual change in steel properties with time after quenching. This is a result of the precipitation of carbon, nitrogen or both from supersaturated interstitial solid solution in ferrite. In carbon steels, carbon is a principal element affecting the quench aging process [2].

Hardness increases more rapidly when aging at above room temperature but higher absolute value of hardness is attained by aging at room temperature as shown in Fig. 7.10. Aging also results in variation of magnetic properties [3].

In order to reduce quench aging, the carbon and nitrogen content in supersaturated solid solutions has to be minimized. It is achieved by cooling the steel relatively slow especially through the temperature range from 540°C to 315°C (1000°F to 600°F). Another method is to form stable carbides and nitrides by alloying the steel with titanium or zirconium.

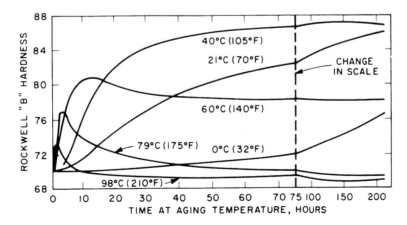

Fig. 7.10 Changes in hardness of 0.06% carbon steel quenched from 720°C (1325°F) after aging at indicated temperatures. (From METALS HANDBOOK, Aging of Iron and Steel, 1948. Copyright ASM International. Reprinted with permission).

REFERENCES

1. G. Krauss, Principles of Heat Treatment of Steel, American Society for Metals, Metals Park, Ohio, pp. 30-33, 103-273 (1980).

2. The Making, Shaping and Treating of Steel, 10th Edition, eds. W.T. Lankford, Jr., et al, Association of Iron and Steel Engineers, Pittsburgh, Pennsylvania, pp. 1258-1287 (1985).

3. Metals Handbook, Aging of Iron and Steel, American Society for Metals, Metals Park, Ohio, p. 439 (1948).

Part III

Making and Casting of Steel for Flat Products

8
Primary Steelmaking Processes

8.1 RAW MATERIALS

The following three primary steelmaking processes are known:

1. Open-hearth steelmaking
2. Oxygen steelmaking
3. Electric-furnace steelmaking.

A comparison of these processes is shown in Table 8.1.

There are two main raw materials which are used in making of steel [1]:

1. Hot metal which is a molten blast furnace iron.

2. Scrap which consists of the by-products of steel fabrication and also of the worn out, broken, or discarded items containing iron or steel.

The basic oxygen steelmaking process uses 10 to 30% scrap while the open-hearth process utilizes 25 to 100%. The electric furnace usually is charged almost entirely with cold scrap. As follows from Table 8.2, there is a difference between the composition of a charge mix and that of a steel specification.

In order to achieve a desired chemical composition of a steel, the unwanted elements, or impurities, are removed and the necessary elements are added to a melt.

8.2 OPEN-HEARTH STEELMAKING

In an open-hearth furnace (Fig. 8.1), a charge is melted on a shallow refractory hearth by a flame passing over the charge so that the charge and the roof are heated by the flame. The depth of molten metal is between 305 and 457 mm (12 to 18 inches) [1, 2].

The open-hearth process allows high flexibility in regard to a composition of raw materials. Metallic materials for the furnace charge can range from 5% cold pig iron plus 95% steel scrap to as high as 70% molten blast furnace iron (hot metal) plus 30% steel scrap. The exceptions are the high-chromium nickel stainless and heat-resisting steels.

Table 8.1 Comparison of primary steelmaking processes. Adapted from
 Peters (1982).

Parameter	Basic Open Hearth	Basic Oxygen	Basic Electric Arc
Original investment	High	Low to medium	Overall low, if continuous caster is included
Operating cost	Medium	Low	Medium to high, low when electric power is cheap
Raw material flexibility	High	Low	Low
Fuel	Gaseous or/and liquid	None (O_2+ metalloids)	None (arc heating)
Electric power consumption	Low	Low	High
Oxygen usage	Medium	High	Low for nonalloy steels, medium for alloy steels
Productivity	low (cold metal) medium (hot metal)	High	Medium
Product range	Carbon and low alloy steels	Carbon and low alloy steels	Carbon, low alloy and high alloy steels, except low N steels
Nitrogen content, %	0.003 to 0.008	0.002 to 0.006	0.008 to 0.016 0.004 to 0.007 for special N practices
Mill or caster supply	Intermittent, unless many furnaces	Every hour or less	Intermittent, unless many furnaces

Table 8.2 Typical composition of hot metal and scrap for making plain carbon steel. Data from Peters (1982).

Composition, %					
C	Mn	P	Si	S	Tramps
Steel specification ranges					
0.04/1.00	0.04/1.65	0.02 max*	0.0/0.3*	0.025 max*	0.2/0.4 max**
Assumed composition of hot metal					
4.0	1.0	0.15	1.0	0.03	0.06
Assumed composition of scrap					
0.2	0.6	0.02	0.04	0.06	0.20
Theoretical charge mix for 30% scrap (BOF)					
2.9	0.9	0.04	0.71	0.04	0.10
Theoretical charge mix for 50% scrap (BOH)					
2.1	0.8	0.085	0.52	0.05	0.13
Theoretical charge mix for 100% scrap (BEAF)					
0.2	0.6	0.02	0.04	0.06	0.20
Typical melt clear (BOF)					
0.2	0.2	0.01/0.02	0.01	0.02	0.11**
Typical melt clear (BOH)					
0.3/0.9	0.1/0.3	0.01/0.03	0.01	0.03	0.15**
Typical melt clear (BEAF)					
0.2	0.3	0.01/0.02	0.02	0.03	0.22**

* Except rephosphorized, silicon-bearing or resulfurized steels.
** Except copper-bearing steels.
BOF – basic oxygen furnace.
BOH – basic open-hearth furnace.
BEAF – basic electric-arc furnace (often abbreviated to EF).

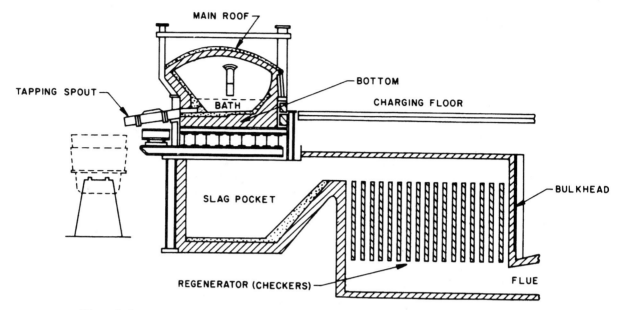

Fig. 8.1 Vertical section across the width of an open-hearth furnace, not to scale, indicating names and relative locations of principal parts. Upper section is through taphole of furnace; bottom section is through slag pocket and regenerator. (From THE MAKING, SHAPING AND TREATING OF STEEL, by W.T. Lankford, et al, Tenth Edition. Copyright AISE, Pittsburgh, Pennsylvania. Reprinted with permission).

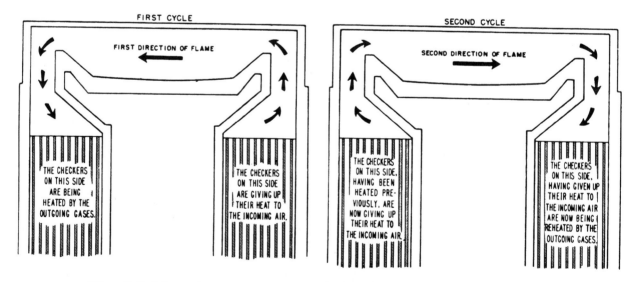

Fig. 8.2 Schematic presentation of preheating the air for combustion in regenerative furnaces. (From THE MAKING, SHAPING AND TREATING OF STEEL, by W.T. Lankford, et al, Tenth Edition. Copyright AISE, Pittsburgh, Pennsylvania. Reprinted with permission).

The furnace can use any available gaseous or liquid fuel as well as pulverized coal. Two methods are used to increase the furnace efficiency. The first method utilizes regeneration of the hot combustion products leaving the furnace by passing them through the checker chambers containing fire brick. The flow of outgoing gases and incoming combustion air is periodically reversed as shown in Fig. 8.2. The second method utilizes direct injection of oxygen into the molten metal by using the roof-mounted, water-cooled lances. Using roughly 1400 cubic feet of oxygen per ton of steel, this method doubles production and reduces the consumption of fuel.

Typical productivity of open-hearth process is:

1. Cold metal   8 to 20 net tons per hour
2. Hot metal   25 to 70 net tons per hour.

Open-hearth furnaces are open for a visual inspection of the hearth after every heat of steel is tapped from the furnace. Any lost refractory material can be replaced promptly. Their production diminishes only moderately when a sudden change from a hot-metal charge to a high-scrap charge is reqiured.

Major disadvantages of the open-hearth process are low productivity and high installation and maintanance cost. These are the main reasons for eventual replacement of this process by a balanced combination of the basic oxygen and electric-arc steelmaking processes.

8.3 OXYGEN STEELMAKING

Oxygen steelmaking process provides refining of a metallic charge through the use of high-purity oxygen and allows one to rapidly produce steel of the desired carbon content and temperature. The metallic charge consists of hot metal (molten pig iron) and scrap. The sulfur and phosphorus contents of a metal bath are reduced to a desired level by adding various steelmaking fluxes during the refining process [1, 2].

The oxygen process does not use fuel. The energy required to melt the scrap, form the slag and raise the temperature of the bath to a desired level is provided by the heat generated during oxidation of silicon, carbon, manganese, phosphorus and iron.

Typical performance of the oxygen process is:

1. Productivity   150 to 550 net ton per hour
2. Time to produce one heat   45 to 60 minutes.

In the oxygen steelmaking process, high purity oxygen is blown under pressure through, onto or over a bath containing hot metal, steel scrap, and fluxes.

The following three processes are presently used:

1. **Top-blown (BOP) process** - In this process oxygen of commercial purity at a high pressure and velocity is blown downward vertically into the bath through a single water-cooled pipe or lance (Fig. 8.3a).

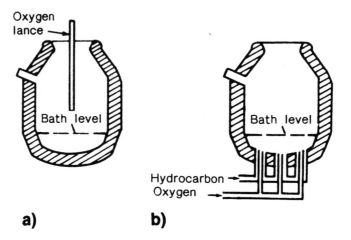

Fig. 8.3 Basic Oxygen Process (BOP):
a) Top-blown process, b) Bottom-blown (Q-BOP) process. Adapted from
THE MAKING, SHAPING AND TREATING OF STEEL (1985).

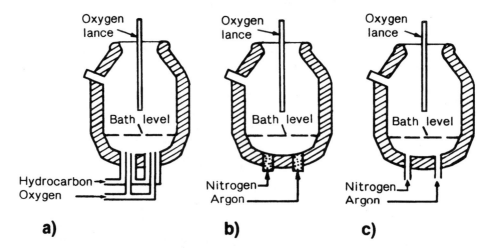

Fig. 8.4 Combination-Blown Basic Oxygen Processes. Adapted from
THE MAKING, SHAPING AND TREATING OF STEEL (1985).

2. Bottom-blown (Q-BOP) process - In this process oxygen of commercial purity at high pressure and velocity is blown upward vertically into the bath through tuyers surrounded by pipes carrying a hydrocarbon coolant such as a natural gas as shown in Fig. 8.3b.

3. Combination-blown processes - One type combines the top-blown (BOP) process with the bottom-blown (Q-BOP) process (Fig. 8.4a). In another type, stirring the bath is accomplished by blowing inert gases upward through either permeable elements (Fig. 8.4b) or uncooled bottom tuyers (Fig. 8.4c).

8.4 ELECTRIC-FURNACE STEELMAKING

Numerous types of electric furnaces have been developed, but only the following three types have been practically, utilized for steelmaking [1, 2]:

1. A.C. (alternating current) direct-arc electric furnace.
2. D.C. (direct current) direct-arc electric furnace.
3. Induction electric furnace.

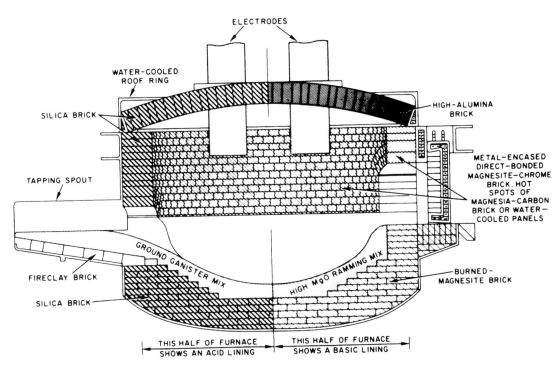

Fig. 8.5 Schematic cross-section of a Heroult electric-arc furnace with a dished-bottom shell and stadium-type sub-hearth construction, indicating typical refractories employed in an acid lining (left), and in a basic lining (right). Although only two electrodes are shown here, furnaces of this type (which operate on three-phase current) have three electrodes. (From Harbison-Walker Refractories Co. Publication, Pittsburgh, Pennsylvania. Reprinted with permission).

In **direct-arc furnace,** electric arcs pass from the electrodes to a metal charge. It creates electric current through the metal charge resulting in generation of heat due to the electric resistance of the metal. This heat is supplemented by the heat radiated from the arcs (Fig. 8.5).

A.C. direct-arc furnaces are designed with non-conducting bottoms, i.e. the current passes from one electrode down through an arc to the metal charge, then from charge up through an arc to another electrode. Furnaces employing three-phase current are used almost exclusively.

D.C. direct-arc furnaces are designed with conducting bottoms, i.e. the current passes from one electrode through an arc and the metal charge to an electrode installed in the bottom of the furnace.

In **induction electric furnace,** electric current is induced in the metal charge by an oscillating magnetic field. One or more inductors are attached to a vessel forming a primary winding of a transformer whereby the secondary winding is formed by a loop of liquid metal confined in a closed refractory channel.

Both the basic and the acid processes are used for making steel in electric furnaces. However, almost all furnaces, used for continuous cast and ingot steel production and a large percentage of the foundry furnaces, are now basic-lined furnaces. They use combinations of high-alloy steel scrap, lower grades of alloy scrap, and plain-carbon steel scrap to produce steels that will meet strict chemical, mechanical-property, and cleanliness requirements.

Electric-furnace process can be used for making almost all steel grades. When production is less than 1,500,000 tons per year, it is often a more economical process than that utilizing a combination of a blast and basic oxygen furnaces. It is especially true in the industrial areas of high steel scrap availability and low-cost power.

Typical productivity of the basic electric-arc furnace steelmaking is:

1. Cold metal   20 to 60 net tons per hour
2. Hot metal   50 to 80 net tons per hour.

Main disadvantages of electric-arc furnace steelmaking include:

1. Inability to produce low residual steel from high residual scrap.

2. The nitrogen contents of the steels produced are usually twice as high as those produced in BOP or Q-BOP furnaces.

3. The necessaty to have at least two furnaces in order to keep pace with a continuous caster.

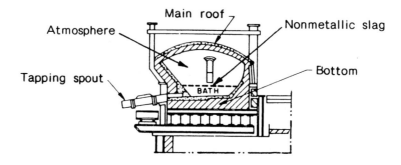

Fig. 8.6 Three phases in steelmaking furnaces.

8.5 CHEMICAL FORMULAS OF STEELMAKING

The following three phases (Fig. 8.6) are usually considered to describe the chemistry of steelmaking process [1]:

1. Metallic bath.
2. Nonmetallic slag floating on the surface of the metallic bath or momentarily intermixed with it.
3. Gas, or atmosphere above the slag.

Removal of unwanted chemical elements from the metallic bath is called **refining**. The main chemical process which is employed during the refining is an **oxidation**. The resulting oxides, except that of carbon, are subsequently neutralized by lime in the slag. The slag serves as a refining medium as well as a receptacle for unwanted impurities.

In chemical formulas describing the steelmaking process, concentrations in the metallic bath are denoted by square brackets, slashes, or underlines; concentrations in the slag are denoted by parentheses whereas no brackets are used to denote concentrations in the gaseous phase.

Hence the reaction:

$$(FeO) + [C] = CO + [Fe]$$

can be interpreted as an iron oxide (FeO), dissolved in the slag, reacts with carbon [C], dissolved in the metallic bath, to form gaseous carbon monoxide CO and iron [Fe], dissolved in the bath.

Reversible reactions are indicated in chemical formulas by bi-directional arrows replacing an equal sign as shown below:

$$[FeS] + (CaO) \rightleftarrows (FeO) + (CaS).$$

8.6 CHEMISTRY OF REFINING

Oxygen is usually present in a slag in the form of oxides such as FeO and CaO. Table 8.3 illustrates the main chemical reactions which take place during the process of refining of the molten metal [1].

The limit of carbon elimination in the open hearth furnace is 0.03% and it is about 0.02% in the basic oxygen and electric-arc furnaces.

Table 8.3 Chemical reaction during refining. Adapted from Peters (1982).

Process	Chemical formula
Decarburization	$(FeO) + [C] = CO + [Fe]$
Desiliconization	$2(FeO) + [Si] = (SiO_2) + 2[Fe]$
Dephosphorization	$5(Fe) + 2[P] = 5[Fe] + (P_2O_5)$
Desulfurization	$[FeS] + (CaO) \rightleftarrows (FeO) + (CaS)$
Loss of manganese	$(FeO) + [Mn] = (MnO) + [Fe]$

There are two main reasons which limit desulfurization:

1. It is due to a limited solubility of CaS in slag.

2. It is due to a chemical incompatibility of desulfurization with other refining processes. Removal of sulfur by oxidation is not possible because no oxides of sulfur exist at steelmaking temperatures.

8.7 AFFINITY FOR OXYGEN

The term 'affinity' is used to describe the ease with which an impurity will become oxidized (Table 8.4). Elements with affinity for oxygen higher than iron will become oxidized before the oxidation of an iron whereas those with affinity lower than iron will become oxidized after oxidation of iron. The elements with low affinities carry over or ride through the steelmaking furnaces; they are known as **tramps** or residuals [1].

It should be realized that there are no clear demarcation lines between formation points of the oxides. These reactions, although proceeding generally in sequence of Si-P-C-Cr-Mn-Fe, always overlap. With the exception of carbon which escapes as carbon monoxide, the oxides of these elements enter the slag that floats on the steel bath.

8.8 BASIC AND ACID STEELMAKING

In the basic steelmaking furnaces, the acid oxides created during an oxidation are neutralized by a basic oxide, lime, which is the main flux of the basic steelmaking process. Also lime is needed for

Table 8.4 Affinities of some chemical elements. Adapted from Peters (1982).

Elements	Effects of oxidation	Remarks
Ca, Mg Al, Ti	Very high affinities	Affinities are very high if elements are in metallic form.
Si	High affinity	Resulting SiO_2 is a strong acid oxide.
P	High affinity	Resulting P_2O_5 is a strong acid oxide. It is unstable except in the presence of FeO and CaO.
C, Cr, V, Cb, Mn	Partially removed by oxidation	
Pb	Tramp, not removed by oxidation	It partly volatilizes and partly seeps into the furnace bottom.
Ni, Cu, Sn, Mo, Sb, As, Co	Tramps, not removed by oxidation	

PRIMARY STEELMAKING PROCESSES

desulfurization. Furnace linings must be basic in order to prevent their dissolving by the lime. Magnesia (MgO) is usually used as a main substance in all linings in the steelmaking furnaces utilizing the basic process [2].

Acid steelmaking furnaces have acid roofs and hearths made of silica. The process is carried out under siliceous slag in which the carbon and manganese contents of steel can be adjusted. Also alloying additions can be made as required. However, it is difficult to obtain low silicon steels. No sulfur or phosphorus refining can be done. This acid steelmaking process is almost completely abandoned in North America.

8.9 DEOXIDATION OF STEEL

Deoxidation of steel is the process of converting oxygen active toward carbon, into inactive form, followed by the floating out of the resulting stable oxides. The most common deoxidizers are aluminum and silicon [2].

The deoxidizers reduce iron oxide to their own oxides which are stable with respect to carbon monoxide. Therefore, decarburization and the resulting generation of carbon monoxide slow down or even stop if a deoxidizer is sufficiently strong.

The following types of steel are known in relation to a degree of suppression of gas evolution:

1. **Killed steel** - It is deoxidized to such an extent that there is no gas evolution during a solidification.

2. **Semi-killed steel** - It is deoxidized less than a killed steel.

3. **Rimmed steel** - It is usually tapped without having made the additions of deoxidizers to the steel in the furnace. However, small additions are made in the ladle.

4. **Capped steel** - It is produced by terminating the rimming action at the end of a minute or more by the application of a metal cap at the top of a mold.

Reoxidation is the oxidation of an excess of deoxidizer present in the steel by the oxygen present in the air. In an aluminum and silicon killed steel, the main resulting inclusions are: alumna which appears first, silica and silicates which appear second, and manganese oxides which make their appearance last [1].

A summary of behavior of individual elements during steelmaking process and their main uses are shown in Table 8.5.

8.10 OPTIMIZATION OF PRIMARY STEELMAKING PROCESS

A major prerequisite for optimization of the primary steelmaking process is **slag-free tapping** [3]. The slag-free tapping has a major influence on the quality of continuously cast product with respect to cleanliness and surface quality. Figure 8.7 illustrates schematically the slag-free tapping system for basic oxygen furnace (BOF), electric arc furnace (EAF), and open hearth furnace (OHF).

The bottom tapping system of an electric arc furnace, developed by Mannesmann Demag in collaboration with Thyssen Edelstahlwerke [4], is shown in Figs. 8.8 and 8.9. The operation of the system is as follows. A ladle is moved under the furnace with alloying material placed in the bottom of the ladle. The tapping is initiated by opening the bottom valve. During tapping the furnace is slightly tilted to insure that the level of liquid steel over the taphole is kept constant. Maximum tilt angle is 10°. After the desired level of steel in the ladle is achieved, the furnace is tilted back to the original position, leaving the taphole open. This allows for cleaning the taphole prior to the next tapping cycle.

Table 8.5 Behavior in furnace and use of elements. Adapted from Peters (1982).

Element	Behavior in furnace			Use*		
	Not removed	Partially removed	Removed	Deoxi- dation	Sulfur control	Alloying
Aluminum			X	X		O
Boron			X			X
Calcium			X	O	X	
Carbon		X		O		X
Cerium**			X		X	
Chromium		X				X
Columbium			X			X
Copper	X					X
Lead			X			X
Manganese		X			X	X
Magnesium			X		X	
Molybdenum	X					X
Nickel	X					X
Nitrogen		X				X
Phosphorus		X				X
Silicon			X	X		X
Sulfur		X				X
Tellurium			X			X
Tin	X					
Titanium			X	X	O	O
Tungsten		X				X
Vanadium			X			X
Zinc			X			
Zirconium			X	O	O	O

* X = major use, O = minor use.
** And other rare earth elements associated with Ce.

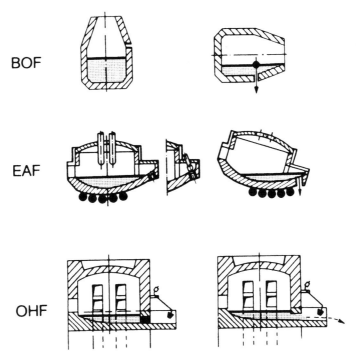

Fig. 8.7 Slag-free tapping of basic oxygen furnace (BOF), electric arc furnace (EAF), and open hearth furnace (OHF). (From Kruger et al, IRON AND STEEL ENGINEER, March 1984. Copyright AISE, Pittsburgh, Pennsylvania. Reprinted with permission).

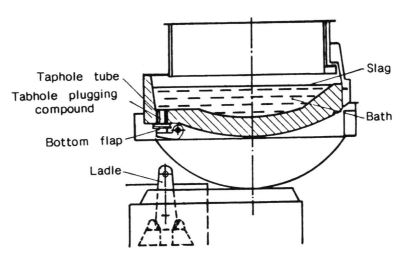

Fig. 8.8 Mannesman-Demag/Thyssen Edelstahlwerke eccentric bottom tapping system. (From Baare et al, IRON AND STEEL ENGINEER, Sept. 1985. Copyright AISE, Pittsburgh, Pennsylvania. Reprinted with permission).

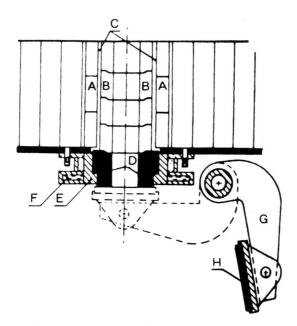

Fig. 8.9 Bottom tapping system (Mannesmann-Demag/Thyssen Edelstahlwerke):
A - nozzle brick (magnesite); B - wear tube (magnesite); C - castable
(magnesite); D - end brick (graphite); E - end ring; F - water cooling;
G - closure flap; and H - graphite plate. (From Baare et al, IRON AND
STEEL ENGINEER, Sept. 1985. Copyright AISE, Pittsburgh, Pennsylvania.
Reprinted with permission).

REFERENCES

1. A.T. Peters, Ferrous Production Metallurgy, John Wiley & Sons, New York, pp. 75-167 (1982).

2. The Making, Shaping and Treating of Steel, 10th Edition, eds. W.T. Lankford, Jr., et al, Association of Iron and Steel Engineers, Pittsburgh, Pennsylvania, pp. 599-701 (1985).

3. B. Kruger, et al, "Continuous Casting of Steel: The Process for Quality Improvement", Iron and Steel Engineer, March 1984, pp. 45-52.

4. R.D. Baare, et al, "An Eccentric Bottom Tapping System - 18-months' Experience", Iron and Steel Engineer, July 1984, pp. 27-32.

9
Secondary Steelmaking Processes

9.1 PURPOSE OF SECONDARY STEELMAKING

Secondary steelmaking process (also referred to as 'ladle metallurgy') is introduced to achieve the following main objectives [1]:

1. Composition control in steel.
2. Teeming temperature control, especially for continuous casting operations.
3. Degassing (decreasing the concentration of oxygen and hydrogen in steel).
4. Decarburization (to less than 0.03%).
5. Desulfurization (to as low as 0.002%).
6. Microcleanliness (removal of undesirable nonmetallics, primarily oxides and sulfides).
7. Inclusion morphology (changing the composition and/or shape of the undesired matter left in the steel to make it compatible with the mechanical properties of the finished steel).
8. Increasing the production rates by decreasing refining time in the furnace.

Main principle of operation utilized during secondary steelmaking is based on exposure of molten steel to a low-pressure environment to remove gases (chiefly hydrogen and oxygen) from steel. Most of vacuum degassing systems provide for adding deoxidizers and alloy additions to the molten steel while maintaining low pressure. Principles of operation and capabilities of different steelmaking processes are described below.

These processes can be divided into the following four groups:

A. Vacuum stream degassing
B. Recirculation degassing
C. Nonvacuum ladle or vessel refining
D. Vacuum degassing with supplemental heating.

9.2 VACUUM STREAM DEGASSING

The following types of vacuum stream degassing systems are implemented in the industry [1]:

A. Ladle-to-mold degassing

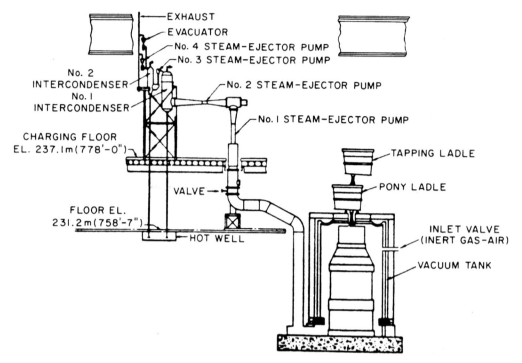

Fig. 9.1 Schematic arrangement of vacuum-casting installation for casting large forging ingots. (From THE MAKING, SHAPING AND TREATING OF STEEL, by W.T. Lankford, et al, Tenth Edition. Copyright AISE, Pittsburgh, Pennsylvania. Reprinted with permission).

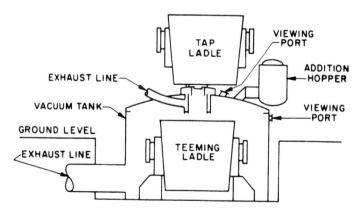

Fig. 9.2 Schematic arrangement of equipment used in the ladle-to-ladle stream degassing process. (From THE MAKING, SHAPING AND TREATING OF STEEL, Tenth Edition. Reprinted with permission).

    B. Ladle-to-ladle degassing

    C. Tap degassing.

In the **Ladle-to-mold degassing system** (Fig. 9.1), a mold is placed inside of a vacuum tank. A heat of steel is tapped from steelmaking furnace into the tapping ladle. Then steel is bottom-poured from the tapping ladle into the pony ladle. As the stream of molten steel enters the evacuated space in the vacuum tank, it breaks up into tiny droplets. It enormously increases the exposed surface of the molten steel resulting in more efficient degassing process.

In the **Ladle-to-ladle degassing system** (Fig. 9.2), a teeming ladle is placed inside of a vacuum tank. In this system, besides providing vacuum degassing, oxygen can be removed by vacuum carbon deoxidation. It is achieved by adding to the steel the required amount of ferrosilicon and aluminum. A special alloy addition hopper is utilized for this purpose so that loss of vacuum does not occur.

In the **Tap degassing system** (Fig. 9.3), a special sealed ladle is provided in order to eliminate a need for a vacuum tank. During degassing, the required deoxidizers are added to the steel in the ladle.

9.3 RECIRCULATION DEGASSING

In the recirculation degassing process the liquid metal in a large ladle is forced by atmospheric pressure into an evacuated degassing chamber where it is exposed to a low pressure and then flows back into the ladle. This cycle may be repeated 40 to 50 times in order to achieve the desired level of degassing.

    Three major processes are known [1]:

    A. D-H (Dortmund-Horder) process

    B. R-H (Ruhrstahl-Heraes) process

    C. RH-OB process.

In the **D-H process** (Fig. 9.4), as the vacuum vessel is lowered, atmospheric pressure causes molten steel to rise into the vacuum chamber where the steel is degassed. Then the vacuum vessel is raised while keeping the lower end of the suction nozzle immersed in steel, thus causing the degassed steel to flow back into the ladle. During the later stages of degassing, the desired additions are made. Means are provided in some D-H systems to introduce argon and/or oxygen inside the vacuum vessel in order to lower the hydrogen or carbon content.

In the **R-H process** (Fig. 9.5), two 'snorkel' tubes are provided at the bottom of a vacuum vessel. An inert gas is introduced into one of the tubes and the gas-steel mixture flows from the ladle up that tube into the vacuum chamber where it is degassed and then flows by means of gravity through the other tube back into the ladle.

RH-OB process (Fig. 9.6) is similar to the R-H process. The difference lies in the gases injected into the system. Whereas the R-H process utilizes an inert gas, the RH-OB process utilizes both the inert gas and oxygen. The inert gas aids in stirring the metal and the oxygen assists in the decarburization of the liquid steel. The RH-OB system is used primarily for the removal of hydrogen and a certain amount of carbon from the steel.

9.4 VACUUM LADLE DEGASSING

The following two processes are utilized [1]:

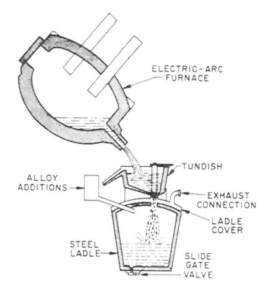

Fig. 9.3 Diagrammatic arrangement of equipment used in the tap-degassing process. (From THE MAKING, SHAPING AND TREATING OF STEEL, Tenth Edition. Reprinted with permission).

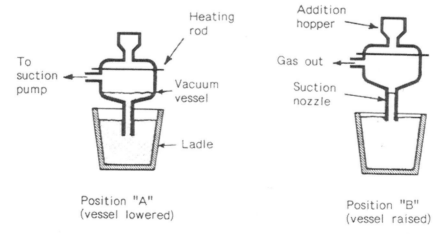

Fig. 9.4 Schematic diagram showing principle of operation of the D-H (Dortmund-Horder) process. (From THE MAKING, SHAPING AND TREATING OF STEEL, Tenth Edition. Reprinted with permission).

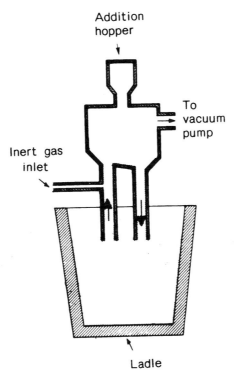

Fig. 9.5 Schematic diagram of the R-H vacuum degassing system. (From THE MAKING, SHAPING AND TREATING OF STEEL, Tenth Edition. Reprinted with permission).

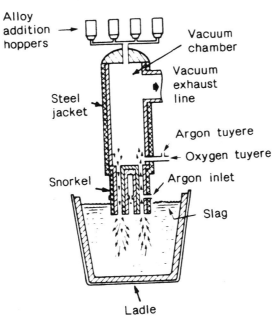

Fig. 9.6 Schematic diagram of the RH-OB vacuum degassing system. (From THE MAKING, SHAPING AND TREATING OF STEEL, Tenth Edition. Reprinted with permission).

A. Induction stirring process

B. Vacuum oxygen decarburization (VOD) process.

In the **Induction stirring process** (Fig. 9.7), stirring effect is produced by eddy current which is induced in the molten steel by the induction coils. The steel is contained in a ladle made of nonmagnetic stainless steel. In order to provide degassing, the ladle is installed inside of a vacuum chamber.

Vacuum-oxygen decarburization (VOD) process (Fig. 9.8) is used in production of stainless steel. In this process, a heat is tapped into a preheated ladle which is placed inside a vacuum chamber. Oxygen is blown through a lance above the bath of steel and argon is blown through the porous plugs in the bottom of the ladle.

9.5 ARGON-OXYGEN DECARBURIZATION

Argon-oxygen decarburization (AOD) process is primarily designed for stainless steel production (Fig. 9.9). The process is conducted at atmospheric pressure. It consists of the following steps [1]:

1. Charging the electric-arc furnace
2. Melting, tapping, deslagging and weighing the heat
3. Pouring the heat into the AOD vessel
4. Decarburization by blowing argon and oxygen with fluxes to be added during blowing
5. Adding ferrosilicon and stirring
6. Sample inspection with necessary alloy adjustment
7. Tapping.

The AOD process allows one to reduce hydrogen content to a level less than 2 ppm and also reduce both nitrogen and sulfur contents to a level less than 0.005%. This process is capable of performing all the preferred metallurgical functions, except for the sulfide shape-control, in a single vessel operation while the quality of AOD refined steel is claimed to be equivalent to that of other secondary steelmaking processes.

9.6 NONVACUUM ARGON BUBBLING

The processes described below are conducted at the atmospheric pressure without supplemental heating during the refining process. They include:

A. Capped argon bubbling (CAB) process

B. Composition adjustment by sealed argon bubbling (CAS) process.

There are two following main methods of using the argon-stirring process:

1. Lance method (Fig. 9.10a) in which argon is blown through a refractory protected lance lowered into a ladle of steel
2. Porous-plug method (Fig. 9.10b) in which argon is blown through a porous refractory plug.

In the **CAB process** (Fig. 9.11), a conventional ladle with a cover is utilized. A synthetic refining slag is added to a heat. The slag absorbs the undesirable nonmetallic inclusions.

In the **CAS process** (Fig. 9.12), a refractory-coated snorkel is lowered into a ladle of steel during the argon stirring so that the steel inside the snorkel will be completely free of furnace slag.

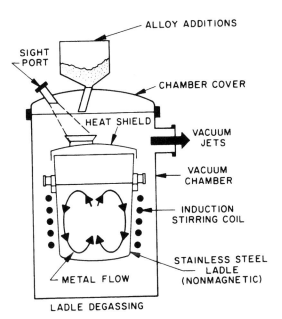

Fig. 9.7 Schematic arrangement of equipment used in the induction-stirring ladle-degassing process. (From THE MAKING, SHAPING AND TREATING OF STEEL, Tenth Edition. Reprinted with permission).

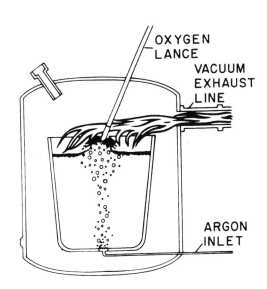

Fig. 9.8 Schematic diagram of the vacuum-oxygen decarburization (VOD) process. (From THE MAKING, SHAPING AND TREATING OF STEEL. Tenth Edition. Reprinted with permission).

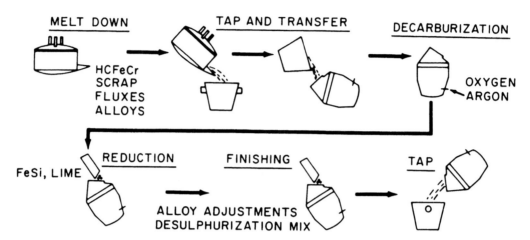

Fig. 9.9 Processing steps for refining stainless steel by the AOD process. (From THE MAKING, SHAPING AND TREATING OF STEEL, Tenth Edition. Reprinted with permission).

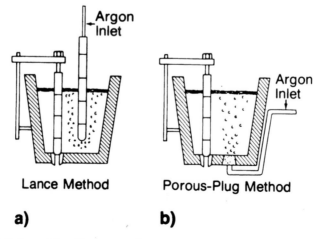

Fig. 9.10 Schematic diagram of non-vacuum argon bubbling process: a) Lance method, b) Porous-plug method. (From THE MAKING, SHAPING AND TREATING OF STEEL, Tenth Edition. Reprinted with permission).

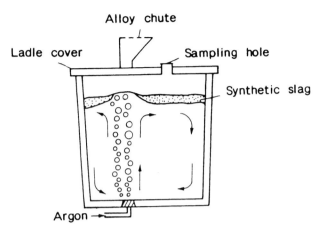

Fig. 9.11 Schematic illustration of Capped argon bubbling (CAB) system. (From THE MAKING, SHAPING AND TREATING OF STEEL, Tenth Edition. Reprinted with permission).

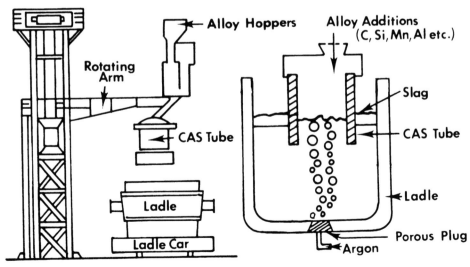

Fig. 9.12 Schematic diagram of Composition adjusted by sealed argon bubbling (CAS) system. (From THE MAKING, SHAPING AND TREATING OF STEEL, Tenth Edition. Reprinted with permission).

In the **MPE/MLD process** with subsequent calcium treatment [2], single alumina crystals from deoxidation process are transformed into liquid particles of the composition $xCaO.yAl_2O_3$ (Fig. 9.13). This process is claimed to have the potential for producing a steel almost free of macroscopic inclusions and with a reduced content of alumina deoxidation products (super-clean steel).

9.7 ELECTRO-SLAG REMELTING

Electro-slag remelting (ESR) process is similar to a vacuum-arc remelting (VAR) process except it is conducted at the atmospheric pressure. In this process (Fig. 9.14), one or more electrodes having a chemical composition about the same as that of the desired product are drip-melted through molten slag into a water-cooled copper mold at the atmospheric pressure [1].

The ESR process cannot eliminate hydrogen; however, its production rate is greater than that of a VAR process. The ESR process allows one to obtain larger size ingots, up to 1520 mm (60 inches) in diameter for round ingots. Round, square, rectangular, and hollow-shaped ingots can be produced. Steel can be desulfurized to a level as low as 0.002%. The ingot surface quality is excellent, requiring little or no conditioning. However, the electro-slag remelting process is costly and usually limited to specialty products.

9.8 LADLE INJECTION

In the ladle injection process, deoxidizers and rare-earth metals (REM) are added to ladles of steel at atmospheric pressure. Two methods of injection are used [1]:

    A. Lance powder injection method
    B. REM canister injection method

The **Lance powder injection method** (Fig. 9.15a) produces an excellent deoxidation and microcleanliness.

The **REM canister injection method** (Fig. 9.15b) is used for adding the rare-earth metals. According to this method, a large steel bloom fastened with a required amount of REM is lowered into a ladle of steel which was alloyed and deoxidized. This method is very effective in changing the morphology of the sulfide inclusions from the stringer type to the globular sulfides resulting in more isotropic steel.

9.9 VACUUM DEGASSING WITH HEATING

The following vacuum degassing processes with a supplemental heating are reviewed below [1]:

    A. Gas stirring-arc reheating (Finkl-Mohr VAD) process
    B. Induction stirring-arc reheating (ASEA-SKF) process
    C. Vacuum-arc remelting (VAR) process
    D. Modified vacuum induction (Therm-I-Vac) process

Gas stirring-arc reheating (Finkl-Mohr VAD) process is designed for desulfurization and additional steelmaking refinement purposes as well as for the degassing of the liquid steel. This process is similar to the vacuum ladle degassing previously described, except that the heat is supplied to the steel before or after degassing operation by electric arcs (Fig. 9.16). Electrodes are inserted through the top of the vacuum-tank cover. Argon is usually used for stirring.

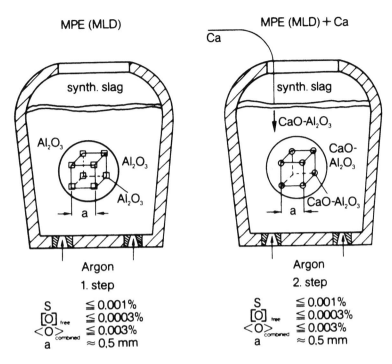

Fig. 9.13 Mannesmann-Demag MPE/MLD process for ladle steel desulfurization. (From IRON AND STEEL ENGINEER, March 1984. Copyright AISE, Pittsburgh, Pennsylvania. Reprinted with permission).

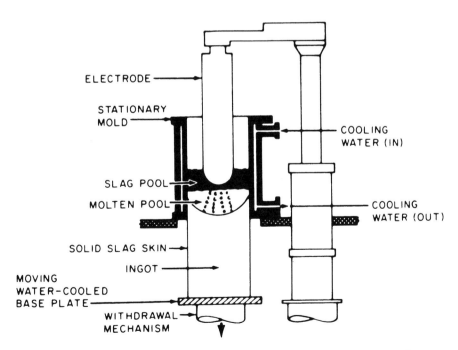

Fig. 9.14 Sketch showing principle of operation of the electroslag-remelting (ESR) process. (From THE MAKING, SHAPING AND TREATING OF STEEL, Tenth Edition. Reprinted with permission).

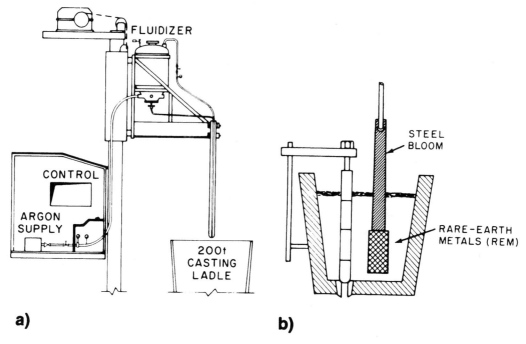

a)

b)

Fig. 9.15 Schematic diagram of the ladle injection process:
a) Lance powder injection method, b) REM canister injection method.
(From THE MAKING, SHAPING AND TREATING OF STEEL, Tenth Edition.
Reprinted with permission).

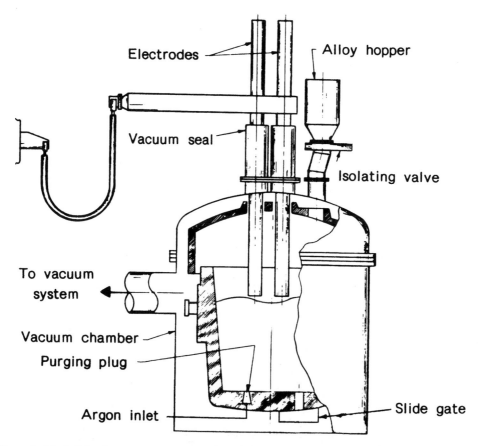

Fig. 9.16 Schematic arrangement of equipment used in the gas-stirring
arc-reheating process. (From THE MAKING, SHAPING AND TREATING OF STEEL,
Tenth Edition. Reprinted with permission).

In the **Induction stirring-arc reheating (ASEA-SKF) process**, stirring is produced in a manner similar to that of the induction stirring which was previously described, except that the heat is supplied to the steel during refining by the electric arcs. This process allows one to reduce sulfur content to less than 0.005%.

In the **Vacuum-arc remelting (VAR) process** a steel electrode having a chemical composition about the same as that of the desired product is drip-melted into a water-cooled copper mold (Fig. 9.17). Because of the high arc temperature and the small pool of liquid metal, it is possible to produce ingots of dense crystal structure, low hydrogen and oxygen contents and with minimal chemical and nonmetallic segregation.

In the **Modified vacuum induction process**, the equipment can be used either as a stream-degassing unit or for an induction-furnace melting of steel under low pressure. The system provides utilization of both cold and hot charging steel into the induction furnace. In the case when cold charging steel is utilized, the system provides charging the steel into the induction furnace with subsequent melting, refining, and teeming the steel into ingots under low pressure. When hot charging steel is used, the system provides charging the steel under low pressure, reheating the steel in the induction furnace, adding alloys for final adjustment of chemical composition, and teeming the heat under low pressure.

9.10 COMPARISON OF SECONDARY STEELMAKING PROCESSES

Comparative capability of the secondary steelmaking processes is summarized in Table 9.1 [1].

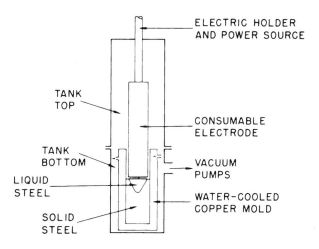

Fig. 9.17 Schematic diagram of a consumable-electrode arc furnace used in the vacuum-arc remelting (VAR) process. (From THE MAKING, SHAPING AND TREATING OF STEEL, Tenth Edition. Reprinted with permission).

Table 9.1 Capability of secondary steelmaking processes.

Secondary steelmaking processes	Metallurgical functions							
Symbols for process performance: – = none C = good B = better A = best	Composition control	Temperature control	Deoxidation (O_2)	Degassing (H_2)	Decarburization	Desulphurization	Microcleanliness	Inclusion morphology
Vacuum stream degassing	C	–	C	A	C	–	C	–
Recirculation degassing:								
* D-H process	B	B	B	B	B	–	B	C
* R-H process	B	B	B	B	B	–	B	C
* RH-OB process	B	B	B	B	A	–	B	C
Nonvacuum ladle or vessel refining:								
* Argon-oxygen decarburization	C	B	B	C	A	B	C	C
* Argon bubbling CAS	B	C	C	–	–	–	C	–
* Argon bubbling CAB	B	C	B	–	–	B	B	B
* Lance powder injection	C	–	B	–	–	A	B	A
* REM canister injection	C	–	C	–	–	C	C	A
Vacuum degassing with heating	A	A	A	B	B	A	A	C

REFERENCES

1. The Making, Shaping and Treating of Steel, 10th Edition, eds. W.T. Lankford, Jr., et al, Association of Iron and Steel Engineers, Pittsburgh, Pennsylvania, pp. 671-690 (1985).

2. B. Kruger, et al, "Continuous Casting of Steel: The Process for Quality Improvement", Iron and Steel Engineer, March 1984, pp. 45-52.

10

Casting of Steel for Flat Products

10.1 TYPE OF CAST PRODUCTS

Cast products utilized for flat rolling can be produced in the following forms:

1. **Ingots** - These are castings of simple shape. Slab ingots range in weight from 9 to 36 metric tons (10 to 40 net tons). In order to roll strip from the ingots, the latter are usually first rolled down to the size of a slab with thickness range from 150 to 350 mm (6 to 14 in.). Then they are further reduced in thickness at the roughing stands of hot strip mill down to 25-65 mm (1.0-2.5 in.) with subsequent reduction to the desired hot rolled thickness at the finishing mill. Final reduction in thickness may be done by rolling at the cold mill (Fig. 10.1a).

2. **Thick cast slabs** - These castings are usually from 150 to 350 mm (6 to 14 in.) in thickness. Utilization of the thick cast slabs allows one to eliminate reduction at the slabbing mill (Fig. 10.1b).

3. **Thin cast slabs** - These castings may be from 25 to 64 mm (1 to 2.5 in.) thick. Utilization of the thin slabs allows the elimination of both the slabbing mill and the roughing mill (Fig. 10.1c).

4. **Cast strip** - The thickness of the cast strip can be as thin as 1.3 mm (0.05 in.). It allows one to eliminate entirely the hot rolling process (Fig. 10.1d).

10.2 CASTING OF INGOT

After the steelmaking operation is completed, the liquid steel is poured into a steel ladle. Additional alloying materials and deoxidizers may be added during the tapping of heat. The steel is then poured or teemed into a series of molds of the designed dimensions [1].

The ingot molds are tall box-like containers made of cast iron with the internal cavity that is usually tapered from the top to the bottom of the mold. There are two principal types of molds (Fig. 10.2):

a) big-end-down molds

b) big-end-up molds.

The inner walls of the molds may be plain sided, cambered, corrugated, or fluted. The last two shapes of the wall promote faster cooling and therefore minimize surface cracking during solidification.

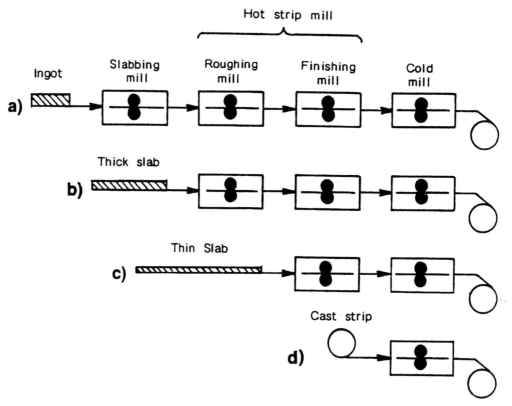

Fig. 10.1 Rolling mill equipment required to produce cold-rolled strip coils from: a) ingot, b) thick cast slab, c) thin cast slab and d) cast strip.

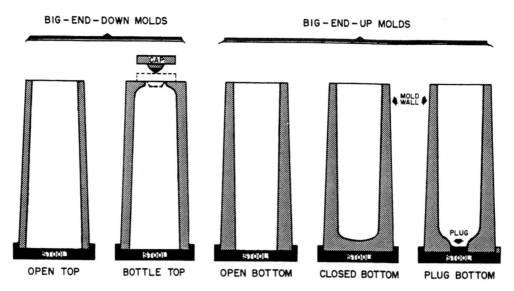

Fig. 10.2 Cross-sections (not to scale) of the five principal types of ingot molds. Molds usually are cast from molten pig iron directly from the blast furnace. (From THE MAKING, SHAPING AND TREATING OF STEEL, by W.T. Lankford, et al, Tenth Edition. Copyright AISE, Pittsburgh, Pennsylvania. Reprinted with permission).

There are two methods of teeming the ingots:

a) top-pouring method

b) bottom-pouring method.

The use of the bottom-pouring method is found especially beneficial for high quality steels.

10.3 TYPES OF INGOTS

Molten steel solidifies first at the regions close to the mold walls, so the gases, chiefly oxygen, evolved from still-liquid portions may be trapped to produce **blowholes.**

Depending on the amount of gases released during solidification, the following types of ingots are known:

1. **Fully-killed ingot** - It is fully deoxidized and therefore it evolves no gas, its top is slightly concave, and below the top there is a shrinkage cavity (see ingot No. 1 in Fig. 10.3), that is commonly called **pipe** [2].

2. **Semi-killed ingot** - This ingot is deoxidized less than fully-killed. As a result, a small amount of carbon monoxide evolves producing a domed top. The blowhole formation in the lower half of the ingot is prevented due to ferrostatic pressure (see ingots No. 2-4 in Fig. 10.3).

3. **Capped ingot** - It is produced by pouring steel into big-end-down bottle-top molds (Fig. 10.2), in which the constructed top or mouth of the mold facilitates the capping operation. The rimming action is allowed to begin normally but is then terminated at the end by sealing the mold with a cast-iron cap. In capped ingot (see ingot No. 5 in Fig. 10.3), the gas bubbles in upper half are swept away due to the strong rimming action. An ingot of this type does not have the interiors of its blowholes exposed to oxidation during heating and soaking.

4. **Rimmed ingot** - This type of ingots is usually tapped without addition of deoxidizers to the steel in the furnace, and with only small additions to the molten steel in the ladle. The evolution of gas produces a boiling action that is commonly known as **rimming action.** Ingot No. 7 in Fig. 10.3 is a typical rimmed ingot in which gas evolution was so strong that the formation of blowholes was confined to only lower part of the ingot.

There are two types of design for the ingots (Fig. 10.4):

a) hot-topped ingots

b) non hot-topped ingots.

The big-end-up, hot-topped killed steel ingots are used in order to provide a complete freedom from pipe.

10.4 METHODS OF CONTINUOUS CASTING OF THICK SLABS

A number of methods have been proposed for continuous casting of steel. Below are some of the methods that have been practically implemented [3].

Vertical or "stick" casting - In this method (Fig. 10.5a), a straight mold, a vertical cooling chamber and a flame cut-off are used. A tilting receiving mechanism transfers the continuously cast slabs onto horizontal run-out table.

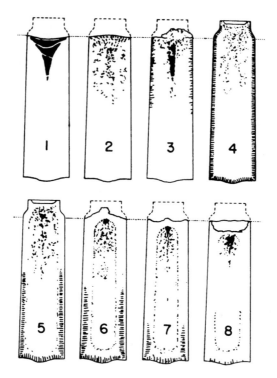

Fig. 10.3 Series of ingot structures. (From THE MAKING, SHAPING AND
TREATING OF STEEL, by W.T. Lankford, et al, Tenth Edition.
Copyright AISE, Pittsburgh, Pennsylvania. Reprinted with permission).

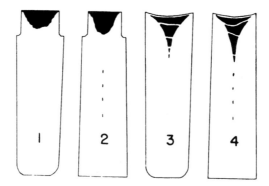

Fig. 10.4 Types of killed ingots:
 1) big-end-up, hot-topped
 2) big-end-down, hot-topped
 3) big-end-up, not hot-topped
 4) big-end-down, not hot-topped.
(From THE MAKING, SHAPING AND TREATING OF STEEL, Tenth Edition.
Reprinted with permission).

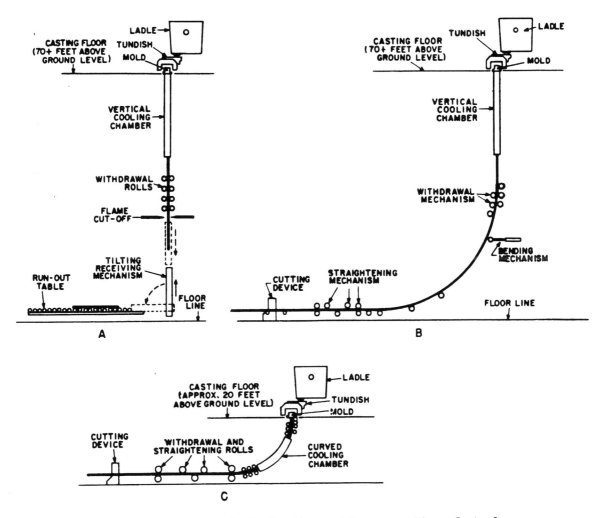

Fig. 10.5 Three methods for the continuous casting of steel. From Roberts (1983).

Vertical plus bending casting - In this method (Fig. 10.5b), the casting direction is smoothly changed from vertical to horizontal as soon as the cast steel emerges from vertical cooling chamber.

Semi-horizontal or curved mold casting - This method (Fig. 10.5c) allows one to simplify design and to substantially reduce dimensions of the continuous casting machines.

Horizontal continuous casting - Schematic representation of a typical horizontal casting machine is shown in Fig. 10.6 [4]. Some horizontal casting machines provide continuous movement of strand with oscillation of either both tundish and mold (Fig. 10.7a) or mold only (Fig. 10.7b). However, the most reliable operation was achieved by providing an intermittent strand movement as shown in Figs. 10.7c-10.7e [5].

10.5 CONTINUOUS CASTING OF THICK SLABS

The most common method for continuous casting of thick slabs is vertical plus bending casting (Fig. 10.5b). Below is a brief description of the casting process that utilizes this method [1].

In order to start the casting process, the dummy bar is inserted in the mold so that its top closes the bottom of the mold. The insertion of the dummy bar is made either from the top of the machine or through entire machine in the bottom of the mold. Liquid steel is then poured at a controlled rate from ladle into the tundish and then the metal flows through nozzles in the bottom of tundish and fills the mold (Fig. 10.8).

There are two methods of pouring the steel from ladle to tundish and from tundish to mold:

a) open stream casting

b) close stream or shrouded casting.

In an open stream casting the liquid metal flows through the air and therefore it picks up oxygen and some nitrogen from the air. It results in formation of undesired inclusions in the liquid steel. Shrouded casting allows to avoid this problem. In this method, steel is protected from contact with the air either by refractory tubes or by gas shrouding [6], as shown in Fig. 10.9.

After the mold is filled, withdrawal of the dummy bar is initiated. The gradually solidifying metal would follow the dummy bar head. At certain position, the dummy bar head is mechanically disassociated from solidified metal being cast and then the dummy bar is removed.

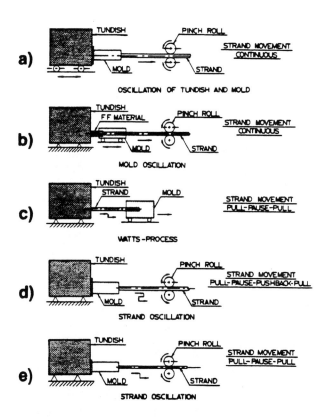

Fig. 10.6 Development of horizontal continuous casting. (From Haissig, IRON AND STEEL ENGINEER, June 1984. Copyright AISE, Pittsburgh, Pennsylvania. Reprinted with permission).

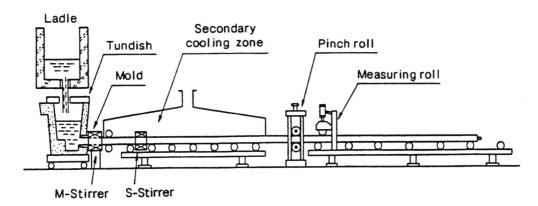

Fig. 10.7 Schematic representation of Kobe Steel pilot plant for horizontal casting. (From Nozaki, et al, IRON AND STEEL ENGINEER, Oct. 1985. Copyright AISE, Pittsburgh, Pennsylvania. Reprinted with permission).

Liquid steel starts to solidify in the water-cooled mold and the solidification of steel continues progressively along its path (Fig. 10.10). The rate of solidification is controlled by secondary cooling water sprays. The distance from the meniscus level in the mold to the point of complete solidification is called **metallurgical length**. The point of complete solidification is usually ahead of straightener. Electromagnetic stirring of liquid steel during solidification may be implemented in order to improve steel quality and increase casting rate.

The mold is oscillated in a vertical direction in order to prevent sticking of the solidified shell to the mold. Also, lubricants such as oils or fluxes are used to reduce friction. Support rolls are installed to guide the metal and to prevent bulging of the solidifying shell from internal ferrostatic pressure. Cutting of the cast section is done after straightening either by shears or by torches.

Table 10.1 shows main characteristics of one of the continuous casting machines installed at the Indiana Harbor Works for casting of thick slabs.

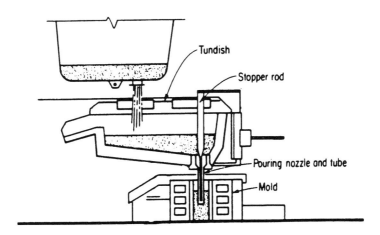

Fig. 10.8 Liquid steel flows from the ladle into the tundish and from the tundish into the mold. (From THE MAKING, SHAPING AND TREATING OF STEEL, Tenth Edition. Reprinted with permission).

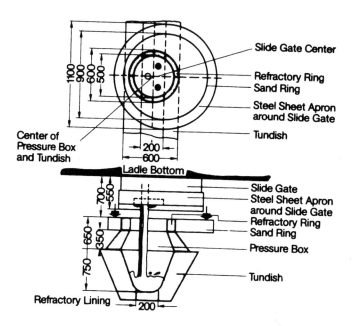

Fig. 10.9 Gas shrouding with Mannesmann-Demag pressure box. (From Kruger, et al, IRON AND STEEL ENGINEER, March 1984. Copyright AISE, Pittsburgh, Pennsylvania. Reprinted with permission).

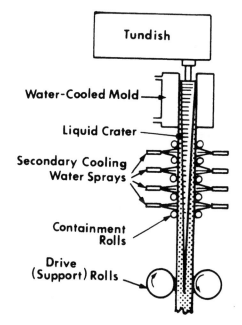

Fig. 10.10 Major components of a continuous casting machine. (From THE MAKING, SHAPING AND TREATING OF STEEL, Tenth Edition. Reprinted with permission).

Table 10.1 Characteristics of thick slab casting facility at LTV Steel Co., Indiana Harbor Works. Data from The Making, Shaping and Treating of Steel (1985).

Caster type...............................	Curved Mold
Number of strands.........................	1
Casting radius, metres (ft)...............	12.2 (40)
Section sizes, mm (in.)...................	1015 to 1980 wide x 250 (40 to 78 wide x 10)
Slab length, m (in.)......................	5.5 to 11.3 (219 to 444)
Heat size, metric tons (net tons).........	260 (285)
Tundish capacity, metric tons (net tons)...	43 (47)
Ladle stream shrouding....................	Argon and nitrogen
Tundish stream shrouding..................	Immersion nozzle
Mold length, mm (in.).....................	900 (35.4)
Mold liner...............................	Silver bearing copper with chrome plating
Metallurgical length, m (ft)..............	39.859 (130.8)
Casting speed (max), m/min (in./min).......	1.8 (71)
Straightener rolls........................	49 pairs (20 driven rolls)
Total cooling water, m^3/s (gpm):	
mold......................	0.31 (4940)
internal machine............	0.77 (12200)
spray.......................	0.64 (10100)

10.6 SLAB WIDTH CONTROL

Desired slab width is usually achieved by using one of the following three methods [1]:

1. Slab slitting - This method allows one to cast a small number of 'master' slab sizes with the slab product being slit longitudinally in a separate operation using either oxy-natural gas torches or rolling machines.

2. Adjustable mold width - This method allows one to minimize the time required to replace a mold. Various designs for changing the mold width are utilized. In the continuous casting machines of earlier designs, the mold width adjustment can be made while the previously cast slab is being removed from the machine. In the latest designs, the mold taper can be changed during the actual casting operation.

3. Divided molds - According to this method, a divider installed in the mold permits the casting of two narrow slabs simultaneously in a single strand machine.

10.7 CONTINUOUS CASTING OF THIN SLABS AND STRIP

Continuous casting of thin slabs and strip is a practical implementation of a modern concept known as **near-net-shape casting process.**

The continuous casting processes which utilize this concept can be divided into six major groups [7]:

1. Stationary-mold casting process

2. Traveling-mold casting process

3. Traveling twin-belt casting process

4. Roller-casting process

5. Roller-belt casting process

6. Spray deposition process.

Stationary-mold casting process utilizes technology that was developed for continuous casting of thick slabs (Fig. 10.11). The process is designed for casting all grades of steel.

Traveling-mold casting process minimizes the interaction between mold wall and cast product. This allows one to increase casting speed. In the process developed by British Steel Corporation [8], steel is poured from a tundish into a moving mold car (Figs. 10.11 and 10.12). The steel is then poured onto the belt formed by horizontally moving cars with cooling plates of cast iron. The cars move back to their starting position as soon as the strip is solidified and taken over by a conveyor belt.

Traveling twin-belt casting process was first implemented in the Hazelett casting machine [9]. In this machine (Fig. 10.13) the mold is formed by two parallel moving metal belts wrapped about pulleys which move both belts at synchronized speed. The two belts are separated by parallel strings of articulated dam blocks whose horizontal separation, easily adjusted, determines the slab width.

The traveling twin-belt casting method, implemented in the Alusuisse Caster II machine [10], utilizes two sets of rotating chilling caterpillar molds. Similar casting machine (Fig. 10.14) was introduced by Kobe Steel [7]. Twin-belt caster of somewhat different designs had been developed by Nippon Steel [7] and Kawasaki Steel [8].

Roller-casting process utilizes either a single-roller casting method or a twin-roller casting method (Figs. 10.15 and 10.16).

A single-roller casting method was implemented by Allegheny Ludlum Steel Corporation [11] for casting thin strip. In this process a tundish conducts the liquid metal stream of the appropriate width to a rotating casting wheel, on which the metal solidifies. The solidified strip is removed from the wheel and carried to a coiler where it is loosely coiled for subsequent recoiling.

A twin-roller casting method was patented by Bessemer in 1865. However, its practical realization has been delayed for more than a century. This method was recently implemented by many steel producers for casting of thin strip of carbon, silicon and stainless steel.

One of modifications of the roller-casting process is the Inside-The-Ring process [7], developed by J & L (now LTV) Steel Co. In this process, the molten steel solidifies between the outer face of a small rotating cylinder and the inner face of a rotating hollow cylinder as shown in Fig. 10.16. The cast strip is withdrawn in a spiral fashion.

Roller-belt casting process was implemented by Hitachi in the casting machine shown in Fig. 10.17. In this machine, the mold cavity is formed by enclosing a groove in the periphery of a rotary wheel with an endless moving belt. The cast steel is free of oscillation marks because the steel travels with the mold. Also, it is reported that this method provides more uniform inclusion distribution as compared with some conventional continuous casters [1].

Spray deposition process was developed by Osprey Metals Ltd. and adapted and refined by Mannesmann Demag Huttentechnick (MDH) for production of flat products [7]. In this process, the steel issuing from a tundish is atomized by a nozzle (Fig. 10.18). The droplets deposit and solidify on a collection surface or substrate. The use of several nozzles permits the production of multilayer materials.

Process	Manufacturer/ operator	Technical data	Product
Stationary mould	SMS Buschhütten	Thickn.: 40–60 mm	Thin slab
	SMS/Nucor Corp. Crawfordsville (II. 1989)	Width: 1600 mm v_G : 6 m/min 750,000 t/a	Usual hot strip grades
Stationary mould 	MDH/MRW Huckingen	Thickn.: 40–70 mm Width: 1600 mm v_G : 6 m/min	Thin slab Usual hot strip grades
Stationary mould	Danieli Udine	Thi.: 28 50 mm Wi. : 1600–1750 mm v_G : 6 m/min	Thin slab Strip
	Danieli Feng Lung Steel Factory Taiwan	Thickness: 75 mm Width: 1220 mm	Thin slab
Stationary mould in horizontal line HCC 	MDH/Bosch-gotthardshütte Siegen	Thickn.: 40–120 mm Width: 450 mm v_G : 4 m/min 100,000 t/a	Thin slab All steel grades
Mould car 	British Steel Corp. Teesside Labs	Thickness: 75 mm Width: 500 mm v_G : 10–20 m/min	Thin slab Plain carbon steel

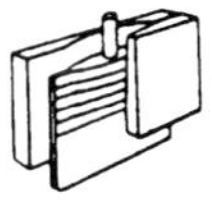

Fig. 10.11 Continuous casting processes for thin slabs and strip. (From Reichelt and Kapellner, METALLURGICAL PLANT AND TECHNOLOGY, Feb. 1988. Copyright Verlag Stahleisen mbH, Dusseldorf, W. Germany. Reprinted with permission).

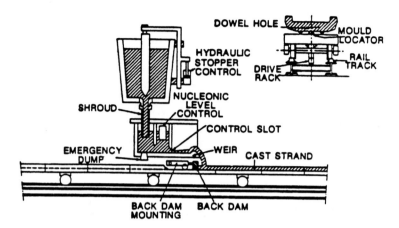

Fig. 10.12 Single-belt caster designed by British Steel Corporation. (From Cygler and Wolf, IRON AND STEELMAKER, Aug. 1986. Copyright Metallurgical Society of AIME. Reprinted with permission).

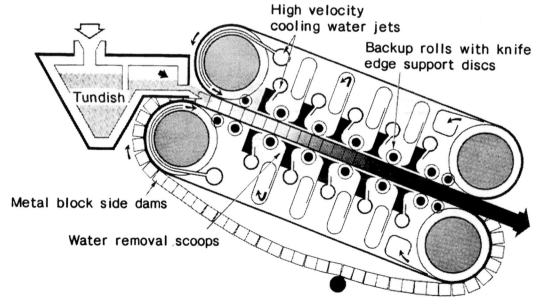

Fig. 10.13 Hazelett twin-belt casting machine. (From Hazelett, AISE Year Book, 1966. Copyright AISE, Pittsburgh, Pennsylvania. Reprinted with permission).

Process	Manufacturer/operator	Technical data	Product
2 travelling belt moulds, vertical 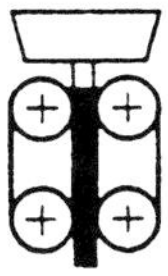	Nippon Steel Corp.	Thickness: 50–80 mm Width: 600 mm v_G: 4 m/min	Thin slab
2 travelling belt moulds, horizontal 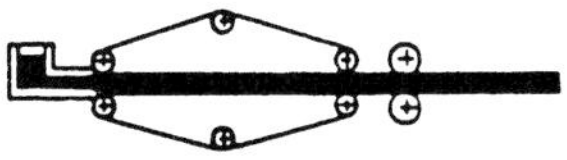 KCC = Kawasaki Horizontal Continuous Caster	Kawasaki Steel Corp. Chiba Research Centre	Thickness: 10–30 mm Width: 100–150 mm v_G: 0.7–12.5 m/min	Strip Low-carbon steels Si-steels Stainless steels
2 travelling belt moulds, inclined 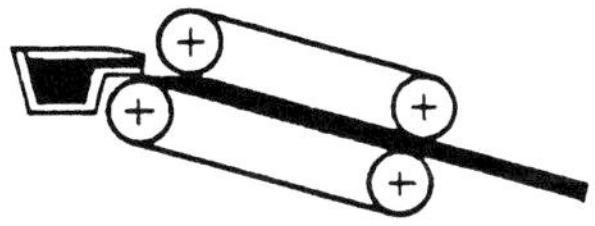 Hazelett	Hazelett/ Sumitomo MI+ Sumitomo HI Kashima	Thickness: 40 mm Width: 600–1300 mm v_G: 2–8 m/min 600,000 t/a (1300 mm)	Thin slab Al-killed steels Stainless steels
	Hazelett/ Nucor Corp. Darlington, S.C.	Thickness: 25–38 mm Width: 1300 mm v_G: 1.5–3 m/min 500 000 t/a (1300 mm)	Strip Carbon steels
	Hazelett/ Krupp Industrietechnik Bochum	Thickness: 70 mm Width: 180 mm abandoned	Thin slab
	Hazelett/ Bethlehem Steel + US Steel Universal, Pa.	Thickness: 12–25 mm Width: 1830 mm interrupted/abandoned	Strip
2 travelling caterpillar moulds, horizontal 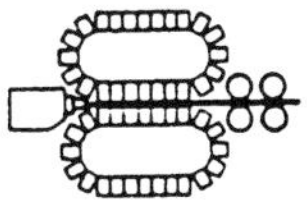	Kobe Steel Nippon Kokan	no details known	

Fig. 10.14 Continuous casting processes for production of thin slabs and strip. (From Reichelt and Kapellner, METALLURGICAL PLANT AND TECHNOLOGY, Feb. 1988. Copyright Verlag Stahleisen mbH, Dusseldorf, W. Germany. Reprinted with permission).

Process	Manufacturer/operator	Technical data	Product
Twin roller 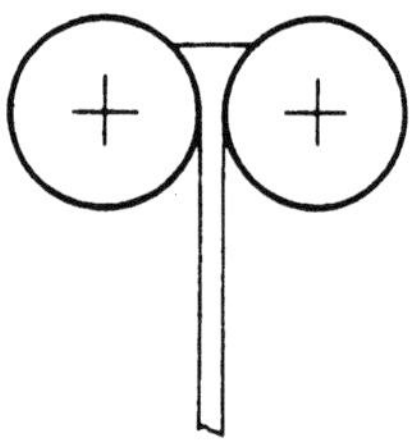	Nippon Steel Corp.	Thickness: 1 mm Width: 200 mm	Thin strip Stainless steels
	Nisshin Steel + Hitachi Corp.	Thickness: < 5 mm Width: 200 mm	Thin strip Stainless steels
	Kawasaki Steel Corp.	Thickness: 0,5 mm Width: 100 mm	Thin strip Si-steels
	Armco + Inland Steel + Weirton Steel + Bethlehem Steel	Thickness: < 5 mm Width: 300 mm	Thin strip
	Clecim/Irsid	Thickness: 2–10 mm Width: 200 + 850 mm	Strip Thin strip Stainless steels
	DEC/British Steel Corp.	Thickness: 3 mm Width: 400 mm	Thin strip Stainless steels
	CSM	Thickness: Width: 300 mm	Thin strip Unalloyed steels
	Voest-Alpine	Thickness: 2–8 mm Width:	Strip Thin strip
	Thyssen Grillo Funke + IBF Aachen	Thickness: 0.1–2 mm Width: 150 mm	Thin strip Si-steels
	MPI für Eisenforschung Dusseldorf	Thickness: 1 mm Width: 100 mm	Thin strip Si-steels
Single roller 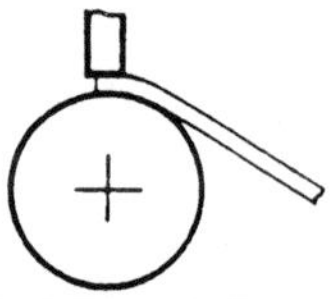	Armco + Westinghouse Middletown	Thickness: 0.8–3 mm Width: 75 mm	Thin strip
	Allegheny Ludlum	Thickness: 1–2 mm Width: 300 mm	Thin strip Stainless steels

Fig. 10.15 Continuous casting processes for production of thin slabs and strip. (From Reichelt and Kapellner, METALLURGICAL PLANT AND TECHNOLOGY, Feb. 1988. Copyright Verlag Stahleisen mbH, Dusseldorf, W. Germany. Reprinted with permission).

Process	Manufacturer/ operator	Technical data	Product
1 travelling casting belt and 1 roller Beltroller DeSC-Process **Demag Strip Casting**	MDH/MSA Belo Horizonte (I. 1988)	Thickness: 5–10 mm Width: 900 mm v_G: 25–50 m/min 750,000 t/a	Strip All steel grades
Inside-The-Ring 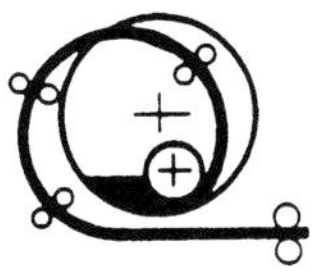	Jones + Laughlin (now LTV Steel) Pittsburgh (1967–1975)	Thickness: 5 mm Width: 380 mm v_G: 7.5 m/min	Strip
Twin roller (top roller) 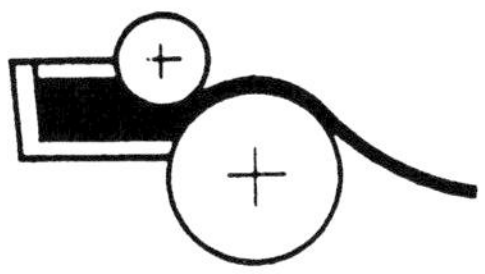	Kobe Steel Amagasaki	Thickness: 1–2 mm Width: 260 mm	Thin strip Stainless steels
	Nippon Metal	Thickness: 1–4 mm Width: 315/650 mm	Thin strip Stainless steels
	Krupp Stahl AG	Thickness: 1–4 mm Width:	Thin strip Stainless steels
	Nippon Yakin	Thickness: 6 mm Width: 150 mm	Strip Stainless steels
Twin roller (hot top) 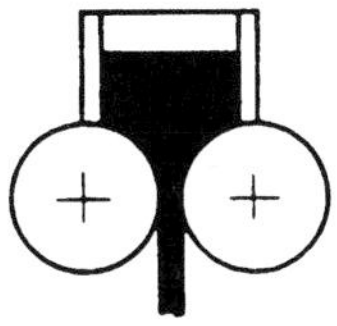	Ishihari HI/ Nippon Kokan	Thickness: 2–6 mm Width: 400 mm v_G: 25 m/min	Strip Thin strip Carbon steels Stainless steels
	Hitachi Zosen Corp.	Thickness: 5–10 mm Width: 300 mm	Strip

Fig. 10.16 Continuous casting processes for production of thin slabs and strip. (From Reichelt and Kapellner, METALLURGICAL PLANT AND TECHNOLOGY, Feb. 1988. Copyright Verlag Stahleisen mbH, Dusseldorf, W. Germany. Reprinted with permission).

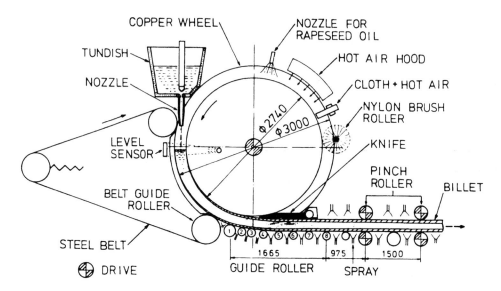

Fig. 10.17 The Hitachi wheel-belt caster. (From THE MAKING, SHAPING AND TREATING OF STEEL, by W.T. Lankford, et al, Tenth Edition, 1985. Copyright AISE, Pittsburgh, Pennsylvania. Reprinted with permission).

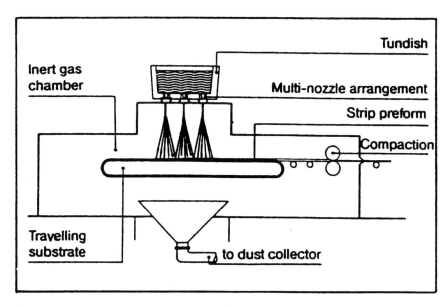

Fig. 10.18 Spray deposition processes for strip production. (From Reichelt and Kapellner, METALLURGICAL PLANT AND TECHNOLOGY, Febr. 1988. Copyright Verlag Stahleisen mbH, Dusseldorf, W. Germany. Reprinted with permission).

10.8 REQUIREMENTS FOR CONTINUOUSLY CAST STEELS

There are three special requirements regarding to steelmaking practices for continuously cast steels [1]:

1. Temperature control
2. Deoxidation practice
3. Desulfurization practice.

Temperature control - The tapping temperature is higher than in ingot production in order to compensate for increased heat losses during continuous casting. In addition, temperature homogenization practices, such as argon stirring, are employed.

Deoxidation practice - Only fully deoxidized (killed) steel can be used for continuous casting in order to avoid formation of blowholes and pinholes. Silicon deoxidation is usually used for coarse grain steels and aluminum deoxidation for fine grain steels.

Desulfurization practice - It is necessary to achieve low sulfur levels including sulfide shape control (primarily in aluminum-killed steels) in order to minimize inclusion deposition in the tundish nozzle.

10.9 OXIDE INCLUSIONS IN CONCAST STEEL

Two distribution patterns of oxide inclusions (Fig. 10.19) are usually identified in continuously cast aluminum-killed steel [6, 12].

First oxide distribution pattern - The first distribution pattern relates to the single crystals of alumina which are deoxidation products remaining in steel. These crystals are uniformly distributed in both solid and liquid steel. Their size can be up to 100 μm (0.004 in.).

The suspension of alumina crystals in liquid steel is caused by inability of the particles to float due to their size and convection currents in the liquid phase. In a steel with a combined oxygen content of 30 ppm and an average particle diameter of 20μm (0.0008 in.), the suspension consists of approximately 10^{10} particles/ton of steel, with a particle interval of approximately 500μm (0.020 in.). It appears that alumina suspension is not affected by the radius of casting machine.

Second oxide distribution pattern - The second distribution pattern relates to the macroscopic inclusions consisting of alumina clusters which are agglomerates of single crystals of alumina. The size of the microscopic inclusions is from 40 to 1300 μm (0.0016 to 0.052 in.).

Unlike single crystals of alumina, the macroscopic inclusions are not uniformly distributed in steel. Since buoyancy of these inclusions is greater than resistance of the flow, they move relative to the liquid phase. In case of curved caster this movement slows down and a partial enrichment of steel with inclusions takes place along the inner bow face of the strand.

As shown in Fig. 10.20, the concentration of the inclusions along the inner bow face of the strand decreases with increase in the casting machine radius. It also decreases with decrease in the particle size.

10.10 FORMATION OF OXIDE PHASES

Mechanism of formation of oxide phases during continuous casting of steel can be illustrated by using an Evler-Venn diagram as shown in Fig. 10.21 [6, 12-14]. The evenly distributed alumina suspension leads to

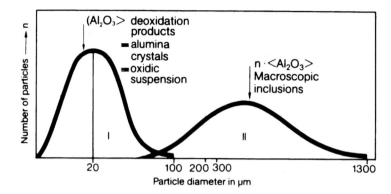

Fig. 10.19 Oxide inclusions in continuously cast aluminum-killed steel (Al 0.010 to 0.060%). (From Kruger, et al, IRON AND STEEL ENGINEER, March 1984. Copyright AISE, Pittsburgh, Pennsylvania. Reprinted with permission).

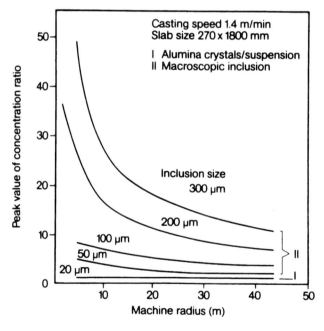

Fig. 10.20 Distribution of suspended particles and macroscopic inclusion as a function of casting radius. (From Kruger, et al, IRON AND STEEL ENGINEER, March 1984. Copyright AISE, Pittsburgh, Pennsylvania. Reprinted with permission).

precipitation as follows:

    a) quantity B precipitates in the submerged nozzle

    b) portion C of quantity B is carried into the mold

    c) portion D of quantity C floats out into and combines with the casting slag

    d) portion E remains in steel and forms the macroscopic inclusions or alumina clusters.

The following measures may prevent the deposition of alumina on submerged nozzles and the reduction of the macroscopic inclusions in steel [6]:

    1. Production of a steel which is free of deoxidation products. This measure, however, does not appear feasible on a large industrial scale in the near future.

    2. Creation of a gaseous phase between the liquid steel and refractory material. This can be achieved by development of a submerged nozzle in which a homogeneous stable gas can be formed between the refractory wall and the liquid steel along the entire length of the shroud tube.

    3. Transformation of solid alumina suspensions, which are solid at casting temperature, into liquid deoxidation particles.

10.11 INFLUENCE OF CASTER TYPE ON STEEL QUALITY

In regard to a ferrostatic pressure height and casting radius, the continuous casting machines can be divided into the following five principal types [6], as shown in Fig. 10.22:

    1. Vertical caster (1)

    2. Bow type caster with one strengthening point (2, 3)

    3. Bow type caster with multiple strengthening points (4)

    4. Oval bow-type, or super low head, caster (5)

    5. Horizontal caster (6).

The overall height and, hence, the ferrostatic pressure decrease starting from a vertical caster through the bow-type to the oval bow-type and horizontal casters. As shown in Fig. 10.22, the danger of internal cracking from bulging decreases with a decrease in the ferrostatic pressure. The decrease in the caster height, however, results in deterioration of the steel cleanliness.

These two opposing quality features play a major role in selection of an optimum type caster. A majority (62%) of the continuous casting machines built worldwide are in the form of a bow-type caster with casting radius between 8 and 15 m (26 to 49 ft.). However, with the introduction of the measures for preventing the formation of microscopic inclusions, the degree of the steel cleanliness becomes less dependent on the caster height. This allows one to substantially reduce the size (Fig. 10.23) and cost of the continuous casting machines.

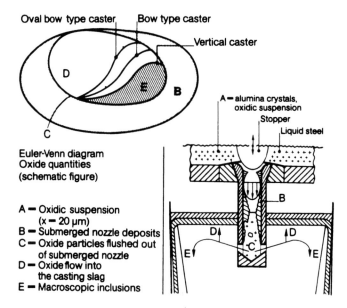

Fig. 10.21 Mechanism of formation of oxide phases in steel. (From Kruger, et al, IRON AND STEEL ENGINEER, March 1984. Copyright AISE, Pittsburgh, Pennsylvania. Reprinted with permission).

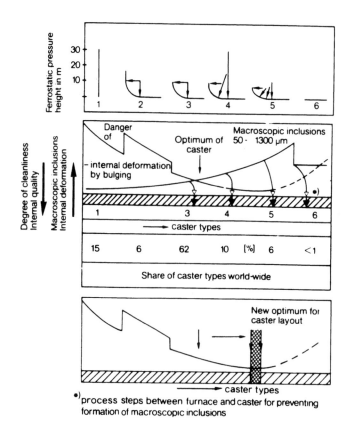

Fig. 10.22 Evolution of new optimum caster type based on new clean steel technology. (From Kruger, et al, IRON AND STEEL ENGINEER, March 1984. Copyright AISE, Pittsburgh, Pennsylvania. Reprinted with permission).

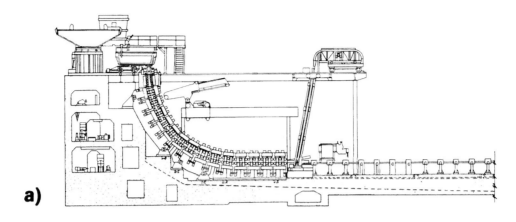

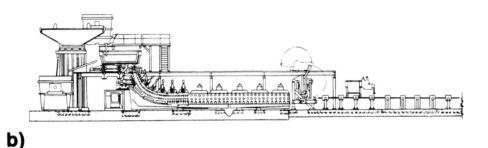

Fig. 10.23 A comparison of a conventional caster (a) with a low overall height modular type caster developed by United Engineering, Inc./L.A.I. (b). (From UNITED ENGINEERING, Inc. PAMPHLET UE875031M. Reprinted with permission).

REFERENCES

1. The Making, Shaping and Treating of Steel, 10th Edition, eds. W.T. Lankford, Jr., et al, Association of Iron and Steel Engineers, Pittsburgh, Pennsylvania, pp. 691-770 (1985).

2. ASM Metals Reference Book, American Society for Metals, Metals Park, Ohio, p. 53 (1981).

3. W.L. Roberts, Hot Rolling of Steel, Marcel Dekker, Inc., New York, pp. 105,106 (1983).

4. T. Nozaki, et al, "Horizontal Continuous Casting of Stainless Steel at Kobe Steel", Iron and Steel Engineer, October 1985, pp. 50-56.

5. M. Haissig, "Horizontal Continuous Casting: A Technology for the Future", Iron and Steel Engineer, June 1984, pp. 65-71.

6. B. Kruger, et al, "Continuous Casting of Steel: The Process for Quality Improvement", Iron and Steel Engineer, March 1984, pp. 45-52.

7. W. Reichelt and W. Kapellner, "Near-Net-Shape Casting of Flat Products", Metallurgical Plant and Technology, No. 2, 1988, pp. 18-35.

8. M. Cygler and M. Wolf, "Continuous Strip and Thin Slab Casting of Steel-An Overview", Iron and Steelmaker, August 1986, pp. 27-33.

9. R.W. Hazelett, "The Present Status of Continuous Casting Between Moving Flexible Belts", AISE Year Book, 1966, pp. 435-440.

10. "Introduction to the Alusuisse Caster II Process", Light Metal Age, Volume 41, Nos. 5 & 6, June 1983, pp. 10-12.

11. R.K. Pitler, "Direct Casting of Coilable Ferrous Alloy Strip", Advanced High-Temperature Alloys: Processing and Properties, eds. S.M. Allen, et al, pp. 1-10, American Society for Metals, Metals Park, Ohio, (1986).

12. H. Schrewe and F.P. Pleschiutschnigg, "Equipment and Process Improvement to Eliminate Oxide Impurities", Mannesmann Continuous Casting Symposium, Oct. 21-22, 1981, Colorado Springs, Colorado.

13. G.M. Faulring, et al, "Steel Flow Through Nozzles: Influence of Calcium", Iron and Steelmaker, IIS-AIME, Vol.7, pp. 14-20.

14. K. Schwerdtfeger and H. Schrewe, "Reoxidation of Aluminum-Killed Steel by Contact with Refractory Materials", Proceedings Electric Furnace Conference, Vol.28, Dec. 9-11, 1970, pp. 95-103.

11

Defects in Ingots and Slabs

11.1 INTERNAL DEFECTS OF INGOTS AND SLABS

Below is a brief description of the most common internal defects of ingots and continuously cast slabs [1-3].

Pipe - The central cavity formed by contraction in metal, especially ingots (Fig. 11.1) during solidification.

Secondary pipe - This is a pipe which is formed due to the shrinkage of the molten metal trapped inside casting during its solidification (Fig. 11.1).

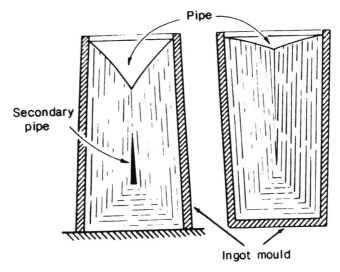

Fig. 11.1 The influence of the shape of the mould on the extent of piping in a steel ingot. (From ENGINEERING METALLURGY, Part 1, by R.A. Higgins, 1983. Copyright Hodder & Stoughton Ltd., England. Reprinted with permission).

Segregation - It is a concentration of dissolved impurities in that portion of the metal which solidifies last. The dendrites which form are almost pure metal, and therefore the impurities become progressively more concentrated in the remaining liquid.

Minor segregation - It is a local concentration of impurities at the crystal boundaries as shown in Fig. 11.2a. It causes overall brittleness of the castings.

Major segregation - It is a concentration of impurities in the central portion of an ingot as V-shaped markings as shown in Fig. 11.2b.

Inverse-V segregation - This type of segregation may occur in very large ingots. In case when the temperature gradient is very slight, the metal may solidify last in the intermediate portions of the ingot causing inverted V-shaped markings as shown in Fig. 11.2c.

Non-metallic inclusions - These are the oxidized materials and sulfides in various combinations with each other. They are a result of oxidizing reactions which take place during refining process. Some of the reactions may be associated with the erosion of ladle or other refractories.

Diagonal crack - It is a pronounced crack following the interface of two different planes of crystallization (Fig. 11.3a). These cracks are caused by thermal stresses, which have been attributed to severe or uneven secondary cooling of slabs during their solidification.

Halfway crack - It is an intercolumnar crack occurring in positions approximately midway between the outside and center of the slab (Fig. 11.3b).

Central unsoundness - It is a cavity or porosity inside the cast product (Fig. 11.4a). Severity of

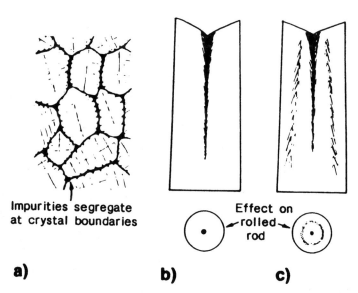

Impurities segregate
at crystal boundaries

Effect on
rolled
rod

a)   b)   c)

Fig. 11.2 Types of segregation which may be encountered in steel ingot: a) minor segregation, b) major segregation, c) major and inverse-V segregation. (From ENGINEERING METALLURGY, Part 1, by R.A. Higgins, 1983. Copyright Hodder & Stoughton Ltd., England. Reprinted with permission).

this defect depends upon the composition, size, and shape of the product being cast, and on casting conditions, particularly, teeming temperature or variation of the withdrawal speed.

Star cracks - These are radial cracks in the form of a star originating from the center of the product (Fig. 11.4b), caused by a severe secondary cooling.

Withdrawal roll cracks - These are parallel transverse internal cracks, normal to the axes of the rolls. They are caused by reduction of the product while the center is still in a semi-molten state.

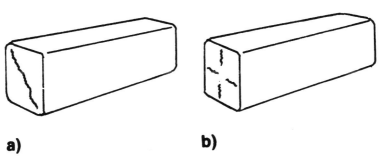

a)   **b)**

Fig. 11.3 Internal cracks in cast products: a) diagonal cracks, b) halfway cracks.

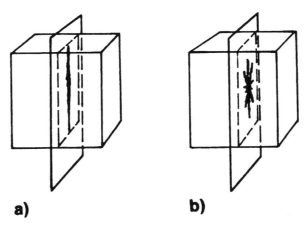

a)   **b)**

Fig. 11.4 Internal cracks in cast products: a) central unsoundness, b) star cracks. (From ISI Publication 106, 1967. Copyright Iron and Steel Institute, England. Reprinted with permission).

11.2 EXTERNAL CRACKS IN INGOTS AND SLABS

Cracks or ruptures in ingots may be caused by a number of reasons such as the following [3]:

1. Restriction to the ingot skin during cooling

2. Inability of the skin to withstand the stresses resulting from ferrostatic pressure exerted by the liquid steel

3. A too high teeming temperature

4. A too high teeming rate

5. A too high mold temperature

6. A too large corner radius on the ingot

7. A too small flute depth on the ingot

8. Steel entering the mold towards one side instead of the center.

In addition to some of the causes listed above, the cracks in continuously cast slabs may be also due to the following reasons [2]:

1. Inadequate mold design

2. Distortion or wear of the mold

3. Fast or uneven cooling in the mold

4. Fast or uneven secondary cooling of different faces of the slabs

5. Bending or straightening of the slab at a low temperature.

Below is a brief description of the most common types of external cracks.

Hanger crack - It occurs when the ingot is suspended in the mold due to a worn or improperly fitted feeder head, or due to overfilling of the mold (Fig. 11.5a).

Fin cracks - It is caused by restrictions resulting from a fin (Fig. 11.5b).

Basal cracks - It occurs when free contraction of the ingot is limited by fins or flashes (Fig. 11.6a).

Double skin crack - It is a crack that is usually associated with 'double skin' as shown in Fig. 11.6b.

Transverse facial crack - It is a crack that runs from the corner of the cast product across its face (Fig. 11.7a).

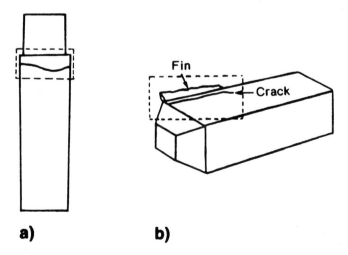

a)   b)

Fig. 11.5 External cracks in cast products: a) hanger crack, b) fin crack. (From ISI Special Report No. 63, 1958. Copyright Iron and Steel Institute, England. Reprinted with permission).

Teeming arrest - It appears as a transverse break in the continuity of the skin extending completely around the product (Fig. 11.7b). This is due to a temporary stoppage in teeming.

Longitudinal facial crack - It is a crack that runs along the ingot face. It may appear either as a corner crack (Fig. 11.8a) or as a facial crack (Fig. 11.8b).

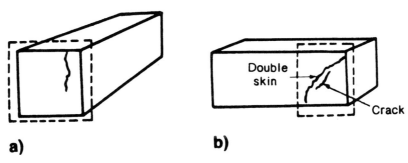

Fig. 11.6 External cracks in cast products: a) basal crack, b) double skin crack. (From ISI Special Report No. 63, 1958. Copyright Iron and Steel Institute, England. Reprinted with permission).

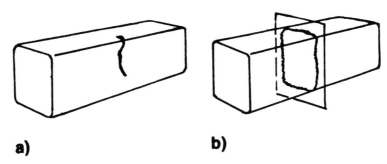

Fig. 11.7 External cracks in cast products: a) transverse facial crack, b) teeming arrest.

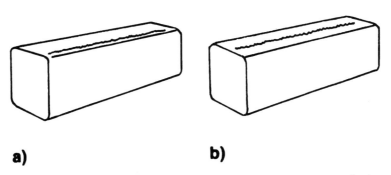

Fig. 11.8 External longitudinal cracks in cast products: a) corner crack, b) facial crack.

Clinks - These are the ruptures which may appear if the ingot is stripped early and exposed to a cold atmosphere. This defect is most prone to occur in high-carbon and alloy steels.

Hot shortness - This is a localized cracking normally associated with concentration of copper and tin in the grain boundaries.

11.3 SURFACE DEFECTS TYPICAL FOR INGOTS

In this method the surface defects in ingots, other than cracks, will be described [3].

Spongy top - It is irregular eruption of steel on the ingot top occurring after the filling of the mold (Fig. 11.9a). It is caused by a late gas evolution during which the metal has become too cold to allow easy escape of the gas.

Bootleg - It is a deeply sunken top in a rimming-steel ingot (Fig. 11.9b). This shape is formed due to excessive gas evolution during the teeming period.

Crazing - This defect usually appears as the 'crocodile-skin' markings (Fig. 11.10a) which are found on the surfaces of ingots being cast into a worn mold.

Rippled surface - This is a series of slight peripheral undulations on the ingot surface (Fig. 11.10b) caused by a rapid succession of layers of frozen metal forming near the mold walls, and becoming engulfed in turn as teeming proceeds. This may be a result of either low teeming temperature or low teeming rate.

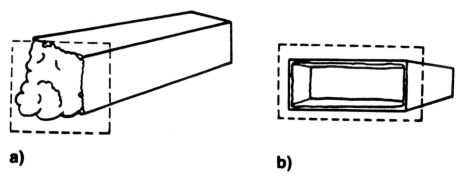

a)   b)

Fig. 11.9 Ingot surface defects: a) spongy top, b) bootleg. (From ISI Special Report No. 63, 1958. Copyright Iron and Steel Institute, England. Reprinted with permission).

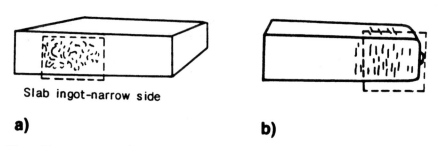

Slab ingot-narrow side

a)   b)

Fig. 11.10 Surface defects in cast products: a) crazing, b) rippled surface. (From ISI Special Report No. 63, 1958. Copyright Iron and Steel Institute, England. Reprinted with permission).

Flash - It is a plate formed by molten steel running up behind loose bricks used to form the feeder head, or between the mold and a superimposed head, or by overfilling the mold.

Scab - This is an irregular bulge on the ingot skin caused by a depression in the mold, or by mold metal adhering to the ingot skin.

11.4 SURFACE DEFECTS FOR CONCAST SLABS

Continuously cast slab may have surface defects which are attributed to the certain specifics of the casting process. The most typical surface defects of this type are listed below [2].

Longitudinal depression - This is a channel-shaped depression on the face of the slab, running in the direction of the axis (Fig. 11.11a).

Transverse depression - This is a localized depression of the slab surface, normal to the axis of the slab (Fig. 11.11b). This defect is attributed to a number of factors including uneven lubrication, too rapid cooling, or fluctuating level of metal in the mold.

Heavy reciprocation marks - Distances between these marks are usually related to the amount of the product that has descended in one cycle of reciprocation (Fig. 11.12a).

False wall - This is a surface irregularity that completely rounds the product (Fig. 11.12b). It is caused by temporary separation of the skin from the descending strand.

Guide marks - This is a mechanical damage arising from irregularities on the guides, support, bending or straightening rolls. The guide marks can also be caused by extraneous metal adhering to the parts of the guiding mechanism.

Carburization - This is a localized surface pick-up of carbon from oil lubrication. It is encountered especially in the casting of low-carbon stainless-steel grades.

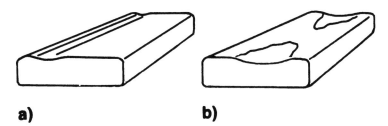

a)   **b)**

Fig. 11.11 Surface defects in concast slabs: a) longitudinal depression, b) transverse depression.

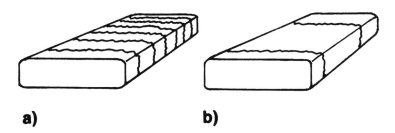

a)   **b)**

Fig. 11.12 Surface defects in concast slabs: a) heavy reciprocation marks, b) false wall.

11.5 SURFACE DEFECTS COMMON FOR INGOTS AND CONCAST SLABS

A number of surface defects are common for both ingots and continuously cast slabs. The most frequent defects are described below [1, 2].

Bleed - it is a result of exudation of molten metal through a rupture in the skin (Fig. 11.13a).

Lap - This is a fold in the skin caused by the rising molten metal that engulfs a layer that is already solidifying against the mold wall (Fig. 11.13b).

Surge - This defect occurs during casting of rimming steel. The rimming action triggers a collapse of the liquid metal level. Consequently, the produced shell envelops the ingot or the slab as the casting proceeds (Fig. 11.14a).

Plating - It appears as a shallow indent on the surface (Fig. 11.14b). The origin of this defect is as follows. Metal that has solidified on the surface of the meniscus is drawn toward the mold walls and subsequently entrapped.

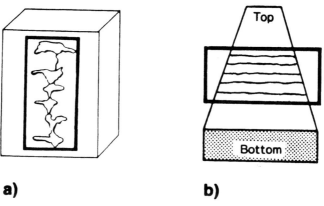

a)   **b)**

Fig. 11.13 Surface defects of cast products: a) bleed, b) lap. (From ISI Publication No. 106, 1967. Copyright Iron and Steel Institute, England. Reprinted with permission).

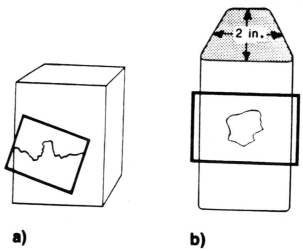

a)   **b)**

Fig. 11.14 Surface defects in cast products: a) surge, b) plating. (From ISI Publication No. 106, 1967. Copyright Iron and Steel Institute, England. Reprinted with permission).

Fin - This is a thin strip of metal protruding approximately at right angle to the surface of the ingot or the slab (Fig. 11.5b). It is caused by molten steel having run into the open cracks in the mold.

Blowholes - These are large holes that are elongated in the direction of columnar growth (Fig. 11.15a). Blowholes may break the surface. This defect is generally due to incorrect level of oxidation. In stainless steel grades, hydrogen can be the cause.

Pinholes - These are small holes close to the surface and often in clusters. When they appear on the surface, they are known as skin holes (Fig. 11.15b). Pinholes are frequently located under the skin of the cast product. They are attributed to high levels of oxygen or hydrogen in steels.

Splash - It is a generally spattered surface caused by small particles of metal trapped between the product skin and mold wall.

Entrapped scum - These are the patches on the ingot or slab surfaces. They are formed by the products of deoxidation or refractory erosion that are trapped at the meniscus.

Bruises - This is a mechanical injury to the ingot or slab skin such as dog marks from crane or indentations caused by sharp objects.

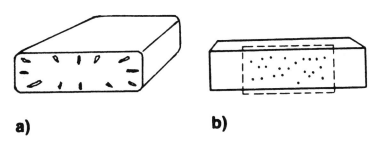

a)   **b)**

Fig. 11.15 Defects of cast products: a) blowholes, b) pinholes. (From ISI Special Report No. 63, 1958. Copyright Iron and Steel Institute, England. Reprinted with permission).

11.6 SLAB SHAPE DEFECTS

The following three defects may be found in continuous cast slabs of rectangular cross-section [2].

Rhomboidity - It is defined as a relative difference between diagonals of a rectangular section of the slab (Fig. 11.16a) and may be expressed as

$$R = (d_1/d_2-1) \times 100, \qquad (11-1)$$

where d_1 = longer diagonal

d_2 = shorter diagonal.

Concavity - It is a distortion of the slab which appears as a concave surface (Fig. 11.16b). This defect is usually attributed to an incorrect spray-cooling pattern.

Bulging - It is a distortion of the slab giving rise to a convex surface or surfaces (Fig. 11.16c). This defect is usually due to inadequate support of the skin against the effects of ferrostatic pressure.

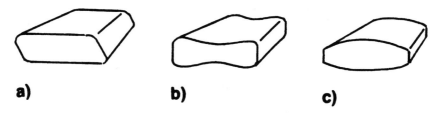

Fig. 11.16 Slab shape defects: a) rhomboidity, b) concavity, c) bulging.

REFERENCES

1. R.A. Higgins, Engineering Metallurgy, Part 1, Applied Physical Metallurgy, Robert E. Krieger Publishing Company, Melbourne, Florida, pp. 68-77, (1983).

2. Definitions and Causes of Continuous Casting Defects, BISRA, ISI Publication 106, The Iron and Steel Institute, London, pp. 1-39 (1967).

3. Surface Defects in Ingots and Their Products, BISRA Special Report No. 63, The Iron and Steel Institute, London, pp. 2-24 (1958).

Part IV

Theory of
Plastic Deformation

12

Principles of Microscopic Plasticity

12.1 ELASTIC AND PLASTIC DEFORMATION

The microscopic viewpoint deals with a physical explanation of plasticity and establishes relationship between plastic behavior and the interatomic forces acting within the crystal structure of a metal.

Elastic deformation is usually defined as a change in dimensions directly proportional to and in phase with an increase or decrease in applied force [1]. In elastic deformation a limited distortion of the crystal lattice occurs, and as soon as the force is removed, the distortion disappears (Fig. 12.1).

Plastic deformation is commonly defined as a change in dimensions that does or will remain permanent after removal of the load that caused it. In plastic deformation an extensive rearrangement of atoms within the lattice structure takes place resulting in this permanent distortion.

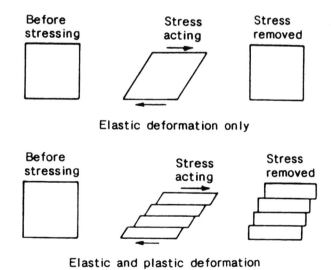

Fig. 12.1 Diagrams illustrating the difference, in action and effect, of deformation by elastic and plastic means. (From ENGINEERING PLASTICITY, Part 1, by R.A. Higgins, 1983. Copyright Hodder & Stoughton Ltd., England. Reprinted with permission).

12.2 DEFORMATION BY SLIP

Plastic deformation proceed in metals by a process known as 'slip'. This is an irreversible shear displacement of one part of a crystal relative to another in a definite crystallographic direction in which the translation of slip takes place [1, 2].

Slip occurs in directions in which atoms are most closely packed since this requires the least amount of energy. In a face-centered cubic lattice (Fig. 12.2) the (111) plane of densest atomic population intersects the (001) plane in line ac. When the (001) plane is assumed to be the plane of the paper (Fig. 12.3), slip is seen as a movement along the (111) plane in the close-packed [110] direction [3, 4].

The result of slip in a polycrystalline mass of metal may be observed by a microscopical examination and is known as 'slip bands'.

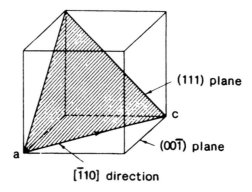

Fig. 12.2 Slip plane and slip direction in face-centered cubic lattice. (From INTRODUCTION TO PHYSICAL METALLURGY, by S. Avner, Second Edition, 1974. Copyright McGraw-Hill Book Co. Inc., New York. Reprinted with permission).

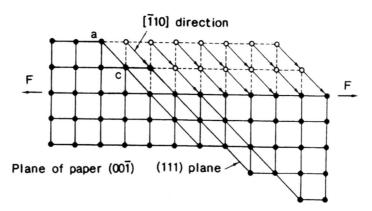

Fig. 12.3 Schematic diagram of slip in a face-centered cubic crystal. (From INTRODUCTION TO PHYSICAL METALLURGY, by S. Avner, Second Edition, 1974. Copyright McGraw-Hill Co. Inc., New York. Reprinted with permission).

12.3 MECHANISM OF SLIP

Modern theory of plastic deformation considers slip as a step by step movement of so-called 'dislocation' within a crystal [2, 3].

Dislocations are linear imperfections in a crystalline array of atoms. It is assumed that the majority of dislocations are formed during the original solidification process. During any subsequent cold-working process the dislocations may reproduce themselves and consequently their number greatly increases.

It is shown in Fig. 12.4a that, by application of the shear force, an extra plane of atoms (dislocation) has been formed above the slip plane. This dislocation moves across the slip plane (Fig. 12.4b,c) and leaves a step when it comes out at the surface of the crystal (Fig. 12.4d). This type of dislocation is known as an **edge dislocation.**

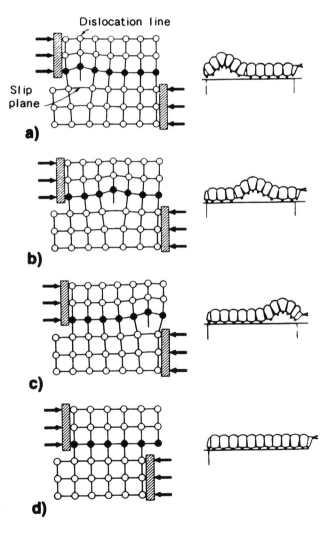

Fig. 12.4 Analogy between the movement of a dislocation through a crystal and the movement of an inchworm as it arches its back while going forward. (From INTRODUCTION TO PHYSICAL METALLURGY, by S. Avner, Second Edition, 1974. Copyright McGraw–Hill Co. Inc., New York. Reprinted with permission).

Slip can also take place by the movement of **screw dislocations**. This differs from edge dislocations in a way that the direction of movement of the dislocation is normal to the direction of the slip step as shown in Fig. 12.5.

When a slip takes place by a combination of a screw dislocation with an edge dislocation, the resultant curved dislocation evolves as shown in Fig. 12.6 [5].

12.4 DEFORMATION BY TWINNING

In certain metals such as zinc, tin and pure iron deformation occurs by a process known as 'twinning' [2, 3].

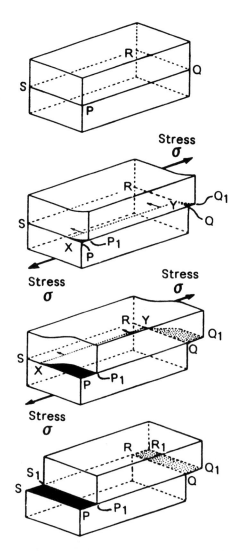

Fig. 12.5 The movement of a screw dislocation. (From ENGINEERING METALLURGY, Part 1, by R.A. Higgins, 1983. Copyright Hodder & Stoughton Ltd., England. Reprinted with permission).

As was mentioned above, in deformation by slip all atoms in one block move the same distance. In deformation by twinning, atoms in each successive plane within a block (Fig. 12.7a) will move different distances (Fig. 12.7b). As a result the direction of the lattice will be altered so that each half of the crystal becomes a mirror image of the other half along a twinning plane (Fig. 12.7c) [6].

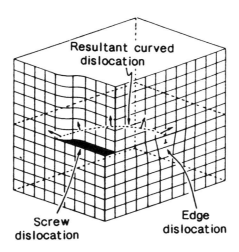

Fig. 12.6 The combination of a screw dislocation with an edge dislocation. (From THE MAKING, SHAPING AND TREATING OF STEEL, by W.T. Lankford, et al, Tenth Edition, 1985. Copyright AISE, Pittsburgh, Pennsylvania. Reprinted with permission.)

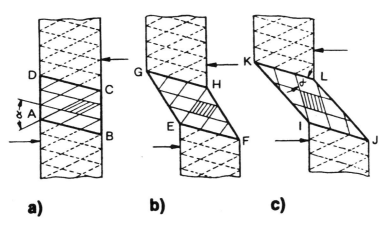

Fig. 12.7 The formation of 'mechanical twins': a) initial state, b) transitional state, and c) final state. (From FUNDAMENTALS OF ROLLING, by Z. Wusatowski. Copyright Wydawnictwo Slask, Katowice, Poland. Reprinted with permission).

12.5 ENERGY OF MECHANICAL DEFORMATION

One of the remarkable features of the plastic deformation is that the stress required to initiate slip is lower than that required to continue deformation on subsequent planes. It is because the dislocations present at the start of stress application move into 'jammed' positions. The material is undergoing **strain hardening** or **work hardening**.

At the point when no further slip by movement of dislocations is possible the material reaches its maximum strength and further increase in stress would give rise to **fracture**.

Energy of mechanical deformation employed during cold-working process consists of the following components:

1. Energy converted to heat as internal forces acting within the metal are overcome (90%)

2. Stored (potential) energy associated with number of new dislocations generated during deformation (9%)

3. Remaining potential energy associated with locked-up residual stresses arising from elastic strains internally balanced (1%).

12.6 GRAIN RESTORATION PROCESS

Restoration of crystalline structure in a cold-worked metal is possible by the heat-treatment process known as **annealing**. This process may proceed in three following phases [2, 3]:

1. Recovery
2. Recrystallization
3. Grain growth.

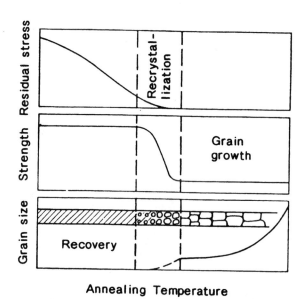

Fig. 12.8 Schematic representation of recovery, recrystallization and grain growth. (From METALS HANDBOOK, 1948. Copyright ASM International, USA. Reprinted with permission).

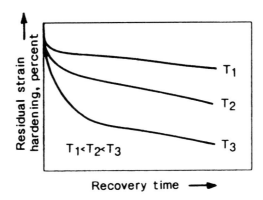

Fig. 12.9 Effect of temperature on recovery. (From INTRODUCTION TO PHYSICAL METALLURGY, by S. Avner, Second Edition, 1974. Copyright McGraw-Hill Book Co. Inc., New York. Reprinted with permission).

Recovery, recrystallization, and grain growth are all processes leading to softening (Fig. 12.8). These three processes all depend on time; they are not instantaneous. The rate at which each of them proceeds increases with increasing temperature, following an exponential law. When the temperature to which a cold worked metal is subjected is sufficiently high, these processes operate with such rapidity as to appear to be instantaneous, but at lower temperatures the time dependence is readily detected. The rates, however, vary widely from metal to metal and depend greatly on composition, purity, grain size of the sample before deformation, and the amount of deformation. Generally, metals with low melting points exhibit high rates at low temperature, as in case of tin. Furthermore, the three processes sometimes overlap and are difficult to distinguish, while in other cases they can be clearly separated [7].

12.7 RECOVERY

Recovery is a mechanism by which some of the crystal imperfections are eliminated or rearranged into new configurations [8].

The principal effect of recovery is the relief of internal stresses accumulated during cold working process [2, 3]. It may be achieved by heating to a relatively low temperature.

The amount of reduction in residual stress that occurs during given time increases with increase in temperature. Also at a given temperature the rate of decrease in residual strain hardening is fastest at the beginning and substantially slows down with time as shown in Fig. 12.9.

12.8 RECRYSTALLIZATION

The recrystallization process begins when annealing temperature reaches certain level. The process commences with formation of new crystals (Fig. 12.10a,b). The new crystals generally appear at the most drastically deformed portions of the grain, usually the grain boundaries and slip planes. The cluster of atoms from which the new grains are formed is called a **nucleus.**

The new crystals are equiaxed in form. These crystals grow at the expense of the old crystals (Fig. 12.10c,d,e) until recrystallization is complete (Fig. 12.10f).

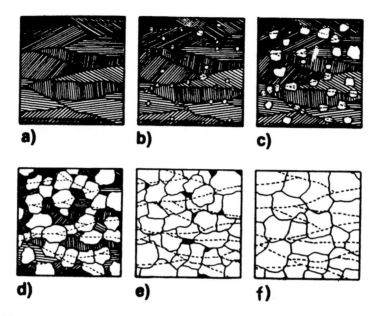

Fig. 12.10 Stages in recrystallization of a metal. (From ENGINEERING METALLURGY, Part 1, by R.A. Higgins, 1983. Copyright Hodder & Stoughton Ltd., England. Reprinted with permission).

The approximate minimum temperature at which complete recrystallization of a cold-worked metal occurs within a specified time is called **recrystallization temperature** [1]. A highly cold-worked low-carbon steel completely recrystallizes in one hour at the temperature approximately 538°C (1000°F) [3].

12.9 GRAIN GROWTH

Two oppositely acting forces determine the character of grain growth. The driving force for the grain growth is associated with the reduction of the amount of grain boundary and therefore with the reduction of free energy in a crystalline body. Opposing this force is the rigidity of the lattice.

If the annealing temperature is maintained above the recrystallization temperature of the metal, the newly formed crystals will continue to grow by absorbing each other.

The following four major factors determine the extent of grain growth [2, 3]:

1. **Annealing temperature** - Grain size increases with increase in temperature as shown in Fig. 12.11.

2. **Duration of annealing process** - At a given temperature the rate of increase in the grain growth is fastest at the beginning and slows down at longer times (Fig. 12.11).

3. **Degree of previous cold work** - In general, increasing the amount of cold deformation gives rise to the production of many nuclei on recrystallization and therefore the grain size will be small. Conversely, light deformation will give rise to a few nuclei and the resulting grain size will be large (Fig. 12.12).

There is a minimum amount of cold-work deformation referred to as **critical** that results in the extremely coarse grain.

4. **Insoluble impurities** - They increase nucleation and act as barriers to the grain growth. The greater the amount and the finer distribution of insoluble impurities the finer the final grain size.

12.10 SUPERPLASTICITY

Superplasticity is the ability of certain metals to undergo unusually large amounts of plastic elongation (up to 2000%) under the action of relatively low tensile stresses [1, 2]. This phenomenon may take place if temperature during deformation is raised above that necessary for recrystallization. In that case the stress required to produce plastic flow falls not only because metals weaken when hot but also due to the recrystallization and the recovery which effectively reduce work-hardening.

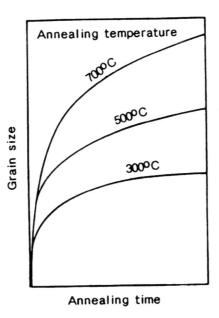

Fig. 12.11 Effect of temperature and time on grain size. (From ENGINEERING METALLURGY, Part 1, by R.A. Higgins, 1983. Copyright Hodder & Stoughton Ltd., England. Reprinted with permission).

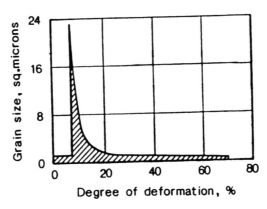

Fig. 12.12 Effect of cold working on grain size developed in a low-carbon steel at 950°C (1742°F). (From METALS HANDBOOK, 1948. Copyright ASM International, USA. Reprinted with permission).

REFERENCES

1. ASM Metals Reference Book, American Society for Metals, Metals Park, Ohio, pp. 1-80, (1981).

2. R.A. Higgins, Engineering Metallurgy, Part I, Applied Physical Metallurgy, Robert E. Krieger Publishing Company, Melbourne, Florida, pp. 78-102, (1983).

3. S.H. Avner, Introduction to Physical Metallurgy, Second Edition, McGraw-Hill Book Company, New York, pp. 107-137, (1974).

4. G.E. Doan, E.M. Mahla, Principles of Physical Metallurgy, McGraw-Hill Book Company, New York, (1941).

5. The Making, Shaping and Treating of Steel, 10th Edition, eds. W.T. Lankford, Jr., et al, Association of Iron and Steel Engineers, Pittsburgh, Pennsylvania, p. 774 (1985).

6. Z. Wusatowski, Fundamentals of Rolling, Pergamon Press, Oxford, p. 5 (1969).

7. Metals Handbook, American Society for Metals, Metals Park, Ohio, pp. 260-262 (1948).

8. G. Krauss, Principles of Heat Treatment of Steel, American Society for Metals, Metals Park, Ohio, pp. 115-118 (1980).

13

Principles of Macroscopic Plasticity

13.1 MACROSCOPIC NATURE OF PLASTICITY

The **macroscopic** viewpoint deals with a phenomenological explanation of plasticity that is based on observations of plastic deformation of a polycrystalline metal. According to this viewpoint the metal is considered as a continuum and its plastic behavior can be described with such parameters as density, stress and velocity at all points within its boundaries [1-4].

13.2 STRESSES IN UNIAXIAL DEFORMATION

Let us consider stresses in a prismatic bar submitted to an axial tensile force P [2]. The stress over cross-section **m-m** (Fig. 13.1a), perpendicular to the axis of the bar, will be equal to

$$\sigma_x = \frac{P}{A}, \tag{13-1}$$

where A = area of the cross-section normal to the axis of the bar.

The stress over cross-section **m-m** (Fig. 13.1b), perpendicular to the plane of the figure and inclined to the axis, will be defined as a ratio of force P over area of the cross-section **m-m**, i.e.

$$s = \frac{P\cos\phi}{A} = \sigma_x\cos\phi \tag{13-2}$$

The stress component σ_ϕ, perpendicular to the cross-section, is called the **normal stress**. Its magnitude is:

$$\sigma_\phi = s\cos\phi = \sigma_x\cos^2\phi \tag{13-3}$$

In the plane **n-n** (Fig. 13.1c), perpendicular to cross-section **m-m**, normal stress will be equal to

$$\sigma_\phi' = \sigma_x\cos^2(\pi/2 + \phi) = \sigma_x\sin^2\phi \tag{13-4}$$

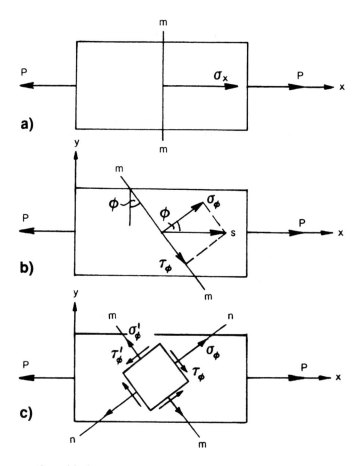

Fig. 13.1 Stresses in uniaxial deformation.

Comparing this result with the Eq. (13-3), we find

$$\sigma_\phi + \sigma'_\phi = \sigma_x\cos^2\phi + \sigma_x\sin^2\phi = \sigma_x \tag{13-5}$$

Thus, the sum of the normal stresses, acting on two perpendicular planes, remains constant and equal to σ_x.

From Eqs. (13-3) and (13-4), it is seen that when $\phi = 0$, normal stress reaches its maximum value

$$\sigma_{max} = \sigma_x, \tag{13-6}$$

whereas σ_ϕ is equal to zero.

In order to apply Eqs. (13-1)-(13-6) to the case of axial compression, the sign of P has to be changed, i.e., P = -P.

The tangential component τ_ϕ (Fig. 13.1b) is called **shear stress** and has the value:

$$\tau_\phi = s\sin\phi = \frac{\sigma_x}{2}\sin2\phi \tag{13-7}$$

In the plane **n-n** (Fig. 13.1c), perpendicular to the cross-section m-m, shear stress will have the value:

$$\tau_\phi' = \frac{\sigma_x}{2}\sin2(\pi/2 + \phi) = -\frac{\sigma_x}{2}\sin2\phi \qquad (13-8)$$

This indicates that the shear stress acting on two perpendicular planes have the same absolute value but opposite sign, i.e.

$$\tau_\phi = -\tau_\phi' \qquad (13-9)$$

From Eqs. (13-7) and (13-8), it is seen that when $\phi = 45^o$, shear stresses reach their maximum value:

$$\tau_{max} = \frac{\sigma_x}{2} \qquad (13-10)$$

13.3 MOHR'S CIRCLE FOR AXIAL DEFORMATION

Equations (13-3) and (13-7) can be represented graphically with Mohr's circle [2], shown in Fig. 13.2. An orthogonal system of coordinates is originated at point O with normal stress σ along horizontal axis and shear stresses τ along vertical axis. A shear stress that induces clockwise rotation of the stress element is plotted, as if it were positive, while one causing counterclockwise rotation is plotted, as if it were negative.

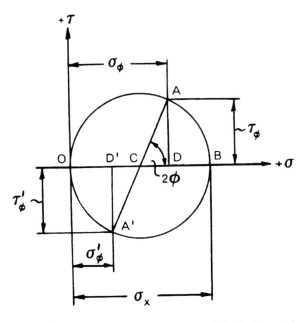

Fig. 13.2 Mohr's circle for uniaxial deformation.

To determine the stress components on an inclined plane, defined by an angle ϕ in Fig. 13.1, the circle has to be constructed on OB = σ_x as a diameter (Fig. 13.2) and the radius CA has to be drawn making the angle ACB equal to 2ϕ.

Then from simple trigonometrical considerations, we obtain that OD = σ_ϕ, OD' = σ_ϕ', AD = τ_ϕ, and A'D' = τ_ϕ'.

Thus, the coordinates of point A define both normal stress σ_ϕ and shear stress τ_ϕ, acting on the plane **m–m**. Similarly, the coordinates of point A' define normal stress σ_ϕ and shear stress σ_ϕ', acting on the plane **n–n**.

13.4 STRESSES IN BIAXIAL DEFORMATION

Now we consider the case when two mutually perpendicular tensile stresses σ_x and σ_y and shear stresses τ_{xy}, produced by external forces (Fig. 13.3), act in the plane of the figure.

From equilibrium consideration, it can be shown that the normal and shear stress components, σ_ϕ and τ_ϕ, vary with orientation in x-y plane as follows:

$$\sigma_\phi = \sigma_x \cos^2\phi + \sigma_y \sin^2\phi + \tau_{xy} \sin^2 2\phi \qquad (13\text{-}11)$$

$$\sigma_\phi' = \sigma_x \sin^2\phi + \sigma_y \cos^2\phi - \tau_{xy} \sin^2 2\phi \qquad (13\text{-}12)$$

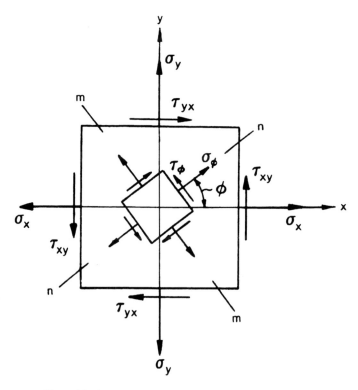

Fig. 13.3 Stresses in biaxial deformation.

$$\tau_\phi = \frac{1}{2}(\sigma_x - \sigma_y)\sin 2\phi + \tau_{xy}\cos 2\phi \tag{13-13}$$

$$\tau'_\phi = -\tau \tag{13-14}$$

Shear stress τ_ϕ is equal to zero on the plane located at the angle $\phi = \phi_0$, which is given by

$$\tan 2\phi_0 = \frac{2\tau_{xy}}{\sigma_x - \sigma_y} \tag{13-15}$$

If $\phi = \phi_0$, then the normal stresses σ_ϕ and σ'_ϕ reach their limiting values σ_1 and σ_2 respectively which are known as **principal stresses** and are equal to

$$\sigma_1 = \frac{\sigma_x + \sigma_y}{2} + \tau_{mxy} \tag{13-16}$$

$$\sigma_2 = \frac{\sigma_x + \sigma_y}{2} - \tau_{mxy}, \tag{13-17}$$

where τ_{mxy} = maximum shear stress in the x-y plane

$$\tau_{mxy} = \frac{1}{2}[(\sigma_x - \sigma_y)^2 + 4\tau_{xy}^2]^{1/2} \tag{13-18}$$

Depending upon the magnitude and sign of the applied stresses, an even larger shear stress, τ_{max}, can exist on other planes.

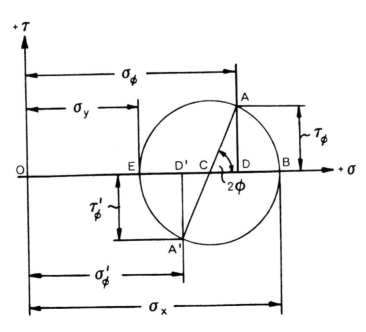

Fig. 13.4 Mohr's circle for biaxial deformation without external shear stress.

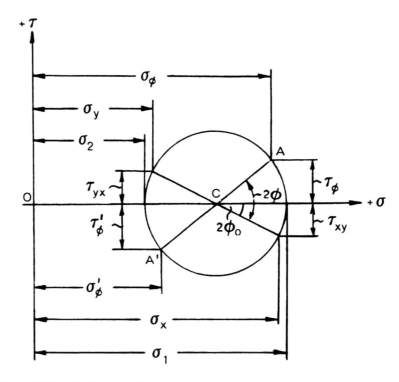

Fig. 13.5 Mohr's circle for biaxial deformation with external shear stress.

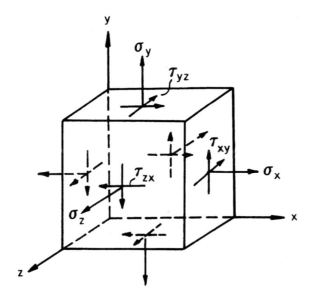

Fig. 13.6 Stresses in triaxial deformation.

13.5 MOHR'S CIRCLE FOR BIAXIAL DEFORMATION

Let us first consider the simplest case when $\tau_{xy} = 0$. Then Eqs. (13-11)-(13-13) can be reduced to the forms:

$$\sigma_\phi = \sigma_x\cos^2\phi + \sigma_y\sin^2\phi \qquad (13\text{-}19)$$

$$\sigma_\phi = \sigma_x\sin^2\phi + \sigma_y\cos^2\phi \qquad (13\text{-}20)$$

$$\tau_\phi = \frac{1}{2}(\sigma_x - \sigma_y)\sin 2\phi \qquad (13\text{-}21)$$

Equations (13-19)-(13-21) can be represented graphically with Mohr's circle shown in Fig. 13.4. Lines OB and OE define respectively the stresses σ_x and σ_y, acting on the side of the element in Fig. 13.3. To obtain the stress components on any inclined plane, defined by an angle ϕ in Fig. 13.3, the circle has to be constructed on EB as a diameter (Fig. 13.4) and the radius AC has to be drawn making the angle ACE equal to 2ϕ. Then from simple trigonometrical considerations we conclude that OD $= \sigma_\phi$, OD' $= \sigma_\phi'$, AD $= \tau_\phi$ and A'D' $= \tau_\phi'$.

Thus, the coordinates of point A define total normal stress σ_ϕ and shear stress τ_ϕ, acting on the plane **m-m**. Similarly, the coordinates of point A' define total normal stress σ_ϕ' and shear stress τ_ϕ', acting on the plane **n-n**.

It follows from Eqs. (13-16)-(13-18) that, if $\tau_{xy} = 0$, then $\sigma_x = \sigma_1$ and $\sigma_y = \sigma_2$; i.e., σ_x and σ_y are the principal stresses.

Figure 13.5 illustrates a Mohr's circle for the case when $\tau_{xy} = 0$. Comparing Fig. 13.5 with Fig. 13.4, it is seen that in both figures the center of the circle, point C, is located at the distance OC $= (\sigma_x + \sigma_y)/2$. However, the initial plane for the angle 2ϕ in Fig. 13.5 is no longer coincident with σ direction as it is in Fig. 13.4, but is at the angle $2\phi_0$ with this direction. The radius of the circle CA $= \tau_{mxy}$.

If $\phi = \phi_0$, then $\sigma_\phi = \sigma_1$ and $\sigma_\phi' = \sigma_2$ as expressed by Eqs. (13-16) and (13-17).

13.6 STRESSES AND STRAINS IN TRIAXIAL DEFORMATION

Figure 13.6 illustrates an elemental cube of metal subjected to a system of external normal and shear stresses, acting on the faces of the cube. These stresses produce in the metal principal stresses σ_1, σ_2 and σ_3. The triaxial deformation may be considered as a general case, whereas both uniaxial and biaxial deformations can be viewed as the particular cases of the triaxial deformation when one or two principal stresses are equal to zero, as shown in Fig. 13.7.

The strains corresponding to the principal stresses σ_1, σ_2 and σ_3 may be calculated through the use of **generalized Hooke's law** [2-5]:

$$\begin{aligned}
\epsilon_1 &= [\sigma_1 - \nu(\sigma_2 + \sigma_3)]/E \\
\epsilon_2 &= [\sigma_2 - \nu(\sigma_1 + \sigma_3)]/E \\
\epsilon_3 &= [\sigma_3 - \nu(\sigma_2 + \sigma_1)]/E
\end{aligned} \qquad (13\text{-}22)$$

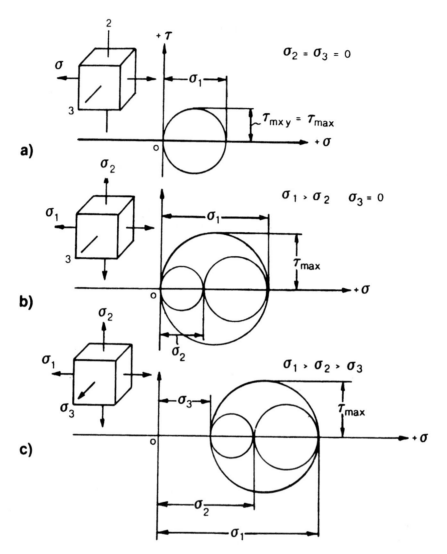

Fig. 13.7 Mohr's circle for three types of deformation: a) uniaxial, b) biaxial, and c) triaxial.

Prior to reaching the elastic limit, the Poisson's ratio ν is equal to 0.3. However, after the yield point is exceeded, the volume of metal remains constant, i.e.

$$\epsilon_1 + \epsilon_2 + \epsilon_3 = 0 \qquad\qquad (13-23)$$

After substituting Eq. (13-22) into Eq. (13-23), we obtain that at the yielding the Poisson's ratio ν is equal to 0.5.

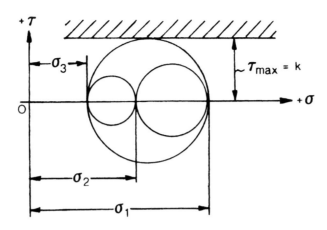

Fig. 13.8 Yield criteria in triaxial deformation.

13.7 YIELD CRITERIA

The yield criteria is a mathematical presentation of the states of stress, that will induce yielding or the onset of plastic deformation. In the most general form (Fig. 13.6), it is expressed as [3]:

$$f(\sigma_x, \sigma_y, \sigma_z, \tau_{xy}, \tau_{yz}, \tau_{zx}) = C_1 \qquad (13\text{-}24)$$

In terms of principal stresses, it reduces to the form:

$$f(\sigma_1, \sigma_2, \sigma_3) = C_2, \qquad (13\text{-}25)$$

where C_1, C_2 = constants.

Since yielding occurs when maximum shear stress τ_{max} reaches the value of the yield shear stress k (Fig. 13.8), the constants C_1 and C_2 are usually expressed as a function of k, or yield stress Y, that is determined in a simple tension test.

Below is a brief description of two mostly used yield criteria. In their derivation it is assumed that the material is continuous, homogeneous and isotropic, i.e., it has the same properties in all directions. Also, the tensile stress is positive and compressive stress is negative, and the yield stress in tension and compression are essentially equal [4].

Maximum-shear-stress criterion - It is also known as the Tresca criterion and can be expressed by

$$\sigma_{max} - \sigma_{min} = \sigma_1 - \sigma_3 = 2k = Y \qquad \text{if } \sigma_1 > \sigma_2 > \sigma_3 \qquad (13\text{-}26)$$

where $\sigma_{max}, \sigma_{min}$ = maximum and minimum principal applied stresses respectively

k = shear yield stress.

Distortion-energy criterion - It is also known as the von Mises criterion and is given by

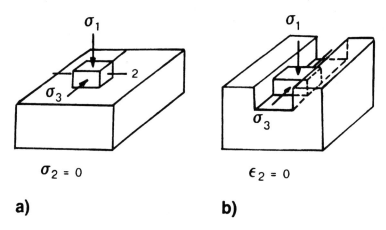

Fig. 13.9 Two states of stress: a) plane stress and b) plane strain.

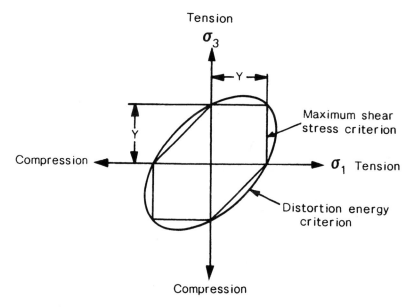

Fig. 13.10 Plane stress diagrams for maximum-shear-stress and distortion energy criteria. (From Serope Kalpakjian, MANUFACTURING PROCESSES FOR ENGINEERING MATERIALS, Copyright 1985 Addison-Wesley Publishing Co., Inc., Reading, Massachusetts. Reprinted with permission.)

$$(\sigma_1 - \sigma_2)^2 + (\sigma_2 - \sigma_3)^2 + (\sigma_3 - \sigma_1)^2 = 2Y^2, \tag{13-27}$$

This criterion considers not only maximum and minimum principal stresses as the maximum-shear-stress criterion does, but also the intermediate principal stresses.

In a more general form, this criterion may be written as:

$$(\sigma_x - \sigma_y)^2 + (\sigma_y - \sigma_z)^2 + (\sigma_z - \sigma_x)^2 + 6(\tau_{xy}^2 + \tau_{yx}^2 + \tau_{zx}^2) = 2Y^2 \tag{13-28}$$

13.8 PLANE STRESS AND PLANE STRAIN

These two states of stress are important in the application to the plastic deformation by rolling.

Plane stress is the state of stress in which one or two of the pairs of faces on an elemental cube are free of any stresses as shown in Figure 13.9a [3, 4].

Figure 13.10 illustrates two yield envelopes. The stress applied to the elemental cube should fall on the outside of these envelopes to cause yielding. The envelope of straight lines is obtained from the maximum-shear-stress criterion (Eq. 13-26). The second envelope is derived from the distortion energy criterion (Eq. 13-27), which for plane stress ($\sigma_2 = 0$) reduces to

$$(\sigma_1)^2 + (\sigma_3)^2 - \sigma_1\sigma_3 = Y^2 \tag{13-29}$$

Plane strain is the state of stress in which one of the pairs of faces on an elemental cube undergoes zero strain as shown in Fig. 13.9b.

It follows from Eq. (13-22), for plane strain ($\epsilon_2 = 0$) in yielding state ($\nu = 0.5$) that

$$\sigma_2 = (\sigma_1 + \sigma_3)/2 \tag{13-30}$$

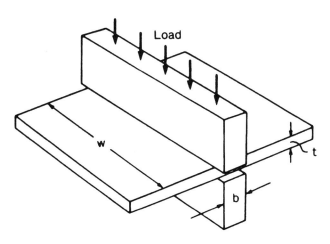

Fig. 13.11 Schematic presentation of compression test. (From William F. Hosford/Robert M. Caddell, METAL FORMING: Mechanics and Metallurgy, Copyright 1983, Reprinted by permission of Prentice-Hall, Inc., Englewood Cliffs, NJ.)

After substituting Eq. (13-30) into Eq. (13-27), we obtain that for the plane strain compression (Fig. 13.9b), the distortion-energy criterion reduces to

$$\sigma_1 - \sigma_3 = (2/\sqrt{3})Y = 1.15Y = S,$$ (13-31)

where S = constrained yield stress for the plane strain compression.

13.9 PLAIN STRAIN COMPRESSION TEST

Figure 13.11 illustrates schematically the plane strain compression test, which simulates processes such as rolling [3, 4]. In this test the following relationships between dimensions of a workpiece and an indenter are recommended: $w/b > 6$ and $2 < b/t < 4$. As a load is applied, the metal between the indenters is constrained from moving in the w direction by the unstressed material adjacent to the indenter region. Thus, $\epsilon_y = \epsilon_w = 0$.

The best results are obtained when incremental loading is employed with 2 to 5% strain per increment. Furthermore, caution should be taken in preparing the indenter surfaces, lubricating the contacting surfaces, etc. It was found that for ductile materials, the true stress-strain curves obtained from the compression test and from tension test coincide. However, it is not true in respect to brittle material.

13.10 EFFECTIVE STRESS AND EFFECTIVE STRAIN

Effective stress $\bar{\sigma}$ and effective strain $\bar{\epsilon}$ are convenient parameters for expressing the state of stress on an element. If the magnitude of the effective strain $\bar{\epsilon}$ reaches a critical value, then the applied stress state will cause yielding [3, 4].

For the maximum-shear-stress criterion

$$\bar{\sigma} = \sigma_1 - \sigma_3$$ (13-32)

and for the distortion-energy criterion

$$\bar{\sigma} = (1/\sqrt{2})[(\sigma_1 - \sigma_2)^2 + (\sigma_2 - \sigma_3)^2 + (\sigma_3 - \sigma_1)^2]^{1/2}$$ (13-33)

When $\bar{\sigma} = Y$, either criterion predicts yielding.

For the maximum-shear-stress criterion, the effective strain is equal to

$$\bar{\epsilon} = (2/3)(\epsilon_1 - \epsilon_3)$$ (13-34)

and for the distortion-energy criterion, it is equal to

$$\bar{\epsilon} = (\sqrt{2}/3)[(\epsilon_1 - \epsilon_2)^2 + (\epsilon_2 - \epsilon_3)^2 + (\epsilon_3 - \epsilon_1)^2]^{1/2}$$ (13-35)

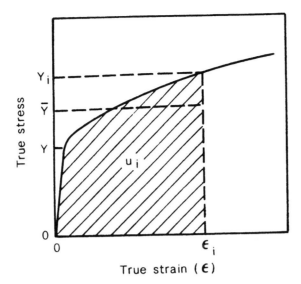

Fig. 13.12 True stress-true strain curve. (From Serope Kalpakjian, MANUFACTURING PROCESSES FOR ENGINEERING MATERIALS, Copyright 1985 Addison-Wesley Publishing Co., Inc., Reading, Massachusetts. Reprinted with permission).

Thus, by using the effective stress and effective strain, the stress-strain curve for triaxial state of strain may be replaced with the stress-strain curve for uniaxial state of strain.

13.11 IDEAL WORK OF DEFORMATION

The work of deformation is considered **ideal** when a homogeneous material is deformed uniformly, internally and externally, and also there is no effects of friction or tool geometry on the deformation process.

The ideal work per unit volume or specific energy can be presented in terms of principal components as

$$d_{ui} = \sigma_x d\epsilon_x + \sigma_y d\epsilon_y + \sigma_z d\epsilon_z \tag{13-36}$$

For a material subjected to uniaxial stresses and strains, the ideal work per unit volume for any strain ϵ_x is equal to the area under the true stress-true strain curve (Fig. 13.12). It is expressed as [4]

$$u_i = \int_0^{\epsilon_x} \sigma\, d\epsilon = \overline{Y}\epsilon_x, \tag{13-37}$$

where $\overline{Y}$ = average flow stress of the material.

For a material subjected to triaxial stresses and strains, the ideal work may be expressed in terms of effective stress and effective strain as follows

$$u_i = \int_0^{\overline{\epsilon}_x} \sigma\, d\epsilon \tag{13-38}$$

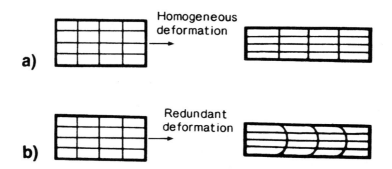

Fig. 13.13 Comparison of ideal and actual deformation to illustrate the meaning of redundant deformation. (From William F. Hosford/Robert M. Caddell, METAL FORMING: Mechanics and Metallurgy, Copyright 1983, Reprinted by permission of Prentice-Hall, Inc., Englewood Cliffs, NJ.)

13.12 TOTAL WORK OF DEFORMATION

The total work of deformation per unit of volume is equal to [3, 4]

$$u_t = u_i + u_f + u_r, \tag{13-39}$$

where u_f = frictional work per unit of volume

u_r = redundant work per unit of volume.

The frictional work is consumed at the interface between the deforming metal and tool faces that constrain the metal.

The redundant work is due to internal distortion in excess of that needed to produce the desired shape. It is affected by such factors as the tool and workpiece geometry and lubrication at the toll-workpiece interface.

Figure 13.13a illustrates an ideal deformation process during which the plane sections remain plane. In reality, internal shearing causes distortion of plane sections (Fig. 13.13b). As a consequence, the metal experiences a strain which is greater than in the case of its ideal deformation.

13.13 TEMPERATURE RISE IN PLASTIC WORKING

During plastic working the mechanical work of deformation is converted into:

a) heat

b) elastic energy stored within the deformed material.

Heat represents the largest component of the converted energy. Stored energy is generally 5 to 10% of the total energy input. For some alloys it can be as high as 30% [4].

Elastic energy stored within the deformed material is due to the fact that a dislocation distorts and strains the crystal lattice. In ideal frictional process when all energy is completely converted into heat, the temperature rise may be defined as

$$T = u_t/(\rho c), \tag{13-40}$$

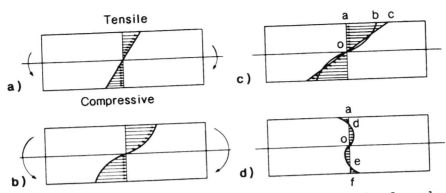

Fig. 13.14 Residual stresses developed in bending a beam made of an elastic, strain-hardening material. (From Serope Kalpakjian, MANUFACTURING PROCESSES FOR ENGINEERING MATERIALS, Copyright 1985 Addison-Wesley Publishing Co., Inc., Reading, Massachusetts. Reprinted with permission).

where u_t = total energy from Eq. (13-39)

ρ, c = density and specific heat of the material respectively.

The theoretical temperature rise during compression of specimens with strain $\epsilon = \ln(L_0/L) = 1$ has been calculated as follows:

Aluminum	75°C (165°F)
Copper	140°C (285°F)
Low-carbon steel	280°C (535°F)
Titanium	570°C (1060°F)

13.14 RESIDUAL STRESSES

Residual stresses are usually defined as the stresses that are present in a body that is free of external forces or thermal gradients [6]. These stresses are the result of inhomogeneous deformation such as the bending of a beam as shown in Fig. 13.14 [4].

When stresses in all fibers of the beam are within elastic range the stress distribution within the beam is linear (Fig. 13.14a). As the moment is increased, the outer fibers of the beam begin to yield and the non-linear stress distribution is obtained (Fig. 13.14b). Unloading may be considered as an equivalent to applying an equal and opposite moment to the beam (Fig.13.14c). The resulting stresses that remain in the beam after unloading are the residual stresses (Fig. 13.14d).

Residual stresses can also be caused by **temperature gradients** within a body. They are also caused by **phase changes** in metals during or after processing. The residual stresses may produce warping, stress cracking, and change in lubricant chemical reactivity as defined below.

Warping - This is a distortion of parts caused by the disturbances of the residual stresses. These disturbances may be either due to relaxation of these stresses over period of time or due to cutting or slitting the parts.

Stress cracking or **stress-corrosion cracking** - This may be a result of tensile residual stresses in the metal.

Change in lubricant chemical reactivity - Surfaces with tensile residual stresses are more reactive than those with compressive stresses.

REFERENCES

1. The Making, Shaping and Treating of Steel, 10th Edition, eds. W.T. Lankford, Jr., et al, Association of Iron and Steel Engineers, Pittsburgh, Pennsylvania, pp. 773-777 (1985).

2. S. Timoshenko, Strength of Materials, Part I, Third Edition, D. Van Nostrand Company, Princeton, pp. 37-61 (1968).

3. W.F. Hosford and R.M. Caddell, Metal Forming: Mechanics and Metallurgy, Prentice-Hall, Inc., Englewood Cliffs, New Jersey, pp. 1-249 (1983).

4. S. Kalpakjian, Manufacturing Processes for Engineering Materials, Addison-Wesley Publishing Company, Inc., Reading, Massachusetts, pp. 71-96, 226 (1984).

5. Z. Wusatowski, Fundamentals of Rolling, Pergamon Press, Oxford, pp. 1-68 (1969).

6. ASM Metals Reference Book, American Society for Metals, Metals Park, Ohio, p. 59 (1981).

14
Slab Analysis of Plastic Deformation

14.1 BASIC PRINCIPLE

This method is also known as the **free-body-equilibrium** approach [1, 2]. It requires the selection of an element in the workpiece and identifying all normal and frictional forces acting on this element. The differential equation is then derived where its variations are considered in one direction only. A solution is provided by an integration of this equation using appropriate boundary conditions.

The following main assumptions are utilized in this method:

1. The direction of the applied load and planes perpendicular to this direction define principal directions.

2. The principal stresses do not vary on these planes.

3. The coefficient of the external friction at the tool-workpiece interfaces is constant at all points of the arc contact.

4. The frictional forces do not produce the internal distortion of the metal and do not change the orientation of principal directions.

5. Plane vertical sections remain plane; thus the deformation is **homogeneous** in regard to determination of induced strain.

6. Elastic deformation of the workpiece is negligible in comparison with the plastic deformation.

7. There is no elastic deformation of the tool in the contact zone.

8. The compressive strength is constant throughout the contact length.

14.2 PLANE-STRAIN COMPRESSION WITH SLIDING FRICTION

Figure 14.1 illustrates a simple compression process with friction. The purpose of this analysis is to determine the pressure distribution at the die-workpiece interfaces. This would then allow one to calculate the total load required for compression [1, 2].

The force balance on the slab in the x direction may be expressed as

$$(\sigma_x + d\sigma_x)h + 2\mu\sigma_y dx - \sigma_x h = 0 \qquad (14\text{-}1)$$

or

$$d\sigma_x + (2\mu\sigma_y/h)dx = 0, \tag{14-2}$$

where μ = coefficient of friction

   $\mu\sigma_y$ = frictional force.

Since σ_x and σ_y are considered as principal stresses, therefore, the distortion-energy criterion for plane strain compression expressed by Eq. (13-31) can be applied

$$\sigma_y - \sigma_x = 1.15Y = S \tag{14-3}$$

In strict terms, Eq. (14-3) is valid when the shear stresses acting at the interfaces are small in comparison to principal stresses which is true with low coefficient of friction.

Differentiation of both sides of Eq. (14-3) gives

$$d\sigma_y = d\sigma_x \tag{14-4}$$

After substituting Eq. (14-4) into Eq. (14-2), we obtain

$$(1/\sigma_y)d\sigma_y = - (2\mu/h)dx \tag{14-5}$$

General solution of this differential equation is obtained by integration of both sides

$$\sigma_y = C\cdot\exp(-2\mu x/h) \tag{14-6}$$

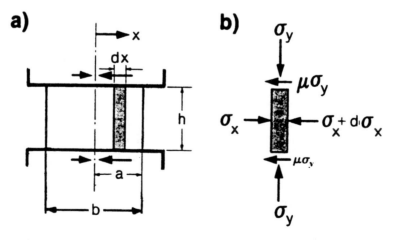

Fig. 14.1 Stresses on an element in plane-strain compression. (From Serope Kalpakjian, MANUFACTURING PROCESSES FOR ENGINEERING MATERIALS, Copyright 1985 Addison-Wesley Publishing Co., Inc., Reading, Massachusetts. Reprinted with permission).

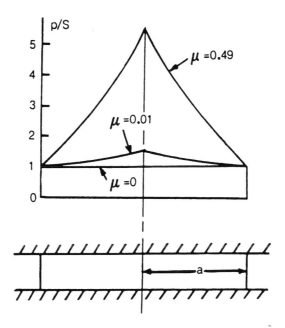

Fig. 14.2 Friction hill in plane-strain compression for different values of coefficient of friction. (From William F. Hosford/Robert M. Caddell, METAL FORMING: Mechanics and Metallurgy, Copyright 1983, Reprinted by permission of Prentice-Hall, Inc., Englewood Cliffs, NJ.)

Constant of integration C may be determined from the boundary conditions at the edges of workpiece such as

$$\text{at } x = a, \quad \sigma_x = 0, \quad \sigma_y = S \tag{14-7}$$

Thus, the normal pressure distribution at the interfaces is equal to

$$p = \sigma_y = S \cdot \exp[2\mu(a - x)/h] \tag{14-8}$$

A plot of p/S versus x is shown in Fig. 14.2, where the rise in p towards the centerline is known as the **friction hill**.

The maximum value for p occurs at the centerline (x = 0), where

$$\frac{p_m}{S} = \exp\left(\frac{\mu b}{h}\right) \tag{14-9}$$

The average pressure may be expressed as

$$\frac{p_a}{S} = \frac{h}{\mu b} \exp\left(\frac{\mu b}{h} - 1\right) \tag{14-10}$$

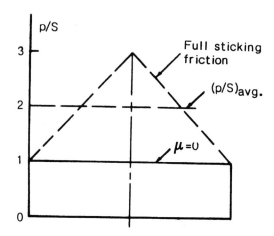

Fig. 14.3 The friction hill in plane-strain compression for sticking friction. (From William F. Hosford/Robert M. Caddell, METAL FORMING: Mechanics and Metallurgy, Copyright 1983, Reprinted by permission of Prentice-Hall, Inc., Englewood Cliffs, NJ.)

14.3 PLANE-STRAIN COMPRESSION WITH STICKING FRICTION

There is a limit at which sliding friction can exist at the tool-workpiece interface. Indeed, in order to avoid shearing at the interface the following condition must be met [1]:

$$\mu\sigma_y < S/2, \tag{14-11}$$

Therefore, the sliding takes place when $\sigma_y > S$, $\mu < 0.5$.

If the shear limit is reached, then interfacial shear of the workpiece occurs and the frictional forces indicated as $\mu\sigma_y$ in Fig. 14.1b have to be replaced by S/2.

Following the analysis that produced Eq. (14-8), the result for sticking friction is

$$p = \sigma_y = S[1 + (a - x)/h] \tag{14-12}$$

which is plotted in Fig. 14.3 for the case when a = 2h.

The maximum value for p occurs at the centerline (x = 0), where

$$\frac{p_m}{S} = 1 + \frac{a}{h} \tag{14-13}$$

The average pressure can be expressed as

$$\frac{p_a}{S} = 1 + \frac{a}{2h} \tag{14-14}$$

14.4 STANDARD TERMS USED IN THEORIES OF FLAT ROLLING

The following standard terms are commonly used in theories of flat rolling:

Average workpiece thickness - It is equal to

$$h_a = (h_1 + h_2)/2, \qquad (14-15)$$

where h_1, h_2 = entry and exit thicknesses of the workpiece respectively.

Mean workpiece thickness - It is given by

$$h_m = (h_1 h_2)^{1/2} \qquad (14-16)$$

Draft - It is expressed as

$$\Delta = h_1 - h_2 \qquad (14-17)$$

Relative reduction - It is given by

$$r = \Delta/h_1 = 1 - h_2/h_1 \qquad (14-18)$$

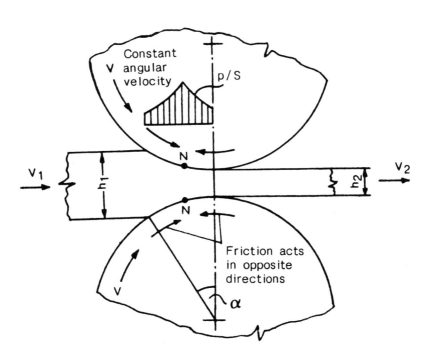

Fig.14.4 Schematic presentation of the deformation zone in flat rolling.

Elongation - It is expressed as

$$e = h_1/h_2 \tag{14-19}$$

Roll bite angle - It is the angle that is formed between two lines crossing at the roll center whereas the first line passes through the entry point and the second line passes through the exit point of the deformation zone (Fig. 14.4). When the roll flattening can be neglected, the roll bite angle is expressed by

$$\alpha = \arccos[1 - \Delta/(2R)] \tag{14-20}$$

Roll contact length - This is a horizontal projection of the arc of contact as shown in Fig. 14.5. When the roll flattening is negligible, it can be presented as

$$L = (R\Delta - \frac{\Delta^2}{4})^{1/2}, \tag{14-21}$$

where R = roll radius.

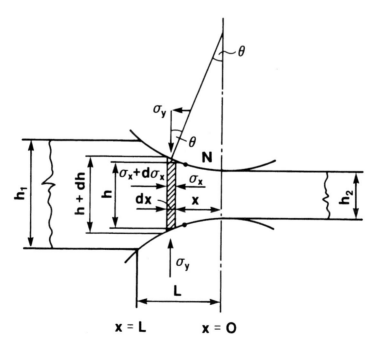

Fig. 14.5 Schematic presentation of the deformation zone in flat rolling.

14.5 FRICTION HILL EFFECT IN FLAT ROLLING

The effect of the deformation zone geometry and friction on flat rolling can be understood in terms of the friction-hill effect [1]. At some point N, in the roll gap (Fig. 14.4), the surface velocities of the rolls and workpiece are equal. This point is known as a **neutral point.**

To the left of this point the surface velocity of the metal V_1 is lower than roll velocity V. This difference in speeds produces friction between the metal and the rolls which tends to draw the metal into the roll gap. To the right of N the metal velocity V_2 is higher than the roll velocity V, so the friction tends to retain the metal in the roll bite.

Because of this roll-metal velocity relationship, the produced friction hill will be somewhat different in comparison with that for the plane-strain compression where both the workpiece and the tool are stationary.

14.6 PLASTIC DEFORMATION IN FLAT ROLLING

In this analysis the following assumptions are made in addition to those described in section 14.1 in respect to the plane-strain compression [3]:

a) the workpiece does not spread laterally

b) there is no roll flattening in the arc of contact

c) the peripheral velocity of the rolls is constant

d) material does not undergo work-hardening during its passage between the rolls

e) the compression rate from point along the arc of contact does not have any effect on the magnitude of the compression strength

f) the vertical component of frictional force is negligible.

Figure 14.5 illustrates stresses acting upon an elemental vertical section of the flat workpiece between the rolls. The horizontal forces acting on vertical faces of the section dx produce compressive stresses $\sigma_x + d\sigma_x$ acting on the face of the section of height h + dh and compressive stresses σ_x acting on the face of height h.

The equilibrium of the horizontal forces acting on section dx may be expressed as

$$(\sigma_x + d\sigma_x)(h + dh) \pm 2\mu\sigma_y dx - 2\sigma_y \tan\theta dx - \sigma_x h_x = 0 \tag{14-22}$$

or

$$2\sigma_y(\tan\theta \mp \mu)dx = d(h\sigma_x) \tag{14-23}$$

where sign (-) before μ corresponds to the sections dx located between the entry and neutral planes and sign (+) corresponds to the sections dx located between the neutral and exit planes.

Since,

$$\tan\theta = \frac{1}{2}\frac{dh}{dx}, \tag{14-24}$$

Eq. (14-23) can be rewritten as

$$\sigma_y \frac{dh}{dx} \stackrel{-}{+} 2\mu\sigma_y = \frac{d(h\sigma_x)}{dx} \tag{14-25}$$

Taking into account the yield criterion given by Eq. (14-3) and that $p = \sigma_y$ we produce the differential equation (14-25) in its final form

$$p\frac{dh}{dx} \stackrel{-}{+} 2\mu p = \frac{d[h(p - S)]}{dx} \tag{14-26}$$

14.7 THEORIES OF HOMOGENEOUS DEFORMATION IN FLAT ROLLING

The differential Eq. (14-26), expressed as above, or in one of its alternative forms, represents the starting point in the analysis known as the **theory of homogeneous deformation** which was first introduced by Von Karman in 1925 [6].

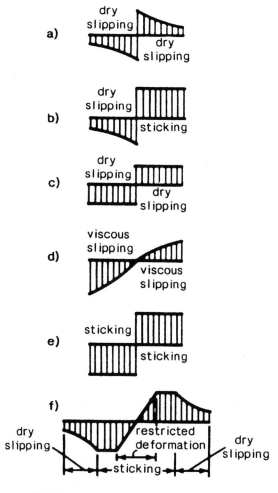

Fig. 14.6 Types of frictional force and its distribution along the arc of contact according to different theories of rolling.

A number of solutions for this equation have been proposed [3-13] as will be described below. These solutions differ mainly in respect to the assumed nature of the frictional forces in the contact zone as shown in Fig. 14.6.

Von Karman's solution [6] is based on the assumption that the **dry slipping** would occur over the whole arc of the contact between the rolls and the rolled material (Fig.14.6a). Also that the frictional force is directly proportional to the value of local normal pressure, i.e.,

$$\tau = \mu p \qquad\qquad\qquad (14\text{-}27)$$

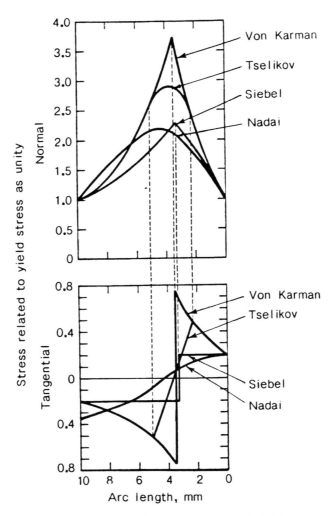

Fig. 14.7 Distribution of normal pressure and frictional force along arc of contact is quite different for the four theories shown. Rolling of wide strip: $h_1 = 2$ mm, $h_2 = 1$ mm, $R = 100$ mm, $\mu = 0.2$. (From Waziri, AISE Year Book, 1963. Copyright AISE, Pittsburgh, Pennsylvania. Reprinted with permission).

Ekelund's solution [7] is given with the assumption that the **dry slipping** would occur over the whole entry side, and **sticking** over the whole exit side of the arc of contact (Fig. 14.6b).

Siebel's solution [8] is obtained for the case when the **dry slipping** occurs over the whole arc of contact between the rolls and the rolled material (Fig. 14.6c). Besides that it is assumed that the frictional force is constant along the arc of contact. Therefore,

$$\tau = \text{constant} \tag{14-28}$$

Nadai's solution [9] is based on the assumption that the **viscous slipping** exists in the roll contact zones (Fig. 14.6d). Also that the frictional force is proportional to the relative velocity of the slip. Thus,

$$\tau_X = \mu(V_X - V)/h, \tag{14-29}$$

where V_X = velocity of metal being rolled at the section dx (Fig. 14.5)

   V = peripheral roll velocity

   h = oil-film thickness.

Orowan and Pascoe's solution [10] is derived for the case when **sticking** occurs over the whole arc of the contact (Fig. 14.6e). Similar assumption is made by Sims [11] and Alexander [12].

Tselikov's solution [13] is provided for the case when there is a zone of **restricted plastic deformation** in the middle of the sticking zone. It is also assumed that **dry slipping** occurs at the entry and the exit of the arc of contact as shown in Fig. 14.6f.

Figure 14.7 shows the distribution of normal pressure and frictional force along the arc of contact as calculated by Tselikov according to the four theories discussed: Von Karman's, Siebel's, Nadai's and Tselikov's.

14.8 PRESSURE DISTRIBUTION ALONG THE ARC OF CONTACT

In order to derive the equation for calculating the pressure distribution along the arc of contact, the solution for the Von Karman's differential equation (14-26) has to be found. One of the simplified solutions of this equation was proposed by Tselikov [14] as described below.

For the case of dry slipping friction (Fig. 14.6a), the simplification was achieved by replacing the variable parameter θ with a constant value, such as

$$\theta = \alpha/2, \tag{14-30}$$

where α = roll bite angle (Fig. 14.4).

This assumption allows one to reduce Eq. (14-26) to a form which can be integrated. However, the obtained results are valid only when the bite angle is small.

If the material is rolled with entry tension s_1, the solution for normal pressure distribution in the arc of contact between the plane of entry and the neutral plane is given by

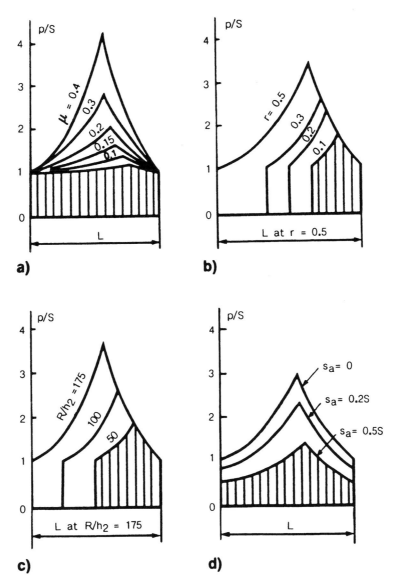

Fig. 14.8 Distribution of normal pressure along the arc of contact in rolling wide strip (r = 0.3, α = 5°40', h₂/D = 0.0116) as a function of: a) coefficient of friction, b) reduction, c) roll radius to exit thickness ratio, d) average strip tension. Adapted from Tselikov (1939).

$$p = \frac{S'}{\xi + 1} \left\{ \frac{p_1}{S'} [(\xi + 1) - 1](\frac{h_1}{h})^{\xi+1} + 1 \right\} \tag{14-31}$$

where

$$S' = \frac{S}{1 + \mu \tan\theta} \tag{14-32}$$

$$p_1 = \frac{S - s_1}{1 + \mu \tan\theta} \tag{14-33}$$

$$\xi = \frac{\tan(f - \theta)}{\tan\theta} \tag{14-34}$$

$$f = \arctan\mu \tag{14-35}$$

If the material is rolled with exit tension s_2, the solution for normal pressure distribution in the arc of contact between the neutral plane and the exit plane is given by

$$p = \frac{S''}{\eta - 1} \left\{ [\frac{p_2}{S''}(\eta - 1) + 1](\frac{h}{h_2})^{\eta-1} - 1 \right\} \tag{14-36}$$

where

$$S'' = \frac{S}{1 - \mu \tan\theta} \tag{14-37}$$

$$p_2 = \frac{S - s_z}{1 - \mu \tan\theta} \tag{14-38}$$

$$\eta = \frac{\tan(f + \theta)}{\tan\theta} \tag{14-39}$$

Figure 14.8 illustrates the distribution of normal pressure over the projected arc of contact as calculated from Eq. (14-31) and (14-36).

Figure 14.8a shows that, theoretically, the coefficient of friction μ exerts a marked influence on the distribution of normal pressure and as μ increases, both the peak pressure and the average pressure increase. The position of neutral point, which lies in the vertical plane through the peak, shifts toward the entry as μ increases.

Figure 14.8b shows that as the reduction r increases the peak normal pressure and average pressure increase. Similar effect is produced when the ratio of the roll radius R to the exit thickness h_2 increases as shown in Fig. 14.8c.

Figure 14.8d illustrates the effect of entry and exit tensions: the greater average tension $s_a = (s_1 + s_2)/2$, the lower normal pressure in the arc of contact.

14.9 LIMITATIONS OF THE THEORIES OF HOMOGENEOUS DEFORMATION

The theories of homogeneous deformation are based on the common assumptions that a plane vertical section of the rolled material remains plane during rolling. However, this contradicts to the facts obtained from numerous experiments.

An attempt to overcome this problem was made by Orowan in his theory of **nonhomogeneous** deformation [15]. According to this theory, the plane vertical sections of sheet before rolling are

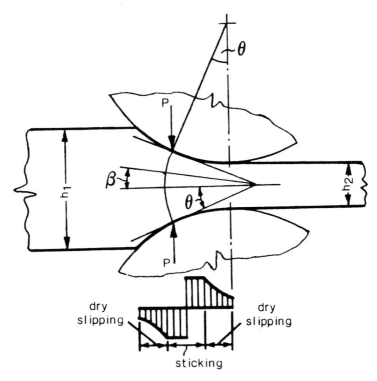

Fig. 14.9 Stress distribution in the roll bite according to the theory of nonhomogeneous deformation.

deformed during the rolling (Fig. 14.9). Also, it is assumed that the stress distribution in vertical plane is not homogeneous. Particularly, the distribution of the shear stresses is described as

$$\tau = - (\mu p \theta)/\beta \tag{14-40}$$

and the distribution of horizontal stress is expressed as

$$\sigma = p - S \sqrt{1 - (\frac{2\tau}{S})^2} \tag{14-41}$$

The Orowan's theory provides a solution for normal pressure distribution along the arc of contact for the case when sticking friction exists in vicinity of neutral plane and dry slipping friction prevails at the entry and exit planes.

Another limitation of the theories of homogeneous deformation arises from the assumptions that the yield stress of the material and the coefficient of friction remain constant. However, in most cases of cold rolling the constrained yield strength S increases due to work-hardening and there is a strong possibility that the coefficient of friction is not constant in all points in the arc of contact especially in dry rolling [3].

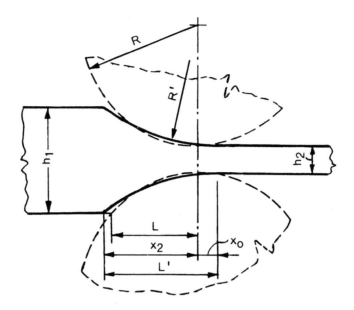

Fig. 14.10 Schematic presentation of the roll flattening effect.

Given a curve showing the variation of the yield stress with reduction, the normal pressure distribution in the arc of contact can be calculated by means of von Karman's equation and a point to point summation. This process, however, is laborious, and for practical purposes a more efficient method is desirable.

Both homogeneous and nonhomogeneous theories of rolling give a simplistic description of the deformation zone. It was found that the accuracy of the calculations based on these theories can be improved if such factors as roll flattening and the elastic recovery of the rolled strip are taken into consideration.

14.10 ROLL FLATTENING EFFECT

Since all rolls are elastic, some deformation of the rolls must occur during rolling. The general nature of this deformation is indicated in Fig. 14.10, where the shape of the rigid roll with radius R is compared with that of deformed roll with radius R'.

It will be noted that the main effects produced by roll flattening are:

a) the arc of contact, compared with rigid roll, is lengthened for the same draft

b) the planes of entry and exit are shifted outward from the centerline of the rolls.

Assuming that the pressure distribution along the arc of contact is elliptical, Hitchcock [16] has derived the following equation for the roll contact length of the flattened roll

$$L' = x_0 + x_2,$$ (14-42)

where

$$x_o = \frac{8R(1 - \nu_r^2)}{\pi E_r} p_a \tag{14-43}$$

$$x_2 = (L^2 + x_o^2)^{1/2}, \tag{14-44}$$

where E_r = modulus of elasticity for roll material

ν_r = Poisson's ratio for roll material.

The radius of flattened roll is equal to

$$R' = R[1 + \frac{16P(1 - \nu_r^2)}{\pi E_r w \Delta}], \tag{14-45}$$

where P = roll separating force

w = width of the rolled strip.

The roll separating force can be expressed by

$$P = p_a w (R'\Delta)^{1/2} \tag{14-46}$$

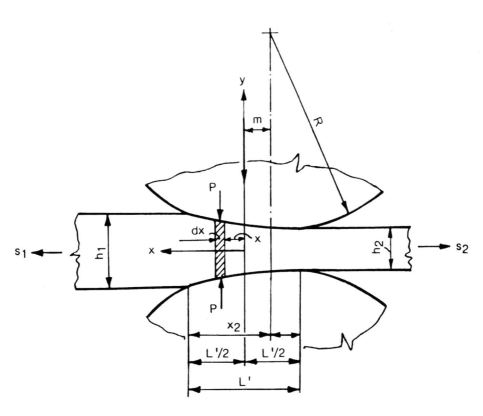

Fig. 14.11 Schematic diagram used for derivation of the Stone's theory of rolling.

14.11 SIMPLIFIED ANALYSIS OF ROLLING WITH DRY SLIPPING FRICTION

One of the most practical analyses of rolling with dry slipping friction has been proposed by Stone [17]. In this analysis, the effects of both the strip tension and the roll flattening have been taken into consideration. The produced results are proved to be applicable for calculations of rolling loads in cold strip mills.

Let us assume that the forces acting on an element dx in the roll bite are described with the following linear differential equation (Fig. 14.11)

$$\frac{dp}{dx} = -\frac{2\mu p}{h_a} \tag{14-47}$$

or

$$\frac{dp}{p} = -\frac{2\mu}{h_a} dx \tag{14-48}$$

After integrating both parts of the equation, we obtain

$$\ln(p) = -\frac{2\mu}{h_a} x + C \tag{14-49}$$

at $x = L'/2$, the yield criterion is expressed as

$$p = S - s_a, \tag{14-50}$$

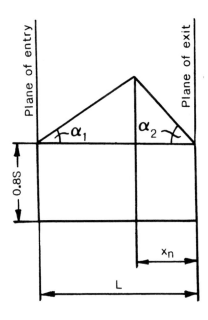

Fig. 14.12 Simplified friction hill for rolling with sticking friction.

where s_a = average strip tension.

Therefore, the integration constant is equal to

$$C = \ln(S - s_a) + \frac{\mu L'}{h_a} \qquad (14\text{-}51)$$

and Eq. (14-49) transforms into

$$p = (S - s_a) \exp \left[\frac{2\mu}{h_a} \left(\frac{L'}{2} - x\right)\right] \qquad (14\text{-}52)$$

Integration of Eq. (14-52) allows one to obtain the average normal pressure

$$\frac{p_a}{S} = \left(1 - \frac{s_a}{S}\right)(PMF), \qquad (14\text{-}53)$$

where (PMF) = pressure multiplication factor

$$PMF = \frac{h_a}{\mu L'} \left[\exp\left(\frac{\mu L'}{h_a}\right) - 1\right] \qquad (14\text{-}54)$$

Then the roll separating force can be calculated from Eq. (14-46)
The roll torque is given by

$$M = 2Pm = 2P(x_2 - L'/2) = 2P(L'/2 - x_0), \qquad (14\text{-}55)$$

where m = lever arm (Fig. 14.11).

14.12 SIMPLIFIED ANALYSIS OF ROLLING WITH STICKING FRICTION

A number of the simplified solutions for rolling with sticking friction have been proposed. These solutions have a practical value for calculations of rolling loads in hot strip mills.

Orowan and Pascoe [10, 15] assumed a simple triangular hill on a rectangular base (Fig. 14.12). The slope α_2 at the exit side of the friction hill is assumed to be equal to

$$\alpha_2 = S/h_2 \qquad (14\text{-}56)$$

Position of neutral plane is described as a distance from the plane of exit

$$x_n = nL, \qquad (14\text{-}57)$$

where n varies somewhere between 0.4 and 0.6.

Then average pressure in the arc of contact can be found by calculating the total area that includes both the friction hill and its base

$$\frac{p_a}{S} = 0.8 + \frac{nL}{2h_2} \qquad (14\text{-}58)$$

If the neutral plane is assumed to be in the middle of the projected arc of contact (i.e., n = 0.5), last equation reduces to

$$\frac{p_a}{S} = 0.8 + \frac{L}{4h_2} \qquad (14\text{-}59)$$

Alexander and Ford [18] modified the analysis described above by assuming that the slope at the entry of the friction hill should be equal to

$$\alpha_1 = S/h_1 \qquad (14\text{-}60)$$

Therefore, the position of neutral plane will be described as (Fig. 14.12)

$$x_n = \frac{h_2 L}{h_1 + h_2} \qquad (14\text{-}61)$$

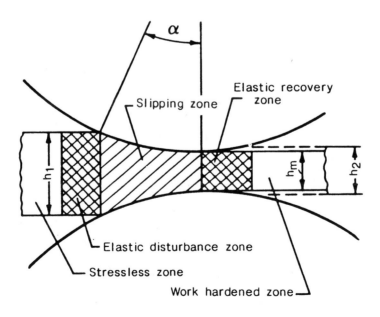

Fig. 14.13 Schematic illustration of the strip elastic recovery.

This allows one to express the average pressure as a function of the average strip thickness h_a

$$\frac{P_a}{S} = 0.8 + \frac{L}{4h_a} \qquad (14\text{-}62)$$

Gupta and Ford [19] showed that Eq. (14-62) is also valid in the presence of roll flattening as described by Eq. (14-45).

Taking into account Eq. (14-15) through Eq. (14-20), Eq. (14-62) can be rewritten in the following form [20]:

$$\frac{P_a}{S} = 0.8 + \frac{\sqrt{Rrh_1}}{2h_1(2-r)} \qquad (14\text{-}63)$$

More sophisticated expression for average rolling pressure was proposed by Sims [21].

14.13 EFFECT OF THE STRIP ELASTIC RECOVERY

Bland and Ford [22] have derived the following equation for the radius of the flattened roll taking into account the strip elastic recovery as shown in Fig. 14.13.

$$R'' = R\left\{1 + \frac{16P(1-\nu_r^2)/(\pi E_r w)}{[(\Delta + \Delta_t + \Delta_2)^{0.5} + (\Delta_2)^{0.5}]^2}\right\} \qquad (14\text{-}64)$$

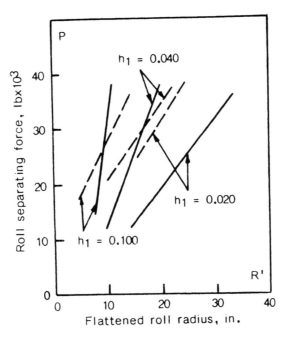

Fig. 14.14 Variation of the roll separating force per unit of width with the flattened roll radius: solid lines - from Eq. (14-45), dotted lines - from Eq. (14-46). (From William F. Hosford/Robert M. Caddell, METAL FORMING: Mechanics and Metallurgy, Copyright 1983, Reprinted by permission of Prentice-Hall, Inc., Englewood Cliffs, NJ.)

where Δ = draft as expressed by Eq. (14-17).

$$\Delta_2 = \frac{h_2}{E_r}(1 - \nu_r^2)(S - s_2) \tag{14-65}$$

$$\Delta_t = \frac{\nu}{E_r}(1 + \nu)(h_2 s_2 - h_1 s_1), \tag{14-66}$$

where h_2 = strip thickness when the tension has been released.

When the elastic recovery is neglected, $\Delta_2 = \Delta_t = 0$ and Eq. (14-64) reduces to the well-known form described by the Eq. (14-45).

14.14 LIMIT OF ROLLABILITY

Roll flattening can become so severe that it is impossible to reduce the thickness of a sheet below some limiting value h_{min}.

Indeed, the effect of roll flattening is to increase the separating force P, since both average normal pressure p_a and the contact length L' increase. Both the roll separating force P and the flattened roll radius R' can be found by solving the equations (14-45) and (14-46) simultaneously [1].

Figure 14.14 is a graphical solution for heavily rolled steel where S = 100,000 psi, $\mu = 0.2$, $E_r = 30,000$ ksi, r = 0.05, and R = 5 in.. Initial thicknesses (h_1) of 0.100, 0.040, and 0.020 in. were used in the calculations. The plots $R' = f_1(P)$ were obtained from Eq. (14-45) whereas the plots $P = f_2(R')$ were obtained from Eq. (14-46). The intersection of these two plots satisfies both equations. There is no intersection point for the sheet where $h_1 = 0.020$ in. It indicates that this thickness can not be reduced under the rolling conditions specified above.

The minimum thickness that can be rolled is given by [17]

$$h_{min} = \frac{A\mu R}{E_r}(1 - \nu_r^2)(S - s_a), \tag{14-67}$$

with a value for A between 7 and 8 [1].

Equation (14-67) suggests a number of ways to reduce minimum rolled thickness, such as:

a) increase of average strip tension s_a

b) reduction of the constrained yield strength of the strip S by annealing

c) reducing the coefficient of friction μ by lubrication and by roll polishing

d) reducing the roll radius R

e) use of the roll materials with higher modulus of elasticity E, such as sintered carbide

f) rolling thin strip between layers of softer material.

Ford and Alexander [23] have derived an equation for minimum rolled thickness which takes into account both modulus of elasticity and Poisson's ratio for strip material. It is expressed as

$$h_{min} = [\frac{14.22R\mu(1 - \nu_s^2)}{E_s} + \frac{9.06R\mu(1 - \nu_r^2)}{E_r}](S_o - s_a) \tag{14-68}$$

where ν_s = Poisson's ratio for strip material

E_s = modulus of elasticity for strip material.

When $E_s = E_r$ and $\nu_s = \nu_r$, Eq. (14-68) reduces to

$$h_m = (9.06 + 14.22\mu) \frac{\mu R(1 - \nu_r^2)}{E_r} (S_o - s_a) \qquad (14-69)$$

This equation produces more conservative estimates of the minimum rolled thickness in comparison with the results obtained from Eq. (14-67) for the same conditions.

REFERENCES

1. W.F. Hosford and R.M. Caddell, Metal Forming: Mechanics and Metallurgy, Prentice Hall, Inc., Englewood Cliffs, New Jersey, pp. 115-142, 1983.

2. S. Kalpakjian, Manufacturing Processes for Engineering Materials, Addison-Wesly Publishing Company, Inc., Reading, Massachusetts, pp. 51-90, (1984).

3. L.R. Underwood, The Rolling of Metals:Theory and Experiment, Volume 1, John Wiley and Sons Inc. New York, pp. 203-241, 1950.

4. V.B. Ginzburg, "Basic Principles of Customized Computer Models for Cold and Hot Strip Mills," Iron and Steel Engineer, September 1985, pp. 21-35.

5. A.H. El-Waziri, "An Up-to-Date Examination of Rolling Theory," AISE Yearly Proceedings, 1963, pp. 753-760.

6. Th. Von Karman, "On the Theory of Rolling", Journal for Applied Mathematics and Mechanics (German), Vol. 5, 1925, pp. 139-141.

7. S. Ekelund, "Analysis of Factors Influencing Rolling Pressure and Power Consumption in the Hot Rolling of Steel", Steel, Vol. 93, August 21, 1933, pp. 27-29.

8. E. Siebel, "Resistance and Deformation and the Flow Material During Rolling", Stahl und Eisen, Vol. 50, 1930, p. 1769.

9. A. Nadai, "The Forces Required for Rolling Steel Strip Under Tension," Journal of Applied Mechanics, June 1939, pp. A54-A62.

10. E. Orowan and K.J. Pascoe, "A Simple Method of Calculating Roll Pressure and Power Consumption in Flat Hot Rolling," Iron and Steel Institute (London), No. 34, 1946, pp. 124-126.

11. R.B Sims, "The Calculation of Roll Force and Torque in Hot Rolling", Proceedings of the Institution of Mechanical Engineers, No. 168, 1954, pp. 191-200.

12. J.M. Alexander, "A Slip Line Field for the Hot Rolling Process," Proceedings for the Institution of Mechanical Engineers, No. 169, pp. 61-76.

13. A.I. Tselikov, "Present State of Theory of Metal Pressure upon Rolls in Longitudinal Rolling," Stahl, Vol. 18, No. 5, May 1958, pp. 434-441.

14. A.I.Tselikov, "Effect of External Friction and Tension on the Pressure of the Metal on the Rolls in Rolling," Metallurg (USSR), No.6, 1939, pp. 61-76.

15. E. Orowan, "The Calculation of Roll Pressure in Hot and Cold Flat Rolling", Proceedings Institute of Mech. Eng., Vol. 50, No. 4, 1943, pp. 140-167.

16. W. Trinks and J.H. Hitchcock, Roll Neck Bearings, American Society of Mechanical Engineers, New York, 1935.

17. M.D., Stone, "Rolling Thin Strip", AISE Yearly Proceedings, 1953, pp. 115-128.

18. J.M. Alexander, H. Ford, "On the Limit Analysis of Hot Rolling", Progress in Applied Mechanics, 1963, pp. 191-203.

19. S. Gupta, H. Ford, "Calculation Method for Hot Rolling of Steel Sheet and Strip", Journal of the Iron and Steel Institute, February 1967, pp. 186-190.

20. M. Tarokh, F. Seredinski, "Roll-Force Estimation in Plate Rolling", Journal of the Iron and Steel Institute, July 1970, pp. 695-697.

21. R.B. Sims, "The Calculation of Roll Force and Torque in Hot Rolling", Proceedings of the Institution of Mechanical Engineers, No. 168, 1954, pp. 191-200.

22. D.R. Bland and H. Ford, "Cold Rolling with Strip Tension, Part 3 - An Approximate Treatment of the Elastic Compression of the Strip in Cold Rolling", Journal of the Iron and Steel Institute, Vol. 171, July 1952, pp. 245-249.

23. H. Ford, J.M. Alexander, "Rolling Hard Materials in Thin Gauges", Journal of the Institute of Metals, Vol. 88, 1959-60, pp. 193-199.

15

Upper-Bound Analysis of Plastic Deformation

15.1 LOWER-BOUND AND UPPER-BOUND ANALYSES

The following two methods are used for approximate evaluation of forces that cause plastic flow of metals [1]:

1. Lower-bound analysis
2. Upper-bound analysis.

The **lower-bound** analysis predicts a load that is less than or equal to the exact load needed to produce full plastic deformation. This analysis is based upon satisfying a yield criterion and stress equilibrium, while ignoring the possible shape change of the body. The lower-bound analysis is applied for a 'safe' calculation in the design of structures that are not intended to deform plastically.

The **upper-bound** analysis predicts a load that is at least equal to or greater than the exact load needed to cause plastic flow. This analysis is based upon satisfying a yield criterion and assuring that shape changes are geometrically self-consistent, whereas no attention is paid to stress equilibrium. The upper-bound analysis is applied to the metal-forming operations where it is required to predict a force that will surely cause the desired shape change.

15.2 PRINCIPLE OF UPPER-BOUND ANALYSIS

In the upper-bound analysis, the internal flow field, or deformation field, is assumed. This field must account for the required shape change. The predicted force that causes plastic deformation in this field is calculated by equating the internal rate of energy dissipation to the rate of work done by external forces [1]. The energy consumed internally is calculated by using the approximate strength properties of the work material.

The assumed deformation field can be checked for complete consistency by drawing a velocity vector diagram, which is commonly known as **hodograph**.

In applying the upper-bound analysis to plastic deformation the following assumptions are used:

1. The workpiece material is isotropic and homogeneous.
2. The effects of the strain hardening and strain rate are neglected.

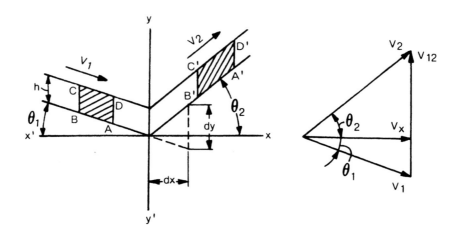

Fig. 15.1 Basis for analysis of energy dissipation along a plane of intense shear discontinuity and the hodograph or velocity vector diagram. (From William F. Hosford/Robert M. Caddell, METAL FORMING: Mechanics and Metallurgy, Copyright 1983, Reprinted by permission of Prentice-Hall, Inc., Englewood Cliffs, NJ.)

There are two basic approaches in constructing the deformation fields.

According to the first approach, the deformation fields are the polygons which are viewed as rigid blocks. These blocks are separated by the planes upon which discrete shear occurs.

According to the second approach, other deformation fields are suitable as long as they are **kinematically admissible.** For example, an acceptable field may include regions undergoing homogeneous deformation.

15.3 ENERGY DISSIPATION ON THE PLANE OF DISCRETE SHEAR

Let us consider an element of rigid metal, ABCD, moving at unit velocity V_1 in the plane of paper at an angle θ_1 to the horizontal (Fig. 15.1), and having unit width in the direction perpendicular to the paper [1]. AD is set parallel to yy'. At the plane yy' the element is forced to have new shape A'B'C'D', new velocity V_2 and new direction described by an angle θ_2.

Horizontal components of both velocities V_1 and V_2 must be equal in order to maintain the same volume of metal approaching and leaving the plane yy'. Velocity V_{12} is the **velocity discontinuity** and is equal to the vector difference between V_1 and V_2.

The rate of energy dissipation on yy' is equal to

$$\frac{dW}{dt} = q\,\frac{v}{dt} \qquad\qquad (15\text{-}1)$$

where q = work of deformation per volume

   v = volume

   dt = increment of time.

Because deformation is due to shear, the work of deformation may be presented as

$$q = k\gamma = k \frac{dy}{dx} \tag{15-2}$$

where k = shear yield stress

- = shear strain.

Taking into account that for unit

$$\frac{v}{dt} = hV_x \tag{15-3}$$

and also that $dy/dx = V_{12}/V_x$ (Fig. 15.1), we obtain after substitution of Eq. (15-2) and Eq. (15-3) into Eq. (15-1) that

$$\frac{dW}{dt} = khV_{12} \tag{15-4}$$

For deformation fields involving more than one plane of discrete shear,

$$\frac{dW}{dt} = \sum_1^{i=n} kh_i V_i, \tag{15-5}$$

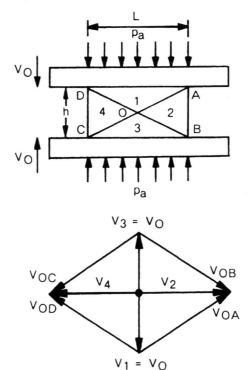

Fig. 15.2 Plane-strain compression with sticking friction for small values of L/h. (From Caddell and Hosford. Reprinted by permission of the Council of the Institution of Mechanical Engineers from International Journal of Mechanical Engineering Education, Vol. 8, 1980.)

where h_i and V_i pertain to each individual plane.

15.4 UPPER-BOUND SOLUTION FOR PLANE-STRAIN COMPRESSION

Figure 15.2 illustrates a deformation field for plane-strain compression of a body with height h and length L. This type of deformation field is usually selected for small values of L/h [1]. Sticking friction is assumed to be at the workpiece-platen interfaces. It is also assumed that the regions AOD and COB are rigid blocks that move with the same velocity V_O, as the compression platens.

Discrete shear occurs on lines $\overline{OA}$, $\overline{OB}$, $\overline{OC}$, and $\overline{OD}$ with corresponding velocities discontinuity V_{OA}, V_{OB}, V_{OC}, and V_{OD}.

Equating the rate of external work per unit of width to internal energy dissipation at shear planes gives

$$2p_a L V_O = 4k(\overline{OA})V_{OA} \tag{15-6}$$

From geometrical considerations

$$V_{OA} = 2V_O(\overline{AO})/h \tag{15-7}$$

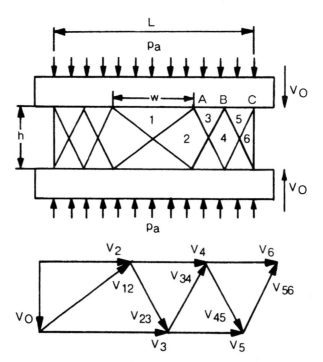

Fig. 15.3 Plane-strain compression with sticking friction for large values of L/h. (From Caddell and Hosford. Reprinted by permission of the Council of the Institution of Mechanical Engineers from International Journal of Mechanical Engineering Education, Vol. 8, 1980.)

and

$$(\overline{AO})^2 = (h^2 + L^2)/4 \qquad (15\text{-}8)$$

After substituting $\overline{AO}$ and V_{OA} into Eq. (15-6) and taking into account that $2k = S$ we obtain that the upper-bound solution for normal pressure is

$$\frac{P_a}{S} = \frac{1}{2}(\frac{h}{L} + \frac{L}{h}) \qquad (15\text{-}9)$$

Figure 15.3 illustrates a deformation field consisting of more than one triangle along the workpiece-platen interface. This type of deformation field gives better upper-bound solutions for large values of L/h.

For this class of upper bounds a general solution corresponding to a minimum value of p_a/S is

$$\frac{P_a}{S} = \frac{1}{hL}[\frac{nh^2}{2} + C(\frac{L}{n+1})^2] \qquad (15\text{-}10)$$

where

$$C = \frac{3n+1}{2} + \sum_{i=1}^{(n-1)/2} (2i - 1) \qquad (15\text{-}11)$$

and n is an odd integer that is equal or greater than 3.

The minimum value of p_a/S occurs when

$$w = \frac{2L}{n+1} \qquad (15\text{-}12)$$

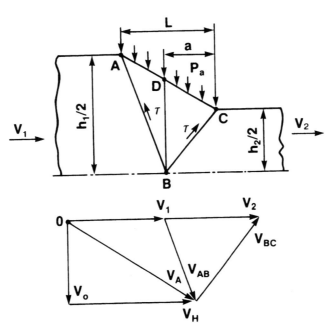

Fig. 15.4 Simplified deformation field for rolling with sticking friction.

When n = 3, w = L/2 and Eq. (15-10) reduces to

$$\frac{p_a}{S} = \frac{3h}{2L} + \frac{3L}{8H} \qquad\qquad (15\text{-}13)$$

When n = 5, w = L/3 and Eq. (15-10) reduces to

$$\frac{p_a}{S} = \frac{5H}{2L} + \frac{L}{3h} \qquad\qquad (15\text{-}14)$$

15.5   UPPER-BOUND SOLUTION FOR FLAT ROLLING

The upper-bound solutions for flat rolling with sticking friction were developed by Green and Wallace [2] and Piispanen [3]. In order to simplify the solution, the problem was reduced to a case of wedge compression as shown in Fig. 15.4. In this drawing:

V_1 = velocity of the material before entering a die

V_2 = velocity of the material after exiting the die

V_{AB}, V_{BC} = velocity discontinuities at the entry and exit of the die respectively.

V_A = velocity of the rigid block ABC

V_O, V_H = vertical and horizontal components of the velocity V_A respectively

h_1, h_2 = entry and exit material thicknesses respectively.

Since the material in triangle ABC moves as a rigid block, velocity V_A must be drawn parallel to AC. Because velocity discontinuity V_{AB} is parallel to AB, this allows one to establish the magnitudes of both V_A and V_{AB} if the magnitude of V_1 is known.

Similarly, at the exit side velocity discontinuity V_{BC} is drawn parallel to BC. Its magnitude is established by the fact that the resultant velocity V_2 (drawn from point O) must be parallel to V_1.

The obtained hodograph meets the requirements of self-consistency. Indeed, the mass-flow relationship

$$V_1 h_1 = V_2 h_2 \qquad\qquad (15\text{-}15)$$

is maintained.

Let us consider the energy balance for two parts, ABD and DBC, of the rigid block ABC.

The energy balance for the part ABD is obtained by equating the rate of external work per unit of width to internal energy dissipation at the shear plane AB

$$P_a(L - a)V_O = k(\overline{AB})V_{AB} \qquad\qquad (15\text{-}16)$$

From geometrical consideration

$$V_{AB} = 2V_O(\overline{AB})/h_1 \qquad\qquad (15\text{-}17)$$

and
$$(\overline{AB})^2 = (L - a)^2 + \frac{h_1^2}{4} \tag{15-18}$$

After substituting $\overline{AB}$ and V_{AB} into Eq. (15-16) we obtain the vertical force that is applied to the part ABD

$$p_a(L - a) = \frac{2k}{h_1} [(L - a)^2 + \frac{h_1^2}{4}] \tag{15-19}$$

The energy balance for the part DBC can be expressed by equating the rate of external work per unit of width to internal energy dissipation at the shear plane BC

$$p_a a V_O = k(\overline{BC})V_{BC} \tag{15-20}$$

From geometrical consideration

$$V_{BC} = 2V_O(\overline{BC})/h_2 \tag{15-21}$$

and
$$(\overline{BC})^2 = a^2 + \frac{h_2^2}{4} \tag{15-22}$$

After substituting $\overline{BC}$ and V_{BC} into Eq. (15-20) we obtain the vertical force that is applied to the part DBC

$$p_a a = \frac{2k}{h_2} (a^2 + \frac{h_2^2}{4}) \tag{15-23}$$

Total vertical force applied to the rigid block ABC is equal to a sum of two components of the force expressed by Eqs. (15-19) and (15-23). Taking into account that $2k = S$ we obtain

$$\frac{p_a}{S} = \frac{1}{L} [\frac{(L - a)^2}{h_1} + \frac{a^2}{h_2} + \frac{h_1 + h_2}{4}] \tag{15-24}$$

Ratio p_a/S reaches minimum value when

$$a = \frac{Lh_2}{h_1 + h_2} \tag{15-25}$$

and Eq. (15-24) reduces to

$$\frac{p_a}{S} = \frac{1}{2} (\frac{L}{h_a} + \frac{h_a}{L}) \tag{15-26}$$

REFERENCES

1. W.F. Hosford and R.M. Caddell, Mechanics and Metallurgy, Prentice-Hall, Inc., Englewood Cliffs, New Jersey, pp. 143–238, 1983.

2. J.W. Green and J.F. Wallace, "Estimation of Load and Torque in the Hot Rolling Process", Journal of Mechanical Engineering Science, Vol. 4, No. 2, 1962, pp. 136–142.

3. V. Piispanen, "Plastic Deformation of Metal: Theory of Simulated Sliding", Wear, Vol. 38, No. 1, pp. 43–72.

16

Slip-Line Field Analysis of Plastic Deformation

16.1 PRINCIPLE OF SLIP-LINE FIELD ANALYSIS

The upper-bound analysis portrays flow of metal as a movement of a solid block inside any polygon. The slip-line field analysis allows one to introduce a more realistic picture of metal flow. It utilizes a graphical approach which presents the flow pattern from point to point in the deforming metal [1-7].

The technique is based on construction of a family of straight or curvilinear lines that intersect each other orthogonally. These lines correspond to the directions of yield stress of the material in shear, k. A network of such orthogonal lines describes a slip-line field.

The developed slip-line field depends largely on intuition and experience and must satisfy certain conditions such as static equilibrium of forces, yield criteria, and boundary conditions.

The slip-line field analysis is based on the following assumptions:

1. The metal is isotropic and homogeneous
2. The metal is rigid-perfectly plastic; this implies the neglect of work hardening
3. Deformation occurs by plain strain
4. There is a constant shear stress at the interfacial boundary
5. Effects of temperature, strain rate and time are not taken into consideration.

16.2 STRESSES AND STRAINS IN THE SLIP-LINE FIELD ANALYSIS

In application to the slip-line field analysis, the assumption of plane strain implies that metal flow occurs in planes parallel to the x-y plane (Fig.16.1) and no movement of metal would occur in the z direction. Taking into account that the volume of the metal remains constant during deformation, we can write [1]:

$$d\epsilon_x = -d\epsilon_y, \quad d\epsilon_z = 0 \qquad (16\text{-}1)$$

$$\dot{\epsilon}_x = -\dot{\epsilon}_y, \quad \dot{\epsilon}_z = 0 \qquad (16\text{-}2)$$

$$d\gamma_{xy} = 0, \quad d\gamma_{yz} = d\gamma_{zx} = 0 \qquad (16\text{-}3)$$

$$\dot{\gamma}_{xy} = 0, \quad \dot{\gamma}_{yz} = \dot{\gamma}_{zx} = 0 \qquad (16\text{-}4)$$

and therefore,

$$\tau_{yz} = \tau_{zx} = 0 \tag{16-5}$$

Since the shear stresses in the x-y plane are equal to zero, so σ_z must be a principal stress

$$\sigma_z = \sigma_2 = (\sigma_y + \sigma_x)/2 \tag{16-6}$$

If the x-and y-axes define principal directions, it follows that:

$$\sigma_2 = (\sigma_1 + \sigma_2 + \sigma_3)/3 = \sigma_{mean} \tag{16-7}$$
$$\sigma_1 = \sigma_2 + k \tag{16-8}$$
$$\sigma_3 = \sigma_2 - k \tag{16-9}$$

Thus, σ_2 may be considered as the **mean hydrostatic stress** at any point in the field of deformation and has no influence upon yielding. It follows from Eqs. (16-7)-(16-9) that the intermediate stress σ_2 is always equal to the mean stress.

Therefore, plane strain deformation causes a state of stress, which is equivalent to pure shear deformation with a superimposed hydrostatic stress. The hydrostatic stress can vary from one point to another in the region of deformation.

Figure 16.2 illustrates an elemental section in the slip-line field. This section is located at an angle ϕ from the x-axis. The shear stress k acts along the slip lines, whereas the mean stress σ_2 acts perpendicular to those lines and to the face of the element as shown. The magnitude of σ_2 can vary from point to point along a slip line. This change is of great importance to this analysis.

16.3 NETWORK OF THE SLIP LINES

Families of orthogonal slip lines are usually designated as α- and β-lines (Fig. 16.3). The angle of rotation ϕ is considered positive for counterclockwise rotation from a given datum such as x-axis here.

The direction of largest algebraic principal stress σ_1 lies in the first and third quadrants of the α-β system. The β-lines are located in the first and third quadrants formed by the σ_1 and σ_3 stress axes.

The following **equilibrium equations** are valid along the slip lines [1]:

$$\sigma_2 - 2k\phi = C_1 = \text{constant along an } \alpha\text{-line} \tag{16-10}$$
$$\sigma_2 + 2k\phi = C_2 = \text{constant along an } \beta\text{-line} \tag{16-11}$$

It means that movement along an α- or β-line causes σ_2 change by

$$\Delta\sigma_2 = 2k\cdot\phi \quad \text{on an } \alpha\text{-line} \tag{16-12}$$
$$\Delta\sigma_2 = -2k\cdot\phi \quad \text{on an } \beta\text{-line} \tag{16-13}$$

Eqs. (16-12) and (16-13) allow one to establish an additional restriction on the shape of a statistically admissible field. This is illustrated in Fig. 16.4 which shows a region bounded by the two α and β lines. The difference between σ_2 at A and C can be calculated by using two paths, A-D-C and A-B-C. In moving on an α-line from A to D,

$$\sigma_{2D} = \sigma_{2A} + 2k(\phi_D - \phi_A) \tag{16-14}$$

In further moving on an α-line from D to C,

$$\sigma_{2C} = \sigma_{2D} - 2k(\phi_C - \phi_D) \tag{16-15}$$

After combining Eqs. (16-14) and (16-15) we obtain

$$\sigma_{2C} - \sigma_{2A} = 2k(2\phi_D - \phi_A - \phi_C) \tag{16-16}$$

Considering the alternative path A-B-C, we obtain

$$\sigma_{2C} - \sigma_{2A} = -2k(2\phi_B - \phi_A - \phi_C) \tag{16-17}$$

Equating Eqs. (16-16) and (16-17) yields

$$\phi_A - \phi_B = \phi_D - \phi_C \tag{16-18}$$

$$\phi_A - \phi_D = \phi_B - \phi_C \tag{16-19}$$

The Eqs. (16-18) and (16-19) indicate that in constructing the net of the α- and β-lines the change of ϕ at the intersections of the families of α-and β-lines must be the same when moving along the lines of the same family.

There are two simple fields that meet these requirements. Figure 16.5a shows one of these fields in which σ_2 remains the same throughout the field. Indeed, since there is no curvature in the net of straight lines ($\phi_\alpha = \phi_\beta = 0$), Eqs. (16-12) and (16-13) would yield $\Delta\sigma_2 = 0$.

Another simple field is the centered fan as shown in Fig. 16.5b. In this field the value for σ_2 is the same along the same radial line. However, this value changes from one radial line to another and become undeterminate at the singular point O.

16.4 BOUNDARY CONDITIONS IN A SLIP-LINE FIELD

The following boundary conditions are usually used in the slip-line field analysis [1]:

1. **Free surface** - In this case the plastic zone extends beyond tool. The α- and β-lines will meet such a surface at 45° since there are no normal or tangential stresses at a free surface (Fig. 16.6).

2. Frictionless interface - As with free surface, when there is no friction between the workpiece and tool surface, the α- and β-lines meet the surface at 45°.

3. Sticking friction at interface - Since in this case the shear yield stress k is reached at interface, one slip line, say α, meets the interface tangentially while the other, β, meets it at 90° [5] as shown in Fig. 16.7.

16.5 VELOCITIES IN A SLIP-LINE FIELD

Hodographs are usually constructed in slip-line field analyses in order to assure that the field is kinematically admissible. It also allows one to determine the relative percentage of energy dissipated by gradual deformation as opposed to intense shear. In the cases when the initial shape of a field is presented with straight lines, the construction of the hodograph may be helpful to predict the approximate change in the shape of these lines as they proceed through the deformation process.

In constructing the hodograph, the following rules are observed [1]:

1. Within any individual zone of constant σ_2, the absolute velocity is constant.

2. The velocity of discontinuity along a single slip line has constant magnitude. Along straight lines, such as a fan radial line in a centered fan (Fig. 16.5b), both the magnitude and direction of the velocity discontinuity are constant.

3. As a particle moves through a region of changing σ_2, its absolute velocity changes in direction and, in many cases, magnitude.

4. Both shear and most of energy dissipation always occur at the boundary between the field and undeformed metal outside the field.

Figure 16.8a illustrates the right half of the slip-line field for plane-strain indentation of a semi-infinite slab (Fig. 16.6). Corresponding hodograph is shown in Fig. 16.8b. In Figure 16.8:

V_O = indentation speed of region OAD.

V_A, V_E, V_B = velocities of the particles located inside AOB at the points A, E, and B respectively.

V_{OA} = velocity discontinuity at OA, parallel to OA.

V_{BC} = velocity discontinuity at BC, parallel to BC.

V_e = exit velocity.

The following observations can be made from the hodograph described above.

1. Intense shear occurs along:

a) AO (Velocity = V_{OA})

b) AEB ($V_A = V_E = V_B$)

c) BC ($V_e = V_B$).

2. Gradual deformation occurring in AOB is accompanied by dissipation of energy.

3. In order to preserve the conservation of mass, the flow rate across OD must be equal to the flow rate across OC.

16.6 INDENTATION

A simple two-dimensional example for simulation of hardness test with a flat rectangular indenter is shown in Fig. 16.9. The material is assumed to deform along shear planes. The triangular blocks slide along each other without separation.

Equating the work done by the indenter to the energy dissipation at shear planes gives [4]:

$$Pd = 2k[(A/\sqrt{2})d/\sqrt{2} + Ad + 2(A/\sqrt{2})(d/\sqrt{2})] = 6kAd, \tag{16-20}$$

where d = displacement of a triangular block

A = area of one side of the triangular block

P = compressive force.

Since the average pressure p_a = P/A and taking into account that S = 2k, the relative average indentation may be obtained from Eq. (16-20)

$$\frac{p_a}{S} = \frac{P}{2kA} = 3 \tag{16-21}$$

16.7 COMPRESSION

Let us consider a case when two flat anvils are loaded to compress a plate of width w and thickness H (Fig. 16.10). The length of the anvils is not less than w and their width L = H. If w ≫ H, then the application of load P produces practically no change in dimension w beneath the anvils, so the deformation zone approximates conditions of plane strain [1].

The horizontal stress, σ_x, must be zero, since no restraint forces act in the x direction. If the interface surfaces between the workpiece and indenter are frictionless, then the vertical or loading stress σ_y must be a principal stress. For now, let us assume that σ_z is intermediate between these other stresses in magnitude. Since all axes coincide with directions of principal stresses, we obtain that $\sigma_1 = \sigma_x$, $\sigma_2 = \sigma_z$ and $\sigma_3 = \sigma_y$ and also that:

$$\sigma_1 > \sigma_2 > \sigma_3 \tag{16-22}$$

It is seen from Mohr's circle (Fig. 16.11), that the compressive stress, required to cause yielding, is σ_3 = 2k. Since the average pressure is equal to p_a = P/(LH) = σ_3, we obtain that

$$\frac{p_a}{S} = 1 \tag{16-23}$$

Figure 16.12 illustrates the slip-line field solutions for compression with sticking friction for the workpieces with different length-to-height ratios L/h. It is shown that the average pressure p_a increases with increase in L/h. It is also shown that the slip-line solution predicts lower values for the average pressure in comparison with those obtained from the slab solution [1].

16.8 ROLLING

Figure 16.13 shows a slip-line field for the hot rolling process as developed by Ford and Alexander [6, 7]. In their solution the problem was reduced to a construction of the slip-line field for a workpiece compressed between two inclined platens with sticking friction.

To further simplify the calculations, it was assumed that the slip line at the entrance to the roll gap is straight meeting the centerline of the workpiece at the 45 degrees and the roll tangentially.

Having constructed a slip-line network, the vertical force applied to the base of each triangle coincident with the centerline of the workpiece may be calculated. The resultant distribution of pressure along the arc of contact is shown in Fig. 16.13 assuming sticking friction between the strip and the roll surface.

Similarly, the resultant of the shearing forces along the bases of the triangles coincident with the roll surface may be used for calculation of the rolling torque.

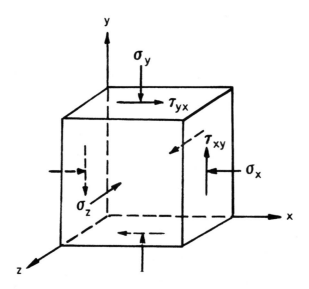

Fig. 16.1 Stress element for plane-strain deformation.

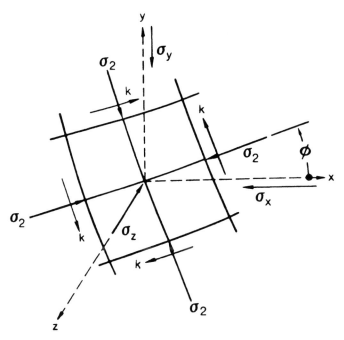

Fig. 16.2 Stresses acting on a small curvilinear element. (From William F. Hosford/Robert M. Caddell, METAL FORMING: Mechanics and Metallurgy, Copyright 1983, Reprinted by permission of Prentice-Hall, Inc., Englewood Cliffs, NJ.)

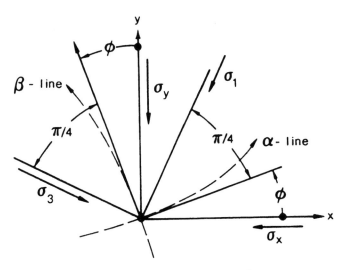

Fig 16.3 Convention for defining α- and β-lines. (From William F. Hosford/Robert M. Caddell, METAL FORMING: Mechanics and Metallurgy, Copyright 1983, Reprinted by permission of Prentice-Hall, Inc., Englewood Cliffs, NJ.)

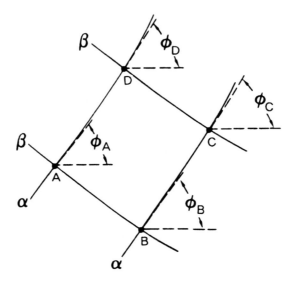

Fig. 16.4 Two pairs of α- and β-lines for analyzing the change
in the mean normal stress by traversing two different paths.
(From William F. Hosford/Robert M. Caddell, METAL FORMING:
Mechanics and Metallurgy, Copyright 1983, Reprinted by
permission of Prentice-Hall, Inc., Englewood Cliffs, NJ.)

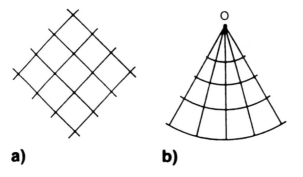

Fig. 16.5 a) A net of straight lines and b) a centered fan.
(From William F. Hosford/Robert M. Caddell, METAL FORMING:
Mechanics and Metallurgy, Copyright 1983, Reprinted by
permission of Prentice-Hall, Inc., Englewood Cliffs, NJ.)

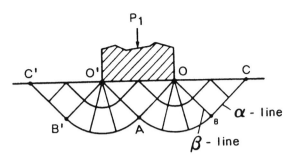

Fig. 16.6 A slip-line field for plane-strain indentation of a semi-infinite slab. (From William F. Hosford/Robert M. Caddell, METAL FORMING: Mechanics and Metallurgy, Copyright 1983, Reprinted by permission of Prentice-Hall, Inc., Englewood Cliffs, NJ.)

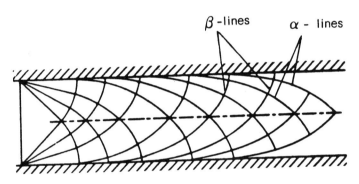

Fig. 16.7 A slip-line field for compression with sticking friction. Adapted from Johnson and Mellor (1973).

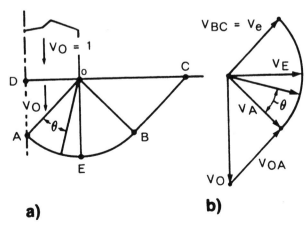

Fig. 16.8 a) A partial field and b) associated hodograph for Fig. 16.6. (From William F. Hosford/Robert M. Caddell, METAL FORMING: Mechanics and Metallurgy, Copyright 1983, Reprinted by permission of Prentice-Hall, Inc., Englewood Cliffs, NJ.)

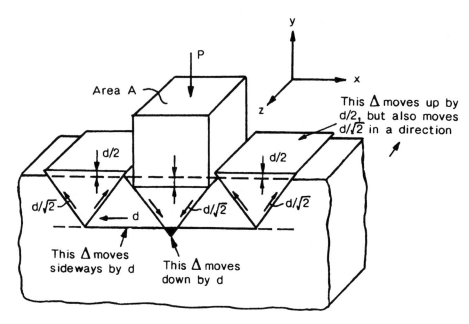

Fig. 16.9 A simple two-dimensional model of slip-line analysis for indentation of a solid body with a flat rectangular punch. Adapted from Kalpakjian (1985).

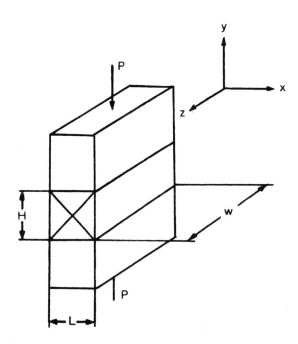

Fig. 16.10 Slip-line field for frictionless compression. Adapted from Hosford and Caddell (1983).

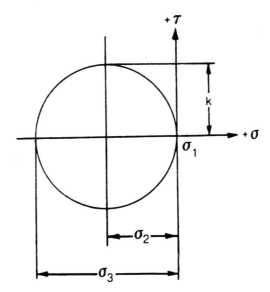

Fig. 16.11 Mohr's circle for frictionless compression.

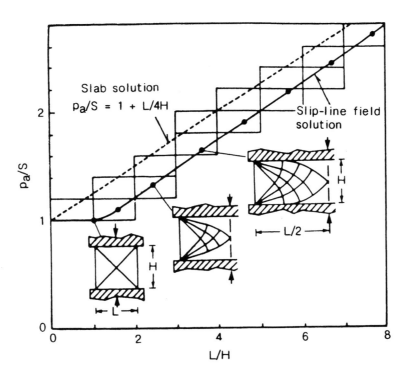

Fig. 16.12 Average pressure versus L/h for compression with sticking friction. (From William F. Hosford/Robert M. Caddell, METAL FORMING: Mechanics and Metallurgy, Copyright 1983, Reprinted by permission of Prentice-Hall, Inc., Englewood Cliffs, NJ.)

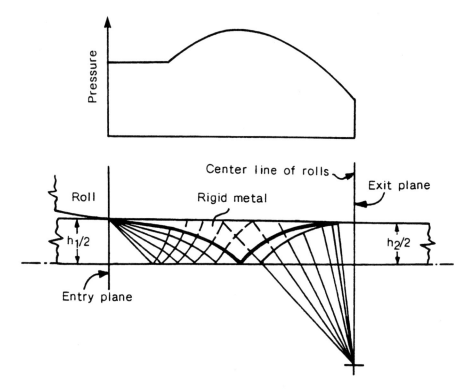

Fig. 16.13 A slip-line field and pressure distribution for hot rolling process (R/h₁ = 80; r = 0.1). Adapted from Ford and Alexander (1963-64).

REFERENCES

1. W.F. Hosford and R.M. Caddell, Mechanics and Metallurgy, Prentice-Hall, Inc., Englewood Cliffs, New Jersey, pp. 143-238 (1983).

2. J.W. Green and J.F. Wallace, "Estimation of Load and Torque in the Hot Rolling Process", Journal of Mechanical Engineering Science, Vol. 4, No. 2, 1962, pp. 136-142.

3. V. Piispanen, "Plastic Deformation of Metal: Theory of Simulated Sliding", Wear, Vol. 38, No. 1, pp. 43-72.

4. S. Kalpakjian, Manufacturing Processes for Engineering Materials, Addison-Wesley Publishing Company, Inc., Reading, Massachusetts, pp. 90, 91 (1984).

5. W. Johnson and P.B. Mellor, Engineering Plasticity, Van Nostrand Reinhold, New York, pp. 381-389, 392. 402-406 (1973).

6. J.M. Alexander, "A Slip Line Field for the Hot Rolling Process", Proceedings of the Institution of Mechanical Engineers, 169, 1955, pp. 1021-1030.

7. H. Ford and J.M. Alexander, "Simplified Hot Rolling Calculations", Journal of the Institute of Metals, Vol. 92 (1963-64), pp. 397-404.

Part V

Calculation of
Rolling Parameters

17

Resistance to Deformation in Hot Rolling

17.1 DEFINITION OF THE RESISTANCE TO DEFORMATION

The resistance to deformation of the material rolled without tension is usually determined as

$$K_w = \frac{P}{F_d} \tag{17-1}$$

where K_w = resistance to deformation

P = roll separating force

F_d = projected area of contact between roll and workpiece.

The roll separating force can be determined if the distribution of normal pressure p_x in the deformation zone (Figs. 17.1 and 17.2) is known.

$$P = \int_0^{l_d} p_x dx = \int_0^{\alpha} R p_\theta \, d\theta, \tag{17-2}$$

where p_x = normal pressure at distance x from the exit plane

p_θ = normal pressure at the roll angle θ

l_d = projected arc of contact between roll and workpiece

α = roll bite angle.

Since both entry and exit strip tensions would reduce the roll separating force, therefore, in order to correctly determine the resistance to deformation of the material rolled with tension, the Eq. (17-1) shall be modified as follows

$$K_w = \frac{P}{F_d} + (\beta_1 S_1 + \beta_2 S_2), \tag{17-3}$$

where S_1, S_2 = entry and exit strip tension respectively

β_1, β_2 = coefficients for entry and exit strip tension respectively.

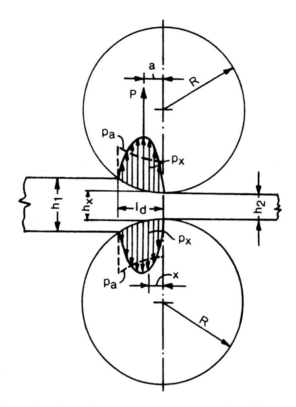

Fig. 17.1 Distribution of normal pressure and roll separating force in rolling.

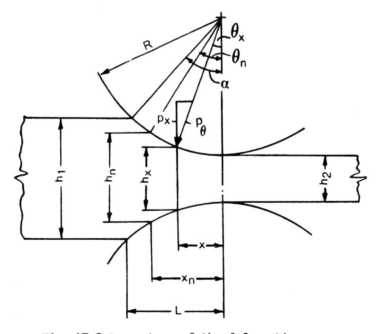

Fig. 17.2 Parameters of the deformation zone.

17.2 PROJECTED AREA OF THE ROLL CONTACT

The projected area of contact between the roll and workpiece is equal to

$$F_d = wl_d,$$ (17-4)

where w = mean width.

When the roll flattening cannot be ignored, the projected arc of contact is expressed as

$$l_d = L' = \sqrt{R'\Delta - \frac{\Delta^2}{4}} \approx \sqrt{R'\Delta},$$ (17-5)

In the case when the roll flattening is negligible, Eq. (17-5) reduces to

$$l_d = L = \sqrt{R\Delta - \frac{\Delta^2}{4}} \approx \sqrt{R\Delta},$$ (17-6)

If the rolls are of unequal diameters, the radius in Eqs. (17-5) and (17-6) has to be substituted with the mean radius that is expressed by [1]

$$R_m = \frac{2R_1R_2}{R_1 + R_2}$$ (17-7)

where R_1, R_2 = radius of the first and second roll.

17.3 MEAN WORKPIECE WIDTH

The mean workpiece width in the rolling deformation zone is usually described in one of the following three terms:

1. Arithmetic average width (w_a)
2. Parabolic mean width (w_p)
3. Geometric mean width (w_g).

Arithmetic average width is determined by assuming that the curve of spread is a straight line, thus

$$w = w_a = \frac{w_1 + w_2}{2},$$ (17-8)

where w_1, w_2 = entry and exit workpiece width respectively.

Parabolic mean width is determined by approximating the curve of spread with a parabola, therefore

$$w = w_p = \frac{w_1 + 2w_2}{3}$$ (17-9)

Geometric mean width is defined as

$$w = w_g = \sqrt{w_1w_2}$$ (17-10)

17.4 PARAMETERS AFFECTING THE RESISTANCE TO DEFORMATION

The following main parameters affect the resistance to deformation in rolling:

1. Material chemical composition
2. Material metallurgical characteristics
3. Material temperature
4. Geometry of the deformation zone
5. External friction in the deformation zone
6. Material work hardening prior to the rolling pass under consideration
7. Strain rate of deformation.

The present state of the art in rolling theory does not allow one to derive a comprehensive analytical relationship between the resistance to deformation and the parameters listed above. A practical solution to the problem was found in an approach that consists of the following two steps:

Step 1 - This step includes determining the yield characteristics of the workpiece material from the test run under controlled conditions.

Step 2 - This step involves determining the correlation between the resistance to deformation corresponding to prerequisite rolling conditions and the yield characteristics of the workpiece material determined during the test in step 1.

Three kinds of tests are mostly common:

1. Laboratory non-rolling tests such as tension, compression and torsion tests
2. Laboratory small scale rolling tests
3. Field real scale rolling tests.

It is obvious that the most difficult task is to establish an accurate relationship between the resistance to deformation K_w and the workpiece yield stress Y_w measured in laboratory non-rolling tests. This is due to a substantial difference in both the characteristics of the deformation process and actual test conditions (temperature, strain rate, friction, etc.).

Better results may be achieved when the yield stress data are obtained from the laboratory small scale rolling tests, and especially, from the field real scale rolling tests. The latter tests are especially advantageous in application for hot rolling due to a possibility to conduct the measurements at the real rolling temperatures, speeds, surface scale conditions, etc.

Before proceeding with a review of different methods for calculating the resistance to deformation, let us briefly describe two parameters which are frequently used in those methods. They are the aspect ratio of the deformation zone and the mean strain rate.

17.5 ASPECT RATIO OF THE DEFORMATION ZONE

The aspect ratio of the rolling deformation zone is described in one of the following three terms:

1. Arithmetic average aspect ratio, Z_a
2. Parabolic mean aspect ratio, Z_p
3. Geometric mean aspect ratio, Z_g.

Arithmetic average aspect ratio is determined by approximating the arc of roll contact with a straight line, thus

$$Z_a = \frac{l_d}{h_a} = \frac{2l_d}{h_1 + h_2} \tag{17-11}$$

where h_a = average workpiece thickness.

Parabolic mean aspect ratio is determined by assuming that the arc of roll contact is a parabola, therefore

$$Z_p = \frac{l_d}{h_p} + \frac{3l_d}{h_1 + 2h_2} \tag{17-12}$$

where h_p = parabolic mean workpiece thickness.

Geometric mean aspect ratio is defined as

$$Z_g = \frac{l_d}{h_g} = \frac{l_d}{\sqrt{h_1 h_2}}, \tag{17-13}$$

where h_g = geometric mean workpiece thickness.

17.6 STRAIN RATE IN FLAT ROLLING

The engineering strain rate was previously defined as

$$\dot{e} = \frac{de}{dt} = \frac{1}{L_o} \frac{dL}{dt}, \tag{17-14}$$

where L_o = initial length of a deformed body

dL = change in length of the body after deformation is completed

dt = time required to deform the body.

Below is a review of some solutions of the Eq. (17-14) in application to flat rolling.

Ford and Alexander's solution - According to the solution proposed by Ford and Alexander [2], the strain rate at any plane of the deformation zone (Fig. 17.2) is expressed by

$$\lambda_x = \frac{1}{h_x} \frac{dh_x}{dt} = \frac{1}{h_x} \frac{dh_x}{dx} \frac{dx}{dt}, \tag{17-15}$$

where x = distance from exit plane in the roll gap

h_x = strip thickness at distance x.

In its turn

$$\frac{dx}{dt} = V_x = V_n \frac{h_n}{h_x} \cos\theta_n, \tag{17-16}$$

where V_x = horizontal velocity of the strip at distance x

V_n = roll peripheral speed

h_n = strip thickness at neutral plane

θ_n = roll contact angle at neutral plane.

Since $\cos\theta_n$ is negligibly different from unity, the Eq. (17-16) reduces to

$$\frac{dx}{dt} = V_n \frac{h_n}{h_x} \tag{17-17}$$

Also, from the geometry of the roll bite we obtain that

$$\frac{dh_x}{dx} = \frac{d}{dx} \left(h_2 + \frac{x^2}{R}\right) = \frac{2x}{R} \tag{17-18}$$

Substitution of Eqs. (17-17) and (17-18) into Eq. (17-15) yields

$$\lambda_x = \frac{2x}{h_x^2} \frac{V_n h_n}{R}, \tag{17-19}$$

V_n and h_n may be expressed as

$$V_n = 2\pi RN/60 \qquad h_n = h_2 + \frac{x_n^2}{R}$$

$$x_n = \frac{h_2 L}{h_1 + h_2}$$

where N = roll peripheral speed, rpm.

Then λ_x can be presented in terms of r, V_n, R, and h_1 as follows

$$\lambda_x = \frac{2V_n}{\sqrt{Rh_1}} \frac{(4 - 3r)(1 - r)\sqrt{r - r_x}}{(2 - r)^2 (1 - r_x)^2} \tag{17-20}$$

where $r_x = \dfrac{h_1 - h_x}{h_1}$

The mean strain rate for any pass is obtained by integrating both parts of Eq. (17-19)

$$\lambda = \frac{1}{L} \int_0^L \lambda_x \, dx = \frac{V_n h_n}{L} \left(\frac{1}{h_2} - \frac{1}{h_1}\right) \tag{17-21}$$

This leads to the final equation for mean strain rate

$$\lambda = \frac{\pi N}{30} \sqrt{\frac{Rr}{h_1}} \frac{4 - 3r}{(2 - r)^2} \tag{17-22}$$

A good approximation of the last equation is

$$\lambda = \frac{\pi N}{30} \sqrt{\frac{R}{h_1}} \left(1 + \frac{r}{4}\right)\sqrt{r} \tag{17-23}$$

Sims' solution - According to Sims [3], the strain rate at any plane of the deformation zone is given by

$$\lambda_\theta = \frac{1}{h_\theta} \frac{dh_\theta}{d\theta} \frac{d\theta}{dt} ,$$

(17-24)

where h_θ = strip thickness corresponding to the roll contact angle θ (Fig. 17.2).

The mean strain rate is obtained by integrating both parts of Eq. (17-24)

$$\lambda = \frac{1}{\alpha} \int_0^\alpha \lambda_\theta \, d\theta = \frac{1}{\alpha} \int_0^a \frac{1}{h_\theta} \frac{dh_\theta}{d\theta} \frac{d\theta}{dt} \, d\theta ,$$

(17-25)

where α = bite angle.

Since

$$\frac{dh_\theta}{d\theta} = \frac{d}{d\theta} (h_2 + R\theta^2) = 2R\theta$$

and $\quad \dfrac{d\theta}{dt} = \dfrac{\pi N}{30} ,$

we obtain the equation for mean strain rate in the form

$$\lambda = \frac{\pi N}{30} \sqrt{\frac{R}{h_1}} \frac{1}{\sqrt{r}} \ln\left(\frac{1}{1-r}\right)$$

(17-26)

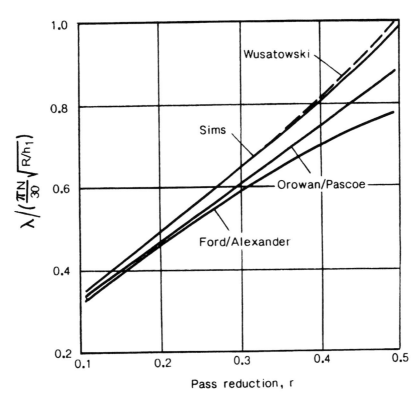

Fig. 17.3 Comparison of the mean strain rates calculated by different formulae.

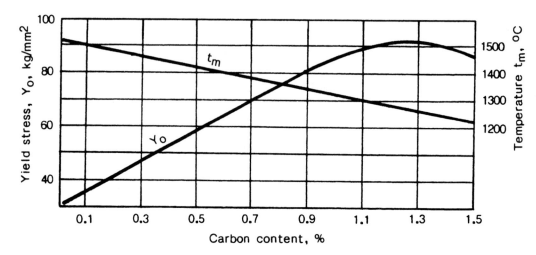

Fig. 17.4 Diagram for determination of yield stress at 20°C and melting point for carbon steels as a function of carbon content. (From FUNDAMENTALS OF ROLLING, by Z. Wusatowski. Copyright Wydawnictwo Slask, Katowice, Poland. Reprinted with permission).

Orowan and Pascoe's solution - The solution for the mean strain rate provided by Orowan and Pascoe [4] is given by

$$\lambda = \frac{\pi N}{30} \sqrt{\frac{R}{h_1}} \left(\frac{1 - 0.75r}{1 - r} \right) \sqrt{r} \qquad (17\text{-}27)$$

Wusatowski's solution - Formula proposed by Wusatowski [1] to calculate the mean strain rate has a form

$$\lambda = \frac{\pi N}{30} \sqrt{\frac{R}{h_1}} \sqrt{\frac{r}{1 - r}} \qquad (17\text{-}28)$$

Figure 17.3 illustrates a comparison of calculation of the relative mean strain rate provided by different authors.

17.7 GOLOVIN-TIAGUNOV'S METHOD OF CALCULATION OF K_w

According to this method the resistance to deformation is determined as [1]

$$K_w = [1 + \mu(Z_a - 1)]k_t Y_0, \qquad (17\text{-}29)$$

where μ = coefficient of external friction

Z_a = arithmetic average aspect ratio of the deformed zone

k_t = temperature effect coefficient

Y_0 = yield stress at 20°C as shown in Fig. 17.4.

Coefficient k_t is obtained from the relations:

$$k_t = \frac{t_m - t - 75}{1500} \quad \text{when } t \geq t_m - 575^oC$$

(17-30)

$$\varsigma_t = \left(\frac{t_m - t}{1000}\right)^2 \quad \text{when } t < t_m - 575^oC$$

where t_m = melting temperature, ^oC (see Fig. 17.4)

$\quad\quad$ t = rolling temperature, ^oC.

17.8 TSELIKOV'S METHOD OF CALCULATION OF K_w

A number of formulas for calculation of the resistance to deformation in hot rolling have been proposed by Tselikov [1, 5]. One of the simplified formulas may be rewritten as

$$K_w = \frac{1.15Y(2 - r)}{r\delta} \left[\left(\frac{1}{1 - r}\right)^{\delta/2} - 1\right]$$

(17-31)

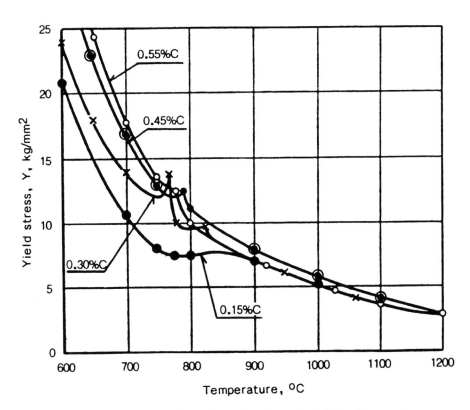

Fig. 17.5 Yield stress of carbon steels related to temperature.
(From FUNDAMENTALS OF ROLLING, by Z. Wusatowski, 1969. Copyright
Wydawnictwo Slask, Katowice, Poland. Reprinted with permission.)

where $\delta = 2\mu L/\Delta$

    r = pass reduction

    μ = coefficient of external friction

    L = contact length of unflattened roll

    Δ = draft

    Y = yield stress of the rolled material corresponding to a given rolling temperature as shown in Fig. 17.5.

Tselikov has also proposed the following general equation for calculating the resistance to deformation [6]:

$$K_w = k_\sigma k_v k_h Y \tag{17-32}$$

where k_σ = stress state effect coefficient

    k_v = speed effect coefficient

    k_h = hardening effect coefficient

    Y = yield stress of the rolled material at a given rolling temperature.

In its turn the stress state effect coefficient is given by:

$$k_\sigma = \gamma k'_\sigma k''_\sigma k'''_\sigma \tag{17-33}$$

where $\gamma = 1.0\text{-}1.15$ ($\gamma = 1.15$ for plane strain conditions, i.e., during rolling without spread; $\gamma = 1.0$ during rolling with free spreading)

    k'_σ = friction effect coefficient

    k''_σ = external zones effect coefficient

    k'''_σ = entry and exit tension effect coefficient.

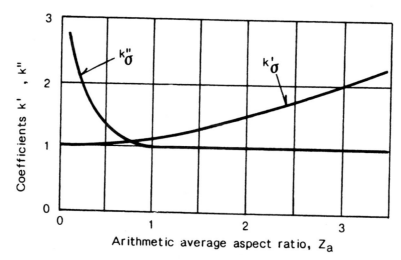

Fig. 17.6 Dependence of coefficients k'_σ and k''_σ on arithmetic average aspect ratio. Adapted from Tselikov (1958).

The coefficients k_σ' and k_σ'' are expressed as the functions of the aspect ratio Z_a as shown in Fig. 17.6. Both coefficients are derived for the case when $k_\sigma''' = 1$ and coefficient k_σ is calculated assuming that $\mu = 0.4$ and $r = 0.2$.

17.9 EKELUND'S METHOD OF CALCULATION OF K_W

The following formula for resistance to deformation may be written using Ekelund's method [1]:

$$K_W = \left(1 + \frac{0.8\mu L - 0.6\Delta}{h_a}\right)\left(Y_c + \frac{\eta V\sqrt{\Delta/R}}{h_a}\right), \tag{17-34}$$

where V = peripheral rolling speed, mm/sec

R = roll radius, mm

η = coefficient of plasticity, kg sec/mm^2

Y_c = yield stress of the rolled material corresponding to a given temperature and chemical composition, kg/mm^2

h_a = average strip thickness, mm.

The coefficient of friction μ is calculated as a function of rolling temperature t, type of the rolls and their surface conditions:

a) for cast iron and rough steel rolls

$\mu = 1.05 - 0.0005t$

b) for chilled and smooth steel rolls $\tag{17-35}$

$\mu = 0.8(1.05-0.0005t)$

c) for ground steel rolls

$\mu = 0.55(1.05-0.0005t)$

The yield stress Y_c is calculated as a function of both rolling temperature and chemical composition of a workpiece

$$Y_c = (14 - 0.01t)(1.4 + C + Mn + 0.3Cr), \tag{17-36}$$

where C, Mn, Cr = content of carbon, manganese, and chromium respectively, %

t = rolling temperature, $^\circ$C.

The coefficient of plasticity of the rolled stock is given by

$$\eta = 0.01(14 - 0.01t) \tag{17-37}$$

The Ekelund's formulas are presumably valid for the following conditions: minimum rolling temperature is 800°C, maximum rolling speed is 7 m/sec, and maximum manganese content is 1%.

17.10 OROWAN-PASCOE'S METHOD OF CALCULATION OF K_W

The following three formulae for resistance to deformation can be obtained using Orowan-Pascoe method [4].

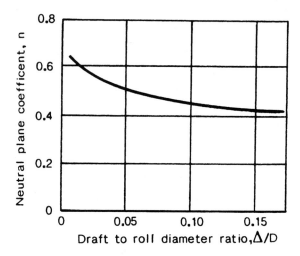

Fig. 17.7 Relationship between the neutral plane location coefficient and the draft to roll diameter ratio. (From FUNDAMENTALS OF ROLLING, by Z. Wusatowski, 1969. Copyright Wydawnictwo Slask, Katowice, Poland. Reprinted with permission.)

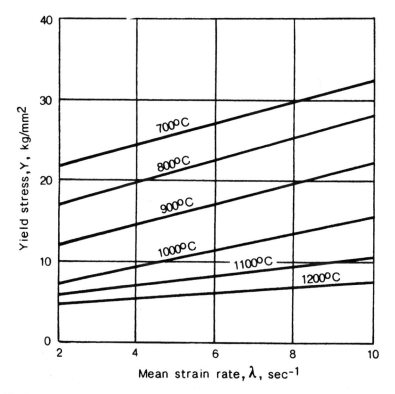

Fig. 17.8 Yield stress of carbon steel (0.28% C) related to rate of deformation and temperature. (From FUNDAMENTALS OF ROLLING, by Z. Wusatowski, 1969. Copyright Wydawnictwo Slask, Katowice, Poland. Reprinted with permission.)

Case 1 - Width of the rolled stock is 6 to 8 times greater than the mean workpiece thickness

$$K_w = 1.15Y \frac{w_1}{w_p} \left(0.8 + \frac{nL}{2h_2} \right) , \qquad (17\text{-}38)$$

where w_1, w_p= entry and parabolic mean width of the workpiece

h_2 = exit thickness of the workpiece

n = neutral plane location coefficient

Y = yield stress of the rolled material corresponding to a given rolling temperature and strain rate.

Case 2 - Width of the rolled stock is more than 1.5 to 2 times greater than the mean workpiece thickness

$$K_w = 1.15Y \left[0.8 + \frac{nL}{2h_2} - \frac{h_2}{3Z_p w_p n} \left(\frac{nL}{h_2} - 0.2 \right)^3 \right] , \qquad (17\text{-}39)$$

where Z_p = parabolic mean aspect ratio of the deformed zone.

Case 3 - Width of the rolled stock is 1.5 to 2 times smaller than the mean thickness

$$K_w = 1.15Y \qquad (17\text{-}40)$$

The neutral plane location coefficient n is expressed as a function of the draft to roll diameter ratio (Fig. 17.7). The values for the yield stress Y are shown in Fig. 17.8 as a function of rolling temperature and strain rate. The strain rate λ is given by Eq. (17-27).

17.11 GELEJI'S METHOD OF CALCULATION OF K_w

By using the Geleji's method [1], the resistance to deformation can be presented as

$$K_w = 1.15Y \left(1 + c\mu Z_a \sqrt[4]{V} \right) , \qquad (17\text{-}41)$$

where Y = yield stress of the rolled material corresponding to a given temperature, kg/mm^2

V = peripheral rolling speed, m/sec

c = geometrical factor

Z_a = arithmetic average aspect ratio of the deformed zone.

The geometrical factor c is given as a function of the average aspect ratio Z_a and is calculated for two ranges of Z_a as follows:

If $0.25 < Z_a \leqslant 1$, then

$$c = 17Z_a^2 - 29.85Z_a + 18.3 \qquad (17\text{-}42)$$

If $1 < Z_a \leqslant 3$, then

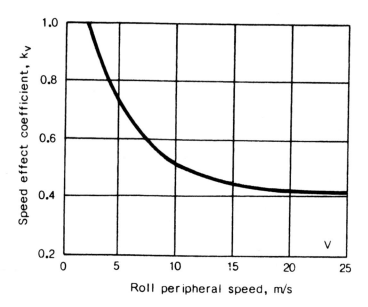

Fig. 17.9 Speed effect coefficient.

$$c = 0.8Z_a^2 - 4.9Z_a + 9.6 \qquad\qquad (17\text{-}43)$$

The coefficient of friction μ is calculated as a function of rolling temperature t, type of rolls and their surface conditions, and rolling speed V:

    a) for steel rolls

      $\mu = (1.05 - 0.0005t)K_V$

    b) for hardened steel rolls

      $\mu = (0.92 - 0.0005t)K_V \qquad\qquad (17\text{-}44)$

    c) for hardened and ground steel rolls

      $\mu = (0.82 - 0.0005t)K_V,$

where K_V = speed effect coefficient (Fig. 17.9).

The yield stress Y is given by

$$Y = 0.015(1400 - t) \qquad\qquad (17\text{-}45)$$

Eq. (17-45) is claimed to be valid for carbon steels of a tensile strength up to 60 kg/mm^2, and a temperature range of 800 to 1300°C.

17.12 SIMS' METHOD OF CALCULATION OF K_W

The Sims' method [3] is based on the slab analysis of rolling deformation with an assumption that only sticking friction exists in the roll contact zone. According to this method, the resistance to deformation can be presented in the form:

$$K_w = Q_p S_p, \tag{17-46}$$

where S_p = mean constrained yield stress corresponding to a given temperature, strain rate and reduction

Q_p = geometrical factor that is given by

$$Q_p = \left[\frac{\pi}{2a} \tan^{-1} a - \frac{\pi}{4} - \frac{1}{a} \sqrt{\frac{R'}{h_2}} \left(\ln \frac{h_n}{h_2} + \frac{1}{2} \ln \frac{a^2}{r} \right) \right], \tag{17-47}$$

where $a = r(1 - r)$ \hfill (17-48)

h_n = strip thickness at neutral point

R' = flattened roll radius.

The values for geometrical factor Q_p are plotted in Fig. 17.10.

The mean constrained yield stress S_p is obtained from the equation:

$$S_p = \frac{1}{\alpha} \int_0^\alpha S \, d\theta, \tag{17-49}$$

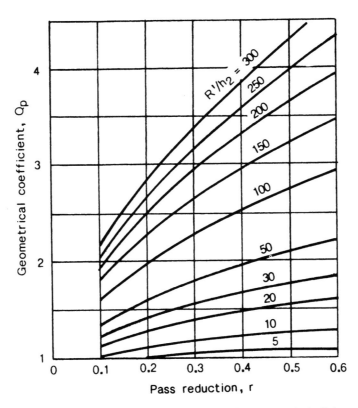

Fig. 17.10 Sims' geometrical factor. (From Sims. Reprinted by permission of the Council of the Institution of Mechanical Engineers from the Proceedings of Institution of Mechanical Engineers, No. 168, 1954.)

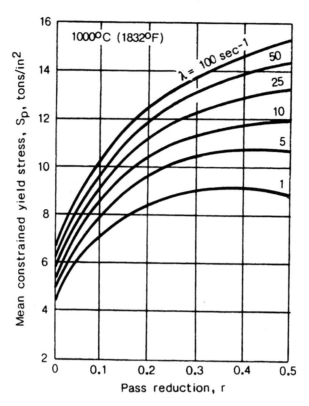

Fig. 17.11 Mean constrained yield stress used for force calculations in rolling of carbon steel (0.17% C) at 1000°C. (From Sims. Reprinted by permission of the Council of the Institution of Mechanical Engineers from the Proceedings of Institution of Mechanical Engineers, No. 168, 1954.)

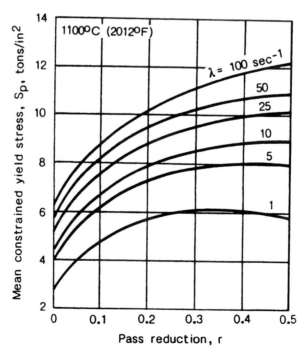

Fig. 17.12 Mean constrained yield stress used for force calculations in rolling of carbon steel (0.17% C) at 1100°C. (From Sims. Reprinted by permission of the Council of the Institution of Mechanical Engineers from the Proceedings of Institution of Mechanical Engineers, No. 168, 1954.)

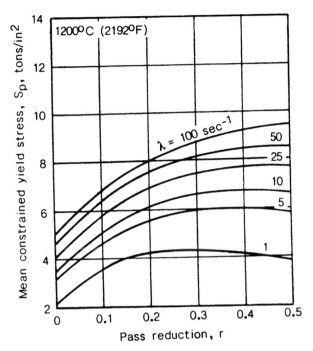

Fig. 17.13 Mean constrained yield stress used for force calculations in rolling of carbon steel (0.17% C) at 1200°C. (From Sims. Reprinted by permission of the Council of the Institution of Mechanical Engineers from the Proceedings of Institution of Mechanical Engineers, No. 168, 1954.)

where S = constrained yield stress in plane strain compression

θ = roll contact angle

α = bite angle.

The values for the mean constrained yield stress S_p are shown in Figs. 17.11-17.13.

17.13 ALEXANDER-FORD'S METHOD OF CALCULATION OF K_w

According to Alexander and Ford method [2], for the case when the roll flattening is negligible, the resistance to deformation can be expressed as

$$K_w = 2k_g k = k_g S, \tag{17-50}$$

where k = shear yield stress of the rolled material

S = constrained yield stress of the rolled material for plane strain compression

k_g = geometrical coefficient that is given by

$$k_g = 0.25(\pi + Z_a), \tag{17-51}$$

where Z_a = arithmetic average aspect ratio of the deformation zone.

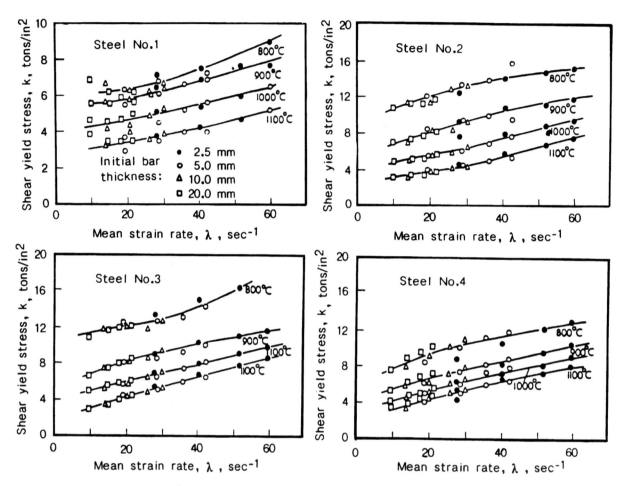

Fig. 17.14 Shear yield stress for some steels (see Table 17.1).
(From Gupta and Ford, Journal of the Iron and Steel Institute,
Vol. 205, Feb. 1967. Reprinted with permission.)

When the roll flattening effect can not be neglected and the flattened roll radius is equal to

$$R' = R\left(1 + \frac{cP}{w\Delta}\right),$$ (17-52)

where $c = \dfrac{16(1 - \nu_r^2)}{\pi E_r}$, (17-53)

then the resistance to deformation may be derived as [7]:

$$K_w' = \frac{K_w}{1 - Xk}, $$ (17-54)

where $X = \dfrac{\pi c}{2\ r}\sqrt{\dfrac{R}{h_1}}\ \dfrac{2c}{2 - r}\ \dfrac{R}{h_1}$ (17-55)

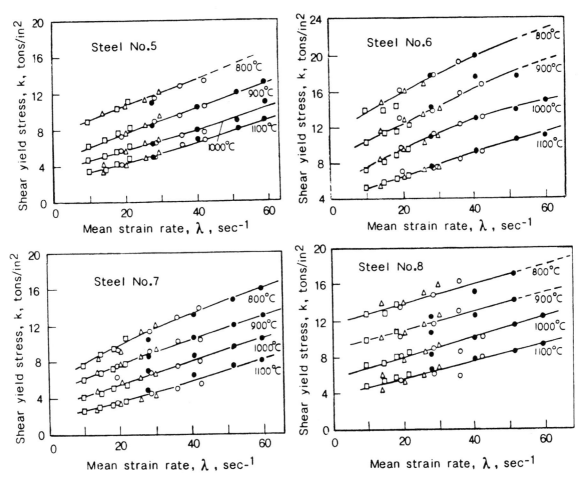

Fig. 17.15 Shear yield stress for some steels. (From Gupta and Ford, Journal of the Iron and Steel Institute, Vol. 205, 1967. Reprinted with permission.)

Table 17.1 Chemical composition of steels (See Figs. 17.14 and 17.15).

Steel No.	Content, %								
	C	Si	Mn	P	S	Sr	Ni	W	Mo
1	0.10	....	0.47	0.063	0.026	....	....	....	....
2	1.03	0.22	0.27	0.030	0.024	....	....	....	....
3	1.00	0.30	0.26	0.029	0.013	1.19	....	....	....
4	0.13	0.25	0.55	0.020	0.018	0.75	2.94	....	....
5	0.55	0.26	0.46	0.017	0.013	0.95	....	....	0.31
6	0.22	0.81	0.50	0.020	0.015	24.0	0.30	....	....
7	0.08	0.45	0.67	0.015	0.015	18.2	9.90	....	0.66
8	0.10	0.50	0.40	0.016	0.017	16.7	20.6	1.18	....

The values for shear yield stress k are plotted in Fig. 17.14 and Fig. 17.15 for the alloys with chemical composition shown in Table 17.1. In the charts, the values for the strain rate λ are calculated by using Eq. (17-23).

17.14 DENTON-CRANE'S METHOD OF CALCULATION OF K_w

Denton and Crane [8] have based their method on examination of the slip line fields solutions which shows that rolling geometry can be conveniently presented by the geometric mean aspect ratio.

According to this method, the equation for resistance to deformation can be expressed in the form

$$K_w = 2k_g k = k_g S, \tag{17-56}$$

where k_g = geometrical coefficient that is given by

$$k_g = 0.655 + 0.265 Z_g, \tag{17-57}$$

where Z_g = geometric mean aspect ratio of the deformed zone.

17.15 GREEN-WALLACE'S METHOD OF CALCULATION OF K_w

According to Green and Wallace's upper-bound analysis of rolling deformation [9], the resistance to deformation can be expressed as

$$K_w = 2k_g k = k_g S, \tag{17-58}$$

where k_g = geometrical coefficient that is given by

$$k_g = 0.5\left(Z_a + \frac{1}{Z_a}\right), \tag{17-59}$$

where Z_a = arithmetic average aspect ratio of the deformed zone.

17.16 YOKOI ET AL'S METHOD OF CALCULATION OF K_w

According to Yokoi et al's method [10], the resistance to deformation in application for slabbing mills is given by

$$k = 1.15 k_g K_m, \tag{17-60}$$

where K_m = mean rolling pressure corresponding to a given rolling temperature, rolling speed, reduction, roll radius, and the carbon content in the rolled material, kg/mm^2

k_g = geometrical coefficient.

The mean rolling pressure is presented as

$$\ln K_m = a_0 + a_1 C + a_2 C^2 + \frac{1}{T_k}(a_3 + a_4 C + a_5 C^2)$$

$$+ a_6 \left[\ln\left(\frac{\pi N}{30}\right) + \frac{1}{2} \ln\left(\frac{R}{h_2}\right) + \frac{1}{2} \ln r \right] + a_7 \left\{ \ln\left[\ln\left(\frac{h_1}{h_2}\right)\right]\right\}, \qquad (17\text{-}61)$$

where h_1, h_2 = entry and exit thicknesses respectively, mm

$\qquad T_k$ = rolling temperature, $^\circ$K

$\qquad N$ = roll peripheral speed, rpm

$\qquad R$ = roll radius, mm

$\qquad r$ = reduction

$\qquad C$ = carbon content in rolled material, %.

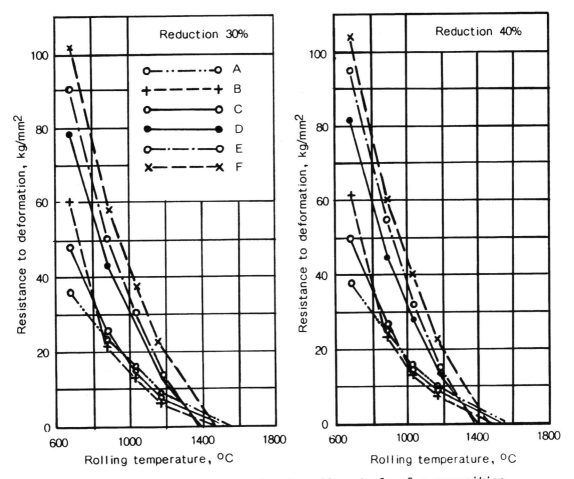

Fig. 17.16 Resistance to deformation for alloy steels of a composition shown in Table 17.2, used in Siebel's method, based on measurements of Pomp and Weddige. (From FUNDAMENTALS OF ROLLING, by Z. Wusatowski, 1969. Copyright Wydawnictwo Slask, Katowice, Poland. Reprinted with permission.)

Table 17.2 Chemical composition of steels (See Fig. 17.16).

Steel type	Content, %								
	C	Si	Mn	P	S	Cr	Ni	W	Al
A	0.11	0.22	0.50	0.020	0.018	....	....	....	....
B	0.88	0.18	0.63	0.014	0.016	....	....	....	....
C	0.06	1.19	0.29	0.010	0.002	22.5	0.14	....	2.25
D	0.11	0.63	0.64	0.015	0.026	18.4	9.10	....	....
E	0.14	1.90	0.09	0.015	0.010	25.0	20.5	.....	....
F	0.47	1.98	0.85	0.015	0.010	15.4	13.1	1.95	....

The constants a_0 through a_7 have the following values:

$$a_0 = 0.126, \quad a_1 = -1.75, \quad a_2 = 0.594, \quad a_3 = 2851$$
$$a_4 = 2968, \quad a_5 = -1120, \quad a_6 = 0.13, \quad a_7 = 0.21$$

The geometrical coefficient k_g is expressed as

$$k_g = \left[1 - b_0 \exp\left(b_1 \frac{w_2}{h_1}\right)\right]\left(b_2 Z_p + \frac{b_3}{Z_p} + b_4\right) , \tag{17-62}$$

where w_2 = exit width, mm

Z_p = parabolic mean aspect ratio.

The constants b_0 through b_4 have the following values:

$b_0 = 0.53$, $b_1 = -1.66$, $b_2 = 0.24$, $b_3 = 0.28$, $b_4 = 0.39$.

17.17 SKF'S METHOD OF CALCULATION OF K_w

According to SKF's method [11], the resistance to deformation can be expressed as

$$K_w = k_v K_m, \tag{17-63}$$

where k_v = speed effect coefficient

K_m = mean rolling pressure corresponding to a given temperature t, roll speed N, reduction r, and entry thickness to roll diameter ratio h_1/D.

In this method, the values for the speed effect coefficient k_v as well as the values for the mean rolling pressure K_m are presented in graphical form.

17.18 SIEBEL'S METHOD OF CALCULATION OF K_w

Siebel [12] utilized the actual rolling data to compute the resistance to deformation. Figure 17.16 presents the resistance to deformation for six alloy steels. The chemical composition of these steels is shown in Table 17.2.

17.19 RIDE'S METHOD OF CALCULATION OF K_w

Ride [13] has utilized a statistical analysis of the rolling mill data in order to derive an empirical formula for calculating the roll separating force. In this formula each of the parameters affecting the roll separating force was included as a parabolic function. The resulting equation is obtained from a correlation by the method of least squares. In relation to the resistance to deformation, it takes the form

$$K_w = (a_0 + a_1V + a_2V^2 + a_3r + a_4r^2 + a_5h_2 + a_6h_2^2 + a_7T + a_8T^2)/l_d, \qquad (17\text{-}64)$$

where a_0 through a_8 = empirical constants

V = roll peripheral speed

T = strip temperature, ^oF

l_d = roll contact length.

For the work roll diameter of approximately 28 in. the constants a_0 through a_8 have the following values:

$a_0 = 102.278$ $a_1 = -0.0012$ $a_2 = -0.000013$

$a_3 = 1.494$ $a_4 = -0.014$ $a_5 = -106.363$

$a_6 = 71.888$ $a_7 = -0.070$ $a_8 = 0.000018.$

17.20 SCHULTZ-SMITH'S METHOD OF CALCULATION OF K_w

Schultz and Smith [14] have derived the empirical equation for the resistance to deformation by applying the multiple regression technique to available test data in order to determine which term in the equation was most significant. It was found that the general equation could be reduced to the following form:

$$\ln K_w = b_0 + b_1 \ln \frac{R}{h_1} + b_2 \ln r + b_3 \ln r \ln \frac{R}{h_1} + b_4 T$$
$$+ b_5 T \ln r + b_6 \ln r \ln\left(\frac{R}{h_1}\right)^2 + b_7 T^2 \quad , \qquad (17\text{-}65)$$

where $b_0 - b_7$ = constants

T = rolling temperature.

17.21 SYKES' METHOD OF CALCULATION OF K_w

The Sykes' method [15] is based on utilization of so-called specific power curves and associated rolling temperature curves which are developed by using the rolling test data.

In developing the specific power curves, the increments in specific power required for rolling of each pass are first calculated

$$\delta H = \frac{(HP)}{0.36 \rho h_2 w_2 V} \quad , \qquad (17\text{-}66)$$

where δH = specific power, (Hp-hours)/ton

(HP) = rolling horse power, Hp

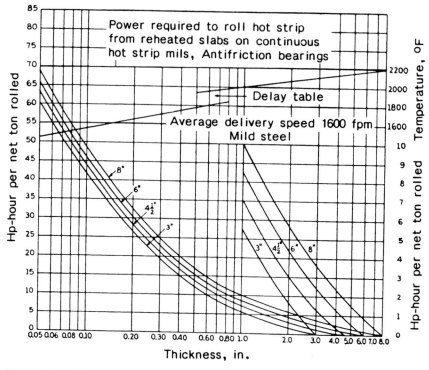

Fig. 17.17 Specific power curve for mild steel. (From Ballenger and Rhea, AISE Yearly Proceedings, 1941. Copyright AISE, Pittsburgh, Pennsylvania. Reprinted with permission.)

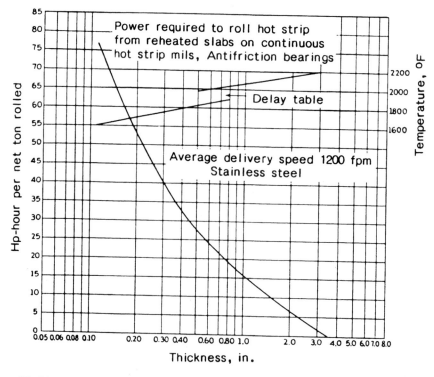

Fig. 17.18 Specific power curve for stainless steel. (From Ballenger and Rhea, AISE Yearly Proceedings, 1941. Copyright AISE, Pittsburgh, Pennsylvania. Reprinted with permission.)

ρ = workpiece density, lb/in.^{3}

h_2, w_2 = exit workpiece thickness and width respectively, in.

V = roll peripheral speed, fpm.

Then the specific power curves are calculated by summing up the increments of the specific power for the previous rolling passes. These curves are usually derived individually for roughing and finishing passes and are usually given with the corresponding rolling temperature curves as shown in Fig. 17.17 and 17.18 [16].

After the specific power curves are developed, then the rolling parameters for the pass reductions different from those used for deriving the power curves can be calculated. The resistance to deformation is computed from the following equation

$$K_w = 11,800 \rho k_t (\delta H) \frac{h_2}{\Delta} \, , \qquad (17\text{-}67)$$

where K_w = resistance to deformation, psi

Δ = draft, in.

k_t = temperature effect coefficient that is given by

$$k_t = 1 + b(T_c - T), \qquad (17\text{-}68)$$

where T_c = temperature read from the rolling temperature curve, $^\circ$F

T = rolling temperature under consideration, $^\circ$F

b = temperature effect constant.

17.22 GINZBURG'S METHOD OF CALCULATION OF K_w

When the restoration process time is shorter than the gap time between the rolling passes, the resistance to deformation depends mainly on the temperature of the rolled material and the geometry of the roll bite [17]. In that case the resistance to deformation can be conveniently expressed in the following form:

$$K_w = K_n k_g k_t, \qquad (17\text{-}69)$$

where K_n = normalized resistance to deformation

k_g = geometrical coefficient

k_t = temperature effect coefficient

The **normalized resistance to deformation** K_n is equal to actual resistance to deformation of the rolled material at selected normalized rolling temperature T_n and normalized aspect ratio Z_n.

For low carbon steel, K_n = 18,000 psi (12.66 kg/mm^2) when T_n = 1800°F (982°C) and $Z_n = Z_a = 1$. In that case, the geometrical coefficient k_g is given by

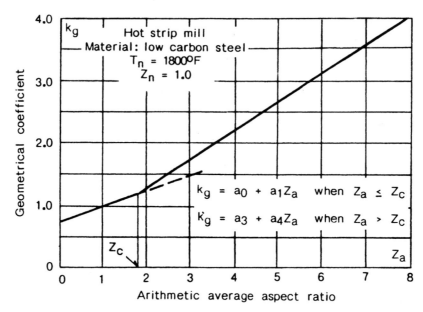

Fig. 17.19 Geometrical coefficient used in calculation of resistance to deformation.

$$k_g = a_0 + a_1 Z_a + a_2 Z_a^2 \qquad \text{when } Z_a \leq Z_c$$
$$k_g = a_3 + a_4 Z_a + a_5 Z_a^2 \qquad \text{when } Z_a > Z_c \qquad\qquad (17\text{-}70)$$

and the temperature effect coefficient k_t is given by

$$k_t = 1 + b_1(T_n - T) + b_2(T_n - T)^2, \qquad\qquad (17\text{-}71)$$

where a_0 through a_5 = geometrical constants

$\qquad\qquad Z_c$ = crossover aspect ratio (Fig. 17.19)

$\qquad\qquad T_n$ = normalized rolling temperature, °F

$\qquad b_1, b_2$ = temperature effect constants

$\qquad\qquad T$ = actual rolling temperature, °F

The crossover aspect ratio Z_c corresponds to the intersection of the curves described by Eq. (17-70).

In the example for low carbon steel shown in Fig. 17.19, the following values for the constants are used:

$a_0 = 0.783 \quad a_1 = 0.217 \quad a_3 = 0.365 \quad a_4 = 0.45 \quad b_1 = 0.001$
$a_2 = a_5 = b_2 = 0.$

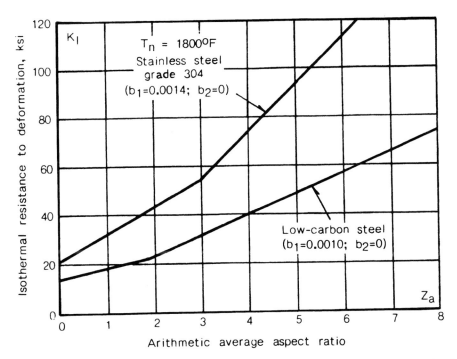

Fig. 17.20 Isothermal resistance to deformation of low-carbon steel and austenitic stainless steel grade 304.

When the resistance to deformation is expressed as a function of a number of variables at a selected normalized rolling temperature, it is called the **isothermal resistance to deformation.**

Application of the isothermal resistance to deformation allows one to standardize and to readily compare the rolling characteristics of different materials. Indeed, at the constant temperature, the main remaining factor affecting the resistance to deformation is the roll bite geometry. Therefore, the isothermal resistance to deformation is given by

$$K_I = K_n k_g \qquad\qquad (17\text{-}72)$$

Figure 17.20 illustrates the isothermal resistance to deformation K_I as a function of the arithmetic average aspect ratio Z_a for two types of materials, low-carbon steel and austenitic stainless steel grade 304.

17.23 EVALUATION OF FORMULAE FOR RESISTANCE TO DEFORMATION

When evaluating the various proposed formulae for the resistance to deformation, a preference shall be given to the ones based on the physics of the deformation process [17].

Utilization of equations based on the physics of the deformation process insures the extrapolative properties of the model. This can be illustrated by calculating the variations of resistance to deformation in hot rolling with the reduction and roll diameter using the following two different methods:

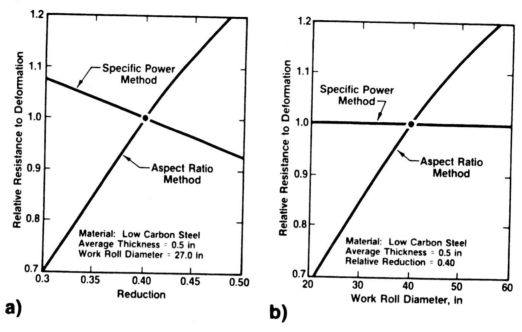

Fig. 17.21 Variation of resistance to deformation with variation in: a) reduction and b) work roll diameter. (From Ginzburg, Iron and Steel Engineer, Sept. 1985. Copyright AISE, Pittsburgh, Pennsylvania. Reprinted with permission.)

1. Specific power method described in Section 17.21, and

2. Aspect ratio method described in Section 17.22.

The specific power method is not based on physics of the rolling process and is not expected to provide adequate extrapolative properties for the model. In fact, this method predicts a decrease in the resistance to deformation with an increase in reduction (Fig. 17.21a). It also predicts no change in the resistance to deformation with an increase in the roll diameter (Fig. 17.21b). Both these predictions contradict those derived from the aspect ratio method and actually observed.

Another important criterion for selecting the equation for resistance to deformation is a capability to verify and calibrate this equation by using actual rolling test data. To provide this capability, the equation shall contain the variables which can be readily measured during rolling test. From this point of view, the aspect ratio method is very convenient since both variables, the aspect ratio and material temperature, can be easily obtained.

The aspect ratio method was also found to be very efficient in simulating the rolling conditions. Indeed, with a minimum number of constants, it allows one to express the resistance to deformation of the rolled products having a wide range of both initial and final thicknesses.

REFERENCES

1. Z. Wusatowski, Fundamentals of Rolling, Pergamon Press, Oxford, pp. 203-386 (1969).

2. H. Ford and J.M. Alexander, "Simplified Hot Rolling Calculations", Journal of the Institute of Metals, Vol. 92, 1963-64, pp. 397-404.

3. R.B. Sims, "The Calculation of Roll Force and Torque in Hot Rolling", Proceedings of the Institution of Mechanical Engineers, No.168, 1954, pp. 191-200.

4. E. Orowan and K.J. Pascoe, "A Simple Method of Calculating Roll Pressure and Power Consumption in Flat Hot Rolling", Iron and Steel Institute (London), No. 34, 1946, pp. 124-126.

5. A.I. Tselikov, Stress and Strain in Metal Rolling, Moscow, 1967.

6. A.I. Tselikov, "Present State of Theory of Metal Pressure Upon Rolls in Longitudinal Rolling", Stahl, Vol. 18, No. 5, May 1958, pp. 434-441.

7. S. Gupta and H. Ford, "Calculation Method for Hot Rolling of Steel Sheet and Strip", Journal of the Iron and Steel Institute, Vol. 205, February 1967, pp. 186-190.

8. B.K. Denton, F.A.A. Crane, "Roll Load and Torque in the Hot Rolling of Steel", Journal of the Iron and Steel Institute, August 1972, pp. 606-617.

9. J.W. Green and J.F. Wallace, "Estimation of Load and Torque in the Hot Rolling Process", Journal of Mechanical Engineering Science, Vol. 4, 1962, pp. 136-142.

10. T. Yokoi, et al, "Model for Calculation of Pass Schedules for Slabbing Mills", Tetsu-to Hagane, Vol. 67, No. 15, 1981, pp. 2356-2364.

11. SKF Industries, Inc., "SKF Calculation of Rolling Mill Loads".

12. E. Siebel, "Calculation of Roll Force Anti-Friction Bearings for Rolling Mills" (German), Schweinfurt, 1941.

13. J.S. Ride, "Analysis of Operational Factors Derived from Hot Strip Mill Tests", AISE Yearly Proceedings, 1960, pp. 867-880.

14. R.G. Schultz and A.W. Smith, "Determination of a Mathematical Model for Rolling Mill Control", AISE Yearly Proceedings, 1965, pp. 461-467.

15. W. Sykes, "Power Requirements of Rolling Mills", Transactions AIEE, 1913, p. 822.

16. W.M. Ballenger and T.R. Rhea, "Power Consumption of Hot Strip Mills", AISE Yearly Proceedings, 1941, pp. 142-150.

17. V.B. Ginzburg, "Basic Principles of Customized Computer Models for Cold and Hot Strip Mills", Iron and Steel Engineer, September 1985, pp. 21-35.

18

Roll Force, Torque and Power in Hot Rolling

18.1 GENERAL EQUATION FOR ROLL SEPARATING FORCE

The following equation is generally used to calculate the roll separating force in rolling of flat products on smooth roll barrels of equal diameter

$$P = K_w F_d = K_w w l_d, \tag{18-1}$$

where K_w = resistance to deformation corresponding to the rolling conditions such as temperature, speed, etc.

F_d, l_d = projected area and arc of contact between roll and workpiece respectively

w = mean workpiece width.

The resistance to deformation K_w is defined by one of the methods described in the previous chapter.

In the presence of strip tension, Eq. (18-1) shall be modified to a form

$$P = w (K_w - \beta_1 s_1 - \beta_2 s_2)\sqrt{R'\Delta}, \tag{18-2}$$

where s_1, s_2 = entry and exit strip tension respectively

β_1, β_2 = entry and exit strip tension coefficients respectively.

18.2 GENERAL EQUATION FOR ROLL TORQUE

The pure torque of deformation is equal to the total torque required to drive both rolls. When both rolls of equal diameter are used, the general equation for pure rolling deformation torque is given by

$$M = 2Pa, \tag{18-3}$$

where a = lever arm as shown in Fig. 17.1.

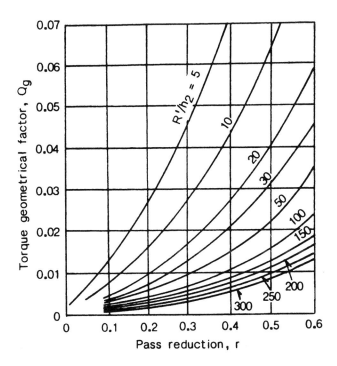

Fig. 18.1 Torque geometrical factor from Sims. (From Sims. Reprinted by permission of the Council of the Institution of Mechanical Engineers from the Proceedings of Institution of Mechanical Engineers, No. 168, 1954.)

The lever arm a is usually expressed as a fraction of the projected arc of contact length l_d

$$a = ml_d = m \sqrt{R'\Delta,} \tag{18-4}$$

where m = lever arm coefficient.

Defining the lever arm coefficients presents the most difficult part in the calculation of the roll torque. From Eqs. (18-3) and (18-4) the lever arm coefficient is equal to

$$m = \frac{M}{2Pl_d} \tag{18-5}$$

18.3 SIMS' FORMULAE FOR FORCE AND TORQUE

Sims' formula for the roll separating force can be obtained from Eqs. (17-46) and (18-1) which yield [1]

$$P = wl_d Q_p S_p, \tag{18-6}$$

where Q_p = geometrical coefficient used for force calculations

S_p = mean constrained yield stress used for force calculations.

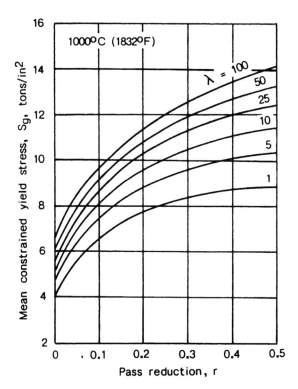

Fig. 18.2 Mean constrained yield stress used for torque calculations in rolling of carbon steel (0.17% C) at 1000°C. (From Sims. Reprinted by permission of the Council of the Institution of Mechanical Engineers from the Proceedings of Institution of Mechanical Engineers, No. 168, 1954.)

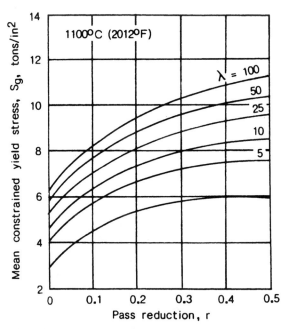

Fig. 18.3 Mean constrained yield stress used for torque calculations in rolling of carbon steel (0.17% C) at 1100°C. (From Sims. Reprinted by permission of the Council of the Institution of Mechanical Engineers from the Proceedings of Institution of Mechanical Engineers, No. 168, 1954.)

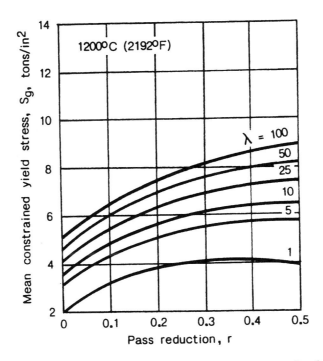

Fig. 18.4 Mean constrained yield stress used for torque calculations in rolling of carbon steel (0.17% C) at 1200°C. (From Sims. Reprinted by permission of the Council of the Institution of Mechanical Engineers from the Proceedings of Institution of Mechanical Engineers, No. 168, 1954.)

The roll torque is presented as

$$M = 2RR'wQ_gS_g,$$
                                                                    (18-7)

where Q_g = geometrical coefficient used for torque calculations

S_g = mean constrained yield stress used for torque calculations.

The geometrical coefficient Q_g is expressed as a function r and the ratio R'/h_2 as shown in Fig. 18.1.

The mean constrained yield stress S_g is obtained from the equation

$$S_g = \frac{1}{r} \int_0^r S\, dr_x \, ,$$

where S = constrained yield stress in plane strain compression

r_x = reduction in the plane located at distance x from the exit of roll bite

r = pass reduction.

The values for the mean constrained yield stress S_g are shown in Figs. 18.2-18.4 for carbon steel with 0.17% C.

The lever arm coefficient m can be derived from Eqs. (18-4)-(18-7) as

$$m = \frac{R}{\Delta}\, \frac{Q_g}{Q_p}\, \frac{S_g}{S_p}$$
                                                                    (18-8)

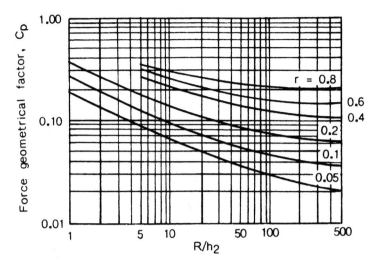

Fig. 18.5 Cook and McCrum's C_p function. (From FUNDAMENTALS OF ROLLING, by Z. Wusatowski, 1969. Copyright Wydawnictwo Slask, Katowice, Poland. Reprinted with permission.)

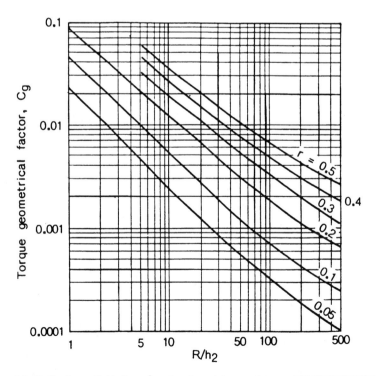

Fig. 18.6 Cook and McCrum's C_g function. (From FUNDAMENTALS OF ROLLING, by Z. Wusatowski, 1969. Copyright Wydawnictwo Slask, Katowice, Poland. Reprinted with permission.)

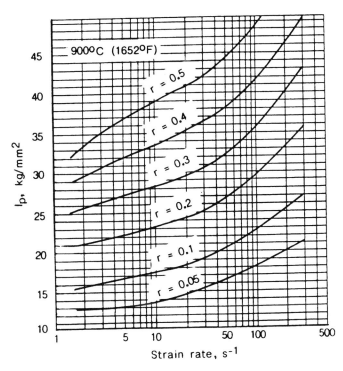

Fig. 18.7 Cook and McCrum's I_p function for low carbon steel at 900°C. (From FUNDAMENTALS OF ROLLING, by Z. Wusatowski, 1969. Copyright Wydawnictwo Slask, Katowice, Poland. Reprinted with permission.)

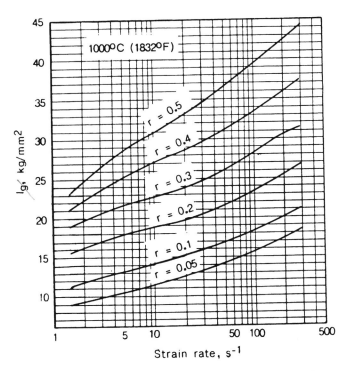

Fig. 18.8 Cook and McCrum's I_p function for low carbon steel at 1000°C. (From FUNDAMENTALS OF ROLLING, by Z. Wusatowski, 1969. Copyright Wydawnictwo Slask, Katowice, Poland. Reprinted with permission.)

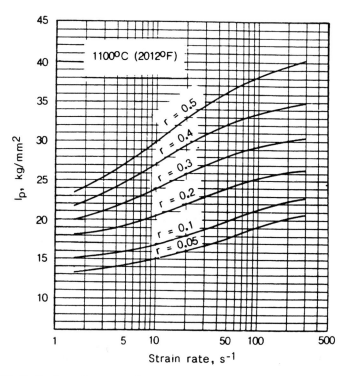

Fig. 18.9 Cook and McCrum's I_p function for low carbon steel at 1100°C. (From FUNDAMENTALS OF ROLLING, by Z. Wusatowski, 1969. Copyright Wydawnictwo Slask, Katowice, Poland. Reprinted with permission.)

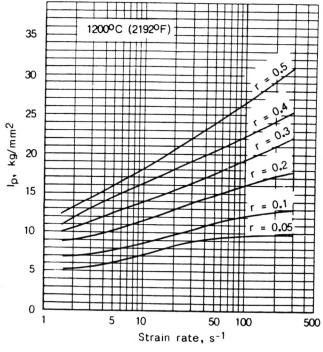

Fig 18.10 Cook and McCrum's I_p function for low carbon steel at 1200°C. (From FUNDAMENTALS OF ROLLING, by Z. Wusatowski, 1969. Copyright Wydawnictwo Slask, Katowice, Poland. Reprinted with permission.)

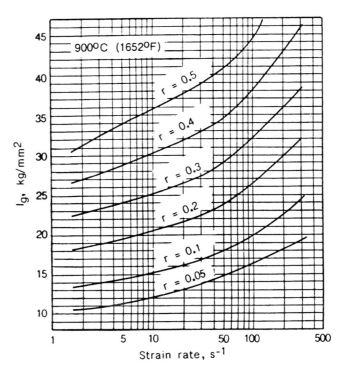

Fig 18.11 Cook and McCrum's I_g function for low carbon steel at 900°C. (From FUNDAMENTALS OF ROLLING, by Z. Wusatowski, 1969. Copyright Wydawnictwo Slask, Katowice, Poland. Reprinted with permission.)

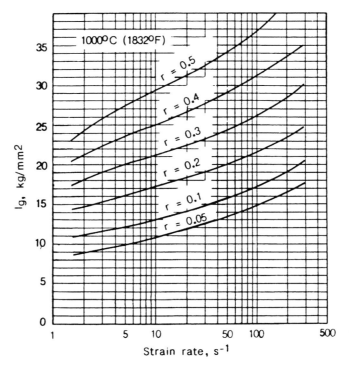

Fig. 18.12 Cook and McCrum's I_g function for low carbon steel at 1000°C. (From FUNDAMENTALS OF ROLLING, by Z. Wusatowski, 1969. Copyright Wydawnictwo Slask, katowice, Poland. Reprinted with permission.)

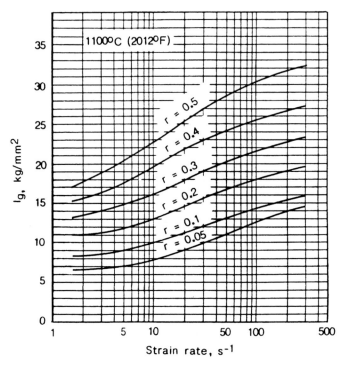

Fig. 18.13 Cook and McCrum's I_g function for low carbon steel at 1100°C. (From FUNDAMENTALS OF ROLLING, by Z. Wusatowski, 1969. Copyright Wydawnictwo Slask, Katowice, Poland. Reprinted with permission.)

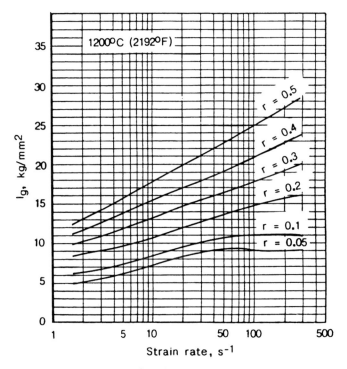

Fig. 18.14 Cook and McCrum's I_g function for low carbon steel at 1200°C. (From FUNDAMENTALS OF ROLLING, by Z. Wusatowski, 1969. Copyright wydawnictwo Slask, Katowice, Poland. Reprinted with permission.)

18.4 COOK-McCRUM'S FORMULAE FOR FORCE AND TORQUE

Cook and McCrum [2] have proposed a graphical method for determination of the roll separating force and torque. The method that is also known as the BISRA method is based on the following formulae:

$$P = R'wC_pI_p \tag{18-9}$$

$$M = 2RR'wC_gI_g \tag{18-10}$$

The geometrical factors shown in Eqs. (18-9) and (18-10) are determined by

$$C_p = Q_p\sqrt{\frac{h_2}{R'}\frac{r}{1+r}} \tag{18-11}$$

$$C_g = Q_g\sqrt{\frac{1-r}{1+r}} \tag{18-12}$$

$$I_p = S_p\sqrt{\frac{1+r}{1-r}} \tag{18-13}$$

$$I_g = S_g\sqrt{\frac{1+r}{1-r}}\ , \tag{18-14}$$

where S_p, S_g = mean constrained yield stresses used for force and torque calculations respectively.

The values for C_p, C_m, I_p and I_m for low carbon steel are shown in Figs. 18.5-18.14. The measurements for determination of these values were carried out on a cam plastometer. The mean strain rate was calculated from Eq. (17-26).

The lever arm coefficient may be obtained by substituting Eqs. (18-9) and (18-10) into Eq. (18-5)

$$m = \frac{R}{2\sqrt{R'\Delta}}\frac{C_gI_g}{C_pI_p} \tag{18-15}$$

18.5 WRIGHT AND HOPE'S FORMULAE FOR FORCE AND TORQUE

Wright and Hope [3] have modified the BISRA method for calculation of the roll separating force when rolling the stainless steel in a hot strip mill.

The modified equation was presented in the form:

Roughing mill, for the temperature range 1250°C < t ≤ 1100°C:

$$P_R = P_B\left(1 - 0.1\frac{1100-t}{100}\right) \tag{18-16}$$

Finishing mill, for the temperature range 950°C < t < 1100°C:

$$P_F = P_B\left(1 + 0.15\frac{1100-t}{100}\right)\ , \tag{18-17}$$

where P_R, P_F = modified roll separating force for roughing and finishing mills at rolling temperature t respectively

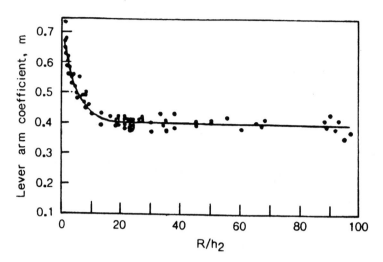

   t   = rolling temperature, $^\circ$C

  P_B = roll separating force determined at the rolling temperature t by BISRA method.

The roll separating force in the roughing mill can be also calculated by the equation

$$P_R = P_{B,1100} \left[1 + \frac{1100 - t}{100} \left(0.25 - 0.01 \frac{R}{h_2} \right) \right] \quad , \tag{18-18}$$

where $P_{B,1100}$ = roll separating force determined at the rolling temperature 1100°C by BISRA method.

    Eq. (18-18) is claimed to be valid for the values of R/h$_2$ between 2 and 27 and for rolling speed between 86 and 215 m/min.

    Wright and Hope have experimentally established values for the lever arm coefficient m from the formula

$$m = \frac{M_s - \mu P_a d_n}{2P_a \sqrt{R'\Delta}} \tag{18-19}$$

where M_s = measured spindle torque, kNm

    P_a = measured roll separating force, MN

    d_n = average backup roll neck diameter, mm

    μ = coefficient of friction.

The coefficient of friction - is defined by.

$$\mu = 0.003 \frac{N}{P_a} + 0.001, \tag{18-20}$$

where N = roll speed, rpm.

    The values of the lever arm coefficients as calculated from Eq. (18-19) are plotted in Fig. 18.15.

18.6 FORD-ALEXANDER'S FORMULAE FOR FORCE AND TORQUE

Ford-Alexander's formula for the roll separating force can be derived from Eqs. (17-50), (17-51) and (18-1). This gives [4]:

$$P = 0.25wl_d(\pi + Z_a)S, \qquad (18\text{-}21)$$

where Z_a = arithmetic average aspect ratio of the deformed zone.

The roll torque is expressed as

$$M = 0.25wl_d^2(\pi + Z_a^2/Z_p)S, \qquad (18\text{-}22)$$

where Z_p = parabolic mean aspect ratio of the deformed zone.

The lever arm coefficient is calculated by substituting Eqs. (18-21) and (18-22) into Eq. (18-5):

$$m = \frac{\pi + Z_a^2/Z_p}{2(\pi + Z_a)} \qquad (18\text{-}23)$$

A simplified form of Eq. (18-23) that includes only one variable Z_a is

$$m = \frac{3.2 + 0.91Z_a}{2(\pi + Z_a)} \qquad (18\text{-}24)$$

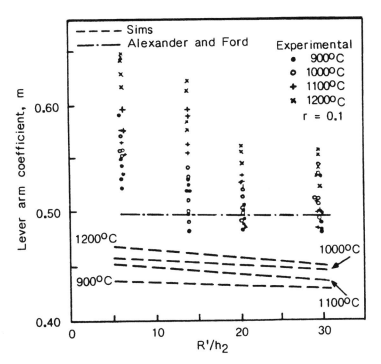

Fig. 18.16 Variation of lever arm coefficient with R'/h₂, r = 0.1. (From Helmi and Alexander, Journal of the Iron and Steel Institute, Sept. 1969. Reprinted with permission.)

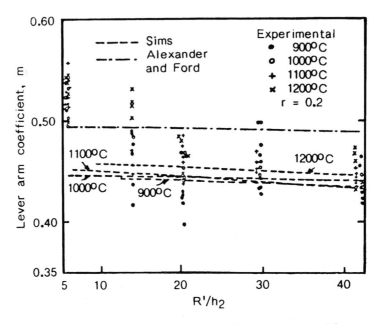

Fig. 18.17 Variation of lever arm coefficient with R'/H_2, $r = 0.2$. (From Helmi and Alexander, Journal of the Iron and Steel Institute, Sept. 1969. Reprinted with permission.)

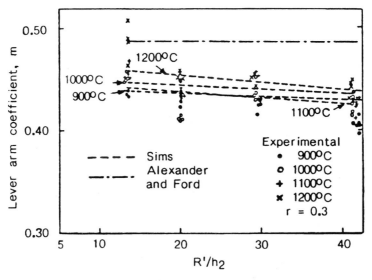

Fig. 18.18 Variation of lever arm coefficient with R'/h_2, $r = 0.3$. (From Helmi and Alexander, Journal of the Iron and Steel Institute, Sept. 1969. Reprinted with permission.)

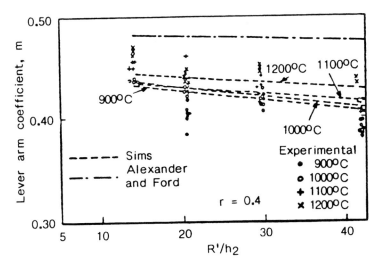

Fig. 18.19 Variation of lever arm coefficient with R'/h$_2$, r = 0.4. (From Helmi and Alexander, Journal of the Iron and Steel Institute, Sept. 1969. Reprinted with permission.)

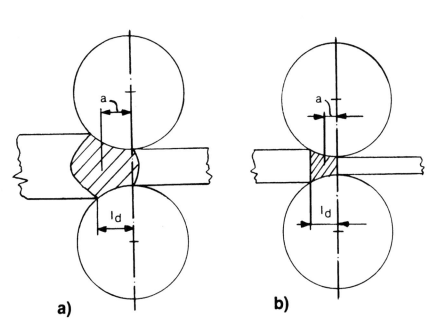

Fig. 18.20 Shape of deformation zone: a) with peening effect, b) without peening effect.

Figures 18.16 through 18.19 give a comparison between theoretical values for the lever arm coefficient m calculated from the Sims' and Ford-Alexander's methods and those calculated from the rolling data obtained by Helmi and Alexander [5] on experimental rolling mill.

A substantial difference can be seen between the theoretical and experimental values for m, when the relative reductions r are less than 0.2 and when the ratios R/h_2 are less than 10.

This discrepancy may be explained by the fact that neither Sims' nor Ford-Alexander's method takes into account a so-called **peening effect**. The peening effect is an expansion of the zone of deformation beyond the arc of roll contact l_d as shown in Fig. 18.20a. In hot rolling this effect becomes more pronounced with decrease in aspect ratio, i.e., when rolled material is thicker and the draft is smaller. The net result is an increase of the lever arm a, as seen from comparison of Fig. 18.20a with Fig. 18.20b.

18.7 DENTON-CRANE'S FORMULAE FOR FORCE AND TORQUE

Denton and Crane's formula for the roll separating force may be found from Eqs. (17-56), (17-57) and (18-1). It gives [6]:

$$P = wl_d(0.655 + 0.265Z_g)S, \tag{18-25}$$

where Z_g = geometrical mean aspect ratio of the deformed zone.

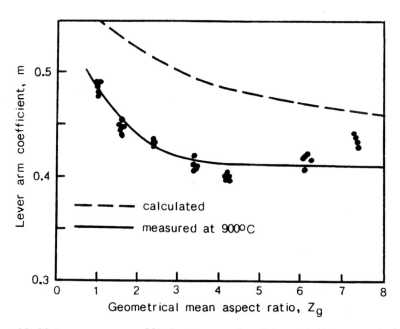

Fig. 18.21 Lever arm coefficient as a function of the geometrical mean aspect ratio. (From Denton and Crane, Journal of the Iron and Steel Institute, Aug. 1972. Reprinted with permission.)

The roll torque is given by

$$M = wl_d(0.795 + 0.22Z_g)S \tag{18-26}$$

After substituting Eqs. (18-25) and (18-25) into Eq. (18-5), we obtain that the lever arm coefficient m is equal to

$$m = \frac{0.795 + 0.22Z_g}{1.31 + 0.53Z_g} \tag{18-27}$$

A comparison of experimental values for m and those calculated from Eq. (18-27) is shown in Fig. 18.21.

18.8 GREEN-WALLACE'S FORMULAE FOR FORCE AND TORQUE

Green and Wallace's formula for the roll separating force can be obtained from Eqs. (17-58), (17-59) and (18-1). It yields [7]:

$$P = 0.5wl_d\left(Z_a + \frac{1}{Z_a}\right)S_p \quad , \tag{18-28}$$

where S_p = mean constrained yield stress used for force calculations by Sims' method [1].

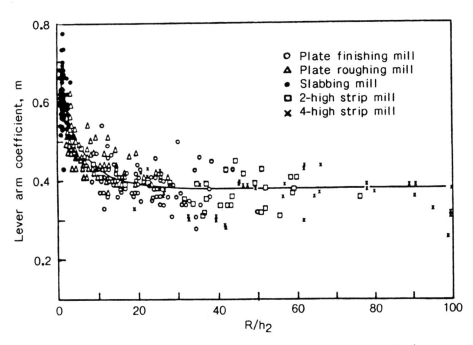

Fig. 18.22 Relationship between lever arm coefficient and R/h_2. (From Sims and Wright, Journal of the Iron and Steel Institute, March 1963. Reprinted with permission.)

The roll torque is given by

$$M = wl_d^2 S_g, \tag{18-29}$$

where S_g = mean constrained yield stress used for torque calculations by Sims' method.

After substituting Eqs. (18-28) and (18-29) into (18-5) we obtain that the lever arm coefficient is equal to

$$m = \frac{0.25Z_a}{1 + Z_a^2} \frac{S_g}{S_p} \tag{18-30}$$

18.9 SIMS-WRIGHT'S FORMULA FOR LEVER ARM COEFFICIENT

Sims and Wright [8] have calculated the lever arm ratio from the actual rolling data obtained from the slabbing, plate and strip mills as shown in Fig. 18.22.

The multiple regression analysis for values of $R/h_2 < 25$ gives an equation for the lever arm coefficient m for a mild steel in the form:

$$m = 0.78 + 0.017 \frac{R}{h_2} - 0.163\sqrt{\frac{R}{h_2}} \tag{18-31}$$

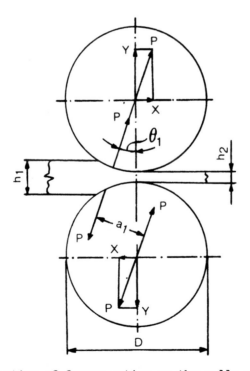

Fig. 18.23 Direction of forces acting on the rolls with bottom driven roll and top dragged roll. (From FUNDAMENTALS OF ROLLING, by Z. Wusatowski, 1969. Copyright Wydawnictwo Slask, Katowice, Poland. Reprinted with permission.)

The equation in the following exponential form allows one to represent the test data for the lever arm coefficient within the range of values for the ratio $R/h_2 < 100$

$$m = 0.39 + 0.295 \exp\left(-0.193 \frac{R}{h_2}\right)$$ (18-32)

18.10 CALCULATION OF TORQUE FOR VARIOUS CONDITIONS

So far we consider the equations for the roll separating force and torque assuming that both top and bottom rolls have had equal diameters and peripheral speeds. Let us consider according to Wusatowski [9] three other cases.

Case 1 - One roll is driven (e.g. the bottom one) and the second roll is dragged by friction between rolled stock and roll surface. Both rolls have the same diameter. In that case, the roll torque will be expressed by (Fig. 18.23):

$$M_w = Pa_1 = P(D + h_2)\sin\theta_1$$ (18-33)

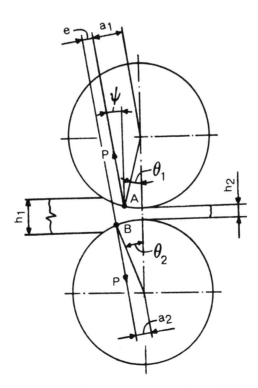

Fig. 18.24 Direction of forces acting on the rolls having different peripheral speeds. (From FUNDAMENTALS OF ROLLING, by Z. Wusatowski, 1969. Copyright Wydawnictwo Slask, katowice, Poland. Reprinted with permission.)

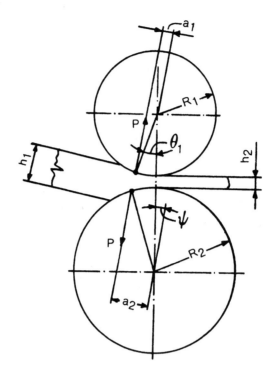

Fig. 18.25 Direction of forces acting on the rolls having different diameters. (From FUNDAMENTALS OF ROLLING, by Z. Wusatowski, 1969. Copyright Wydawnictwo Slask, Katowice, Poland. Reprinted with permission.)

Case 2 - Top and bottom rolls have different peripheral speeds. Both rolls have the same diameter. In that case the roll torques corresponding to each roll will be given by (Fig. 18.24)

$$M_{w1} = Pa_1 = PR_1 \sin(\theta_1 - \psi) \tag{18-34}$$
$$M_{w2} = Pa_2 = PR_2 \sin(\theta_2 - \psi), \tag{18-35}$$

where $\tan \psi = e/h_{AB}$ $\tag{18-36}$

Case 3 - One roll diameter is considerably greater than the other. The peripheral speeds of rolls are equal. In that case, the roll torques for each roll will be presented by Eqs. (18-34) and (18-35) in which the angle ψ is equal to (Fig. 18.25):

$$\psi = \frac{R_2 \sin\theta_2 - R_1 \sin\theta_1}{h_2 + R_1(1 - \cos\theta_1) + R_2(1 - \cos\theta_2)} \tag{18-37}$$

Values of lever arm a or angle θ occuring in Eqs. (18-33) through (18-37) are determined from the distance between point of action of the resultant roll force of the roll surface and the plane passing through the roll axes. The point of action is defined as the center of gravity of the roll pressure distribution curve along the arc of the roll contact.

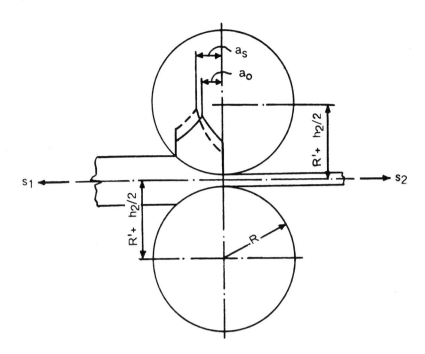

Fig. 18.26 Effect of the entry and exit strip tension on the rolling torque.

18.11 EFFECT OF STRIP TENSION ON ROLL TORQUE

In the previous text, the formulae for roll torque were applied for the cases when the strip tension is negligibly small. In more general case, however, the effects of both entry and exit tension has to be taken into account.

The strip tension affects the roll torque, or the torque applied to the rolls, in the following ways:

Firstly, as follows from Eq. (18-2), both entry and exit tension decreases the roll separating force P. This would proportionally reduce the roll torque. Secondly, it modifies the lever arm a due to redistribution of the normal pressure along the arc of the roll contact. Figure 18.26 shows, in solid lines, the normal pressure distribution and a corresponding lever arm a_0 when no tension applied to the strip. For comparison it also shows in dotted lines the pressure distribution with strip tension and corresponding lever arm a_s.

And lastly, it creates an additional torque that is equal to

$$M'_s = \left(R' + \frac{h_2}{2}\right)(s_1 h_1 w_1 - s_2 h_2 w_2) \quad , \tag{18-38}$$

where s_1, s_2 = specific strip entry and exit tension respectively.

If neglecting the variation in the length of the lever arm due to tension, and assuming that $h_2 < R'$ and $w_1 = w_2 = w$ we can obtain from Eqs. (18-2) through (18-4) and (18-38) the following formula for the roll torque when the strip tension is present:

$$M_s = wR'[2m\Delta(K_w - \beta_1 s_1 - \beta_2 s_2) + (s_1 h_1 - s_2 h_2)] \qquad (18\text{-}39)$$

18.12 TRUE WORK AND POWER IN ROLLING

The formula for determination of the true work of deformation during rolling is [9]:

$$A_w = K_w w_2 h_2 l_2 \ln \frac{h_1}{h_2} \quad , \qquad (18\text{-}40)$$

where l_2 = exit length of the rolled product.

Since by definition power equals to the work done in unit time, the true power N_w required for a pass is equal to

$$N_w = K_w w_2 h_2 V_2 \ln \frac{h_1}{h_2} \quad , \qquad (18\text{-}41)$$

where V_2 = workpiece exit speed.

The true work of deformation can be also expressed as a function of the roll separating force

$$A_w = 2Pa\phi = 2Pa \frac{l_2}{R} \quad , \qquad (18\text{-}42)$$

where ϕ = angle of roll rotation during rolling of the workpiece with the exit length equal to l_2.

Similarly, the pure power required for driving both rolls can be expressed as a function of roll torque

$$N_w = M_w \omega = M_w \frac{V}{R} = \frac{2\pi N}{60} M_w \quad , \qquad (18\text{-}43)$$

where ω = roll angular speed, s^{-1}

$\quad$ V = roll peripheral speed, m/s

$\quad$ N = roll angular speed, rpm.

18.13 TOTAL DRIVING TORQUE

The total driving torque M_D, necessary on the motor shaft, consists of the following component torques:

$$M_D = M_w + M_f + M_d, \qquad (18\text{-}44)$$

where M_w = roll torque required to overcome the resistance to deformation and the consequent frictional resistance between roll surface and workpiece

$\quad$ M_f = frictional torque, i.e., torque necessary to overcome frictional forces in the driving elements and roll bearings

$\quad$ M_d = dynamic torque necessary to overcome inertia forces arising due to varying roll speeds.

The ratio of the roll torque M_w to the required rolled torque is called **rolling efficiency coefficient** that is equal to

$$\eta_w = \frac{M_w}{M_w + M_f + M_d} \tag{18-45}$$

The dynamic torque M_d for a driving system consisting of n elements can be expressed

$$M_d = \sum_{i=1}^{n} I_i \frac{d\omega_i}{dt} = \frac{2\pi}{60} \sum_{i=1}^{n} I_i \frac{dN_i}{dt} \tag{18-46}$$

where l_i = moment of inertia of i-th element

ω_i = angular velocity of i-th element

dN_i/dt = angular acceleration of i-th element.

18.14 ROLL FORCE AND TORQUE IN HEAVY DRAFT ROLLING

Data on rolling with heavy drafts are not readily available. One of the most detailed investigations has been conducted by Wusatowski [9]. The experiments were carried out using 2-high rolling mill stand with

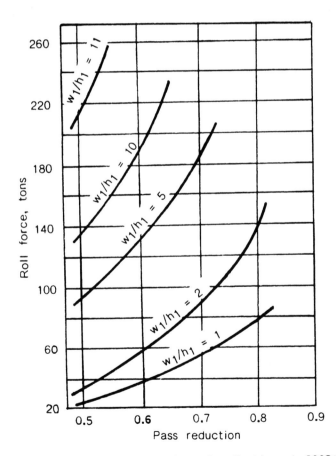

Fig. 18.27 Force measured as a function of reduction at 900°C (1652°F) and fixed form factors. (From FUNDAMENTALS OF ROLLING, by Z. Wusatowski, 1969. Copyright Wydawnictwo Slask, katowice, Poland. Reprinted with permission.)

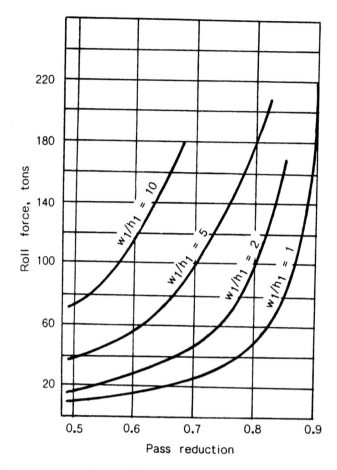

Fig. 18.28 Force measured as a function of reduction at 1100°C (2012°F) and fixed form factors. (From FUNDAMENTALS OF ROLLING, by Z. Wusatowski, 1969. Copyright Wydawnictwo Slask, Katowice, Poland. Reprinted with permission.)

the initial roll diameters 308 mm (12.126 in.). The rolled specimens were made of mild steel with the composition: 0.19% C, 0.31% Mn, 0.10% Si, 0.022% P, 0.031% S.

Initial cross-sections of the specimens were: 20 x 20 mm (0.787 x 0.787 in.), 20 x 40 mm (0.787 x 1.575 in.), 15 x 100 mm (0.591 x 3.937 in.), and 10 x 110 mm (0.394 x 4.331 in.). Thus, the following four different values of the initial width to initial thickness ratio w_1/h_1 were used: 1, 2, 6.67, and 11.

The reduction in longitudinal direction was accompanied by a considerable spread of the metal in lateral direction. When the specimens with initial cross-section 20 x 40 mm were rolled with the relative reductions r equal to 0.8 and 0.9, the ratios of the exit width to entry width w_2/w_1 were equal to 1.6 and 1.82 respectively.

Figure 18.27 shows the results of the roll force measurements as a function of reduction at 900°C (1652°F) and fixed initial width to thickness ratios w_1/h_1. Similar data for the specimen's temperature of 1100°C (2012°F) are shown in Fig. 18.28.

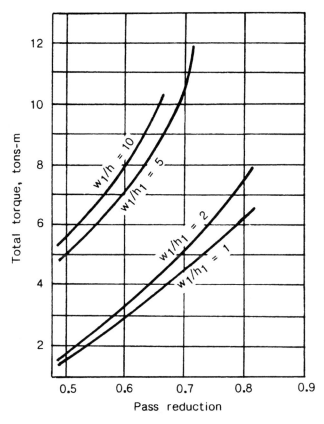

Fig. 18. 29 Total torque as a function of reduction at 900°C (1652°F) and fixed form factors. (From FUNDAMENTALS OF ROLLING, by Z. Wusatowski, 1969. Copyright Wydawnictwo Slask, Katowice, Poland. Reprinted with permission.)

The obtained curves are almost rectilinear for the relative reductions r less than 0.5. The steepness of these curves, however, increases with the increase in reduction above 0.5. With the change of relative reduction from 0.5 to 0.8 at a temperature 1100°C (2012°F) and for the ratio $w_1/h_1 = 1$, the roll force becomes five times greater, and for r = 0.9 it is more than 20 times greater.

Figures 18.29 and 18.30 illustrate the results of the torque measurements. These curves have the relationship between their steepness and reduction similar to the curves for roll force (Fig. 18.27 and 18.28).

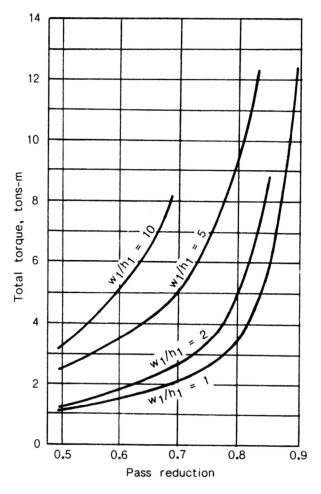

Fig. 18.30 Total torque as a function of reduction at 1100ºC (2012ºF) and fixed form factors. (From FUNDAMENTALS OF ROLLING, by Z. Wusatowski, 1969. Copyright Wydawnictwo Slask, Katowice, Poland. Reprinted with permission.)

REFERENCES

1. R.B. Sims, "The Calculation of Roll Force and Torque in Hot Rolling", Proceedings of the Institution of Mechanical Engineers, No. 168, 1954, pp. 191-200.

2. P.M. Cook and A.W. McCrum, "The Calculation of Load and Torque in Hot Flat Rolling", 1958, BISRA.

3. H. Wright and T. Hope, "Rolling of Stainless Steel in Wide Hot Strip Mills", Metals Technology, December 1975, pp. 565-576.

4. H. Ford and J.M. Alexander, "Simplified Hot Rolling Calculations", Journal of the Institute of Metals, Vol. 92, 1963-64, pp. 397-404.

5. A. Helmi and J.M. Alexander, "Geometric Factors Affecting Roll Force and Torque in the Hot Flat Rolling of Steel", Journal of the Iron and Steel Institute, September 1969, pp. 1219-1231.

6. B.K. Denton, F.A.A. Crane, "Roll Load and Torque in the Hot Rolling of Steel", Journal of the Iron and Steel Institute, August 1972, pp. 606-617.

7. J.W. Green and J.F. Wallace, "Estimation of Load and Torque in the Hot Rolling Process", Journal of Mechanical Engineering Science, Vol.4, 1962, pp. 136-142.

8. R.B. Sims and H. Wright, "Roll Force and Torque in Hot Rolling Mills-A Comparison Between Measurement and Calculation", Journal of the Iron and Steel Institute, Vol. 201, March 1963, pp. 261-269.

9. Z. Wusatowski, Fundamentals of Rolling, Pergamon Press, Oxford, pp. 203-386, 477-493 (1969).

19

Roll Force, Torque and Power in Cold Rolling

19.1 RESISTANCE TO DEFORMATION IN COLD ROLLING

As was discussed earlier, resistance to deformation in hot rolling depends mainly on temperature of a workpiece and its roll bite geometry. Unlike the hot rolling, the principal parameters affecting the resistance to deformation in cold rolling are work-hardening and friction in the roll contact zone. Roll flattening plays a much more important role in cold rolling due to higher resistance to deformation. Also, a contribution of strip tension becomes more substantial as cold rolling is conducted with greater specific tensions in comparison with those used in hot rolling. These and some other features of cold rolling process are usually taken into consideration to a different degree in the methods for calculation of roll force and torque. Some of these methods are briefly discussed below.

19.2 WUSATOWSKI'S METHOD FOR CALCULATION OF ROLL FORCE

In the method proposed by Wusatowski [1], the resistance to deformation is determined from the work-hardening curves (Fig. 19.1) which may be expressed in the following general form:

$$Y = f(r_m) \quad , \tag{19-1}$$

where r_m = mean total percentage reduction

Y = yield stress.

In order to calculate the work-hardening effect in the roll bite more precisely, the bite angle α for each pass is divided into four parts (Fig. 19.2) corresponding to the angles ϕ_i for each point on the arc of contact defined by these angles; the roll gap is then calculated from the formula:

$$h_i = h_1 + R\phi_i^2, \tag{19-2}$$

where h_1 = entry thickness for the pass under consideration

R = work roll radius for the pass under consideration.

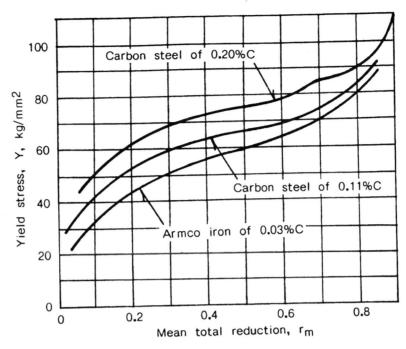

Fig. 19.1 Work-hardening curves for three steels. (From FUNDAMENTALS OF ROLLING, by Z. Wusatowski. Copyright Wydawnictwo Slask, Katowice, Poland. Reprinted with permission.)

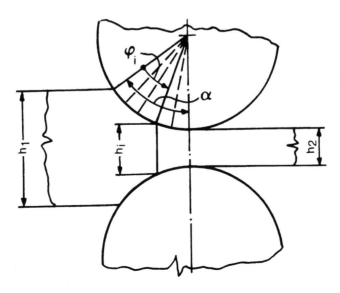

Fig. 19.2 Workpiece thickness in relation to roll angle.

Then for each of these points, the corresponding mean total reduction is calculated as following:

$$r_{mi} = 1 - h_i/h_o, \tag{19-3}$$

where h_o = thickness prior to the first cold rolling pass.

For each value of mean total reduction r_{mi} along the arc of contact the value of yield stress Y_i is then read from an appropriate work-hardening curve (Fig. 19.1). The mean yield stress along the arc of contact is calculated from the formula:

$$Y = \frac{1}{4}\left(\frac{Y_1}{2} + Y_2 + Y_3 + Y_4 + \frac{Y_5}{2}\right) \tag{19-4}$$

This would allow one to calculate the roll force from the well-known formula similar to one used for hot rolling:

$$P = 1.15YL'w, \tag{19-5}$$

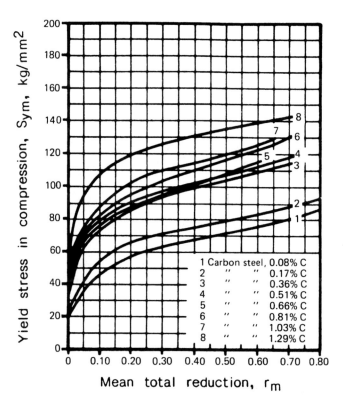

Fig. 19.3 BISRA curves for yield stress in compression for some grades of carbon steel. (From Ginzburg, Iron and Steel Engineer, Sept. 1985. Copyright AISE, Pittsburgh, Pennsylvania. Reprinted with permission.)

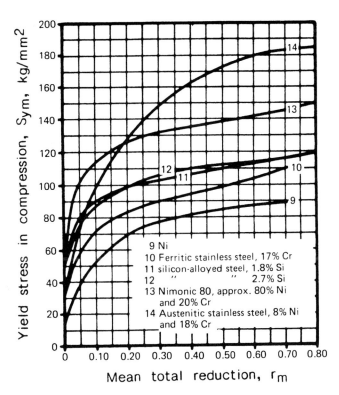

Fig. 19.4 BISRA curves for yield stress in compression for some metals and alloys. (From Ginzburg, Iron and Steel Engineer, Sept. 1985. Copyright AISE, Pittsburgh, Pennsylvania. Reprinted with permission.)

where L' = constant length of the deformed roll

w = mean workpiece width.

19.3 SKF METHOD FOR CALCULATION OF ROLL FORCE

In the method for calculating the roll force proposed by SKF [2], the effect of the strip tension has been taken into account in addition to the work-hardening effect.

According to this method the mean total reduction is equal to:

$$r_m = r_{to} + 0.6r(1 - r_{to}), \tag{19-6}$$

where r_{to} = total reduction prior to the pass under consideration

r = reduction in the pass under consideration.

The total reductions r_{to} and r are given by :

$$r_{to} = 1 - h_1/h_0 \tag{19-7}$$

$$r = 1 - h_2/h_1, \tag{19-8}$$

where h_1 = entry thickness for the n-th pass

h_2 = exit thickness for the n-th pass.

Using the value for mean total r_m obtained from Eq. (19-6), the average yield stress S_{ym} of the material in compression can be read off from the chart developed by BISRA (Figs. 19.3 and 19.4).

Then the roll force can be calculated from the formula:

$$F_s = F \left(1 - \frac{2s_1 + s_2}{3S_{ym}}\right)$$

(19-9)

where s_1 and s_2 = entry and exit specific strip tensions respectively

F_s = roll force with strip tension

F = roll force whithout strip tension

S_{ym} = average yield stress of the material in compression.

19.4 BLAND AND FORD'S SOLUTION FOR ROLL FORCE AND TORQUE

Bland and Ford's general solution [3] for roll force and torque is based on Orowan's general theory of rolling [4]. It was shown that if certain approximations are made, the normal pressure distribution in the roll bite from the entry to the neutral point (Fig. 19.5) can be expressed by

$$p^+ = \frac{kh}{h_2} \left(1 - \frac{s_2}{k_2}\right) \exp(\mu H)$$

(19-10)

and from the exit to the neutral point, the pressure distribution is given by

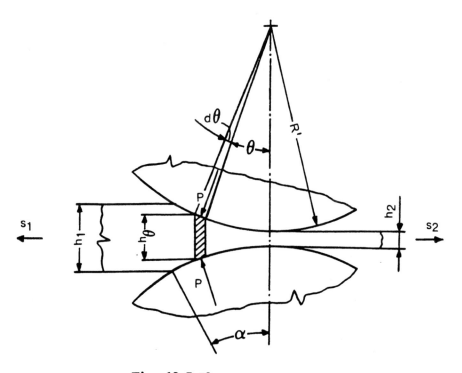

Fig. 19.5 Element in roll gap.

$$p^- = \frac{kh}{h_1}\left(1 - \frac{s_1}{k_1}\right) \exp \mu(H_1 - H),$$

(19-11)

where k_1, k_2 = yield stresses at the entry and exit points respectively

k = yield stress at any point in the arc of contact

μ = coefficient of friction between the strip and the roll

H_1 = value of H at the entry ($\theta = \alpha$)

H = dimensionless quantity which is equal to:

$$H = 2\sqrt{\frac{R'}{h_2}} \, \tan^{-2}\left(\sqrt{\frac{R'}{h_2}}\,\theta\right) \, ,$$

h = strip thickness at any point in the arc of contact

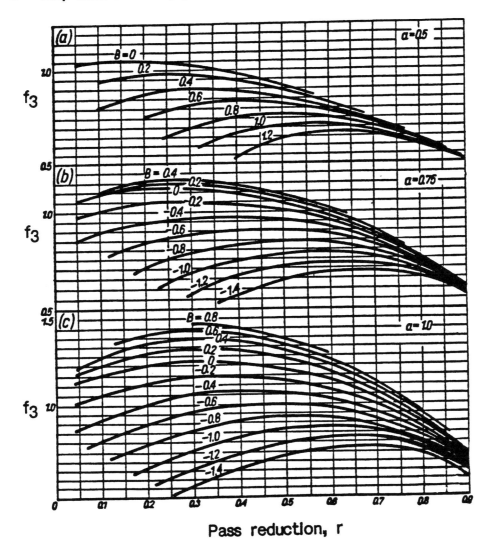

Fig. 19.6 Calculated values of function $f_3(a,r,B)$ used for determination of roll force in cold rolling with a = 0.5; 0.75 and 1.0. (From Ford, et al, Journal of the Iron and Steel Institute, Vol. 168, May 1951. Reprinted with permission.)

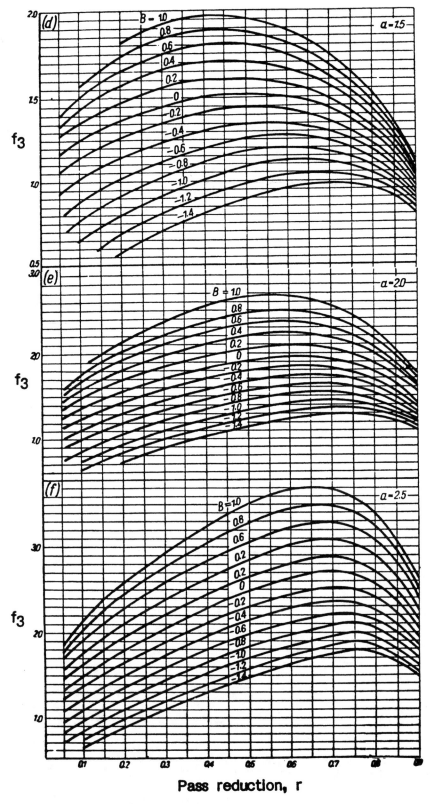

Fig. 19.7 Calculated values of function $f_3(a,r,B)$ used for determination of roll force in cold rolling with a = 1.5; 2.0 and 2.5. (From Ford, et al, Journal of the Iron and Steel Institute, Vol. 168, May 1951. Reprinted with permission.)

where θ = roll angle.

The roll force is obtained from the expression:

$$P = wR' \left(\int_0^{\theta_n} p^+ \, d\theta + \int_{\theta_n}^{\alpha} p^- \, d\theta \right) ,$$ (19-12)

where θ_n = neutral point angle, is given by the equations:

$$\theta_n = \sqrt{\frac{h_2}{R'}} \, \tan \left(\frac{H_n}{2} \sqrt{\frac{h_2}{R'}} \right)$$ (19-13)

and

$$H_n = \frac{H_1}{2} - \frac{1}{2\mu} \ln \frac{h_1}{h_2} \frac{1 - s_2/k}{1 - s_1/k} ,$$ (19-14)

where μ = coefficient of friction between the strip and the roll.

The total roll torque transmitted by both top and bottom work rolls is expressed as follows:

$$M = 2RR'w \left[\int_0^{\alpha} p\theta \, d\theta + \frac{s_1 h_1 - s_2 h_2}{2R'} \right]$$ (19-15)

Evaluating the roll force P and torque M from the above equations presents a difficult task due to the variation of yield stress k along the roll gap.

19.5 FORD ET AL'S METHOD OF CALCULATION OF ROLL FORCE

Ford, Ellis and Bland [5] have proposed a graphical method for calculation of the roll force given by Eq. (19-12). The method uses a constant mean value of yield stress along the roll gap. The mean yield stress is expressed by:

$$Y = \frac{1}{\alpha} \int_0^{\alpha} k \, d\theta$$ (19-16)

The roll force is determined as

$$P = YwL'(1 - s_1/Y)f_3(a,r,B),$$ (19-17)

where $f_3(a,r,B)$ = non-dimensional roll force function found from curves given in Figs. 19.6 and 19.7
        P = roll force, kg
        w = strip width, mm
        L' = contact length of deformed roll, mm
        s_1 = entry strip specific tension, kg/mm^2
        Y = mean yield stress, kg/mm^2
        r = reduction in the pass.

The independent variables a and B are given by:

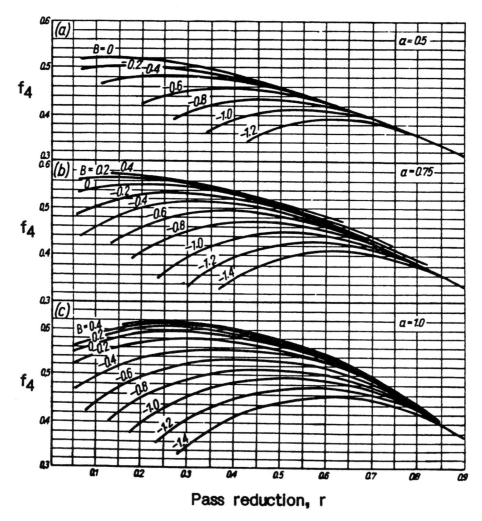

Fig. 19.8 Calculated values of function $f_4(a,r,B)$ used for determination of roll torque in cold rolling with a = 0.5; 0.75 and 1.0. (From Ford, et al, Journal of the Iron and Steel Institute, Vol. 168, May 1951. Reprinted with permission.)

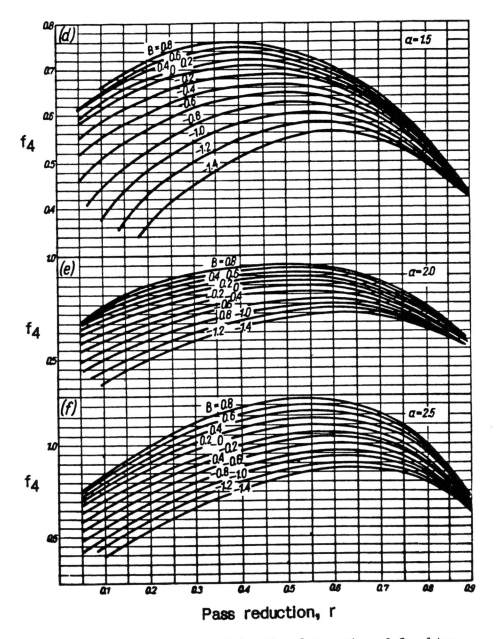

Fig. 19.9 Calculated values of function $f_4(a,r,B)$ used for determination of roll torque in cold rolling with a = 1.5; 2.0 and 2.5. (From Ford, et al, Journal of the Iron and Steel Institute, Vol. 168, May 1951. Reprinted with permission.)

$$a = \mu \sqrt{\frac{R'}{h_2}} \qquad\qquad\qquad (19\text{-}18)$$

and

$$B = \ln \frac{1 - s_2/Y}{1 - s_1/Y} \qquad\qquad\qquad (19\text{-}19)$$

The mean yield stress can be defined from the work-hardening curves as previously described.

Bland and Ford [3] have simplified the procedure for calculating the mean total reduction by introducing the formula:

$$r_m = 0.4 r_{to} + 0.6 r_t, \qquad\qquad\qquad (19\text{-}20)$$

where r_{to} = total reduction prior to the pass under consideration

r_t = total reduction after completion of the pass under consideration.

19.6 FORD ET AL'S METHOD OF CALCULATION OF ROLL TORQUE

Ford, Ellis and Bland's method [5] for calculation of roll torque is based on graphical solution of Eq. (19-15). Similar to a solution for roll force, the assumption has been made that the mean yield stress in plane strain is constant along the roll gap and expressed by Eq. (19-16).

The method is based on solution of the equation:

$$M = 2YRw(h_1 - h_2)(1 - s_1/Y)f_4(a,r,B), \qquad\qquad\qquad (19\text{-}21)$$

where $f_4(a,r,B)$ = dimensionless roll torque function found from curves given in Figs. 19.8 and 19.9.

The dimensionless independent variables a and B are given by Eqs. (19-18) and (19-19).

19.7 STONE'S METHOD OF CALCULATION OF ROLL FORCE

The method developed by Stone [6] is based on a simplified slab analysis of deformation during rolling with dry slipping friction (see Chapter 14).

By assuming that the resistance to deformation K_w is equal to average normal pressure p_a and also that for the plane strain compression the constrained yield stress is given by S = 1.15Y, we obtain from Eq. (14-53) the following formula for the resistance to deformation:

$$K_w = (1.15Y - s_a)(PMF), \qquad\qquad\qquad (19\text{-}22)$$

where s_a = average strip tension

(PMF) = pressure multiplication factor.

The average strip tension is given by

$$s_a = \beta_1 s_1 + \beta_2 s_2 \qquad\qquad\qquad (19\text{-}23)$$

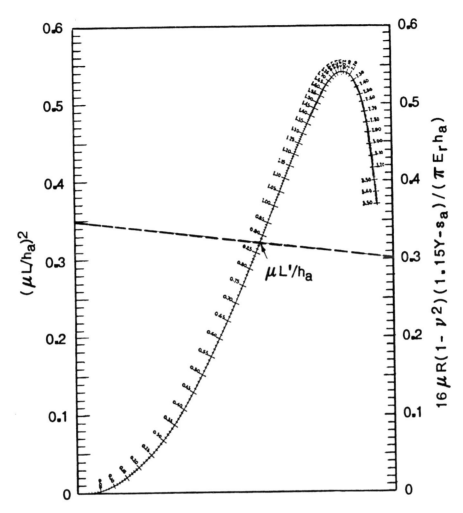

Fig. 19.10 Nomographic chart for use in determining roll force. Adapted from Stone (1953).

The values for the tension effect coefficients used by Stone are: $\beta_1 = \beta_2 = 0.5$. The other values are more frequently used, such as $\beta_1 = 0.7$ and $\beta_2 = 0.3$. The latter values are based on the rolling data confirming that the entry strip tension affects the roll force stronger than the exit strip tension.

Pressure multiplication factor PMF is expressed as

$$\text{PMF} = \frac{\exp(\mu L'/h_a - 1)}{\mu L'/h_a} \tag{19-24}$$

The pressure multiplication factor (PMF) is needed to calculate the roll separating force. Since the PMF, in its turn, depends on the roll separating force (Eqs. 14-45 and 17-5), the calculation of roll force is usually made by using either an iteration or a graphical method.

In the graphical method proposed by Stone, the ratio $\mu L'/h_a$ is found from a nomographic chart shown in Fig. 19.10.

The values for the coefficient of friction μ depend on a number of factors which will be discussed in the following chapter. The yield stress Y can be defined from the work-hardening curves such as shown in Fig. 19.1.

Once the resistance to deformation K_w is found, then the length of the arc of contact of the flattened roll L' can be calculated from the following formula derived from Eqs. (14-41)-(14-43)

$$L' = \sqrt{R\Delta + (aK_w)^2} + aK_w,$$ (19-25)

where

$$a = \frac{8R(1 - \nu^2)}{\pi E_r}$$ (19-26)

And finally, the roll force P is obtained by substituting the values for K_w and L' into the following equation:

$$P = wL'K_w$$ (19-27)

19.8 STONE'S METHOD OF CALCULATION OF ROLL TORQUE

This method is based on the same assumptions as the method of calculation of the roll force previously described [6].

From Eqs. (14-43), (14-55) and (19-26) we can obtain the following equation for roll torque

$$M = 2P \left(\frac{L'}{2} - aK_w \right)$$ (19-28)

Taking into account that the projected length of the arc of contact of the unflattened roll is equal to

$$L = \sqrt{R\Delta},$$ (19-29)

the resistance to deformation K_w can be derived from Eq. (19-25) and expressed as

$$K_w = \frac{1}{2a} (L' - L^2/L')$$ (19-30)

After substituting K_w from Eq. (19-30) into Eq. (19-28), we obtain that

$$M = \frac{PL^2}{L'}$$ (19-31)

Considering the Eqs. (19-27) and (19-29), the roll torque may be expressed in the following most commonly known form:

$$M = K_w wR\Delta$$ (19-32)

19.9 ROBERTS' METHOD OF CALCULATION OF ROLL FORCE AND TORQUE

Based on experimental data, Roberts [7] has derived the following empirical equations for calculation of roll force and torque in temper rolling.

The projected length of the arc of contact of the flattened roll is given by:

$$L'' = 0.5[\mu R_r + \sqrt{(\mu Rr)^2 + 4Rh_1 r}] \qquad (19\text{-}33)$$

For the conditions of the temper rolling, the average strain rate is found to be independent of the pass reduction or draft and is approximated by

$$\lambda_a = \frac{V}{\mu R} , \qquad (19\text{-}34)$$

where V = rolling speed.

The resistance to deformation is given by the expression:

$$K_w = 1.15[Y_t + b\log_{10}(1000\lambda_a)] - s_a, \qquad (19\text{-}35)$$

where Y_t = yield strength of the rolled strip as measured in tension at very low strain rate

b = strain rate factor

s_a = average strip tension.

The strain rate factor b is defined as an increase in the yield strength in tension per tenfold change in strain rate.

The total roll torque for two rolls is proposed to calculate by the formula

$$M = wRh_1 r(S_d - s_a) \left(1 + \frac{\mu L''}{h_a}\right) , \qquad (19\text{-}36)$$

where S_d = constrained dynamic yield strength corrected for strain rate

L'' = length of the arc of contact of the flattened roll as given by Eq. (19-33).

19.10 RESISTANCE TO DEFORMATION METHOD

Roll force and torque can be readily defined if the resistance to deformation of the rolled material is known. The resistance to deformation can be derived from the rolling test data and the following equation:

$$K_w = \frac{P}{L'w} + \frac{\beta_1 S_1}{h_1 w} + \frac{\beta_2 S_2}{h_2 w} \qquad (19\text{-}37)$$

where P, S_1 and S_2 = measured values of the roll force, entry and exit strip tensions respectively

w, h_1 and h_2 = measured values of the strip width, entry and exit thickness respectively.

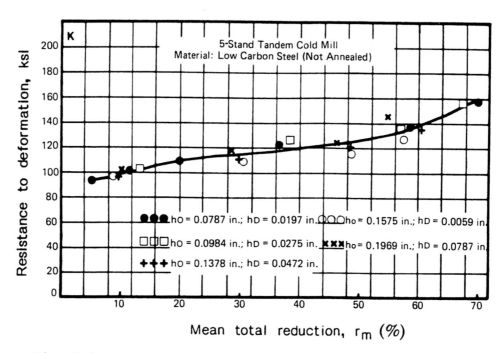

Fig. 19.11 Resistance to deformation derived from the rolling test at 5-stand tandem cold mill. (From Ginzburg, Iron and Steel Engineer, Sept. 1985. Copyright AISE, Pittsburgh, Pennsylvania. Reprinted with permission.)

The length of the arc of contact L' is calculated from the formula:

$$L' = \sqrt{R\left[\Delta + \frac{16P(1 - \nu_r^2)}{\pi E_r w}\right]} \qquad (19\text{-}38)$$

For the cases when the effect of the strain rate is negligible, the resistance to deformation can be assumed as a function of the mean total previous reduction r_m only and be expressed by using the single variable regression analysis:

$$K_w = a_0 + a_1 r_m + a_2 r_m^2 + \ldots + a_n r_m^n, \qquad (19\text{-}39)$$

where a_i = constant coefficients.

Figure 19.11 shows a typical plot of Eq. (19-39) derived from the rolling test at 5-stand tandem cold mill [8].

REFERENCES

1. Z. Wusatowski, Fundamentals of Rolling, Pergamon Press, Oxford, pp. 266-38 (1969).
2. SKF Industries, Inc., "SKF Calculation of Rolling Mill Loads".
3. D.R. Bland and H.Ford, "The Calculation of Roll Force and Torque in Cold Strip Rolling with Tensions", Proceedings Institute of Mech. Eng., Vol. 159, 1948, pp. 144-153.

4. E. Orowan, "The Calculation of Roll Pressure in Hot and Cold Flat Rolling", Proceedings Institute of Mech. Eng., Vol. 150, No. 4, 1943, pp. 140-167.

5. H. Ford, F. Ellis and D.R. Bland, "Cold Rolling with Strip Tension, Part 1 - A New Approximate Method of Calculation and a Comparison with Other Methods", Journal of the Iron and Steel Institute, Vol. 168, May 1951, pp. 57-72.

6. M.D. Stone, "Rolling Thin Strip", AISE Yearly Proceedings, 1953, pp. 115-128.

7. W.L. Roberts, Cold Rolling of Steel, Marcel Dekker, Inc., New York and Basel, (1978).

8. V.B. Ginzburg, "Basic Principles of Customized Computer Models for Cold and Hot Strip Mills", Iron and Steel Engineer, Sept. 1985, pp. 21-35.

Part VI

Tribology in the
Rolling Process

20

Basic Concept of Friction

20.1 DEFINITION OF FRICTION

Rolling process involves contacts between the workpiece and the work rolls. Friction is defined as 'the resisting force tangential to the common boundary between two bodies when, under the action of an external force, one body moves or tends to move to the surface of the other' [1].

Since the frictional force is a result of an interaction between contacting bodies at their interface, the nature of friction cannot be understood without explaining the nature of the interface. Modeling of the interfaces is one of the subjects of **tribology** which is a branch of science that studies friction and friction related phenomena.

Three main viewpoints of the interfaces are usually considered [2]:

1. Mechanical viewpoint
2. Macroscopic viewpoint
3. Microscopic viewpoint.

20.2 MECHANICAL VIEWPOINT OF INTERFACE

From the mechanical viewpoint, the interface is presented as a continuous film of τ_i shear strength, interimposed between a rigid tool (work roll in case of rolling) and a deforming workpiece [2].

The assumption, that the die is rigid, means that the effect of the die deformation on the frictional forces can be ignored. In addition to that, the mechanical viewpoint does not require the knowledge of the substance of interface and assumes that any changes in the interface can be taken into account simply by choosing an appropriate value of τ_i.

This simplified model of interface is often adequate for calculation of main rolling parameters, but it fails when it is necessary to understand the sources of friction and the mechanisms of lubrication and wear.

Although in this model the interface can be described by its shear strength τ_i for convenience of calculation, it is preferable to use the non-dimensional factors.

The most commonly used non-dimensional factors are:

a) coefficient of friction

b) interface shear factor.

20.3 COEFFICIENT OF FRICTION

The coefficient of friction is usually expressed by [2]:

$$\mu = \frac{F}{P} = \frac{\tau_i}{p} \qquad (20\text{-}1)$$

where F = force required to move the body

P = normal force

τ_i = shear strength of the interface

p = normal pressure.

Both τ_i and p are obtained by dividing the corresponding forces F and P by the apparent area of contact A between two bodies (Fig. 20.1), i.e.:

$$\tau_i = \frac{F}{A} \qquad (20\text{-}2)$$

and

$$p = \frac{P}{A} \qquad (20\text{-}3)$$

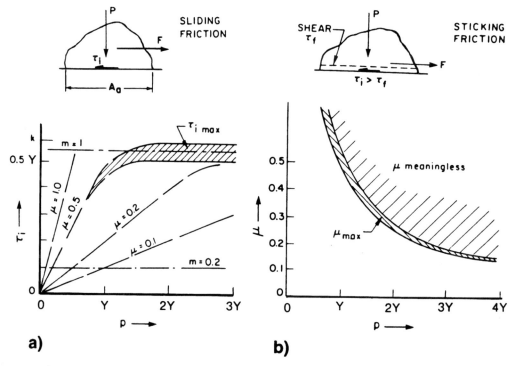

Fig. 20.1 Shear stress (a) and maximum coefficient of friction (b) in sliding at various pressures. (From J.A. Schey, TRIBOLOGY IN METALWORKING: FRICTION, LUBRICATION AND WEAR, American Society for Metals, Metals Park, OH, 1983, p. 15. Reproduced with permission.)

The following two basic laws of friction are stipulated by Eq. (20-1):

1. The frictional force is proportional to normal force

2. The frictional force is independent of the size of the apparent contact area.

It means that for a constant coefficient of friction μ, the frictional force F must increase at the same rate as normal pressure p. This relationship is valid for **sliding friction**, which is often referred to as Coulomb friction. The condition of sliding friction (Fig. 20.1a) is

$$\tau_i = \mu p < k, \tag{20-4}$$

where k = shear yield stress of the interface.

When τ_i reaches the value of k, it will take less energy for the material to shear inside the body of the workpiece (Fig. 20.1b), rather than to slide against the tool surface. In that case it is usual to speak of **sticking friction**, although no actual sticking to the die surface has to occur. The condition of sticking friction is

$$\tau_i = \mu p \geq k \tag{20-5}$$

The maximum value of the coefficient of friction μ_{max} is sometimes defined assuming that full surface conformity is reached at p = Y. Since k = 0.577Y according to von Mises criterion, we obtain from Eq. (20-1) that when $\tau_i = k, \mu_{max} = 0.577$. In many cases, however, the full surface conformity is obtained when the interface pressure p reaches a multiple of the yield stress Y, whereas the shear yield stress k remains constant. Therefore, the calculated values of the coefficient of friction μ actually drop as shown in Fig. 20.1b.

When this condition occurs, it is fair to assume that the term coefficient of friction is no longer applicable since there is no relative sliding at the interface.

20.4 INTERFACE SHEAR FACTOR

The interface shear factor m is another nondimensional quantity that has been proposed by some researchers as an alternative to the coefficient of friction. It is described by:

$$\tau_i = mk, \tag{20-6}$$

where m = interface shear factor.

The value of the interface shear factor varies from m = 0 for a frictionless interface to m = 1 for sticking friction (Fig. 20.1a).

It is assumed that the shear yield stress of the interface k is closely related to that of the workpiece. The possibility of such occurrence, however, is very unlikely, especially in the cases when the lubricants are used.

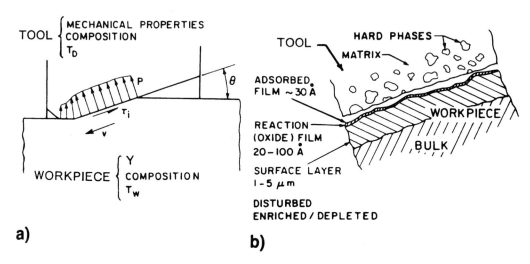

Fig. 20.2 The tool-workpiece interface on the (a) macroscopic and
(b) microscopic scales. (From J.A. Schey, TRIBOLOGY IN METALWORKING:
FRICTION, LUBRICATION AND WEAR, American Society for Metals,
Metals Park, OH, 1983, p. 28. Reproduced with permission.)

20.5 MACROSCOPIC VIEWPOINT OF INTERFACE

The macroscopic viewpoint of interface recognizes the following phenomena [2]:

A. The tool material is not considered to be rigid but to have a finite elastic modulus and strength. In case of rolling it may lead to either roll flattening or, if some critical loading is exceeded, to plastic deformation of the roll surface layer, or perhaps to the roll break.

B. The materials of both, tool and workpiece, are considered to have more or less well-defined compositions which govern the process of mutual attraction or adhesion in the contact zone (Fig. 20.2a).

C. It takes into account the fact that the initial temperatures of the tool and workpiece will change during the deformation process. The work of deformation transformed into heat and work of friction cause a temperature rise that affects the lubrication and properties of the rolled product.

20.6 MICROSCOPIC VIEWPOINT OF INTERFACE

The microscopic viewpoint accounts for the following tribological phenomena which can be revealed on a microscopic scale as shown in Fig. 20.2b [2]:

A. The surface minute peaks (asperities) and valleys of both the tool and workpiece and their magnitude and geometry are considered to affect not only the frictional force but also the stability of the lubricant film.

B. The tool material is viewed as a multiphase structure in which hard, wear-resistant particles (usually intermetallic compounds) are embedded in a softer and more ductile matrix.

C. The tool surface may be different from the bulk as a result of either an intentional surface treatment or a diffusion process during which some alloying elements are removed from or added to the surface.

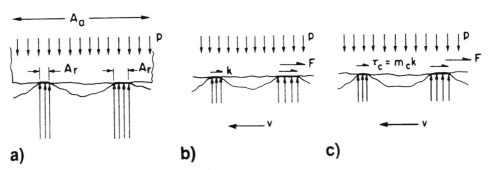

Fig. 20.3 Contact between die and rough workpiece with plastic deformation of asperities. (From J.A. Schey, TRIBOLOGY IN METALWORKING: FRICTION, LUBRICATION AND WEAR, American Society for Metals, Metals Park, OH, 1983, p. 34. Reproduced with permission.)

D. The workpiece material is described as either a single phase (pure metals or solid-solution alloys) or a multiphase structure. The phase structure of the surface may differ from the bulk because of an intentional deposition of some elements on the surface or because of a diffusion of alloying elements.

E. The technical surfaces are considered to be covered with the products of interaction with environment. The films formed from different reactions are usually superimposed. For example, the oxide film may be covered with an absorbed lubricant film. Often, chemical reaction between different films takes place.

F. Lubricants are viewed as chemically active substances whose properties depend not just on elemental composition but also on the molecular forms in which various elements are present.

20.7 THEORIES OF DRY FRICTION

Contact between real surfaces is very complex to be described comprehensively in analytical terms. In order to simplify the analysis the first useful assumption is often made that no lubricant has been intentionally applied at the interface. This condition is usually referred to as dry friction.

A number of theories of dry friction have been proposed. The most well-known of them are [2]:

1. Adhesion theory
2. Junction-growth theory
3. Asperity interaction theory
4. Molecular theory.

A brief description of these theories is given below.

20.8 ADHESION THEORY OF DRY FRICTION

The adhesion theory of friction, proposed by Bowden and Tabor [2, 3], assumes that friction in the contact between tool and rough workpiece is mainly due to pressure welding or adhesion at the asperities.

As shown in Fig. 20.3a, the real area of contact A_r is only a small fraction of the apparent contact area A_a. Under the load P the real contact area A_r will increase until it can support the load, i.e.:

$$P = A_r H, \tag{20-7}$$

where H = hardness of asperities.

The force F required to move the body relative to the tool (Fig. 20.3b), assuming that the shear strength of the junction is roughly equal to the shear yield stress k, is:

$$F = A_r k \tag{20-8}$$

Then the coefficient of friction can be found by substituting P and F in Eq. (20-1):

$$\mu = \frac{k}{H} \tag{20-9}$$

20.9 JUNCTION-GROWTH THEORY OF DRY FRICTION

In the case when welding of asperities is prevented by a contaminant layer of shear strength τ_C (Fig. 20.3c), the latter can be expressed as:

$$\tau_C = m_C k, \tag{20-10}$$

where m_C = interface shear factor related to the real contacting area A_r.

According to the junction-growth theory proposed by Tabor [2, 4], the coefficient of friction is given by:

$$\mu = \frac{m_c}{\sqrt{a(1 - m_c^2)}} \quad , \tag{20-11}$$

where

$$a = (H/k)^2 \tag{20-12}$$

For a three-dimensional junction the average value for a is around 9 and is found to be equal to 4 for brass and 20 for mild steel.

On the very clean surfaces the real junction area is maximum, so $m_C = 1$. This gives infinity for μ in Eq. (20-11). For contaminated surfaces with $m_C < 0.2$, Eq. (20-11) reduces to

$$\mu = \frac{\tau_c}{H} \tag{20-13}$$

Thus, when m_C is small, the effect of junction growth is negligible and friction is governed mainly by the strength of the contaminant film.

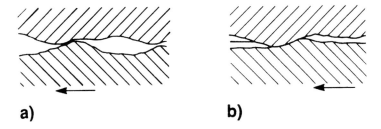

Fig. 20.4 Asperity encounters. (From J.A. Schey, TRIBOLOGY IN METALWORKING: FRICTION, LUBRICATION AND WEAR, American Society for Metals, Metals Park, OH, 1983, p. 35. Reproduced with permission.)

20.10 ASPERITY-INTERACTION THEORY OF DRY FRICTION

In the previous discussion, the junction between tool and workpiece is assumed to be sheared in a plane parallel to the direction of sliding (Fig. 20.3b). In the process of sliding (Fig.20.4a), the asperities may also deform each other so the plastic deformation of the softer asperity takes place [2]. The shear force developed for this type of interaction between asperities is calculated by application of plasticity theory.

The asperity-interaction theory includes the junction-growth theory as a limiting case when the asperity slope approaches zero. In the opposite case, when the tool asperity is large (Fig. 20.4b), the asperity plows through the workpiece surface layer. When adhesion is negligible, this type of asperity encounter may be the only source of friction.

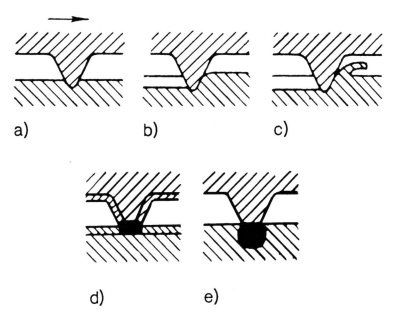

Fig. 20.5 Types of frictional interactions. (From EXTERNAL FRICTION DURING ROLLING, by A.P. Grudev, 1973. Copyright Metallurgia, Moscow, USSR. Reprinted with permission.)

20.11 MOLECULAR THEORY OF DRY FRICTION

The molecular theory of dry friction explains the adhesion between the tool and workpiece by the interatomic forces acting in the zone of contact. In the theory of friction developed by Kragelski [2, 5], the molecular attraction plays an important role along with other sources of friction. The theory considers the following types of interactions [6]:

  1. Elastic deformation of a softer surface by a harder asperity (Fig. 20.5a)

  2. Plastic deformation of a softer surface by a harder asperity (Fig. 20.5b)

  3. Microcutting, scratching (Fig. 20.5c)

  4. Breaking the interface, sticking the film to the tool and/or workpiece surface, weak sticking between the bulk interface surfaces (Fig. 20.5d)

  5. Strong sticking between the bulk interface surfaces accompanied by a break of one of the surfaces (Fig. 20.5e).

20.12 FRICTION IN PLASTICALLY DEFORMED WORKPIECE

Until now we have analyzed the plastic deformation of the individual asperities assuming that the interaction between the deformation zones is negligible. This assumption is valid when the distances between adjacent asperities are sufficiently large and normal pressure at the interface is low. As the normal pressure increases, the stress fields acting on the adjacent asperities start to interact and the plastic flow begins in the substrate [2, 7]. It leads to lower pressure required for asperity deformation. The asperity flattens giving a rise for valley located between them.

Initially, the real contact area A_r increases rapidly with pressure and then approaches the apparent contact area A_a asymptotically. This is accompanied by both the hardening of asperities and increase in the frictional force as follows from Eq. (20-8).

The average frictional stress over entire apparent area of contact may be expressed as [2, 7]:

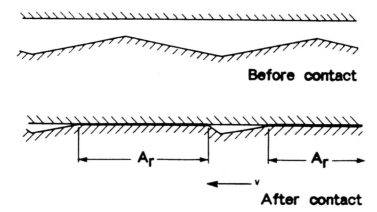

Fig. 20.6 Distortion of asperities on sliding. (From J.A. Schey, TRIBOLOGY IN METALWORKING: FRICTION, LUBRICATION AND WEAR, American Society for Metals, Metals Park, OH, 1983, p. 38. Reproduced with permission.)

$$\tau_i = m_c c k, \tag{20-14}$$

where c = real contact area coefficient (Fig. 20.6) that is equal to

$$c = A_r/A_a \tag{20-15}$$

Taking into account Eq. (20-1), we obtain

$$\mu = \frac{m_c c k}{p} \tag{20-16}$$

Wanheim and Bay [7] showed from slip-line field theory that the real contact area coefficient c depends on both the interface pressure p and the interface shear factor of the real contact zone m_c. When $p/2k < 1.3$, the coefficient c increases linearly with interface pressure p. Since the average frictional stress τ_i also increases linearly, the coefficient of friction μ remains constant. At higher pressure, however, it becomes pressure dependent.

20.13 RUBBING, WEAR AND CUTTING MODELS

In the previous discussion, the tool surface was assumed to be perfectly smooth whereas the workpiece to have some regular roughness features. In practice, however, the tool surface may also be roughened, whether intentionally or through wear. In that case, the asperities of the harder tool material will encounter and plastically deform the asperities of the workpiece, resulting in higher frictional forces.

Mechanics of this phenomena have been studied by several researchers. Challen and Oxley [2, 8] have suggested that the mechanics of interaction between the tool and workpiece asperities may be expressed as a function of the die asperity slope (hard asperity angle) and the interface shear factor m_c.

The following three models have been proposed:

1. Rubbing model that relates to the case when the interactive forces are too weak to produce plastic deformation of the workpiece asperities

2. Wear model that describes the conditions when the workpiece asperities are destroyed by plastic deformation

3. Cutting model that corresponds to the case when the workpiece asperities are machined away.

REFERENCES

1. Glossary of Terms and Definitions in the Field of Friction, Wear and Lubrication: Tribology, OECD, Paris, 1969.

2. J.A. Schey, Tribology in Metalworking: Friction, Lubrication and Wear, American Society for Metals, Metals Park, Ohio, pp. 11-39 (1983).

3. F.P. Bowden and D. Tabor, The Friction and Lubrication of Solids, Clarendon Press, Oxford, Pt.I, 1950, and Pt. II, 1964.

4. D. Tabor, Proc. Roy. Soc. (London), A251, 1959.

5. I.V. Kragelski, Friction and Wear, Butterworth, Washington (1965).

6. A.P. Grudev, External Friction During Rolling, Metallurgia (Russian), Moscow, pp. 52, 53 (1973).

7. T. Wanheim and N. Bay, CIRP, 27, 1978, pp. 189-194.

8. J.M. Challen and P.L.B. Oxley, Wear, 53, pp. 229-243 (1979).

21

Basic Principles of Lubrication and Wear

21.1 SOLID-FILM LUBRICATION

Solid-film lubricant prevents metal-to-metal contact. It may also have a low shear strength, τ_s. If there is a full conformity between die and workpiece surface (Fig. 21.1a), the interface is described either by a coefficient of friction:

$$\mu = \frac{\tau_s}{p} \tag{21-1}$$

or by using the interface shear factor m_s which is valid when real contact area is equal to apparent contact area:

$$\tau_s = m_s k \tag{21-2}$$

At partial conformity (Fig. 21.1b), the average shear strength of the interface is equal to

$$\tau_i = m_{cs} c k, \tag{21-3}$$

where m_{cs} = interface shear factor related to real contact area

c = real contact area coefficient.

Some solid-film lubricants are briefly described below [1].

Oxide films - As will be discussed later, the scale formed on iron at higher temperature consists of three layers. The outermost oxygen-rich Fe_2O_3 and the intermediate layer of Fe_3O_4 are brittle and plastically do not deform. The FeO layer adjacent to the metal, however, becomes softer than iron above $900^{\circ}C$ ($1652^{\circ}F$), where the phase transformation from ferrite into the stronger austenite occurs. Under these conditions the softer oxide may act as a soft-film lubricant.

Metal films - Coating of a tool or workpiece with a continuous film of another metal may provide lubrication by reducing adhesion between die and workpiece. It may reduce friction if the metal film of

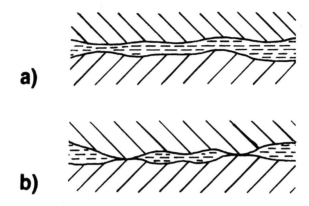

Fig. 21.1 Solid-film lubricated interface with: a) full conformity; b) partial conformity.

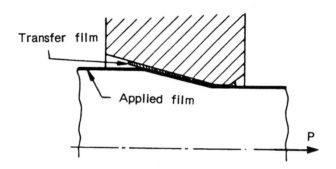

Fig. 21.2 Formation of polymer transfer film in drawing. (From J.A. Schey, TRIBOLOGY IN METALWORKING: FRICTION, LUBRICATION AND WEAR, American Society for Metals, Metals Park, OH, 1983, p. 44. Reproduced with permission.)

low shear strength is used. The metal films may improve the performance of another (usually liquid) lubricant. The efficiency of lubrication is improved when continuous coating is provided with metal films thinner than 1 μm (0.00004 in.).

Polymer films - The most suitable polymer films are thermoplastic. Below glass-transition temperature they are brittle and do not follow the surface deformation. Above this temperature they behave as viscoelastic solids but have no spreading ability.

Strong adhesion between the polymer and the metal substrate can be obtained by rubbing, melting on, fusing of dry powder, or application from a dispersion or solution. A coating deposited on a bar or strip may withstand drawing or rolling (Fig. 21.2). At high interface pressures the shear strength of polymers is proportional to hydrostatic pressure. It is also affected by temperature and strain rate.

Layer-lattice compounds - These compounds have a layered crystal structure with the platelets (lamellae) which comprise strongly bonded atoms. The most commonly used compounds are graphite and molybdenum disulfide.

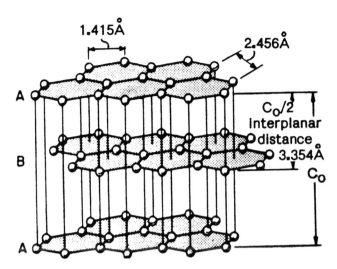

Fig. 21.3 Structure of graphite crystals. (From J.A. Schey, TRIBOLOGY IN METALWORKING: FRICTION, LUBRICATION AND WEAR, American Society for Metals, Metals Park, OH, 1983, p. 46. Reproduced with permission.)

In graphite (Fig. 21.3), bonding in the c direction is substantially weaker than in basal plane, but it is still too strong to provide the lubricating properties. These properties are usually obtained when considerable vapors are absorbed onto the edges of the platelets. In contrast to graphite, molybdenum disulfide MoS_2 does not require considerable vapor for lubrication, but, according to Pawelski [2], the lubricating properties of MoS_2 start to deteriorate at much lower temperature than those of graphite.

21.2 EXTREME PRESSURE (E.P.) LUBRICATION

The E.P. lubrication is a version of solid-film lubrication with the difference that the application of the low-shear-strength film is limited to contact points. Organic E.P. compounds contain phosphorus, chlorine, sulfur or their combination which are applied in a carrier such as mineral oil. The mechanism of their action may involve the following steps [1]:

1. Producing more reactive species as a result of interactions of the additive with the environment such as oxygen, water, other additives, and carrier fluid.

2. Absorbing the additives and reactive species on the metal surface.

3. Forming the polymeric films under intense pressure and temperature conditions of contact.

4. Removing the reaction product by sliding and also by chemical dissolution.

5. Re-establishing the reaction product by steps 1 to 3 described above.

Success of lubrication depends on maintaining a balance between removal and regeneration of the reaction product.

21.3 BOUNDARY LUBRICATION

Boundary lubrication is defined as a condition of lubrication in which the friction and wear between contacting surfaces are determined by the properties of the surfaces, and the properties of the lubricant other than bulk viscosity [3].

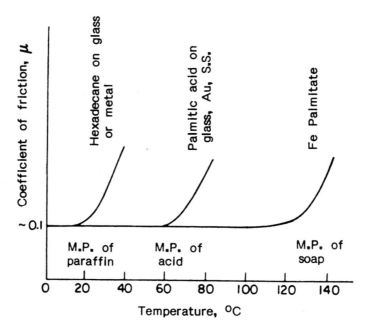

Fig. 21.4 Critical (breakdown) temperatures in boundary lubrication. (From J.A. Schey, TRIBOLOGY IN METALWORKING: FRICTION, LUBRICATION AND WEAR, American Society for Metals, Metals Park, OH, 1983, p. 55. Reproduced with permission.)

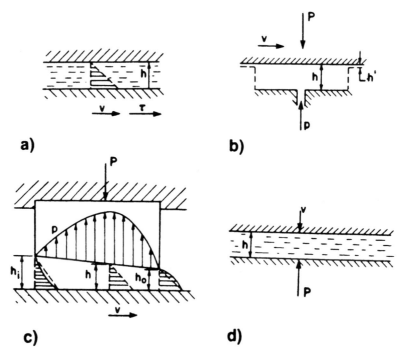

Fig. 21.5 Hydrodynamic lubrication, illustrating (a) measurement of viscosity, (b) hydrostatic lubrication, (c) converging gap, and (d) approaching surfaces. (From J.A. Schey, TRIBOLOGY IN METALWORKING: FRICTION, LUBRICATION AND WEAR, American Society for Metals, Metals Park, OH, 1983, p. 61. Reproduced with permission.)

In case of organic films, the lubrication mechanism involves forming a thin solid film in the interface as a result of adsorption of molecules of the polar liquid, such as primarily derivatives of fatty oils, on the surface. Adsorption is aided by an oxide film and by the presence of water.

At higher temperature, the adsorbed layer becomes increasingly disoriented, and friction rises suddenly at a critical temperature as shown in Fig. 21.4. At yet higher temperature, molecules desorb and the contacting surface becomes unprotected.

21.4 FLUID THICK-FILM LUBRICATION

The fluid thick-film lubrication relates to the simplest case when the film in the interface between two bodies is thick enough so the surface roughness can be ignored, and the contacting bodies are assumed to be rigid [1].

When the lubrication fluid has an excess into the space between two parallel surfaces moving against the other at a relative velocity v, the fluid layer located at the moving surface will be dragged at the same velocity, and is stationary at the nonmoving surface (Fig. 21.5a). The shear stress which is necessary to overcome for movement is given by

$$\tau = \eta \frac{dv}{dh} = \eta \dot{\gamma} \tag{21-4}$$

where v = relative velocity of two surfaces

   h = local film thickness

   η = dynamic viscosity of fluid

   $\dot{\gamma}$ = shear strain rate.

When viscosity is independent of the shear strain rate, the fluid is referred to as **Newtonian.** Many fluids exhibit complex relationship between viscosity and the shear strain rate and are often called **Non-Newtonian.**

To prevent a collapse of the film under the load normal to the lubricant film, one of these methods can be used:

1. Supplying the lubricant under sufficient hydrostatic pressure p (Fig. 21.5b)

2. Generating the hydrodynamic pressure by creation of a converging gap between two nonparallel surfaces (Fig. 21.5c)

3. Generating the hydrodynamic film, a so-called squeeze film, between two surfaces that approach each other at a given velocity v (Fig. 21.5d).

21.5 HYDRODYNAMIC LUBRICATION

Two types of hydrodynamic lubrication are usually considered: elastohydrodynamic (EHD) and plastohydrodynamic (PHD).

Theory of EHD lubrication recognizes the fact that viscosity of practical lubricants increases with pressure and also that the materials of two contacting surfaces deform elastically [1].

Theory of PHD lubrication is an extension of the theory of EHD lubrication to the case when the workpiece material is subjected to plastic deformation.

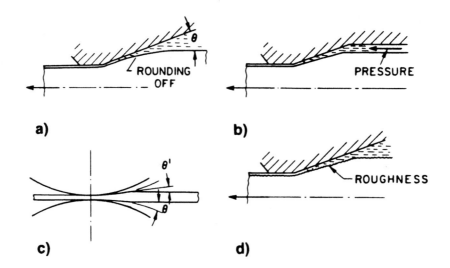

Fig. 21.6 Factors promoting plastohydrodynamic (PHD) lubrication. (From J.A. Schey, TRIBOLOGY IN METALWORKING: FRICTION, LUBRICATION AND WEAR, American Society for Metals, Metals Park, OH, 1983, p. 69. Reproduced with permission.)

In case of steady-state PHD lubrication the entrained film thickness is

$$h = \frac{6V\eta}{Y \tan \theta} \tag{21-5}$$

where V = mean surface speed

Y = yield stress of the workpiece

θ = angle between converging surfaces.

Thus, the entrained film thickness is larger for a more viscous lubricant, for a softer workpiece, and for smaller angle between converging surfaces.

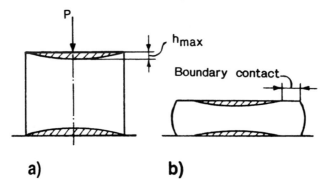

Fig. 21.7 Squeeze-film development and breakdown in forging. Adapted from Schey (1983).

Factors promoting PHD lubrication may include:

a) change of the angle θ between converging surfaces of the entire entry geometry by deformation (rounding) of the workpiece just before entry into the tool (Fig. 21.6a),

b) increase in the fluid pressure at the tool entry (Fig. 21.6b),

c) decrease in the angle θ by elastic deformation of the tool (Fig. 21.6c),

d) enhanced entrainment of the lubricant into the deformation zone due to increased workpiece roughness (Fig. 21.6d).

In other process, such as forging, the thickest part of the squeeze film in the center of the billet is equal to (Fig. 21.7a):

$$h_{max} = \left(\frac{3V\eta r^2}{Y}\right)^{1/3} \tag{21-6}$$

where V = approach velocity

r = radius of the billet.

The entrapped film thins out toward the edges. As the deformation proceeds, the squeezed film expands much less than the workpiece surface and unlubricated (or boundary-lubricated) annulus forms around the periphery (Fig. 21.7b). Lubricant entrapment is promoted by a controlled roughness of the surface.

21.6 MIXED-FILM LUBRICATION

The lubrication mechanism of the mixed-film lubrication can be qualitatively described as follows:

1. A liquid lubricant enters the deformation zone. The film thickness is not sufficient enough to provide a complete separation of contacting surfaces.

2. Local entrapment of a thick fluid film in the surface pockets (Fig. 21.1b) occurs as a result of differential yielding of the workpiece material. This produces roughening of the workpiece surface. As the deformation proceeds, the entrapped lubricant expands less than the workpiece surface and, thus, an increasing portion of the surface makes boundary contact.

3. Asperity slope and radius have little effect on plastic deformation with mixed-film lubrication. Much of the interface is actually in boundary contact, so that the tool surface features are imprinted on the plateaus of the workpiece surface.

4. When an appropriate combination of the tool and workpiece materials is selected, only a few of the metal-to-metal contact points result in adhesion. However, when adhesion (welding) does take place, it may result in some metal pickup.

5. Electrochemical effects play important roll in lubrication, especially when the lubricating fluids are electrolytes. As shown by Waterhouse [4], a virgin metal surface submerged in electrolytes develops an electrical double layer: positively charged metal ions enter into solution, leaving behind a negatively charged metal surface onto which the aqua-cation are back-absorbed. The electrolytic process (hydrolysis) leads to the formation of oxide-hydroxide films. This could be regarded as a form of E.P. lubrication except that the presence of liquid phase may promote some squeeze film or hydrodynamic contribution. This is particularly true in oil-in-water emulsions.

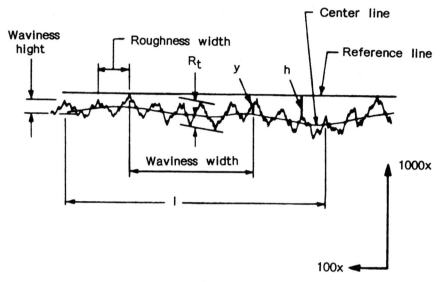

Fig. 21.8 Characterization of surface roughness. Adapted from Schey (1983).

21.7 REGIMES OF LUBRICATION MECHANISM

Regimes of lubrication mechanism can be analyzed as proposed by Schey [1].

   1. When $V\eta$ is small, hydrodynamic effects are discouraged, and only boundary lubricant films separate the surfaces. The tool surface finish is imprinted on the workpiece surface. If the lubrication breakdown occurs, the surface may be smeared.

   2. With increasing $V\eta$, the entrapped film increases and the mixed film hydrodynamic lubrication becomes possible. The coefficient of friction drops with increase of $V\eta$ and reaches its minimum value when entire surface is covered with hydrodynamic film, but the surface will now be substantially roughened.

   3. Further increase of $V\eta$ produces greater values of μ due to increasing drag in the hydrodynamic film. This is a zone of plastohydrodynamic lubrication which occurs very rare in practice.

   4. For any given value of $V\eta$, increasing pressure reduces the entrained film thickness and coefficient of friction usually increases.

21.8 SURFACE ROUGHNESS CHARACTERISTICS

The surface roughness is determined by a number of characteristics (Fig. 21.8):

   1. The centerline average (CLA) or arithmetic average (AA) surface roughness is defined as the average deviation from the mean center line

$$R_a = \frac{1}{l} \int_0^l |y| \ dl \ , \tag{21-7}$$

where l = length of the surface chosen in the record

y = measured deviation of the surface profile.

2. The root mean square (RMS) average surface roughness is calculated from the equation:

$$R_q = \left[\frac{1}{l} \int_0^l y^2 \ dl \right]^{1/2} \tag{21-8}$$

3. The peak-to-peak valley height R_t is determined as shown in Fig. 21.8.

4. The mean distance from the reference line is defined as

$$R_p = \frac{1}{l} \int_0^l h \ dl \ , \tag{21-9}$$

where h = measured deviation of the surface profile from the reference line.

5. Skewness is determined as the degree of asymmetry of the profile about a mean center line and is given by:

$$\text{Skewness} = \frac{1}{R_q^3} \ \frac{1}{n} \ \sum_1^n y^3 \ , \tag{21-10}$$

where n = number of y values chosen in the record length.

6. Fullness of the profile gives a direct information on the ability of the surface to entrap lubricants or accommodate wear particles. If asperities shown in Fig. 21.8 have limited length in the direction perpendicular to the plane of the paper, entrapment volume increases. Thus, ground and, especially, shot-blasted surfaces, entrap much more lubricant than a turned surface of the same roughness.

21.9 SURFACE ROUGHENING DURING DEFORMATION

Roughening of free (nonlubricated) surface of a workpiece depends on deformation mode and also depends on whether the surface is in contact with a tool or it is a side surface [1]. In all cases, however, the roughening is a function of strain. Roughening is greater for coarse-grain materials and for the metals with limited number of slip systems. Thus, materials with face-centered cubic (f.c.c.) lattice will roughen more than those with body-centered cubic (b.c.c.) lattice.

Roughening of lubricated surfaces appears as follows [1]:

a) at very light strains, the termination of the slip planes at the contact surface appear as a fine structure of parallel lines, especially on annealed material

b) surface roughening occurs even during homogeneous deformation because the deformation of individual grains varies as a function of their crystallographic orientations and the constraint imposed by surrounding grains. Thus, some grains will deform earlier; pockets of lubricant form and surface roughens.

c) when a workpiece is completely separated from a tool with liquid film, the surface roughness is the same as it would be in free deformation

d) annealed materials with equiaxed grain structures develop a nondirectional roughness. In two-phase materials, the pockets first develop in the softer phase or in softer zones between hard precipitate particles

e) the pockets of lubricant are generally elongated in the direction of major deformation.

21.10 EFFECT OF WORKPIECE SURFACE ROUGHNESS

The surface roughness enhances the mechanical entrapment of lubricant; thus, for a given peak-to-valley roughness, a less full profile is preferable [1, 5]. The lubrication is greatly influenced by the lay of the surface roughness. Surfaces with a lay parallel to the direction of deformation allow to escape much more lubricant than surfaces with a lay perpendicular to the direction of deformation [1, 6].

The surface roughness, however, is not always beneficial. Since a rougher surface gives more boundary contact, the benefit of lubricant entrapment may be lost. Under certain conditions, the effect of roughness orientation can be reversed too. Reduced friction may cause instabilities in deformation process (such as a failure of the strip to enter the roll gap).

An excessively rough initial surface may result in unacceptable appearance of the final surface. It is very difficult to eliminate the hydrostatic pockets. Therefore, if a bright surface is desired, the starting surface must be smooth and the hydrodynamic effects have to be avoided.

21.11 EFFECTS OF TOOL SURFACE ROUGHNESS

The effect of tool surface roughness on coefficient of friction has been studied by Schey and Myslivy [1] by upsetting smooth rings between anvils finished to three levels of random roughness and one intermediate level of directional roughness as shown below

Surface finish	Roughness, μm		
	R_a	10-point avg.	Peak-to-valley
Lapped	0.12-0.15	0.25	0.5
Fine shot blasted	0.5-0.7	0.8	2.5
Rough shot blasted	2.0-2.5	4.0	11.0
Ground: perpendicular	0.5-0.7	1.0	3.0
parallel	0.18-0.2	0.4	1.0

The study shows that with a low viscosity lubricant, large asperities of rough shot-blasted surface pierce through the film and give higher friction. Transversely oriented grinding also results in higher friction.

With thicker film produced by high viscosity oil, friction is generally lower. Most of the surface finishes, except for shot-blasted and the transversely oriented ground surfaces, give identical coefficient of friction.

21.12 FRICTION INSTABILITIES

The phenomenon known as the stick-slip motion is a typical example of friction instability. It can be observed in bar drawing as described by Schey [1]. In the simplest case, the properties of the tester can be modeled with a damped mass-spring system (Fig. 21.9a). The stick-slip motion consists of the following steps:

1. The draw starts by pulling at a speed v. This gives rise to frictional stress τ_i which results in microdeformation of asperities

2. After pulling force reaches some critical F value, the bar begins to move at speed v_m

3. For boundary lubrication, τ_i is constant irrespective of sliding speed and, thus, the force F will be constant too (Fig. 21.9b). For full fluid lubrication, the force F rises with increasing speed v. Therefore, stability in both these cases is ensured.

4. For mixed-film lubrication the interface shear stress τ_i drops with increasing speed v due to larger contribution of plastohydrodynamic lubrication (Fig. 21.9c). This produces a sudden acceleration

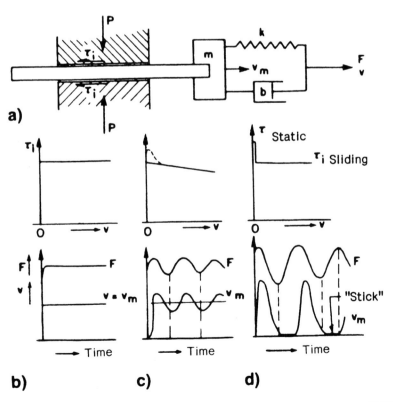

Fig. 21.9 Frictional instabilities. (From J.A. Schey, TRIBOLOGY IN METALWORKING: FRICTION, LUBRICATION AND WEAR, American Society for Metals, Metals Park, OH, 1983, p. 92. Reproduced with permission.)

and oscillating process commences. The amplitude of oscillation and its attenuation depend greatly on magnitude of damping.

5. The extreme case of so-called **static friction,** when τ_i is large prior to beginning of gross sliding (Fig. 21.9d), is usually observed in a system which includes a relatively soft, elastically deformed member. In that case, the asperities deform without breaking away and establish junctions. Once sliding begins, a lower τ_i is reached very suddenly. The oscillations will be similar as for mixed-film lubrication except that the bar may come to a temporary stop.

21.13 CLASSIFICATION OF WEAR

Wear is defined in this context as a progressive loss of the tool material during deformation. Theoretically, deformation may proceed without wear if the tool and workpiece are separated by a thick film of a nonreactive lubricant without a presence of any foreign particles [1]. This is a case of a pure plastohydrodynamic (PHD) lubrication. In practice, however, lubrication is mostly of mixed-film or boundary type. This type of lubrication cannot completely eliminate wear.

The wear process is a very complex phenomenon and usually involves a number of different wear mechanically acting at the same time. It makes classification of wear very difficult. Classification proposed by Burwell [7] is based on an assumption that, at least under certain conditions, a single kind of wear mechanism prevails.

Adhesive wear - This kind of wear may occur when asperities of a workpiece and a tool plow through the interface film. Adhesion may lead to cold (solid-phase) welding. Further relative sliding destroys the junction. Depending on relative strengths of the tool material, workpiece and the junction, the latter separates either in the workpiece or in the tool.

Abrasive wear - This is a removal of material by a hard asperity (two-body wear) or by a particle located between the two surfaces (three-body wear). Resistance to abrasive wear is a function of hardness; the greater hardness the greater wear resistance.

Fatigue wear - This type of wear occurs during cycling loading. The process involves a generation of shear stress below a compressed tool surface. Repeated loading produces microcracks, usually below the surface. The microcracks propagate on subsequent loading and unloading. After the microcrack reaches a critical size, it changes direction and emerges at the surface. This results in detachment of the flat sheet-like particle (delamination wear). When the detached particle is very large, the process is called **spalling.** Small-scale surface fatigue is known as **micropitting.**

A very destructive thermal fatigue process arises during hot rolling. It produces a mosaic network of cracks described as **crazing** or **firecracking.**

Chemical wear - The chemical wear is defined as a material loss due to tribochemical reactions between the lubricant, tool and workpiece materials and environment.

REFERENCES

1. J.A. Schey, Tribology in Metalworking: Friction, Lubrication and Wear, American Society for Metals, Metals Park, Ohio, pp. 27-130 (1983).

2. O. Pawelski, Schmiertechnik, 15, 1968, pp. 129-138 (German).

3. E.R. Booser, Handbook of Lubrication, Theory and Practice of Tribology, Vol. II, Theory and Design, CRC Press Inc., Boca Raton, Florida, p. 49 (1983).

4. R.B. Waterhouse, Tribology, 3, 1970, pp. 158-162.

5. P.R. Lancaster and G.W. Rowe, Wear, 2, 1959, pp. 428-437.

6. L.B. Sargent and Y.H. Tsao, ASLE Trans., 23, 1980, pp. 70-76.

7. J.T. Burwell, Wear, 1, 1957, pp. 119-141.

22

Friction, Lubrication and Wear in Rolling

22.1 FRICTIONAL FORCES IN THE ROLL CONTACT ZONE

The frictional forces are generated in the contact zone as a result of either sliding of the workpiece surface relative to the roll surface (sliding friction) or, in case of sticking friction, due to a potential movement between those surfaces. In general case, deformation in the roll contact zone occurs in longitudinal, transverse, and vertical directions [1].

Thus, the frictional force vector F at any point of the roll contact surface has three components: F_x, F_y, and F_z (Fig. 22.1)

$$F = (F_x + F_y + F_z) \tag{22-1}$$

Tselikov [1, 2] had classified the distribution of frictional forces along the arc of contact zone as a function of the arithmetic average aspect ratio Z_a as follows:

Type	Z_a	Figure
1	Greater than 5	22.2a
2	2 to 5	22.2b
3	0.5 to 2	22.2c
4	Less than 0.5	22.2d

It has been found that the character of distribution of the frictional forces is also affected by other factors such as coefficient of friction, bite angle, strip tension, etc. [1].

The frictional forces that are acting in the lateral direction resist the spread of the workpiece and are distributed in the plan view of the contact zone as shown in Fig. 22.3. In the point a, the lateral component F_y of the frictional force is equal to zero. In the point c, both lateral F_y and vertical F_z components are equal to zero. In the point b, at which the neutral line ed intersects with the

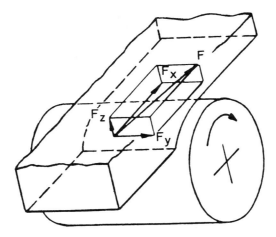

Fig. 22.1 Vector of elementary frictional force and its components.

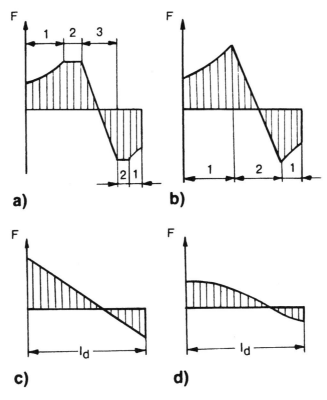

Fig. 22.2 Four types of distribution of frictional forces along the arc of the contact zone: 1 - zone of sliding friction; 2 - zone of maximum friction; 3 - zone of sticking friction. (From EXTERNAL FRICTION DURING ROLLING, by A.P. Grudev, 1973. Copyright Metallurgia, Moscow, USSR. Reprinted with permission.)

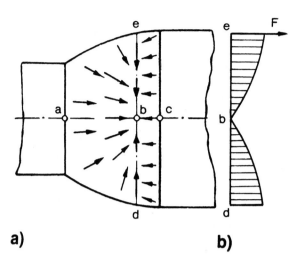

Fig. 22.3 Distribution of frictional forces: a) in plan view of the roll contact zone, b) along neutral plane. Adapted from Grudev (1973).

longitudinal axis ac, all three components F_x, F_y, and F_z of the frictional force are equal to zero.

In calculations of frictional forces in the roll bite it is common to use a mean value of the coefficient of friction that is defined as ratio of the sum of elemental frictional forces F_i, acting in the roll contact to the sum of corresponding normal forces P_i, or as a ratio of the average frictional force F_a in the roll bite to the average normal force P_a:

$$\mu = \frac{\Sigma F_i}{\Sigma P_i} = \frac{F_a}{P_a} \qquad (22\text{-}2)$$

22.2 CLASSIFICATION OF COEFFICIENTS OF FRICTION
In initial stages of development of theory of rolling, the difference in values for the coefficient of friction during different stages of rolling process has not been recognized. The later studies, however, have shown a necessity for a differentiation of the coefficients of friction and their classification as follows [1]:

1. Entry coefficient of friction μ_e that describes friction during initial entry of the workpiece into a roll bite.

2. Transient coefficient of friction μ_t that describes friction during passing of the workpiece leading edge through the roll bite.

3. Steady-state coefficient of friction μ that describes friction during a steady-state rolling process.

The entry coefficient of friction μ_e has the greatest value. During threading of the leading edge of the workpiece through the roll bite the value of the transient coefficient of friction μ_t gradually reduces from the value being equal to μ_e to the value equal to the steady-state friction μ, i.e.:

$$\mu_e \geq \mu_t \geq \mu \qquad (22\text{-}3)$$

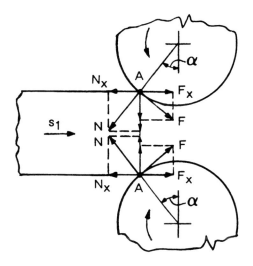

Fig. 22.4 Forces acting in the roll bite during initial entry of the workpiece.

22.3 FRICTION DURING INITIAL ENTRY OF ROLL BITE

Let us consider an equilibrium of forces acting at the points A of initial contact of a workpiece with two rolls. As shown in Fig. 22.4, N are the forces acting in the normal direction to the rolls and F are the frictional forces tangent to the roll. The frictional forces F pull the workpiece into the roll bite whereas the normal forces N resist to the process.

A grip of the workpiece occurs when horizontal components F_x of frictional forces are greater or, at least, equal to horizontal components N_x of normal forces, i.e.:

$$N_x \leq F_x \qquad\qquad (22\text{-}4)$$

According to Amonton's law

$$F = \mu_i N \qquad\qquad (22\text{-}5)$$

From Fig. 22.4 we obtain

$$N_x = N \sin\alpha \qquad\qquad (22\text{-}6)$$
and

$$F_x = F \cos\alpha, \qquad\qquad (22\text{-}7)$$

where α = roll bite angle.

Thus, from Eqs. (22-4)-(22-7), we can find the following condition for 'biting' the workpiece:

$$\tan\alpha < \mu_i \tag{22-8}$$

22.4 ENTRY COEFFICIENT OF FRICTION IN HOT ROLLING

Formulae for the entry coefficient of friction during hot rolling have been derived by a number of research workers. Ekelund [1, 3] has investigated a variation of the entry coefficient of friction with temperature during rolling of 225 mm (8.9 in.)-thick bar of 0.15% C steel. The roll diameters of the mill were equal to 427 mm (16.8 in.). The investigated temperature range was from 700 to 1100°C (1292 to 2012°F).

The obtained equations for the entry coefficient of friction were expressed as a function of the workpiece temperature, type of the rolls and their surface conditions:

    a) for cast iron and rough rolls

$$\mu_e = 1.05 - 0.0005t \tag{22-9}$$

    b) for chilled and smooth iron rolls

$$\mu_e = 0.8(1.05 - 0.0005t), \tag{22-10}$$

where t = workpiece temperature, °C.

    Smirnov and Uk [1, 4] have derived the equation for the entry coefficient of friction as a function of the workpiece temperature, roll surface roughness, and workpiece chemical composition in the form:

$$\mu_e = [0.7935 - 0.000356t + 0.012(R_a)^{1.5}]k_1k_2, \tag{22-11}$$

where R_a = arithmetic average surface roughness, μm.

    Coefficient k_1 is equal to

$$k_1 = 1 - (0.348 + 0.00017t)C, \tag{22-12}$$

where C = carbon content in steel, %.

    Coefficient k_2 depends on rolling speed as shown below.

Rolling speed, V, m/s	0-2	2-3	>3
k_2	(1-0.1V)	(1.44-0.28V)	0.5

Equations (22-11) and (22-12) have been obtained from the results of the test conducted at the experimental mill with the roll diameter 90 mm (3.54 in.) and rolling speed 0.05 m/s. The arithmetic average roughness of the rolls varied from 4 to 74μm. The rolled materials were carbon steel (approx.

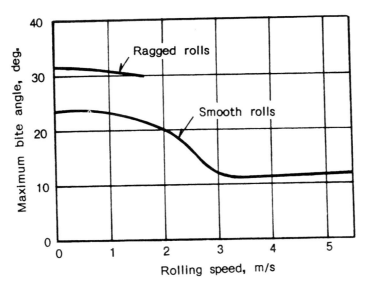

Fig. 22.5 Variation of maximum bite angle with rolling speed in blooming mill. (From EXTERNAL FRICTION DURING ROLLING, by A.P. Grudev, 1973. Copyright Metallurgia, Moscow, USSR. Reprinted with permission.)

Table 22.1 Maximum bite angle and entry coefficient of friction in hot rolling. Data from Wusatowski (1969).

Rolls	Maximum bite angle, degrees	Entry coefficient of friction
Smooth ground rolls	12-15	0.212-0.268
Plate mill rolls	15-22	0.268-0.404
Smooth rolls of small section mills	22-24	0.404-0.445
Rectangular grooves for rolling flats	24-25	0.445-0.466
Box passes	28-30	0.532-0.577
Box passes with ragging	28-34	0.532-0.675

0.3% C) and stainless steels with 20% Cr and 20 to 30% Ni. The investigated workpiece temperature range was from 700 to 1150°C (1292 to 2102°F).

It was found that minimum value of the entry coefficient of friction corresponded to 920°C (1688°F) for carbon steel and to 1030°C (1886°F) for stainless steels. The study also established that with hardened steel rolls the entry coefficient of friction was 1.2 times less than that with the soft steel rolls.

Tafel and Schnider [1, 5] have conducted experiments to determine maximum bite angle for low-carbon steel (0.09-0.12% C) as a function of rolling speed for two cases (Fig. 22.5).

Case 1: Roll surface   smooth
             Material cross-section   180 x 100 mm (7.1 x 4.0 in.)
             Material temperature   1200°C (2192°F)
Case 2: Roll surface   ragged
             Material cross-section   180 x 500 mm (7.1 x 19.7 in.)
             Material temperature   1250°C (2282°F)

In both cases the work roll diameters were equal to 990 mm (39 in.).

Maximum bite and entry coefficient of friction for different types of rolls with various surface roughness are shown in Table 22.1 according to Wusatowski [6]. As expected, ragging of the rolls increases the maximum bite angle.

22.5 ENTRY COEFFICIENT OF FRICTION IN COLD ROLLING

In general, maximum draft in cold rolling is not limited by a biting capability of the work rolls except for the cases when initial strip thickness is greater than 4 mm (0.158 in.). The effects of the workpiece material, type of lubricant and rolling speed on entry coefficient of friction during cold rolling have been a subject of numerous studies [1].

Effect of workpiece material - Table 22.2 summarizes the results obtained during rolling some specimens of carbon steel with the roll RMS roughness from 0.2 to 0.4μm. It was found that change in

Table 22.2 Entry coefficient of friction during cold rolling of some carbon steels. Data from Grudev (1973).

Lubricant	Entry coefficient of friction for steels			
	0.08% C	0.10% C	0.20% C	0.25% C
Without lubricant	0.136	0.131	0.133	0.131
Cotton-seed oil	0.116	0.118	0.116	0.117
Castor oil	0.109	0.109	0.101	0.115

Table 22.3 Effect of lubrication on entry coefficient of friction during cold rolling of low-carbon steel. Data from Grudev (1973).

Lubricant	Entry coefficient of friction	
	Range	Average
Water	0.152-0.160	0.156
Kerosene	0.154-0.157	0.156
Transformer oil	0.148-0.161	0.152
Machine oil	0.128-0.139	0.136
Sunflower oil	0.133-0.138	0.137
Castor oil	0.115-0.124	0.122

carbon content from 0.08 to 0.25% and in manganese content from 0.27 to 0.65% has no practical effect on the entry coefficient of friction. It was also shown that the entry coefficient of friction for stainless steel (18% Cr, 10% Ni) was 5 to 20% greater than that for carbon steel.

Effect of lubrication - Table 22.3 shows the ranges and average values for the entry coefficient of friction with different lubricant. A relatively weak effect of lubrication on μ_e is due to poor conditions for a formation of lubricant film during entry of the strip leading edge into a roll bite.

Effect of rolling speed - Fig. 22.6 illustrates the effect of rolling speed on entry coefficient of friction during rolling of 3.9 mm (0.154 in.)-thick samples of carbon steel (0.3% C) with castor oil as a lubricant. The RMS roughness of the rolls was between 0.2 and 0.4μm. The entry coefficient of friction reduces very steeply within the speed range from 0 to 0.15 m/s. Beyond this range there is a more moderate decrease of the entry coefficient of friction with increase in speed.

Effect of roll material and surface roughness - Table 22.4 shows the values for the maximum bite angle and entry coefficient of friction for some rolls with various surface roughness.

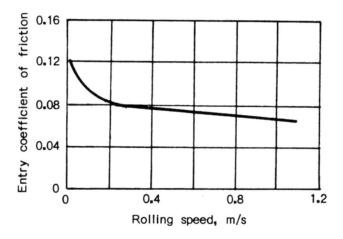

Fig. 22.6 Effect of rolling speed on entry coefficient of friction during cold rolling of carbon steel. (From EXTERNAL FRICTION DURING ROLLING, by A.P. Grudev, 1973. Reprinted with permission.)

Table 22.4 Maximum bite angle and entry coefficient of friction in cold rolling. Data from Wusatowski, (1969).

Rolls and lubricant	Maximum bite angle, degrees	Entry coefficient of friction
Smooth ground rolls with mineral oil	3-4	0.052-0.07
Chromium steel rolls, medium ground, with mineral oil	6-7	0.105-0.12
Dry rough rolls	up to 8	0.150

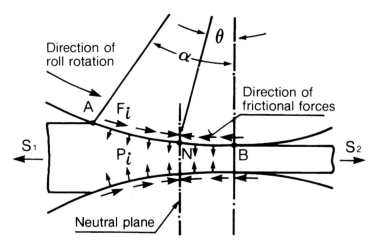

Fig. 22.7 Forces acting along the arc of the roll contact during rolling. (From FUNDAMENTALS OF ROLLING, by Z. Wusatowski, 1969. Copyright Wydawnictwo Slask, Katowice, Poland. Reprinted with permission.)

22.6 FRICTION DURING STEADY-STATE ROLLING CONDITIONS

During steady-state rolling conditions the frictional forces F_i, acting in the entry zone AN of the arc of the roll contact (Fig. 22.7), assist rolling, whereas the frictional forces acting in the zone NB along with normal forces P_i, acting along overall length of the arc of the roll contact hinder rolling. When strip tensile forces S_1 and S_2 are present, they also have to be considered in defining the equilibrium of horizontal components of the forces acting in the roll bite. By considering this equilibrium of forces, Ekelund [3] had derived the following formula for the neutral angle:

$$\phi = \frac{\alpha}{2}\left(1 - \frac{\alpha}{2\mu}\right)$$

(22-13)

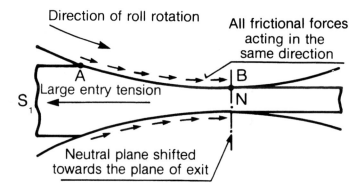

Fig. 22.8 Forces acting along the arc of the roll contact with large entry tension. (From FUNDAMENTALS OF ROLLING, by Z. Wusatowski, 1969. Copyright Wydawnictwo Slask, Katowice, Poland. Reprinted with permission.)

Thus, with increasing friction the neutral plane moves toward the middle of the arc of contact.

For rolling with strip tension, the neutral angle can be defined from the equation proposed by Ford et al [7]:

$$\phi = \frac{\alpha}{2} - \frac{S(h_1 - h_2) + h_1 s_1 - h_2 s_2}{4SR'\mu} \, , \tag{22-14}$$

where s_1, s_2 = entry and exit strip tensile stresses respectively

   S = mean constrained yield stress.

Thus, with increase of exit tension s_2, the neutral plane moves toward the entry side and with increase of the entry tension s_1 and also with increase in draft the neutral plane moves toward the exit side.

Figure 22.8 illustrates the extreme condition when the neutral plane is shifted to the end of the arc of the roll contact, so all frictional forces are started to act in rolling direction.

22.7 EFFECT OF FRICTION ON FORWARD SLIP

At the neutral plane, both the roll and the workpiece move at the same speed. At the entry side, the workpiece surface speed is slower than the peripheral speed of the roll. At the exit side, the workpiece speed is higher. Forward slip is defined as:

$$S_f = \frac{V_w - V_r}{V_r} \tag{22-15}$$

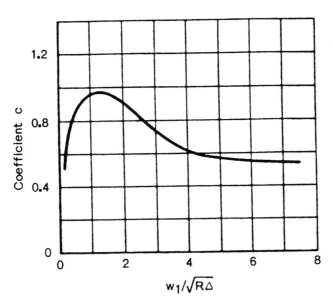

Fig. 22.9 Spread factor. (From FUNDAMENTALS OF ROLLING, by Z. Wusatowski, 1969. Copyright Wydawnictwo Slask, Katowice, Poland. Reprinted with permission.)

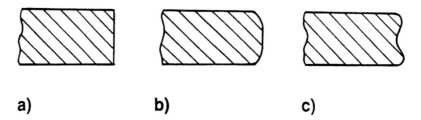

a)   **b)**   **c)**

Fig. 22.10 Edge profiles of rolled slabs. (From J.A. Schey, TRIBOLOGY
IN METALWORKING: FRICTION, LUBRICATION AND WEAR, American Society for
Metals, Metals Park, OH, 1983, p. 266. Reproduced with permission.)

where V_w = workpiece exit speed

$\quad$ V_r = roll peripheral speed.

As was shown by Ford et al [7], the forward slip can be calculated from the roll bite geometry as follows:

$$S_f = \frac{h_n}{h_2} \cos \phi - 1 \cong \frac{R'}{h_2} \phi^2 \tag{22-16}$$

By substituting ϕ from Eq. (22-13) into Eq. (22-16), we obtain the following equation for forward slip which is valid for rolling without strip tension:

$$S_f = \frac{R'\alpha}{2h_2} \left(1 - \frac{\alpha}{2\mu}\right)^2 \tag{22-17}$$

Thus, forward slip is equal to zero when $\mu = \alpha/2$.

22.8 EFFECT OF FRICTION ON LATERAL SPREAD

Relationship between the lateral spread and coefficient of friction has been proposed by Bakhtinov [6] in the form:

$$\Delta w = 1.15 \frac{\Delta}{2h_1} \left(\sqrt{R\Delta} - \frac{\Delta}{2\mu}\right) \tag{22-18}$$

Tselikov [6] had derived a more complicated formulae for lateral spread

$$\Delta w = \frac{2c}{r^2} \left(\sqrt{R\Delta} - \frac{\Delta}{2\mu}\right) \left[(1 - r)^2 \ln \frac{1}{1 - r} - r + 1.5r^2\right] \tag{22-19}$$

where c = spread factor dependent on the ratio $w/\sqrt{R\Delta}$ and varies from 0.5 to 1.0 (Fig. 22.9).

Friction equations indicate that when friction increases, it results in increase of lateral spread.

Friction also affects the contour of the side surfaces of the rolled material [9]. When deformation is homogeneous, the edge contour is straight (Fig. 22.10a). If friction is present it results not only in greater spread but also in barreling of the side faces (Fig. 22.10b). Under sticking conditions, folding over of the side surfaces (Fig. 22.10c) may take place [8].

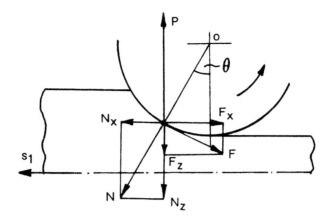

Fig. 22.11 Forces acting in the roll bite during skidding.
Adapted from Grudev (1973).

22.9 MEASUREMENT OF COEFFICIENT OF FRICTION

A number of methods for measurement of the steady-state coefficient of friction have been proposed [1]. The most well-known methods are briefly described below.

Method of forced skidding - This method has been proposed by Pavlov [10]. The method involves an application of entry tension to the rolled workpiece. The tension is gradually increased until the rolls start to skid. At that moment the equilibrium of forces in the roll bite (Fig. 22.11) in horizontal and vertical directions is given by

$$2F\cos\theta - 2N\sin\theta - S_1 = 0 \qquad\qquad (22\text{-}20)$$

$$P - N\cos\theta - F\sin\theta = 0 \qquad\qquad (22\text{-}21)$$

Since $F = \mu N$, and also assuming that the forces P and F are applied in the middle of the arc of the roll contact ($\theta = \alpha/2$), we obtain from Eqs. (22-20) and (22-21) the expression for the steady-state coefficient of friction

$$\mu = \frac{S_1 + 2P\,\tan(\alpha/2)}{2P - S_1\,\tan(\alpha/2)} \qquad\qquad (22\text{-}22)$$

Roll torque method - Whitton and Ford [11] have shown that under skidding conditions the steady-state coefficient of friction can be determined from a simple relationship:

$$\mu = \frac{M}{PR} \,, \qquad\qquad (22\text{-}23)$$

where M = roll torque

R = roll radius.

This method is valid when the back tension is sufficient to move the neutral plane to the exit point as shown in Fig. 22.8.

Method of maximum draft - This method involves a gradual increase of draft during rolling until the roll skidding begins. Under these conditions the steady-state coefficient of friction is equal to

$$\mu = \tan\alpha \tag{22-24}$$

The roll bite angle in Eq. (22-24) is determined from Eq. (14-20).

Forward slip method - This method is based on measuring the forward slip S_f during rolling and calculating the steady-state coefficient of friction μ from the equation that expresses μ as a function of S_f. When strip tension can be neglected, Eq. (22-17) can be used.

Roll force method - According to this method the steady-state coefficient of friction - is determined by using the equations for roll force P in which P is expressed as a function of μ. As an example, see Eqs. (19-22)-(19-27) related to the Stone's method of calculation of roll force.

22.10 STEADY-STATE COEFFICIENT OF FRICTION IN HOT ROLLING

The steady-state coefficient of friction - in hot rolling is affected by a number of factors which are briefly described below.

Workpiece temperature - For low-carbon steel, for the temperatures above 700°C (1292°F), μ decreases with increase in temperature as approximated by [1]

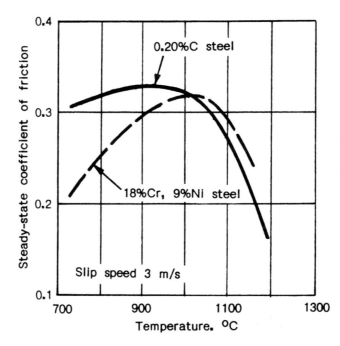

Fig. 22.12 Effect of the workpiece temperature on the steady-state coefficient of friction. (From FUNDAMENTALS OF ROLLING, by Z. Wusatowski, 1969. Copyright Wydawnictwo Slask, Katowice, Poland, Reprinted with permission.)

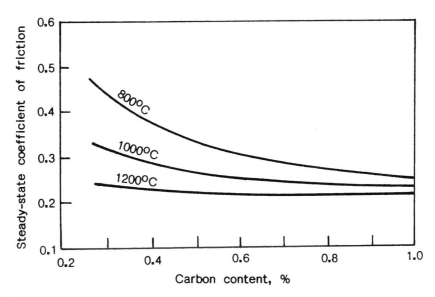

Fig. 22.13 Effect of the workpiece carbon content on the steady-state coefficient of friction in hot rolling. (From EXTERNAL FRICTION DURING ROLLING, by A.P. Grudev, 1973. Copyright Metallurgia, Moscow, USSR. Reprinted with permission.)

Table 22.5 Effect of the roll surface roughness on the steady-state coefficient of friction in hot rolling with quenched steel rolls. Data from Grudev (1973).

Roll diameter, mm	RMS average surface roughness, μm	Steady-state coefficient of friction
193	0.63	0.20-0.28
193	0.8-1.6	0.21-0.31
188	12.5-50	0.51-0.69

$$\mu = 0.55 - 0.00024t \quad , \tag{22-25}$$

where t = workpiece temperature, $^{\circ}$C.

In general, it was found that μ reaches maximum at a certain temperature which value depends on the chemical composition of the workpiece [12], as shown in Fig. 22.12.

Workpiece chemical composition - Effect of chemical composition of the workpiece on coefficient of friction in hot rolling is usually related to a mechanism of formation of scale. The experiments [13] show that μ decreases with increase in carbon content in carbon steel as shown in Fig. 22.13. The effect becomes less pronounced with increase in temperature. This phenomenon is also sometimes explained by a weakening of the forces of molecular attraction between metallic surfaces with increase in carbon content. This is confirmed by the fact that during rolling of austenitic stainless steel, which has a tendency for bonding to the roll surface, the coefficient of friction is from 1.3 to 1.5 times greater than that during rolling of carbon steels [1].

Roll surface roughness - The steady-state coefficient of friction substantially increases with increase in roll surface roughness as shown in Table 22.5 where the values for μ were determined using the method of forced skidding.

Rolling speed - According to Gelej [14], an increase in rolling speed reduces the steady-state coefficient of friction according to the formula

a) for steel rolls

$$\mu = 1.05 - 0.0005t - 0.056V$$

b) for iron rolls

$$\mu = 0.92 - 0.0005t - 0.056V \qquad (22\text{-}26)$$

c) for ground steel or iron rolls

$$\mu = 0.82 - 0.0005t - 0.056V$$

where V = rolling speed, m/s

t = workpiece temperature, $^\circ$C.

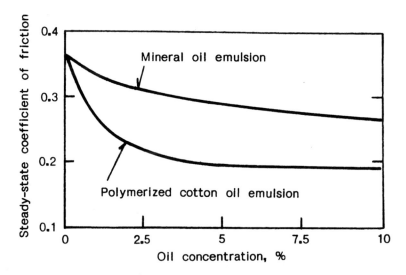

Fig. 22.14 Effect of oil concentration on the steady-state coefficient of friction in hot rolling. (From EXTERNAL FRICTION DURING ROLLING. Reprinted with permission.)

Oil concentration - Generally, the steady-state coefficient of friction decreases with increase in oil concentration. However, after the oil concentration reaches a certain maximum level, further addition of oil does not result in any additional reduction in friction (Fig. 22.14). This maximum level of oil concentration depends on the type of lubricant as shown below [1,8]:

Polymerized cotton oil emulsion	5%
Stearic acid	20%
Rapeseed oil	40%

22.11 LUBRICATION MECHANISM IN HOT ROLLING

There are many uncertainties in identifying the actual mechanism of lubrication in hot rolling. According to Schey [8], these mechanisms may be summarized as presented below.

Hydrodynamic lubrication - This type of lubrication can be provided with glasses which have sufficient viscosity to allow a reasonable reduction in finishing passes. Full-fluid-film lubrication with glass improves surface finish. It also decreases heat loss. The main obstacles for industrial application of hydrodynamic lubrication are the difficulties in the removal and the possible abrasiveness of solidified glass.

Mixed-film lubrication - In most of the instances this type of lubrication occurs when a compounded oil or an emulsion is applied into a roll bite. In hot rolling, the lubricant film usually

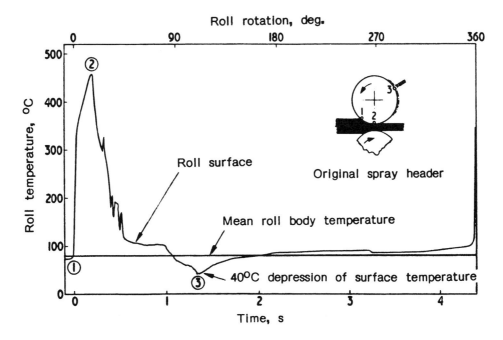

Fig. 22.15 Variation of roll surface temperature during one revolution. (From Robinson and Westlake. Reprinted by permission of the Council of the Institution of Mechanical Engineers from Proceedings of the First European Tribology Congress, 1973.)

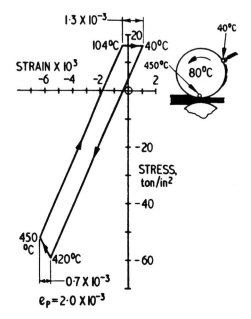

Fig. 22.16 Resulting stress-strain hysteresis loop produced by thermal cycle in Fig. 22.15. (From Robinson and Westlake. Reprinted by permission of the Council of Institution of Mechanical Engineers from Proceedings of the First European Tribology Congress, 1973.)

operates beyond its breakdown point, resulting in metal-to-metal contact. The organic substances, however, remain effective. Even though some of the film burns up, contact time is short and the lubricating action is presumably due to a film which is a mixture of original lubricant, its reaction products, and residues.

Mixed-film lubrication may be provided by application of mineral oils, fatty oils and synthetics. Mineral oils burn readily and pollute atmosphere when applied in the neat form; their efficiency is doubtful. They are also found ineffective in the form of emulsions. Fatty oils, such as tallow, rapeseed oil, or even mineral oil blended with fatty oils or their derivatives, have recently become the basis of steel-rolling lubricants. Synthetic esters have higher temperature resistance and, probably because of that, are found advantageous for steel rolling.

Solid lubrication - A number of solid lubricants have been tested in hot rolling. They are found, however, to have little spreading power, and their industrial application is still questionable.

22.12 ROLL WEAR IN HOT ROLLING

In hot rolling, the work rolls are subjected to periodic loading that is accompanied with abrasion by hard oxide and fluctuations in temperature [15] as shown in Figs. 22.15 and 22.16. The four major causes of roll wear are closely related to these conditions [16]:

   1. Abrasion of the roll surface due to its contact with the rolled material and backup roll
   2. Mechanical fatigue of the roll surface layer as a result of cyclical loading of the rolls

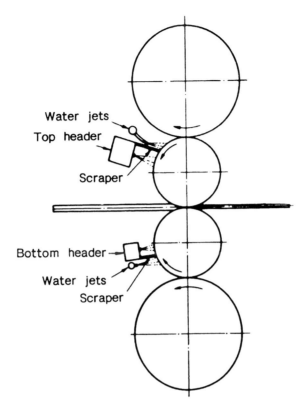

Fig. 22.17 Positions of the water and lubricant spray headers on the 4-high mill stand. (From Robinson and Westlake. Reprinted by permission of the Council of Institution of Mechanical Engineers from Proceedings of the First European Tribology Congress, 1973.)

3. Thermal fatigue of the roll surface layers as they are periodically heated by rolled material and cooled by water sprays

4. Corrosion.

Two types of the roll wear are usually considered: uniform wear and localized wear [8].

Uniform wear - The uniform wear is caused by abrasion in combination with thermal fatigue, with corrosive wear playing a subordinate role.

Localized wear - This is a local wear along bands on the roll surface. It is mainly caused by accumulation of secondary scale on the roll surface. Such accumulation increases with increase in roll surface roughness and rolling temperature and, therefore, is more likely to occur in the early stands of the finishing mill.

The roll surface roughening (banding) is closely related to the metallurgical structure of the cast iron rolls. According to Judd [17], the banding is a result of the following sequence:

a) thermal fatigue initiates cracks perpendicular to the roll surface,

b) at large eutectic carbides the cracks turn parallel to the surface,

c) oxidation in the cracks causes swelling and spalling of cells, and

d) layers of cells are successively removed to form the bands.

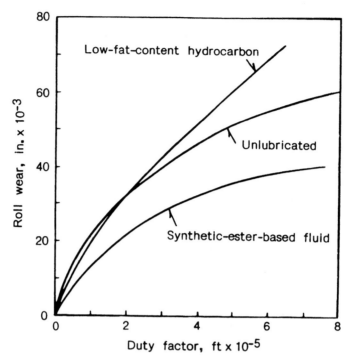

Fig. 22.18 Comparison of the roll wear curves for No. 2 finishing mill stand of 762 mm (30 in.) wide hot strip mill. (From Robinson and Westlake. Reprinted by permission of the Council of Institution of Mechanical Engineers from Proceedings of the First European Tribology Congress, 1973.)

Reduction of roll wear by lubrication is greatly affected by the type of lubricant used. Tests conducted by Robinson and Westlake [15] at No. 2 finishing mill stand of 762 mm (30 in.)-wide hot strip mill (Fig. 22.17) have shown that an application of a synthetic oil substantially reduces roll wear whereas the application of hydrocarbon fluid is detrimental to the roll life as shown in Fig. 22.18.

22.13 APPLICATION OF HOT ROLLING LUBRICANTS

According to data collected by Wandrei [16] from different hot strip and plate mills, the lubrication sites are provided at backup rolls, as well as at work rolls on either entry or exit sides or on both.

The lubricants are applied either in neat form or as emulsions. Dispension and steam atomization are used in several mills. Some attempts of using the solid lubricants are also reported.

Application of lubricants in production hot rolling mills provides a number of benefits. The contributions vary widely from mill to mill as summarized below by using data from 28 rolling mills.

Coefficient of friction reduction	34 to 67%
Roll life increase	20 to 250%
Roll force reduction	4.2 to 30%
Mill power reduction	3 to 35%
Pickling rate increase	10 to 50%
Rolled tonnage increase	1.5 to 22%.

Additional benefits include improvement in surface finish, reduced amount of scale generated during rolling, reduced rolled-in scale. Use of lubricants provides more homogeneous deformation allowing one to improve both strip profile and shape.

The following changes in physical and metallurgical properties of the rolled material are reportedly obtained by application of lubricants:

    a) improved control in grain size

    b) decreased yield point, ultimate tensile strength and notch strength

    c) increased elongation

    d) possible improved surface texture of silicon steel sheet

    e) possible improved precipitation kinetics of niobium carbide.

22.14 LUBRICATION MECHANISM IN COLD ROLLING

Two types of lubrication mechanism are usually considered in cold rolling, full- fluid-film lubrication and mixed-film lubrication.

Full-fluid-film lubrication - Nadai [18] was one of the first researchers who recognized a possibility of plastohydrodynamic (PHD) lubrication in cold rolling. His solution for the shear strength at the interface due to frictional force is based on assumed film thickness h and given as:

$$\tau_x \sim \frac{\mu \, dV_x}{h} \qquad (22\text{-}27)$$

where τ_x = shear strength at the point x along the arc of the roll contact

    dV_x = difference between velocity of metal and peripheral roll velocity at the point x.

In its turn, the film thickness h may be expressed by Eq. (21-5), modified here in application for rolling as follows:

$$h = \frac{6V\eta}{Y \tan \alpha} \qquad (22\text{-}28)$$

where α = roll bite angle.

Full-fluid-film lubrication reduces forward slip. This increases a probability of roll skidding. This type of instability is a main reason why practical rolling is usually conducted in the mixed-film regime. However, Eqs. (22-27) and (22-28) are found to be useful for qualitative analysis of the effects of lubrication on rolling parameters.

Mixed-film lubrication - In application for mixed-film lubrication Eq. (22-28) has been modified by Mizuno [8] as follows:

$$h = \frac{6V\eta}{Y \tan \alpha} (1 - 2r/3) \qquad (22\text{-}29)$$

where V = arithmetic average of roll speed and strip exit speed

    r = pass reduction.

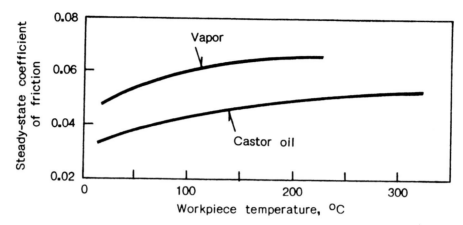

Fig. 22.19 Effect of the workpiece temperature on the steady-state coefficient of friction in cold rolling. (From EXTERNAL FRICTION DURING ROLLING, by A.P. Grudev, 1973. Copyright Metallurgia, Moscow, USSR. Reprinted with permission.)

A number of studies show that the average film thickness increases and lubrication becomes more hydrodynamic in character as the lubricant viscosity increases. This results in lower value of the coefficient of friction as given in general form by the equation:

$$\mu = f(\eta^a), \tag{22-30}$$

where a = exponent, that was found by different studies to be between -0.2 and -0.5.

Rolling speed has great influence on the oil film. At low speed most of the lubricant is squeezed out, leaving a thinner film in the roll gap. As shown in Chapter 21, the boundary lubrication prevails at low product $V\eta$. As the product $V\eta$ increases, the entrained film becomes thicker resulting in predominantly mixed-film lubrication.

22.15 STEADY-STATE COEFFICIENT OF FRICTION IN COLD ROLLING

The steady-state coefficient of friction μ in cold rolling is affected by various factors which are briefly reviewed below.

Workpiece temperature - Increase in workpiece temperature generally increases μ. This increase is more pronounced when the lubricants with higher viscosity are used (Fig. 22.19). The temperature dependence of μ can be approximated by [1]:

$$\mu = \mu_{20} + a(t - 20)^{0.5}, \tag{22-31}$$

where μ_{20} = steady-state coefficient of friction at 20°C (68°F)

      t = workpiece temperature, °C

      μ = steady-state coefficient of friction at temperature t

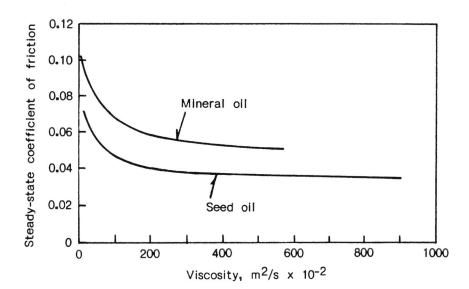

Fig. 22.20 Effect of lubricant viscosity on the steady-state coefficient of friction in cold rolling. (From EXTERNAL FRICTION DURING ROLLING, by A.P. Grudev, 1973. Reprinted with permission.)

a = empirical coefficient depending on the roll surface finish as follows:

Smooth roll surface   0.0011-0.0015

Rough roll surface   0.0035- 0.0073

Roll surface roughness - As would be expected, μ increases with increase in the roll surface roughness. This effect may be expressed by [1]:

$$\mu = \mu_{0.2}[1 + 0.5(R_a - 0.2)], \tag{22-32}$$

where $\mu_{0.2}$ = steady-state coefficient of friction corresponding to the roll surface roughness $R_a = 0.2$ μm

μ = steady-state coefficient of friction corresponding to the roll surface roughness R_a.

Equation (22-28) is valid within the range of R_a between 0.2 to 10 μm.

Workpiece chemical composition - When rolling carbon steels with lubrication, μ was found to be negligibly affected by variations in chemical composition of rolled material. Austenitic stainless steel, however, has a tendency to stick to the rolls and the measured values for μ are usually from 10 to 20% greater than those for carbon steel.

Lubricant viscosity - As follows from Eq. (22-29), the entrained film thickness increases with increase in lubricant viscosity. Therefore, the frictional forces decrease. This is confirmed by a number of experiments. The effect of viscosity on μ for two groups of oils is shown in Fig. 22.20 and is approximated by the equation:

$$\mu = k[0.15(\eta_{50})^{-0.5} + 0.03], \tag{22-33}$$

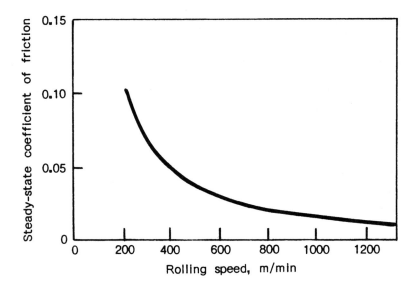

Fig. 22.21 Effect of cold rolling speed on the steady-state
coefficient of friction. Adapted from Pawelski (1964-65).

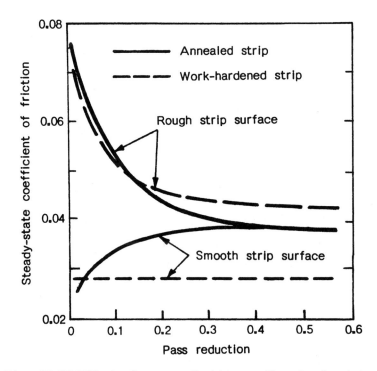

Fig. 22.22 Effect of pass reduction on the steady-state
coefficient of friction in cold rolling. Adapted from
Grudev (1973).

where η_{50} = viscosity at 50°C (122°F), m²/s x 10⁻².

For mineral oil, k = 1.4. For seed oil, k = 1.0.

Rolling speed - In accordance with Eq. (22-29), the entrained film thickness is proportional to rolling speed. Therefore, friction decreases as rolling speed increases. Fig. 22.21 shows typical values of μ as calculated from steel rolling trials [19].

Sims and Arthur [20] have shown that at the speed greater than 0.25 m/s (50 ft/min) the speed effect on the steady-state coefficient of friction can be expressed in the form:

$$\mu = a \exp(bV^{-c}),$$

(22-34)

where a, b = constants

V = rolling speed.

In case of dry rolling, friction increases with speed as shown in the formula proposed by Starchenko et al [21]:

$$\mu = \mu_0 \pm \frac{V}{a + bV},$$

(22-35)

where μ_0 = steady-state coefficient of friction at very low speed

a, b = empirical constants.

In Eq. (22-35), the minus sign corresponds to rolling with lubrication and the plus sign relates to dry rolling condition.

Pass reduction - The effect of pass reduction on friction depends greatly on the surface roughness and a degree of work-hardening of the workpiece. Figure 22.22 shows the results of rolling the samples of low-carbon steel with the castor oil and 10% mineral oil emulsion being used as lubricants [1]. During rolling a strip with rough surface, μ reduces with increase in reduction for both annealed and work-hardened strip. When a strip with smooth surface is rolled, μ increases with increase in reduction for annealed strip and remains constant for work-hardened strip.

Work roll diameter - As follows from Eq. (22-29), the oil film thickness increases when bite angle α decreases. Since increasing the work roll diameter results in smaller bite angle, it is expected that this would also yield a lower value of the coefficient of friction.

22.16 APPLICATION OF COLD ROLLING LUBRICANTS

In cold rolling there are three main methods of lubrication which are described below [8].

Lubrication with separate water - This method includes an application of the lubricant to the strip in neat form prior to rolling, and adding of water on the mill.

Lubrication by dispersion - This method involves mixing water with oil at the mill and application of the mixture in the form of a mechanical dispersion.

Lubrication with emulsions - This method presumes a use of emulsions which are oils or fats emulsified in water.

Table 22.6 Steady-state coefficient of friction in cold rolling. Data from Whitton and Ford (1955).

Lubricant	Viscosity at 37.8°C (100°F) Redwood No. 1	Pass number	Pass reduction, %	Steady-state coefficient of friction
Vac. R.O. 546	128	1	15.0	0.070
Vac. R.O. 950	37	1	15.6	0.069
Vac. R.O. 40A	59	1	17.0	0.061
Shell P.E. 6	120	3	23.0	0.050
Shell P.E. 6	120	4	27.9	0.053
Esso Pale 885	81	3	24.0	0.052
Esso Baywest	5500	4	27.5	0.050
Olive oil	–	2	18.1	0.057
Castor oil	–	4	23.1	0.045
Lanoline	–	4	26.5	0.041
Camphor flowers	–	4	27.2	0.038

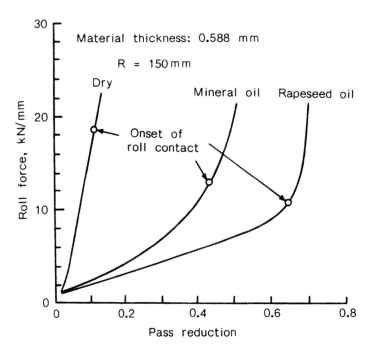

Fig. 22.23 Roll forces measured in rolling of steel strip under conditions of limiting reduction. Adapted from Pawelski and Kading (1967).

The steady-state coefficient of friction is often used for comparison of different types of lubricants and their method of application. Table 22.6 presents the values of the coefficient of friction calculated from the tests conducted by Whitton and Ford [22]. Fully annealed mild steel strip of 1.78 mm (0.070 in.) thick was rolled during the tests using 100 mm (4.0 in.) diameter polished steel rolls. The rolling speed was about 0.16 m/s (32 ft/min). The lubricants were generously applied to both the strip and the rolls.

Roll force is another parameter that is used as a criterion for evaluation of lubricants. Figure 22.23 illustrates the roll forces measured by Pawelski and Kading [23] in rolling of thin steel strip. When rolling thin gauges the rolls make direct contact outside the strip, but this is only part of the reason for reaching a limiting reduction. Another important contributor is roll flattening to a degree that strip deformation becomes all elastic.

The test data show that under conditions of limiting reduction it is advantageous to use the rapeseed oil over mineral oil.

22.17 ROLL WEAR IN COLD ROLLING
Abrasive, adhesive, and corrosive wear each may play a role in cold rolling [8].

Abrasive wear - When a softer adhesive material is rolled, abrasive action of the debris is a prime cause for roll wear. Abrasive wear is enhanced in the presence of hard oxide supported by a strong substrate.

Corrosive wear - This type of roll wear may occur with some lubricants and can become localized when the rolls are left unturned for a prolonged period of time with a lubricant film between them.

Adhesive wear - This type of roll wear is expected when harder adhesive materials are rolled.

The most severe damage of the rolls is caused by **spalling** that is a type of fatique failure, which may be deep enough to destroy a hard surface layer of the rolls. Spalling usually takes the form of shallow flakes or circumferential cracks. The roll spalling is promoted by uneven stress distribution due to localized wear, grinding cracks, residual stresses in the roll, and hydrogen embrittlement in contact with emulsion.

22.18 FRICTION INSTABILITIES
A rolling mill in combination with rolled strip presents a very sophisticated mass-spring system. When this system is being excited, it starts to vibrate. Two types of vibration are possible: roll torsional vibration and roll vibration in vertical plane [8, 24]. Torsional vibrations have frequencies of 10 to 35 Hz.

Vertical vibrations are usually classified into two modes which have the following ranges of frequencies:

1. Third octave mode   120-260 Hz
2. Fifth octave mode   500-700 Hz.

Some phenomena caused by friction instabilities are briefly discussed below.

Herringbone (or chevron) - This is a periodic marking at some angle to the roll axis, occasionally in a V or multiple-V pattern. In rolling of aluminum with relatively poor lubrication, the bands are associated with stick-slip motion initiated presumably by torsional vibration. As was observed by Moller and Hoggart [25], the vibration was initiated by any sudden change in torque. The vibration was self-sustained only when friction decreased with increasing speed, and vibration was damped out if friction increased with increasing speed.

At heavier reductions, up to 85%, the herringbone typically corresponds to frequencies from 140 to 200 Hz and is ascribed to mill vibration in vertical plane.

Third-octave-mode chatter - This type of vibration is observed mostly in cold rolling of strip that is less than 0.2 mm (0.008 in.) in thickness. The chatter imposes on the strip surface the alternating bright and dull bars strictly in line with the roll axes. The resulting periodic gauge variations may be as high as 25%. The phenomenon is attributable to the roll vibration in vertical plane.

Yarita et al [26] have observed that the chatter sets in when rolling starts with new emulsion of very fine particle size, and that the chatter disappears as the emulsion ages and the particle size becomes coarser. The effect, however, reappears again shortly before emulsion starts to break down. Application of new emulsion with organometallic zinc compounds was found helpful when reductions close to the limiting gauge were taken.

Roberts [24] identifies, together with inadequate lubrication, excessively high entry tension as the main source of chatter.

Fifth-octave-mode chatter - This type of vibration occurs during cold rolling of wide strip. The chatter is attributed to periodic wear marks developed on the roll surface. According to Roberts [24], the frequency of the fifth-octave-mode chatter may be approximated by:

$$f_5 = 14200/d, \tag{22-36}$$

where f_5 = frequency, Hz

d = work roll diameter, in.

The fifth-octave-mode chatter does not affect the strip gauge and is only detrimental from an aesthetic or operational viewpoint. Plausible relationships between rolling parameters and a tendency to chatter have been established by Pawelski et al [27]. Their theory is based on interaction of plastic and elastic deformations.

To reduce a probability of the chatter, DOFASCO (Canada) has utilized a so-called 'tuned damper' [28] that is an auxiliary mass attached to a vibrating system with a damping medium. The tuned damper consists of a block of steel bonded with adhesive to the rubber which is, in turn, bonded to the mill housing. The rubber acts as a spring and a damping medium. Application of the damper had reportedly allowed to increase the maximum mill speed up to 15%.

The same principle is used in the friction damper or 'hydraulic mill post liner' which provides damping to the mill stand by increasing the friction between the mill housing post and the backup roll chocks.

REFERENCES

1. A.P. Grudev, External Friction During Rolling, Metallurgia (Russian), Moscow (1973).

2. A.I. Tselikov, Izvestia Vuz. Machinostroenie (Russian), No. 11, 1959.

3. S. Ekelund, Steel 93 (Aug. 21), 1933, pp. 27-29.

4. B.S. Smirnov and H.P. Uk, "Treatment of Metals by Pressure", Machgiz, Moscow, LPI No. 203, 1959, pp. 38-48.

5. W. Tafel, Rolling and Roll Pass Design (German), Dortmund (1923).

6. Z. Wusatowski, Fundamentals of Rolling, Pergamon Press, Oxford, pp. 69-202 (1969).

7. H. Ford, F. Ellis, and D.R. Bland, "Cold Rolling with Strip Tension, Part 1 - A New Approximate Method of Calculalation and Comparison with Other Methods", Journal of the Iron and Steel Institute, Vol. 168, May 1951, pp. 57-72.

8. J.A. Schey, Tribology in Metalworking: Friction, Lubrication and Wear, American Society for Metals, Metals Park, Ohio, pp. 131-341 (1983).

9. T. Sheppard and D.S. Wright, Metals Technology, 8, 1981, pp. 46-57.

10. I.M. Pavlov, Theory of Rolling and Principles of Plastic Deformation of Metals (Russian), GONTI, Moscow (1938).

11. P.W. Whitton and H. Ford, "Surface Friction and Lubrication in Cold Strip Rolling", Proceedings of the Institution of Mechanical Engineers, No. 169, 1955, pp. 123-140.

12. I.M. Pavlov and M.I. Kuprin, "Technological Processes of Treatment of Steels and Alloys", Moscow, Metallurgizdat, MIC, No. 33, 1955, pp. 154-192 (Russian).

13. L.F. Molotkov, "Theory and Practice of Metallurgy", No.3, 1940, pp. 20-22 (Russian).

14. S. Gelej, Calculation of Forces and Energy During Plastic Deformation of Metals, Metallurgizdat (Russian), Moscow (1958).

15. C.L. Robinson and F.J. Westlake, "Roll Lubrication in Hot Strip Mills", Proceedings of the First European Tribology Congress, London, 25-27 Sept. 1973, pp. 389-398.

16. C.L. Wandrei, Review of Hot Rolling Lubricant Technology for Steel, ASLE Special Publication SP-17, Park Ridge, IL, (1984).

17. R.R. Judd, "Surface Deterioration of Grain Iron Work Rolls in the First Stands of a Hot Strip Mill Finishing Train", AISE Yearly Proceedings, 1979, pp. 65-74.

18. A. Nadai, "The Forces Required for Rolling Steel Strip Under Tension", Journal of Applied Mechanics, June 1939, pp. A54-A62.

19. O. Pawelski, Lubrication and Wear, Third Convention, Proc. Inst. Mech. Eng., Pt. 3D, 1964-65, pp. 80-92.

20. R.B. Sims and D.F. Arthur, "Speed-Dependent Variables in Cold Strip Rolling", Journal of Iron and Steel Institute, Vol. 172, 1952, No. 3, pp. 285-295.

21. D.I. Starchenko et al, Izv. Vuz. Chernaya Metallurgia, No. 8, 1967, pp. 86-91.

22. P.W. Whitton and H. Ford, "Surface Friction and Lubrication in Cold Strip Rolling", Proceedings of the Institution of Mechanical Engineers, Vol. 169, 1955, pp. 123-140.

23. O. Pawelski and G. Kading, Stahl von Eisen, 87, 1967, pp. 1340-1355 (German).

24. W.L. Roberts, Flat Processing of Steel, Marcel Dekker, New York and Basel, pp. 625–664 (1988).

25. R.H. Moller and J.S. Hoggart, Journal of the Australian Institute of Metals, 12, 1967, pp. 155–165.

26. I. Yarita et al, Transaction of the Iron and Steel Institute of Japan, 18, 1978, pp. 1-10.

27. O. Pawelski, et al, "Chattering in Cold Rolling. Theory of Interaction of Plastic and Elastic Deformation", Proceedings of the 4th InternationalSteel Rolling Conference: The Science and Technology of Flat Rolling, Vol. 2, Deauville, France, June 1-3, 1987, pp. E.11.1-E.11.5.

28. S. Critchley and D. Paton, "Tandem Mill Vibration", Proceedings of the 4th International Steel Rolling Conference: The Science and Technology of Flat Rolling, Vol. 2, Deauville, France, June 1-3, 1987, pp. E.12.1-E.12.6.

Part VII

Heat Transfer in Rolling Mills

23

Steel Heating for Hot Rolling

23.1 PURPOSE OF HEATING PROCESS

Heating of ingots and slabs accomplishes the following main purposes [1]:

1. Surface scaling for removal of surface defects.

2. Softening the steel for rolling.

3. Providing a sufficiently high initial temperature, so that rolling process is completed in fully austenitic temperature region.

4. Dissolving (where applicable) carbides or nitrides that are to be precipitated at a later stage of processing.

23.2 REQUIREMENTS FOR HEATING PROCESS

Because of its nature, heating process produces coarsing of the austenitic structure with grain diameters in the range of 100 to 1000 microns. In the process of grain coarsing, the larger grains get larger and the smaller grains tend to disappear.

The grain growth may proceed with increase in temperature in two ways (Fig. 23.1):

a) continuous growth that is typical for plain carbon steel

b) discontinuous growth that is typical for aluminum-killed and microalloyed steels.

During continuous grain growth, the grain coarsing of C-Mn steels between 900 and 1200°C (1652 and 2192°F) follows well-known continuous grain-growth law:

$$D = kt^a exp(-b/T), \tag{23-1}$$

where k, a, b = constants

$\qquad$ D = average grain diameter

$\qquad$ t = time

$\qquad$ T = temperature.

In case of discontinuous grain growth, the process is suppressed up to a certain temperature after

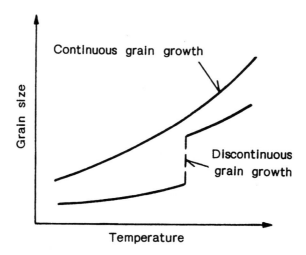

Fig. 23.1 Two types of the grain growth process during heating.

which a sudden increase in grain size takes place. This discontinuous grain growth may also start after holding material at the same temperature during longer time periods. The main reason for discontinuous grain growth is suppression of the grain boundary motion by the precipitates that are dissolved in steel.

Vanadium-bearing steels generally resist coarsing only up to 1000 to 1050°C (1832 to 1922°F). Niobium-bearing steels resist grain coarsing to temperatures up to 1100 to 1150°C (2012 to 2102°F). Steel containing a fine dispersion of the very stable TiN can resist grain coarsing to temperatures above 1200°C (2192°F). Another feature of the heating process is non-uniform increase in temperature of the ingot or slab. This is mainly due to the following factors:

a) penetration of heat from metal surface to its center that produces rapid increase in surface temperature whereas the center of the slab remains cold

b) uneven heating conditions in the zones surrounding the ingot or slab

c) larger surface to volume ratio in a unit volume of the ingot or the slab near their edges in comparison to their middle portion

d) heat sink to the supportive members of the ingots or slabs producing in the slabs so-called 'skid marks'.

In application for hot rolling of flat steel products, the main requirements for heating process are:

1. Achieving the desired material temperature which for plain carbon steel is usually between 1200 to 1320°C (2192 to 2408°F).

2. Reducing the temperature differential between surface and center to a desired level which can be as low as 14°C (25°F) for the slabs of 250 mm (10 in.)-thick.

3. Reducing skid marks effect.

4. Avoiding the overheating ('washing') of the slab surface in order to reduce the grain coarsing as well as the excessive scaling.

5. Reduce energy consumption for heating.

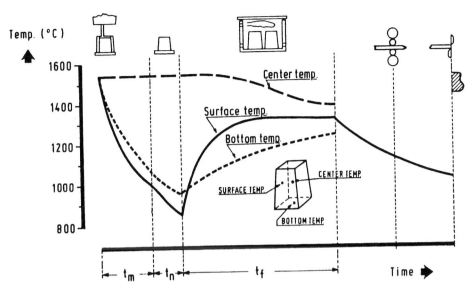

Fig. 23.2 Three periods of the ingot heat transfer cycle. (From Hollander and Zuurbier, AISE Year Book, 1982. Copyright AISE, Pittsburgh, Pennsylvania. Reprinted with permission.)

23.3 HEAT TRANSFER IN INGOTS

Heating of ingots is closely coordinated with teeming process. The ingot heat transfer time prior to rolling may be divided into the following three time periods [2]:

1. **Mold cooling time t_m** (Fig. 23.2) - It starts with beginning of teeming of the ingot and finishes after stripping the mold. This time period includes the transportation time of the ingot in the mold from the steel plant to the stripping bay and the time for stripping the mold.

2. **Air cooling time t_n** - It starts from the moment when the ingot is placed in the soaking pit. The object of heating is to reduce the temperature differences in the ingot and to reach an average temperature level such that the ingot can be rolled.

23.4 TYPES OF SOAKING-PIT FURNACES

Soaking pits are the deep chambers, or furnaces, of square, rectangular or circular shape, into which the ingots are placed in an upright position. Most of the existing soaking pits are fuel-fired in which fuel oil, natural gas or coke-oven gas is burned [1]. Below is a brief description of three types of the soaking-pit furnaces.

1. **Regenerative pits** - In the furnaces of this type the ingots are heated by burning the gas through a port in the pit wall on one side, permitting the products of combustion to pass across the pit and out through the regenerator flues and stack to the atmosphere. Direction of the gas passage is periodically reversed. The air, after each reversal, is passed through the hot regenerations to provide heat for combustion of the fuel. In order to better equalize ingot temperature, the practice of firing and dampering is usually utilized.

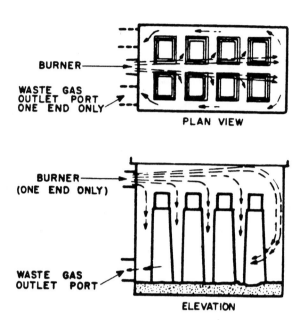

Fig. 23.3 Diagrammatic elevation and plan view illustrating principle
of continuous firing and flow of hot gases in a 'one-way top-fired'
soaking pit. Hot gases from outlet ports pass to recuperators
(not included). (From THE MAKING, SHAPING AND TREATING OF STEEL,
eds. W.T. Lankford et al, 1985. Copyright AISE, Pittsburgh,
Pennsylvania. Reprinted with permission.)

2. **Continuous-fired pits** - In the furnaces of this type a sufficient space was provided for combustion of fuel. These furnaces are usually equipped with recuperators. A number of different designs have been implemented.

In **one-way fired pits** (Fig. 23.3) the combustion space is provided above the ingots. The fuel is fired horizontally and the combustion gas flow is vertical in accordance with hydrostatic principles.

In **bottom-center-fired** or **vertically fired pits,** the fuel is fired vertically through a port, centrally located in the bottom of the pit, around which the ingots are placed.

In **circular pits** the tangential firing of fuel is employed from a series of recessed burners located in the lower periphery of inclined side walls.

In **bottom-two-way fired pits** (Fig. 23.4) the burners are located in opposite endwalls about 600 mm (2 ft) above the bottom of the pit and the waste-gas ports are located in the same endwalls at each of four pit corners. Combustion of the fuel takes place in the center of the pit. This method of firing provides turbulence to the flow of gases in the pit, thus improving the heating of the ingot bottoms.

In **top-two-way fired pits** the fuel is fired from opposite ends into a combustion space above the ingots. A swirling motion of the gases is provided by setting the burners to fire horizontally at an angle to the centerline of the pit.

3. **Electric soaking pits** - In the soaking-pit furnaces of this type the pit itself consists of a rectangular steel casing and a refractory lining and an insulation which form a closed unit. The heating elements run the entire length of the pit. Power is applied to the heating elements by electrodes that

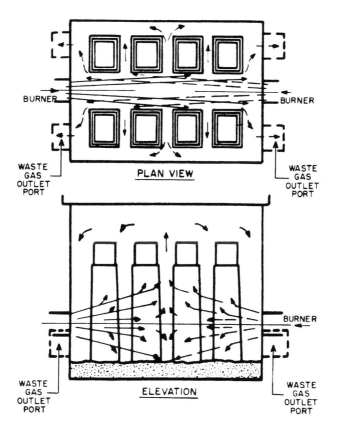

Fig. 23.4 Schematic diagram showing principle of firing and flow
of gases in a side-fired or bottom two-way fired soaking pit.
Gases from waste-gas ports go to a recuperator (not shown).
(From THE MAKING, SHAPING AND TREATING OF STEEL, eds. W.T. Lankford
et al, 1985. Copyright AISE, Pittsburgh, Pennsylvania. Reprinted
with permission.)

are inserted through the pit walls at each end. The electric soaking-pit furnaces allow one to provide an
improved control of the furnace atmosphere during heating of stainless and alloy-steel ingots.

23.5 HEAT TRANSFER IN SOAKING PITS

Heat transfer to the ingot in a soaking pit is affected by the following parameters [1]:

   a) temperature distribution in the ingot

   b) temperature differences between gas, soaking pit wall and ingot

   c) maximum firing rate

   d) combustion air temperature

   e) air-to-gas ratio

   f) fuel quality.

The heat input consists of approximately 85% fuel, 12% combustion air and 3% from oxidation of
ingot surface (Fig. 23.5). The heat content of the waste cases represents 58% of the heat input with 10%
lost through the pit walls and covers.

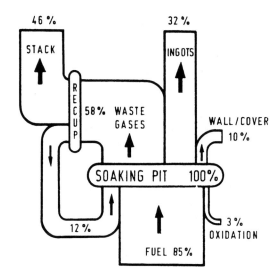

Fig. 23.5 Heat balance of soaking pit. (From Hollander, AISE Year Book, 1983. Copyright AISE, Pittsburgh, Pennsylvania. Reprinted with permission.)

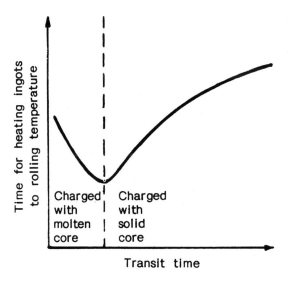

Fig. 23.6 Sketch illustrating the relationship between heating time and transit time in a soaking pit. (From THE MAKING, SHAPING AND TREATING OF STEEL, 1985. Reprinted with permission.)

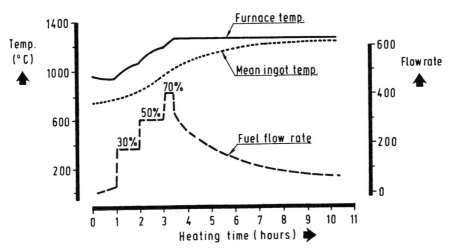

Fig. 23.7 Step heating practice in ingots. (From Hollander, AISE Year Book, 1983. Copyright AISE, Pittsburgh, Pennsylvania. Reprinted with permission.)

In order to improve the heating efficiency, the optimum ingot transit time ($t_t = t_m + t_n$) has to be defined. As follows from Fig. 23.6, when the transit time is shorter than optimum, the heat content of ingot is high, but the ingot residence in soaking pit must be longer to allow the ingot to completely solidify prior to rolling. When the transit time is longer than optimum, the heating time increases with increase in transit time.

When the ingot track time is not excessive, the heating time required for ordinary carbon steel is approximately 1.5 times longer than transit time. The heating time of cold ingots usually requires from 8 to 18 hours depending on the ingot size and type of steel. A common practice in heating of some alloys is known as **step heating** (Fig. 23.7). This practice allows one to better utilize the latent heat in the ingot, reduces stresses due to severe temperature gradients in the ingot and provides more uniform heating.

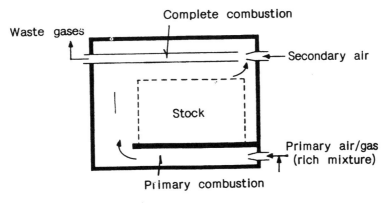

Fig. 23.8 Schematic presentation of fuel-fired batch-type reheating furnace. (From the Journal of Iron and Steel Institute, Publication No. 111, London, 1968. Reprinted with permission.)

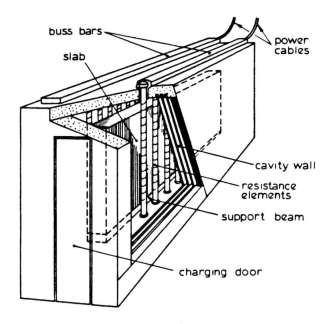

Fig. 23.9 Electrical resistance batch-type reheating furnace. (From Laws, Slab Reheating, Iron and Steel Institute Publication, London, 1973. Reprinted with permission.)

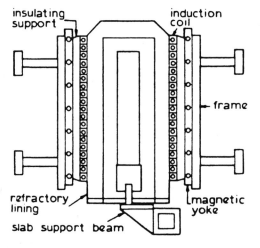

Fig. 23.10 Diagram of induction slab reheating furnace. (From Korey, Slab Reheating, Iron and Steel Institute Publication, London, 1973. Reprinted with permission.)

23.6 BATCH-TYPE SLAB REHEATING FURNACES

In the batch-type furnaces the charged material remains in a fixed position on the hearth during heating.

These furnaces may be divided into the following four groups:

1. Fuel-fired furnaces
2. Electrical resistance furnaces
3. Induction furnaces
4. Dual-fuel furnaces.

In **fuel-fired furnaces** (Fig. 23.8) either gaseous or liquid fuel may be utilized, with preheated or cold combustion air [1, 3].

In **electrical resistance furnaces** (Fig. 23.9) high-intensity radiant energy is produced by silicon carbide elements located near the slab [4].

In **induction furnaces** (Fig. 23.10) the slab heating is provided by eddy-current induced in the slab by electromagnetic field which is generated by the induction coils surrounding the slab [5].

In **dual furnaces** a fuel-fired furnace may be used for preheating of the slab followed by rapid heating to rolling temperature in an induction heating furnace [6].

Main advantages of the electric furnaces are [5]:

a) instant on/off action
b) maximum flexibility for scheduling of both hot and cold slabs
c) minimum product in process
d) reduced space requirements
e) no air-pollution problem.

High power cost is considered a main disadvantage of the electric furnaces.

23.7 CONTINUOUS REHEATING FURNACES

In the continuous-type slab reheating furnace, the charged material moves inside the furnace while it is being heated.

Depending on the method used to move the slabs during heating these furnaces may be classified as following [1]:

1. Pusher furnaces
2. Rotary hearth furnaces
3. Walking beam furnaces
4. Walking hearth furnaces
5. Roller-hearth furnaces.

In **pusher furnaces** (Fig. 23.11) the slabs can be charged either from the end or through a side door. The steel is moved through the furnace by pushing the last slab charged with a pusher at the charging end. A heated slab is simultaneously removed through a discharge door.

In **rotary hearth furnace** (Fig. 23.12) the hearth section of the furnace revolves while the external walls and the roof remain stationary.

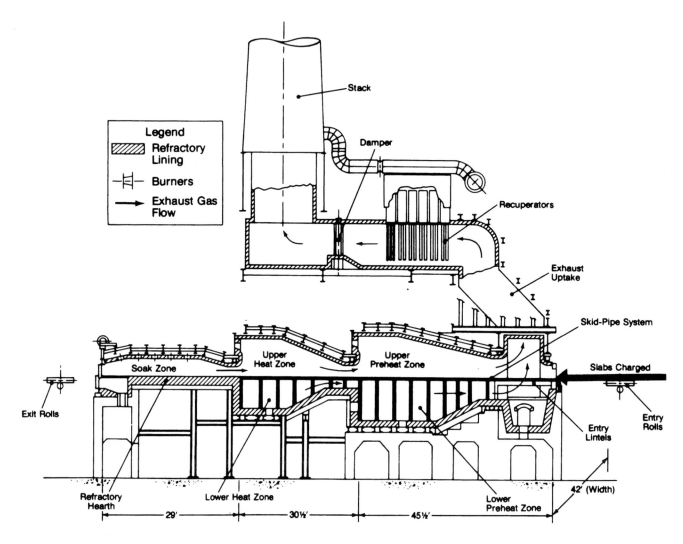

Fig. 23.11 Schematic longitudinal section through a three-zone counter-current fired pusher-type continuous furnace. (From Vance et al, AISE Year Book, 1978. Copyright AISE, Pittsburgh, Pennsylvania. Reprinted with permission.)

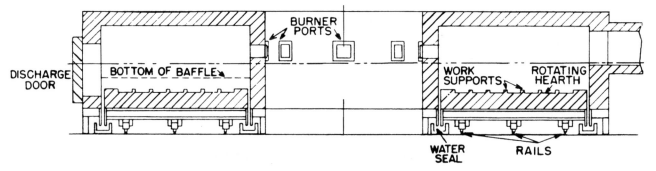

Fig. 23.12 Schematic arrangement of rotary hearth reheating furnace. (From THE MAKING, SHAPING AND TREATING OF STEEL, by W.T. Lankford et al, 1985. Copyright AISE, Pittsburgh, Pennsylvania. Reprinted with permission.)

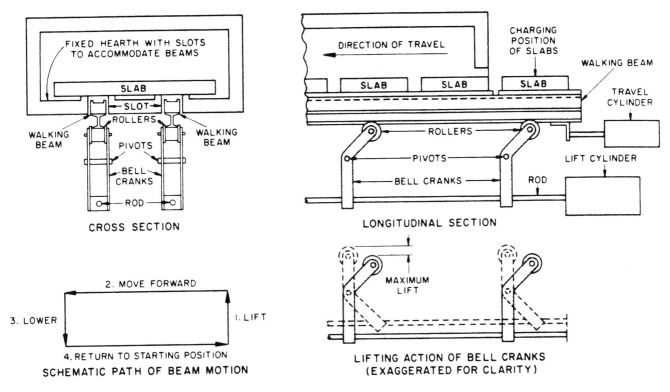

Fig. 23.13 Schematic diagram (not to scale) showing a method by which walking beams can operate to advance slabs along the hearth of a heating furnace. (From THE MAKING, SHAPING AND TREATING OF STEEL, 1985. Reprinted with permission.)

In **walking beam furnaces** (Fig. 23.13) the heated slabs are moved intermittently with so-called 'walking beams'. The slabs at rest are supported on the raised stationary ridges on the hearth. At that time the walking beams are located below the slabs. The walking motion is provided in four steps: (1) the walking beams are raised by pivoting the bell cranks, thus lifting the slabs from the stationary ridges, (2) the walking beams with the slabs are pushed in the direction of travel by means of travel cylinder, (3) the bell cranks are pivoted to lower the walking beams, thus placing the slabs on the stationary ridges in the advanced position, and (4) the walking beams are returned to the initial horizontal position by the travel cylinder.

In **walking hearth furnaces** the slab rests on fixed refractory piers which are extended through the openings in the hearth, thus providing a gap between the slabs and the furnace hearth. The slabs are advanced toward the discharge end by 'walking' the hearth in a manner that is similar to 'walking' the beams in a walking beam furnaces.

Both walking beam and walking hearth furnaces may have two entrances [7]. The second entrance is provided for preheated slabs and is usually located in a midway of the furnace. In order to achieve optimum slab moving rate through the furnace, the walking beam or walking hearth mechanism may be split into two or more independently driven sections.

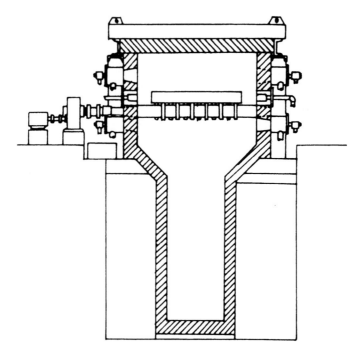

Fig. 23.14 Cross-section of a roller-hearth-type furnace for reheating continuously cast slabs immediately after casting to equalize temperature for rolling. Note over- and under-firing main burners and auxiliary edge-heating burners. (From THE MAKING, SHAPING AND TREATING OF STEEL, 1985. Reprinted with permission.)

In **roller-hearth furnaces** (Fig. 23.14) the heated slabs move through the furnace in the direction of their length on the motor-driven table rolls. This allows for non-intermittent advance of the slabs of much longer length than would be practical in pusher-type or walking-beam-type furnaces.

23.8 FUEL-FIRING IN CONTINUOUS REHEATING FURNACES

Gaseous or liquid fuel is usually used in continuous-type slab reheating furnaces. The fuel is mixed with air in burners. Some burner designs are shown in Fig. 23.15 [8]. In low-velocity burners (Fig. 23.15a), air and gas come out in parallel currents at the same speeds. This produces inefficient mixture of air and gas. In the burner with a disk (Fig. 23.15b) the air speed is increased. However, this design does not provide desired short flames.

Better results have been achieved by utilizing the principle of air rotation. In the burner shown in Figure 23.15c, the central air is forced into rotation by inclined blades and the gas is sent through a thin ring just between the turbulent air and the parallel air current. The flame length is adjusted by the air distribution between two air circuits. The simplified high-velocity burner, based on the same principle but without the flame adjustment, is shown in Fig. 23.15d. The best results (no hot spots and maximum possible flow variation) have been produced with burners (Fig. 23.15e) in which the air is injected at high pressure (above 1 kg/cm^2) through a series of holes perpendicular to the burner axis.

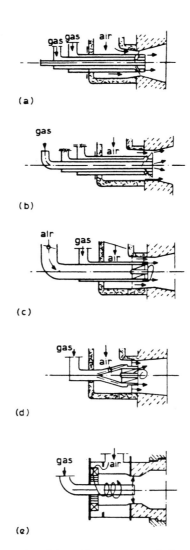

Fig. 23.15 Burners for reheating furnaces: a) low-velocity, b) with disk, c) high-velocity with variable flame, d) simplified high-velocity, e) high-velocity. (From Defise, Slab Reheating, Iron and Steel Institute Publication, London, 1973. Reprinted with permission.)

In order to conserve energy, some furnaces are equipped with recuperators which provide preheating of the combustion air.

Depending on the arrangement of the fuel-firing burners and firing chambers, the continuous reheating furnaces can be divided into the following groups [1, 3]:

1. Single zone furnaces (Fig. 23.16a)

2. Multiple zone furnaces (Fig. 23.16b, c, d)

3. Roof-fired furnaces (Fig. 23.16e)

4. Reverse-fired furnaces (Fig. 23.16f).

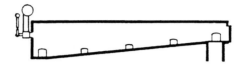

a) Single zone furnace

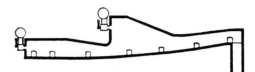

b) Two-zone furnace

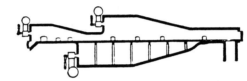

c) Three-zone furnace

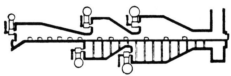

d) Five-zone furnace

e) Roof-fired furnace

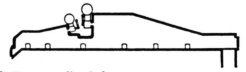

f) Reverse-fired furnace

Fig. 23.16 Types of continuous reheating furnaces. (From Flux, Iron and Steel Institute Publication No. 111, 1968, London. Reprinted with permission.)

23.9 ANALYSIS OF A CONTINUOUS HEATING PROCESS

In a continuous reheating furnace the wall temperature distribution along the furnace is regulated by a zone temperature control system. This temperature distribution may be described by temperature-distance curve. When slab push rate is known this temperature-distance curve can be converted into a temperature-time curve.

Because heating conditions for the top and bottom zones of the furnace are different, both top and bottom zone wall temperature-time curves have to be considered in order to provide accurate simulation of the heating process.

It was determined from mill experiments that the heat transfer in these type of furnaces is mainly due to radiation whereas the heat transfer by convection contributes less than 5% to the total heat transfer [9, 10].

The heat radiated to the slab may be expressed by the equation similar to Eq. (2-29)

$$Q_s = S\xi(1 - \epsilon)A_s[(T_w + 460)^4 - (T_s + 460)^4] \tag{23-2}$$

where A_s = slab surface area

$\quad T_s$ = slab surface temperature, ^oF

$\quad T_w$ = furnace wall temperature, ^oF

$\quad \epsilon$ = portion of the heat flow absorbed by gases.

Heat transfer inside the slab is done mainly by conduction. Neglecting both the edge effect and the skid marks effect, the process is described by the following equation similar to Eq. (2-27)

$$dT/dt = a_s(d^2T/dx^2), \tag{23-3}$$

where a_s = slab thermal diffusivity

$\quad T$ = slab temperature

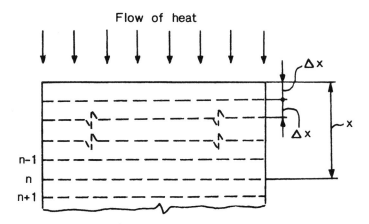

Fig. 23.17 Schematic presentation of finite layers of a slab.

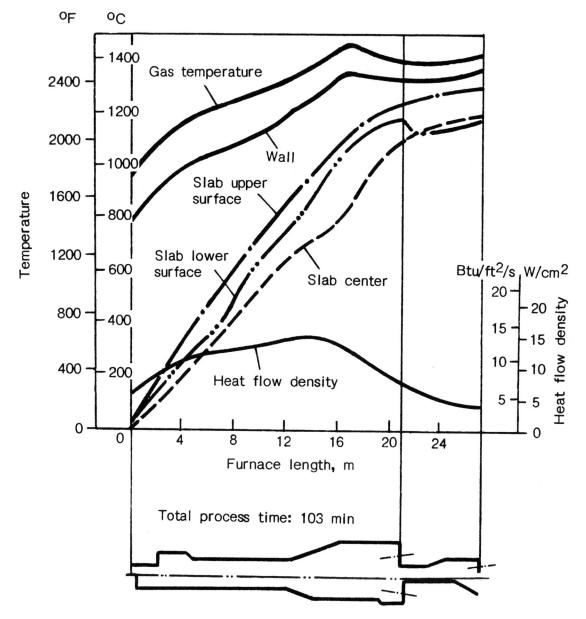

Fig. 23.18 Temperature and heat flow density distribution along furnace length. (From Hollander and Zuurbier, AISE Year Book, 1982. Copyright AISE, Pittsburgh, Pennsylvania. Reprinted with permission.)

x = distance from the slab surface.

The finite difference method proposed by Schmidt [11] allows one to replace Eq. (23-3) by

$$(T_n' - T_n)/(\Delta t) = a_s(T_{n+1} - 2T_n + T_{n-1})/(\Delta x)^2, \tag{23-4}$$

where n = layer number (Fig. 23.17)

Δx = layer thickness

T_n = temperature of the n-th layer at the time t

T_n' = temperature of the n-th layer at the time $t + \Delta t$.

In order to obtain better accuracy of approximation, the time increment Δt and layer thickness Δx are chosen so that [12]

$$\frac{(\Delta x)^2}{a_s(\Delta t)} \geq 2 \tag{23-5}$$

For a given furnace wall temperature distribution along the furnace and a given slab pushrate, the heat transfer analysis allows one to define the following parameters (Fig. 23.18):

a) gas temperature distribution along the furnace length

b) slab temperature at any position in the furnace

c) heat flow density along the furnace length.

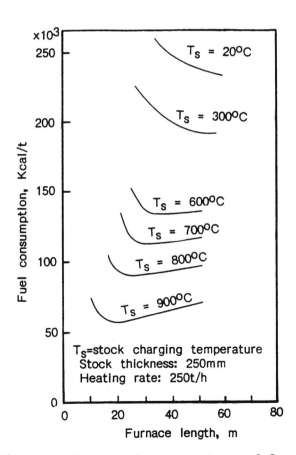

Fig. 23.19 Effect of slab charging temperature and furnace length on fuel consumption. Adapted from Matsukawa and Yoshibe (1982).

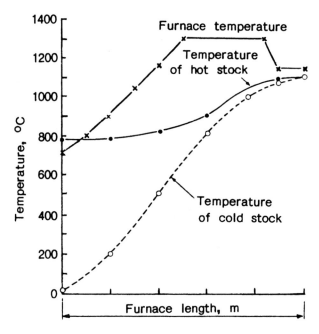

Fig. 23.20 Heating of cold and hot stock in conventional reheating furnace. Adapted from Matsukawa and Yoshibe (1982).

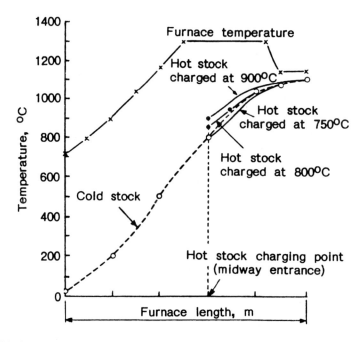

Fig. 23.21 Heating of cold and hot stock in two-entrance reheating furnace. Adapted from Matsukawa and Yoshibe (1982).

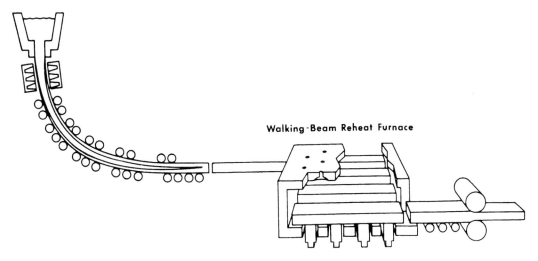

Fig. 23.22 Continuous caster with walking beam furnace. (From Cook and Rasmussen, AISE Year Book, 1970. Copyright AISE, Pittsburgh, Pennsylvania. Reprinted with permission.)

23.10 HOT SLAB CHARGING

Hot slab charging allows one to substantially reduce fuel consumption required for heating of the slab [13]. Amount of fuel saved increases with an increase in the slab charging temperature (Fig. 23.19).

The three most common methods of hot slab charging are described below.

1. Charging both hot and cold slabs into a single-entrance reheating furnace (Fig. 23.20). In that case the furnace design length is selected to receive cold slabs, and therefore, this length is too long for the hot charge slabs. Furthermore, the in-furnace stock traveling speed is matched to the cold stock. As a result, this type of operation impairs not only heat economy but also product quality and productivity because it sometimes causes washing on the surface of the hot stock.

2. Charging both hot and cold slabs into a double-entrance reheating furnace. The second slab charging port is designated for hot slabs and is usually located at a midpoint where the temperature of the traveling cold stock becomes equal to that of the hot stock (Fig. 23.21). This permits the hot stock to be more efficiently heated along with cold stock. The double-entrance furnace also provides some flexibility in synchronizing the production rates of continuous casters and rolling mills.

3. Charging only hot slabs into a single-entrance furnace (Fig. 23.22). This method allows one to substantially reduce the design furnace length as well as to achieve maximum fuel savings.

23.11 EFFECTS OF SKID SYSTEM ON HEATING

Skid systems have two main effects on heating:

    a) they absorb heat from their surroundings

    b) they cast radiation shadows on the stock and thus prevent heat from reaching it.

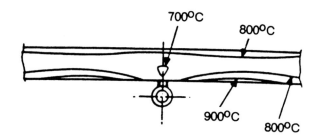

Fig. 23.23 Temperature distribution inside a slab in a continuous pusher-type furnace. (From Salter, Slab Reheating, Iron and Steel Institute Publication, London, 1973. Reprinted with permission.)

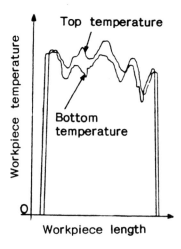

Fig. 23.24 Typical slab temperature distribution along its length after reheating in continuous pusher-type furnace. (From Salter, Slab Reheating, Iron and Steel Institute Publication, London, 1973. Reprinted with permission.)

This results in reduction of fuel efficiency and also in generation of so-called 'skid marks', cooler regions on the slabs, as shown in Figs. 23.23 and 23.24.

The heat absorbed by a skid system may be described by equation [14]:

$$q = hA(T_f - T_c), \tag{23-6}$$

where q = rate of heat loss

   h = heat transfer coefficient

   A = surface area of skid pipes

   T_f = mean temperature of the furnace surrounding the pipe

   T_c = temperature of the skid coolant.

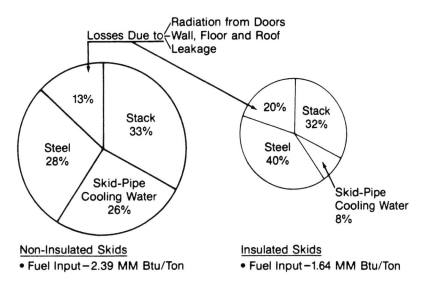

Fig. 23.25 Energy balance for a 5-zone pusher-type slab reheating furnace. (From Vance, et al, AISE Year Book, 1978. Copyright AISE, Pittsburgh, Pennsylvania. Reprinted with permission.)

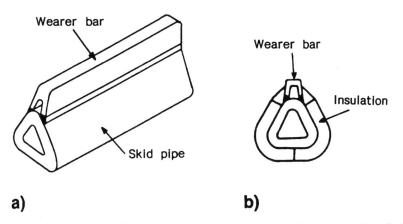

a)   **b)**

Fig. 23.26 Hot skids for pusher-type furnace: a) without insulation, b) with insulation. (From Salter, Slab Reheating, Iron and Steel Institute Publication, London, 1973. Reprinted with permission.)

The skid coolant temperature is usually maintained as high as possible and is limited by structural properties of the steel used to make the skid pipes. Thus, the only two variables available for alteration are the heat-transfer coefficient and the surface area of the skid system. This is usually achieved by the use of well insulated skid systems with minimum possible surface area. The effect of the skid insulation on the furnace energy balance is shown in Fig. 23.25 [15].

In order to reduce the skid marks, the skid system should produce the minimum radiation shadowing on the slab. Figure 23.26 shows one of the designs of the hot skids utilizing the apex-uppermost pipes that reportedly meet the above requirements. The skid marks are also reduced by either staggering the hot skids or by a gradual change of the distance between them.

REFERENCES

1. The Making, Shaping and Treating of Steel, 10th Edition, eds.W.T. Lankford, Jr., et al,Association of Iron and Steel Engineers, Pittsburgh, Pennsylvania, pp. 783–860 (1985).

2. F. Hollander,"Reheating Processes and Modifications to Rolling Mill Operations for Energy Savings", AISE Yearbook, 1983, pp. 252-259.

3. J.H. Flux, "Reheating for Hot Working in the Steel Industry", Reheating for Hot Working, Iron and Steel Institute Publication No. 111, London, 1968, pp. 70-82.

4. W.R. Laws, "Trends in Slab Reheating Furnace Requirements and Design", Slab Reheating, Iron and Steel Institute Publication, London, 1973, pp. 1-10.

5. W.J. Korey, "Induction Heating of Slabs Today", Slab Reheating, Iron and Steel Institute Publication, London, 1973, pp. 13-19.

6. W.R. Laws and F.M. Slater, "Future Trends in Design of Continuous Pusher Furnaces for the Hot Strip Mill", Reheating for Hot Working, Iron and Steel Institute Publication No. 111, London, 1968, pp. 132-139.

7. T. Matsukawa and Y. Yoshibe, "Recent Developments in Reheating Furnaces", Nippon Steel Technical Report, No. 20, December 1982, pp. 119-130.

8. M. Defise, "Walking-Beam Furnaces of the Chartal Works, Cockerill", Slab Reheating, Iron and Steel Institute Publication, London, 1973, pp. 58-67.

9. A. Schack, Industrial Heat Transfer, Chapman and Hall, London (1965).

10. F. Hollander and S.P.A. Zuurbier, "Design, Development and Performance of On-Line Computer Control in a 3-Zone Reheating Furnace", AISE Yearbook, 1982, pp. 58-66.

11. E. Schmidt, Foppls Festschrift, 1924, reported in Heat Transmission, ed. W.H. McAdams, McGraw Hill, New York, 1954.

12. F. Fitzgerald, "Aspects of Furnace Design for Hot Working", Reheating for Hot Working, Iron and Steel Institute Publication No. 111, London, 1968, pp. 123-131.

13. T. Matsukawa and Y. Yoshibe, "Recent Developments in Reheating Furnaces", Nippon Steel Technical Report, No. 20, December 1982, pp. 119-130.

14. F.M. Salter, "Improving Pusher Furnace Skid Systems", Slab Reheating, Iron and Steel Institute Publication, London, 1973, pp. 83-94.

15. M. W. Vance, et al, "Application of Skid Pipe Insulation in Slab Reheating Furnaces", AISE Yearbook, 1978, pp. 72-82.

24

Heat Transfer During the Rolling Process

24.1 WORKPIECE TEMPERATURE CHANGE IN HOT STRIP MILL

After reheating a slab to a desired temperature, it is subjected to rolling. A rolling cycle in a typical hot strip mill includes the following main steps:

1. Descaling of the slab prior to flat rolling by using high-pressure water descaling system in combination, in some cases, with edging.

2. Rough rolling to a transfer bar thickness which may vary from 19 to 40 mm (0.75 to 1.57 in.). The rough rolling is usually accompanied by edging and interpass descaling.

3. Transfer of the transfer bar from roughing mill to a flying shear installed ahead of finishing mill. The shear is usually designed to cut both head and tail ends of the bar.

4. Descaling of the transfer bar prior to entering the finishing mill.

5. Finish rolling to a desired thickness with a possible use of interstand descaling and/or strip cooling.

6. Air and water cooling of the rolled product on run-out table.

7. Coiling of the rolled product.

Various types of heat transfer from the rolled workpiece to its surrounding matter occur during the rolling cycle. Some of the lost heat is recovered by generating heat inside the workpiece during its deformation.

The main components of the workpiece temperature loss and gain in hot strip mill are usually identified as follows:

a) loss due to heat radiation,

b) loss due to heat convection,

c) loss due to water cooling,

b) loss due to heat conduction to the work rolls and table rolls, and

e) gain due to mechanical work and friction.

The analytical aspects of these components are briefly described below.

419

Table 24.1 Temperature loss rate due to radiation [1].

Author's Name	Year	Equations
Tyagunov. B.A.	1944	$\alpha_r = \dfrac{T - 752}{406.4h}$
Vasin, I.I.	1968	$\alpha_r = \dfrac{(T - 32)^2 - 900(T_a - 32)}{457,200h}$
Wusatowski, Z.	1969	$\alpha_r = \dfrac{3.26}{\rho h}(T + 460)^4 \cdot 10^{-14}$
Chernyavski, A.L.	1971	$\alpha_r = \dfrac{6.1}{\rho h}(T - 32)^4 \cdot 10^{-14}$
Seredynski, F.	1973	$\alpha_r = \dfrac{3.48}{\rho h}(T + 460)^4 \cdot 10^{-14}$

24.2 TEMPERATURE LOSS DUE TO RADIATION

Two methods have been employed to derive equations for temperature loss due to radiation [1].

In the first method, the temperature gradient within the material is assumed to be negligible. The amount of heat radiated to the environment is then calculated using the Stefan-Boltzmann law [2]:

$$dq'_r = S\xi A_r[(T + 460)^4 - (T_a + 460)^4]dt, \qquad (24\text{-}1)$$

where A_r = surface area of body subjected to radiation, $in.^2$

dq_r = amount of heat radiated by a body, Btu

S = Stefan-Boltzmann constant

T = temperature of rolled material at time t, °F

T_a = ambient temperature, °F

t = time, s

ξ = emissivity.

The amount of heat lost by a body dq''_r is given by:

$$dq''_r = \rho c v_r dT, \qquad (24\text{-}2)$$

where c = specific heat of rolled material, Btu/lb/°F

v_r = volume of body subjected to radiation, $in.^3$

ρ = density of rolled material, $lb/in.^3$

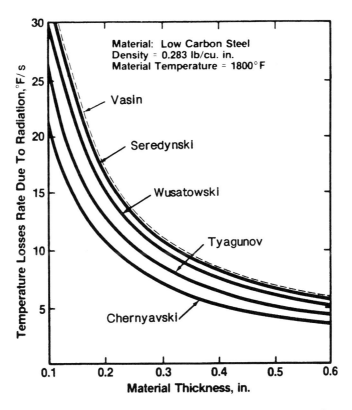

Fig. 24.1 Temperature loss rate due to radiation as a function of material thickness. Data from Ginzburg (1985).

The rate of temperature loss α_r can be calculated by considering the heat balance condition $dq_r' = dq_r''$, and Eqs. (24-1) and (24-2):

$$\alpha_r = \frac{dT}{dt} = \frac{S\xi A_r}{\rho c v_r} [(T + 460)^4 - (T_a + 460)^4] \tag{24-3}$$

Equations for the rate of temperature loss due to radiation which have been obtained by reducing some of the known equations [3-7] to a compatible form with an assumption that $T_a \ll T$ are summarized in Table 24.1 and plotted in Fig. 24.1. In the derivation of these equations, the dependency of the parameters S, ξ, p, and c on temperature is not taken into account (i.e., they are assumed constant). However, the variations of these constants with temperature may be significant and, therefore, the final form of Eq. (24-3) will depend on the average values selected for these constants.

The temperature loss ΔT_r during radiation time t_r can be calculated by integrating the differential equation:

$$\Delta T_r = \int_0^{t_r} \alpha_r dt \tag{24-4}$$

The second method of calculating temperature loss due to radiation takes into account the heat

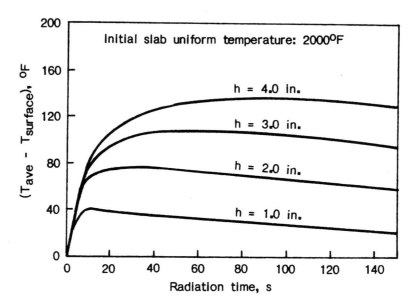

Fig. 24.2 The difference between the average and surface temperatures of a plate as a function of radiation time.

transfer along the thickness of the material. If z is the distance from the center of the body toward its surface, then from a Fourier equation we obtain [8]:

$$\frac{dT}{dt} = a \frac{d^2 T}{dz^2} \tag{24-5}$$

where a= thermal diffusivity of rolled material, in.^{2}/s.

The differential equation (24-5) can be solved numerically by the method of finite differences.

The goal of these calculations is to establish a relationship between the average temperature of the material T_{ave} which would affect the rolling deformation process and the material surface temperature $T_{surface}$ which could be measured (Fig. 24.2).

24.3 TEMPERATURE LOSS DUE TO CONVECTION

In the hot strip mill, heat transfer by convection is related to the motion of air surrounding a workpiece. This motion continuously brings new particles of air into contact with the workpiece [2]. Depending upon whether this internal motion is forced (through an externally generated pressure difference), or free (natural buoyancy motion), the heat transfer is referred to as either **forced** or **free convection**. The latter is a usual case in the hot strip mills.

A key factor in the calculation of temperature losses due to convection is to determine the heat transfer coefficient, which depends on the material temperature, ambient temperature, material specific heat and density, and the dynamic viscosity of the air flow and its characteristic, i.e., free, enforced laminar, turbulent, etc. The known mathematical interpretations of this relationship are too controversial to be recommended for practical calculation [8]. A consensus among some research workers is that the temperature loss due to convection ΔT_{cv} should be expressed as a certain percentage of the temperature loss due to radiation:

$$\Delta T_{cv} = k_{cv}(\Delta T_r) \tag{24-6}$$

Here k_{cv} is the ratio between the temperature loss due to convection and radiation and varies between 0.01 and 0.22 according to different studies.

24.4 TEMPERATURE LOSS DUE TO WATER COOLING

The temperature loss due to water cooling can be calculated by assuming that conduction plays a major role in heat transfer from a workpiece to water. Therefore, when water contacts one side of the workpiece continuously across its width, the amount of heat passing through the outer surface of the workpiece may be expressed by the formula [2]:

$$q'_w = 2kbw(T - T_w)\sqrt{\frac{t_w}{\pi a}} \tag{24-7}$$

where k = thermal conductivity of the surface layer, Btu/in./s/^{o}F.

q'_w = amount of heat passing through outer surface of the workpiece, Btu

b = water contact length, in.

w = workpiece width, in.

T_w = water temperature, ^{o}F

t_w = water contact time, s.

The amount of heat released by a workpiece is given by:

$$q''_w = \rho cv(\Delta T_d), \tag{24-8}$$

where v = volume of workpiece cooled by the water, in.^{3}

ΔT_d = temperature loss due to water cooling, ^{o}F.

From the heat balance condition $q'_w = q''_w$, Eqs. (24-7) and (24-8), and taking into account that

$$t_w = \frac{5b}{V} \tag{24-9}$$

where V = workpiece velocity, fpm

and

$$\frac{bw}{v} \approx \frac{1}{h} \tag{24-10}$$

we obtain that the temperature loss due to water cooling is equal to

$$\Delta T_d = \frac{2k}{\rho ch} (T - T_w) \sqrt{\frac{5b}{\pi a V}}$$
                                                                                                    (24-11)

The amount of heat absorbed by cooling water may be expressed as:

$$q_w''' = \rho_w c_w v_w (\Delta T_w),$$ (24-12)

where ρ_w = density of water, lb/in^3

   c_w = specific heat of water, Btu/lb/^{o}F

   v_w = volume of water absorbing heat, in^3

   ΔT_w = temperature rise of water, ^{o}F.

From heat balance $q_w'' = q_w'''$, Eqs. (24-8), (24-11), and (24-12), and also taking into account that

$$\frac{v_w}{v} = \frac{19.25d}{hV} ,$$ (24-13)

where d = water flow per unit of strip width, gpm/in.,
we obtain the following formula for the temperature rise of water:

$$\Delta T_w = \frac{0.104 \ k}{\rho_w c_w d} (T - T_w) \sqrt{\frac{5bV}{\pi a}}$$ (24-14)

Equation (24-11) does not show an explicit dependence of the temperature loss on the flow rate and pressure of cooling water. The flow rate and pressure, however, may substantially affect the thermal conductivity k of the surface layer that separates the body of workpiece from cooling water. Indeed, the surface layer consists of scale and boiled water, which work as a thermal barrier. This barrier will be weakened to a greater degree with increase of both the flow rate and pressure of cooling water.

24.5   TEMPERATURE LOSS DUE TO CONDUCTION TO WORK ROLLS

Temperature loss due to heat conduction to the work roll can be calculated if it is assumed that two bodies of uniform initial temperature T and T_r are pressed against each other and that, at the interface, considered to be plane, there is contact resistance formed by oxide layer.

Under these assumptions, the process can be described with the following heat balance equations. According to Schack [2], the total amount of heat passing through two outer surfaces of the plate may be calculated from the formula

$$q' = 4kA_c(T - T_r) \sqrt{\frac{t_c}{\pi a}}$$ (24-15)

where A_c = contact area between rolled material and work rolls, in^2

   k = thermal conductivity of the workpiece oxide layer, Btu/in./s/^{o}F

   q_c' = heat gained by work roll or heat lost by body due to thermal conduction, Btu

   T_r = roll temperature, ^{o}F

t_c = contact time of rolled material with work roll, s

a = thermal diffusivity of workpiece, in^2/s.

The amount of heat lost by the rolled metal in the roll bite is given by:

$$q_c'' = \rho c v_c(\Delta T_c), \tag{24-16}$$

where ΔT_c = temperature loss by rolled material due to contact with work rolls, $^\circ$F.

From the heat balance condition $q_c' = q_c''$, Eqs. (26-15) and (24-16), and also taking into account that

$$t_c \cong \frac{5\sqrt{R\Delta}}{V} \tag{24-17}$$

and

$$\frac{A_c}{v_c} = \frac{1}{h_a} \tag{24-18}$$

where R = work roll radius, in.

h_a = average workpiece thickness, in.

we obtain the following formula for the temperature loss due to conduction to work rolls:

$$\Delta T_c = \frac{4k}{\rho c h_a} (T - T_r)\sqrt{\frac{5(R\Delta)^{0.5}}{\pi a V}} \tag{24-19}$$

Equations for temperature loss due to contact with rolls which have been obtained by reducing some of the known equations [6, 9-12] to a compatible form are summarized in the Table 24.2 [1] and

Table 24.2 Temperature loss due to contact to the rolls [1].

Author's Name	Year	Equations
Lee, P.W., Sims, R.B., Wright, H.	1963	$\Delta T_c = 0.321\left[\frac{(\Delta H)}{c} \cdot 98.1\right]$
Ventzel, H.	1965	$\Delta T_c = \frac{0.606}{h_1+h_2}(T-T_r)\sqrt{\frac{\sqrt{R\Delta}}{V} \cdot \frac{h_1}{h_1 \cdot h_2}}$
Zheleznov, Y.D.	1968	$\Delta T_c = \frac{0.051}{h_1+h_2}(T-108)\sqrt{R\cos^{-1}\left(1-\frac{\Delta}{2R}\right)\frac{1-s}{V}}f$
Seredinski, F.	1973	$\Delta T_c = \frac{0.561}{Vh_2}(T-T_r)\sqrt{R\Delta h_1}$
Wright, H., Hope, T.	1975	$\Delta T_c = \frac{0.163}{h_1+h_2}(T-T_r)\sqrt{\frac{\sqrt{R\Delta}}{V}}$

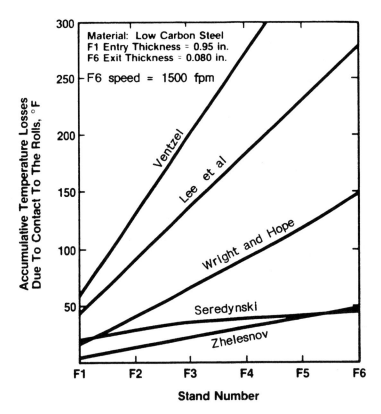

Fig. 24.3 Accumulative temperature losses in the finishing mill due to contact to the work rolls. Data from Ginzburg (1985).

are plotted in Fig. 24.3. The substantial discrepancies in temperature losses calculated from different equations are due mainly to the uncertainty in estimating thermal conductivity k which depends on the contact resistance of the oxide layer between the roll and the rolled material.

24.6 TEMPERATURE RISE DUE TO MECHANICAL WORK

There are two key components to be considered when calculating the temperature rise due to mechanical work [13]: mechanical work dissipated to the interface between the material being rolled and the roll; and mechanical work absorbed by the material being rolled during deformation.

It is assumed, in the case of hot rolling, that there is no sliding at the interface and, therefore, the component of mechanical work to overcome friction is negligible. A major part of the mechanical work is absorbed by the material being rolled during deformation. This energy is not completely transformed into heat: as the deformation process progresses, the displacement of dislocations is being counteracted which causes internal stresses to appear at different points of the grains.

The heat transfer process can be described with the following heat balance equations. Heat generated due to mechanical work is equal to [5]:

$$q'_m = \frac{K_w \eta_m v_c}{9338} \ln \frac{h_1}{h_2}$$

(24-20)

where h_1, h_2 = entry and exit thicknesses of rolled material respectively, in.

$\quad$ k_w = resistance to deformation, psi

$\quad$ q'_m = heat generated in the rolled material due to mechanical work, Btu

$\quad$ v_c = volume of rolled material in roll bite, in.^{3}

$\quad$ η_m = portion of mechanical work transformed into heat.

Table 24.3 Temperature rise due to mechanical work [1].

Author's Name	Year	Equations
Tyagunov, B.A.	1944	$\Delta T_m = \left(\dfrac{2417-T}{28.86}\right)\left(1+\dfrac{\sqrt{R\Delta}}{h_1+h_2}\right)\ln\dfrac{h_1}{h_2}$
Zaikov, M.A.	1960	$\Delta T_m = \dfrac{K_w}{11,690\rho c}\ln\dfrac{h_1}{h_2}$
Lee, P.W., Sims, R.B., Wright, H.	1963	$\Delta T_m = \dfrac{2.96(HP)}{bh_2 V\rho c} = \dfrac{1.07(\Delta H)}{c}$
Zheleznov, Y.D.	1968	$\Delta T_m = \dfrac{K_w}{441}\ln\dfrac{h_1}{h_2}$
Seredynski, F.	1973	$\Delta T_m = \dfrac{K_w}{9,345\rho c}\ln\dfrac{h_1}{h_2}$
Wright, H. Hope, T.	1975	$\Delta T_m = \dfrac{K_w}{790}\dfrac{h_1+h_2}{h_2}\ln\dfrac{h_1}{h_2}$

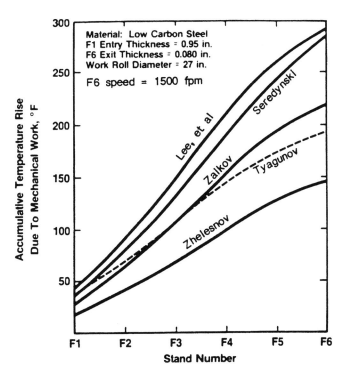

Fig. 24.4 Accumulative temperature rise in the finishing mill due to mechanical work. Data from Ginzburg (1985).

Amount of heat absorbed by the rolled material is given by

$$q_m'' = \rho c v_c (\Delta T_m),$$ (24-21)

where ΔT_m = temperature rise by rolled material due to mechanical work, $^\circ F$.

From the heat balance condition $q_m' = q_m''$, Eqs. (24-20) and (24-21), we obtain the following formula for the temperature rise due to mechanical work:

$$\Delta T_m = \frac{K_w \eta_m}{9338 \rho c} \ln \frac{h_1}{h_2}$$ (24-22)

Equations for the rise in temperature due to mechanical work, which have been obtained by reducing some of the known equations [3, 7, 11, 12, 14] to a compatible form, are summarized in Table 24.3 and are plotted in Fig. 24.4.

24.7 INTERMEDIATE REHEAT FACILITIES

The intermediate reheat facilities are used in hot strip mill for reheating of either entire body of a workpiece or only its edges. The entire body is usually preheated in tunnel furnaces.

When stock passes through a tunnel furnace, the following heat balance equations can be written [15]:

$$q_f' = \eta_f W_f t_f,$$ (24-23)

where q_f' = heat transfer to a body in a tunnel furnace, Btu

  t_f = residence time in tunnel furnace, s

  W_f = energy generated by tunnel furnace, Btu/s

  η_f = efficiency of tunnel furnace.

The heat transferred to the stock is equal to the amount of heat absorbed by a portion of the stock residing in the furnace during the same time

$$q_f'' = \rho c v_f (\Delta T_f),$$ (24-24)

where v_f = volume of stock in tunnel furnace, in^3

  The volume of stock is given by

$$v_f = L_f b h,$$ (24-25)

where L_f = length of tunnel furnace, in.

  b = width of rolled material (stock), in.

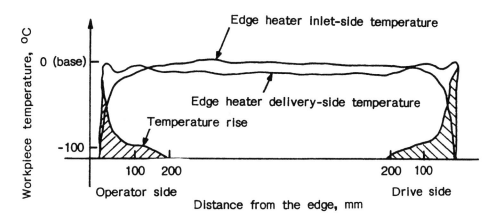

Fig. 24.5 Heat pattern of electromagnetic induction heating. Adapted from Sugita and Oi (1984).

The rate of temperature rise of the stock α_f in the tunnel furnace can be calculated from Eqs. (24-23)-(24-25):

$$\alpha_f = \frac{\Delta T_f}{t_f} = \frac{\eta_f W_f}{\rho c b h V_f} , \qquad (24\text{-}26)$$

where V_f = transfer speed of rolled material in tunnel furnace, fpm.

The purpose of edge preheating is to equalize temperature across the width of a workpiece [16]. Figure 24.5 shows the temperature profiles of a transfer bar before and after entering the induction type edge heater installed upstream of the finishing mill.

24.8 THERMAL COVERS

Thermal cover systems reduce the heat radiation rate by maintaining a higher ambient temperature T_a around the transfer bar, as described in Eq. (24-3). The known thermal cover systems for hot strip mills may be classified as follows:

1. Insulating thermal covers
2. Reflecting thermal covers
3. Reradiating thermal covers.

In insulating thermal covers, the transfer bar is surrounded by insulating material which reduces heat conduction and thus provides higher ambient temperature around the transfer bar. Such systems are relatively inexpensive but their efficiency is low [17]; on a mill with a full length delay table cover, the effective saving in slab temperature has been as little as 13°C (23°F) [18, 19]. According to Laws [20], heat insulating panels covering the top surface of the transfer bar reach an equilibrium temperature of only 700°C (1292°F).

In reflecting thermal covers, the transfer bar is surrounded by covers which reflect heat coming from the bar. In some installations, uninsulated aluminum reflecting covers have been used [21, 22]. Only limited benefits from reflecting thermal panels have been reported; to date, the maximum equilibrium

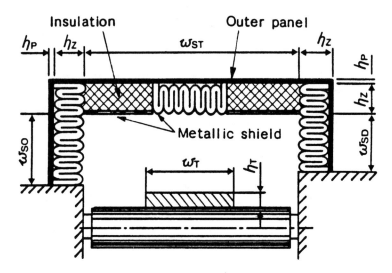

Fig. 24.6 Schematic cross-section of reradiating thermal cover. (From Bakhtar and Ginzburg, Proceedings of 4-th International Steel Rolling Conference, 1987. Copyright IRSID, France. Reprinted with permission.)

temperature achieved has been 300°C (572°F). Also, problems in keeping the reflectors clean were encountered.

In radiating thermal covers, the transfer bar is surrounded by a thin metallic shield covered with insulating material (Fig. 24.6). When the hot transfer bar passes the thermal cover, the metallic shield is heated very quickly and reaches an equilibrium temperature of approximately 1000°C (1832°F). In contrast to reflecting covers, the blacker the shield surfaces, the more effective reradiating thermal covers become.

To achieve first-bar performance, the design of the reradiating thermal covers must provide for almost instantaneous heating of the reradiating shield to a temperature near that of the transfer bar. As follows from the foregoing analysis, this condition is met when the metallic shield thickness is from 200 to 500 times thinner than thickness of the workpiece under the covers.

24.9 HEAT TRANSFER IN RERADIATING HEAT COVERS

The main purpose of the analysis of heat transfer in the reradiating heat covers is to establish a relationship between the workpiece temperature rundown inside the covers and their design parameters (Fig. 24.6). The analysis involves a one-dimensional representation of the heat flow equations using the fundamental theories of radiation and conduction in application to a heat transfer system which includes both the heat panels and the transfer bar [23].

Once all the interacted heat balance relationship are derived, the equations describing the heat transfer rates between adjacent components of the system can be obtained. In order to calculate the temperature of each component of the system, these equations have to be solved simultaneously.

Heat balance for transfer bar - Heat losses by a transfer bar are mainly due to radiation.

Therefore, the heat balance equation can be expressed as:

$$\Delta q_t''' = \Delta q_t' + \Delta q_t'',$$ (24-27)

where $\Delta q_t'''$ = amount of heat lost by transfer bar, Btu/in.

$\Delta q_t'$ = amount of heat radiated from top and side surfaces of the transfer bar

$\Delta q_t''$ = amount of heat radiated from bottom surface of the transfer bar, Btu/in.

Solution of Eq. (24-27) allows to define the transfer bar temperature loss rate $-_t$.

Heat balance for metallic shield - Metallic shield either absorbs heat radiated from transfer bar during transfer time or radiates heat during gap time. It also conducts some heat to the adjacent layers of insulation. Therefore, the heat balance equation can be expressed as:

$$\Delta q_s''' = \Delta q_s' - \Delta q_s'',$$ (24-28)

where $\Delta q_s'''$ = amount of heat absorbed or released by metallic shield, Btu/in.

$\Delta q_s'$ = amount of heat transferred by radiation, Btu/in.

$\Delta q_s''$ = amount of heat transferred to the first adjacent layer of insulation by conduction, Btu/in.

Solution of Eq. (24-28) allows to determine the metallic shield temperature change rate α_s.

Heat transfer through insulation - The heat transfer through insulation can be calculated from Fourier's general law of heat conduction:

$$\frac{\Delta T_z}{\Delta t} = \alpha_z \frac{\Delta^2 T_z}{\Delta x^2}$$ (24-29)

where T_z = change in temperature of the insulation layer, $^\circ$F

Δx = thickness of the finite insulation layer, in.

t = time increment, s

a_z = insulation thermal diffusivity, in^2/s.

Heat balance for outer metallic panel - Outer metallic panel absorbs heat transferred from insulation by conduction and also radiates some heat through its outer surface. Therefore, the heat balance equation can be expressed as:

$$\Delta q_p''' = \Delta q_p' - \Delta q_p'',$$ (24-30)

where $\Delta q_p'''$ = amount of heat absorbed by outer metallic panel, Btu/in.

$\Delta q_p'$ = amount of heat transferred by radiation, Btu/in.

$\Delta q_p''$ = amount of heat transferred by conduction Btu/in.

Solution of Eq. (24-30) allows one to calculate the outer metallic panel heat transfer rate α_p.

The temperature of the heat cover components can be found by simultaneous solution of the following equations using the iterative calculation procedure:

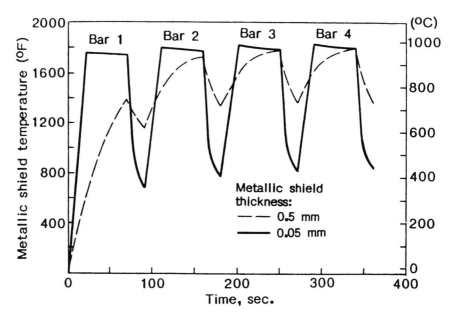

Fig. 24.7 Variation of metallic shield temperature during transfer of the 28 mm (1.1 in.)-thick and 1250 mm (49.2 in.)-wide bars. (From Bakhtar and Ginzburg, Proceedings of 4-th International Steel Rolling Conference, 1987. Copyright IRSID, France. Reprinted with permission.)

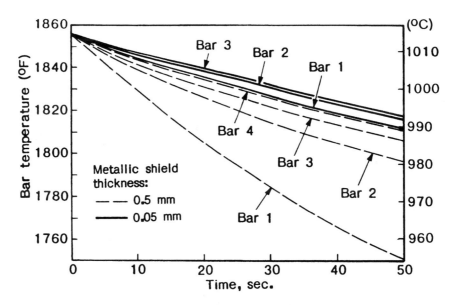

Fig. 24.8 Temperature rundown of the tail end of the transfer bars inside heat retention panels. (From Bakhtar and Ginzburg, Proceedings of 4-th International Steel Rolling Conference, 1987. Copyright IRSID, France. Reprinted with permission.)

$$T'_t = T_t - \alpha_t(\Delta t)$$
$$T'_s = T_s + \alpha_s(\Delta t)$$
$$T'_{z1} = T_{z1} + \Delta T_{z1} \qquad\qquad (24\text{-}31)$$
$$T'_{zn} = T_{zn} + \Delta T_{zn}$$
$$T_p = T_p + \alpha_p(\Delta t)$$

where indexes 1 and n correspond to the first (i.e., closest to the metallic shield) and last (i.e., closest to the panel) layer of insulation.

Figure 24.7 shows the results of a computer simulation of the metallic shield temperature of the heat covers installed between roughing and finishing mills of 80-inch hot strip mill. When the metallic shield thickness is equal to 0.5 mm (0.020 in.), the metallic shield temperature reaches only 1400°F after transferring the first bar. Besides, this temperature is achieved at the end of the transfer time. When the metallic shield thickness is reduced to 0.05 mm (0.002 in.), it requires only 5 seconds for the metallic shield to be heated up to 1760°F after the first bar enters the heat shield.

Figure 24.8 illustrates the temperature rundown of the tail end of the transfer bar inside of the heat panels. When the metallic shield thickness is equal to 0.5 mm (0.020 in.), up to four bars are needed in order to reach the optimum heat conservation, whereas the first bar performance is provided when the metallic shield thickness is reduced to 0.05 mm (0.002 in.)

24.10 HEAT TRANSFER IN COILBOX

The main purpose of the Coilbox is to reduce heat losses of the bar while it is being transferred from roughing mill to finishing mill. Coiling of the transfer bar allows one to increase its equivalent thickness h_t as shown in Fig. 24.9. This reduces the surface to volume ratio A_r/v_r of the workpiece. Since the heat losses in the Coilbox are mainly due to radiation, the rate of temperature loss α_r will also reduce as follows from Eq. (24-3).

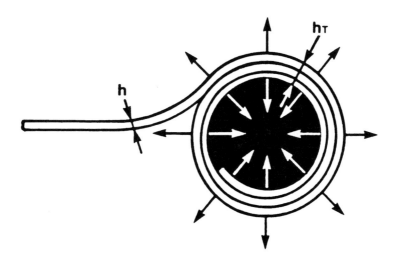

Fig. 24.9 Schematic presentation of the heat transfer in the Coilbox.

Additional heat conservation in the Coilbox are attained due to the fact that the eye of the Coilbox forms a cylindrical reradiating surface so the heat inside the eye is self-maintained. Therefore, the heat losses in the Coilbox can be calculated assuming that the heat transfer occurs only through the outer surfaces of the coil.

REFERENCES

1. V.B. Ginzburg, "Basic Principles of Customized Computer Models for Cold and Hot Strip Mills", Iron and Steel Engineer, Sept. 1985, pp. 21-36.

2. A. Schack, Industrial Heat Transfer, Chapman and Hall, London, pp. 1-216 (1965).

3. B.A. Tyagunov, Rational Settings of Hot Strip Mills, Metallurgizdat (Russian), Moscow (1944).

4. I.I. Vasin and A.S. Gindin, "Ural Metallurgical NII Proceedings" (Russian), 6, 1968, pp. 105-107.

5. Z. Wusatowski, Fundamentals of Rolling, Pergamon Press, Oxford, pp. 136-144 (1969).

6. A.L. Chernyavski, et al, "Rolling Production", Metallurgia (Russian), Moscow, 1971, 35, pp. 5-19.

7. F. Seredynski, "Prediction of Plate Cooling Temperature During Rolling Mill Operations", Journal of Iron and Steel Institute, March 1973, pp. 197-203.

8. Y.B. Konovalov and A.L. Ostapenko, Temperature Conditions in Wide Hot Strip Mills, Metallurgia (Russian), Moscow (1974).

9. H. Ventzel, "Rolling Process and Rolling Equipment", Viniti (USSR), 27, 1965, pp. 8-43.

10. P.W. Lee, R.B. Sims and H. Wright, "A Method for Predicting Temperature in Continuous Hot Strip Mills", Journal of Iron and Steel Institute, (London), 203, March 1963, pp. 270-274.

11. H. Wright and T. Hope, "Rolling of Stainless Steel in Wide Hot Strip Mills", Metals Technology, Dec. 1975, pp. 565-576.

12. Y.D. Zheleznov and B.A. Tsifrinovich, "Problem of the Heat Balance of Sheet in the Continuous Hot Rolling Mill", IZV VUZ Chernaya Met., Vol. 9, 1968, pp. 105-111.

13. S. Wilmotte, et al, "Model of the Evolution of the Temperature of the Strip in the Hot Strip Mill", Centre de Researches Metallurgiques, Belgium, C.R.M. Report No. 36, Sept. 1973, pp. 35-44.

14. M.A. Zaikov, Roll Separating Forces and Deformations in Hot Rolling, Metallurgizdat (Russian), Moscow (1960).

15. V.B. Ginzburg and W.F. Schmiedberg, "Heat Conservation Between Roughing and Finishing Trains of Hot Strip Mills", Iron and Steel Engineer, April 1986, pp. 29-39.

16. K. Sugita and J.Oi, "Maintenance of Temperature in Hot Strip Mill under Continuous Casting - Direct Rolling Process", Nippon Steel Technical Report No. 23, June 1984.

17. W.R. Laws, et al, "New Roller Table Thermal Insulation Systems Increase Product Range and Save Energy in Hot Strip Mills", Proceedings of the 4th International Steel Rolling Conference: The Science and Technology of Flat Rolling, Vol. 1, Deauville, France, June 1-3, 1987, pp. A.6.1-A.6.7.

18. E.C. Hewitt, "Hot Strip Mill Developments", I&SM 17, Sept. 1982.

19. "Energy Savings Through Improvements in Hot Rolling Process", Kobe Steel R&D Report, Nov. 29, 1980.

20. W.R. Laws, "Development of the ENCOPANEL Hot Strip Mill Insulation System", AISE Year Book, 1983, pp. 457-463.

21. P. Polukhin, et al, "Efficiency of Screening the Metal on the Roller Tables of Wide Hot Strip Mills", Steel in the USSR, Aug. 1974, pp. 649-650.

22. W.R. Laws, et al, "Rolling to Low Fuel Cost", Iron and Steel International, June 1984.

23. F. Bakhtar and V.B. Ginzburg, "Mathematical Model for Heat Retention Panels", Proceedings of the 4th International Steel Rolling Conference: The Science and Technology of Flat Rolling, Vol.1, Deauville, France, June 1-3, 1987, pp. A.7.1- A.7.6.

Part VIII

Metallurgical Aspects of the Rolling Process

25

Structural Changes in Steel During Hot Rolling

25.1 STRUCTURAL CHANGES DURING REHEATING

One of the consequences of reheating process is grain coarsing. The control of grain coarsing behavior of steels is an important step in the design of thermomechanical process striving to achieve fine-grained products [1, 2].

For microalloyed steels, the reheating temperature should be high enough to provide solubility of stable particles. If the stable particles remain undissolved, the beneficial precipitation hardening effects can not be obtained.

Addition of aluminum, niobium, vanadium, titanium, etc., produces abnormal type of grain growth (Fig. 25.1) which involves the growth of very few grains in relatively unchanged fine-grain matrix. The abnormal grain growth occurs at the temperatures which are significantly lower than the microalloying solution temperature. The temperature that corresponds to commencing of the abnormal grain growth is sometimes referred to as **grain-coarsing temperature.**

The grain size distribution has a complicated dependence on the reheating temperature as depicted in Fig. 25.2 in application to Nb-V-microalloyed steel. When reheating temperature is equal to 1200°C (2192°F), the maximum area fraction of the steel microstructure corresponds to the grain size of approximately 0.12 mm (0.0048 in.). When the reheating temperature is lowered to 1150°C (2102°F), the grain size occupying the maximum area fraction is reduced to 0.06 mm (0.0024 in.). However, further decrease in reheating temperature to 1050°C (1922°F) produces two pronounced peaks in distribution of the grain size, one of each is at the grain size of about 0.18 mm (0.0072 in.) and the second one is at 0.022 mm (0.0009 in.).

Reheating temperature also affects a formation of so-called **deformation bands** which play an important role during subsequent grain restoration processes [3]. As can be seen from Fig. 25.3, the higher reheating temperature the smaller amount of deformation bands will be formed and with less uniformity after the same reduction.

While it does not appear that the final average austenite grain size after deformation is strongly dependent on the reheated grain size, it is likely that the distribution of the grain sizes above average is much smaller when the reheating temperature is kept below the grain-coarsing temperature [1].

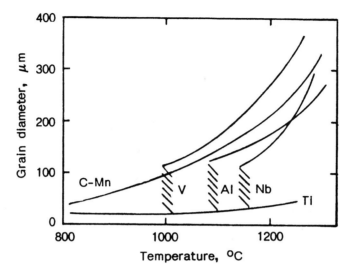

Fig. 25.1 Austenite grain coarsening characteristics in steels containing various microalloying additions. (From Speich, Phase Transformation in Ferrous Alloys, Proceedings of TMS-AIME, 1984. Reprinted with permission.)

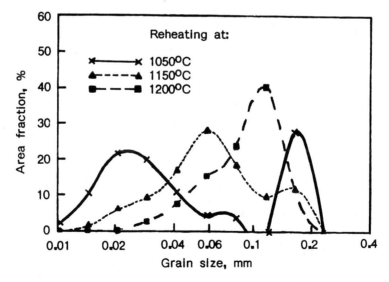

Fig. 25.2 Grain-size distribution in a Nb-V-microalloyed steel after reheating 1 hour at three various temperatures. (From Kaspar and Pawelski, Steel Research 57, 1986. Copyright Verlag Stahleisen mbH. Reprinted with permission.)

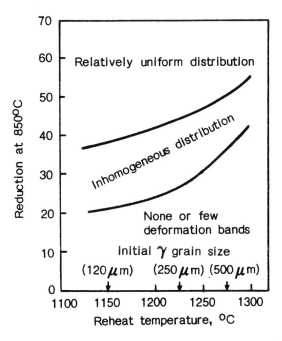

Fig. 25.3 Effect of amount of reduction on the ease of formation of deformation bands. (From Tanaka, et al. Reprinted with permission from Thermomechanical Processing of Microalloyed Austenite, edited by A.J. DeArdo, G.A. Ratz, and P.J. Wray, The Metallurgical Society, 420 Commonwealth Drive, Warrendale, Pennsylvania 15086, 1982.)

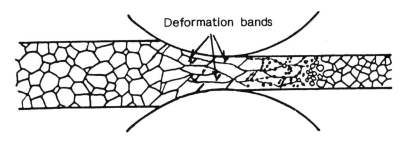

Fig. 25.4 Schematic illustration of static recrystallization during hot rolling. (From Katsumata, et al. Reprinted with permission from Thermomechanical Processing of Microalloyed Austenite, edited by A.J. DeArdo, G.A. Ratz, and P.J. Wray, The Metallurgical Society, 420 Commonwealth Drive, Warrendale, Pennsylvania 15086, 1982.)

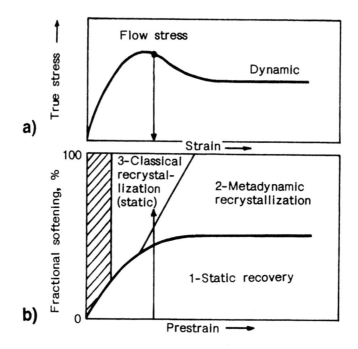

Fig. 25.5 Schematic representation of: a) stress-strain curve and
b) interrelation between three softening mechanisms (hatched area
is forbidden zone). (From Tanaka, International Metals Review,
No. 4, 1981. Copyright ASM International. Reprinted with permission.)

25.2 KINDS OF GRAIN RESTORATION PROCESS

Prior to the start of hot rolling, the steel microstructure consists of coarse equiaxed grains of austenite
(Fig. 25.4). During passing through the rolls, the austenite grains are getting flattened and elongated on
the average, each austenite grain undergoes a dimensional change corresponding to that of the workpiece
as a whole [4]. The deformation bands may also be induced within the grains [5] as illustrated in Fig.
25.4.

The three following kinds of restoration process are associated with hot rolling [6]:

1. Dynamic restoration process - This process starts and completes during deformation.

2. Metadynamic restoration process - This process starts during deformation and completes after
deformation.

3. Static restoration process - This process starts and completes after deformation.

25.3 DYNAMIC RESTORATION PROCESS

When steel is deformed in the austenitic state at high temperature, the flow stress rises to a maximum
and then falls to a steady-state [7, 8] as shown in Fig. 25.5a.

The strain ϵ_p that corresponds to the maximum value of the flow stress is equal to [9]:

$$\epsilon_p = AZ^n\sqrt{d_o}, \qquad\qquad (25\text{-}1)$$

where A, n = constants

$\qquad$ d_0 = initial grain size

$\qquad$ Z = Zener-Hollomon parameter.

The Zener-Hollomon parameter is a temperature compensated strain rate. It is expressed by:

$$Z = \dot{\epsilon}\exp(Q_{def}/RT), \qquad\qquad\qquad (25\text{-}2)$$

where $\qquad \dot{\epsilon}$ = strain rate

$\qquad$ Q_{def} = activation energy

$\qquad$ R = gas constant

$\qquad$ T = absolute temperature.

Dynamic restoration process includes dynamic recovery and dynamic recrystallization.

Dynamic recovery is a reduction of work-hardening effects without motion of large-angle grain boundaries. It occurs in a range of strain less than that for peak stress.

Dynamic recrystallization takes place in the range of strain that corresponds to steady state of flow stress.

The dynamically recrystallized grain size d is usually associated with Zener-Hollomon parameter and is given by the following equations [10]:

$$1/d \sim \sigma_{flow} \sim \ln Z \qquad\qquad\qquad (25\text{-}3)$$

For the grade 304 stainless steel, the grain size was found to be equal to [11]:

$$d = 6.813 \times 10^6 \dot{\epsilon}^{1/6} Z^{-1/9}, \qquad\qquad\qquad (25\text{-}4)$$

where d = grain size, μm

$\qquad$ $\dot{\epsilon}$ = strain rate, sec^{-1}

$\qquad$ Z = Zener-Hollomon parameter, sec^{-1}.

Role of dynamic recrystallization of austenite in practical rolling of C-Mn steels is small. It is due to the fact that a critical strain required for achieving the steady-state of the flow stress is very large, even at high temperatures. The grain refinement of these steels is usually achieved by static recrystallization.

25.4 STATIC RESTORATION PROCESS

The microstructures developed by dynamic restoration are not stable and at the elevated temperatures are modified by metadynamic and static restoration processes. The latter processes may include static recovery, static recrystallization and metadynamic recrystallization as shown in Fig. 25.5b.

In hot rolling, static recrystallization may start spontaneously. Nuclei of recrystallization take place preferentially at elongated grain boundaries and interfaces of deformation bands [4].

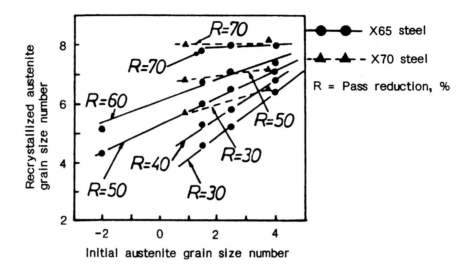

Fig. 25.6 Effect of initial austenite grain sizes and rolling reductions on recrystallized austenite grain size at 1050ºC. (From Katsumata, et al. Reprinted with permission from Thermomechanical Processing of Microalloyed Austenite, edited by A.J. DeArdo, G.A. Ratz, and P.J. Wray, The Metallurgical Society, 420 Commonwealth Drive, Warrendale, Pennsylvania 15086, 1982.)

Softening by static recovery and recrystallization occurs at the rates which depend on the prior deformation conditions and the holding temperature. The recrystallization curves generally follow an Avrami equation of the form [9]:

$$X_t = 1 - \exp[-C(t/t_f)^k, \tag{25-5}$$

where X_t = fraction recrystallized in time t

t_f = time to produce the specified fraction of recrystallization f

$$C = -\ln(1-f) \tag{25-6}$$

The recrystallization rate and size of recrystallized grains are controlled by the following three major factors [4]:

1. Austenite grain size prior deformation which is a function of heating temperature.
2. Temperature of recrystallization.
3. Amount of deformation prior to any recrystallization.

The effect of each of these factors is considered below in more detail.

25.5 EFFECT OF INITIAL GRAIN SIZE ON STATIC RECRYSTALLIZATION

Since nucleation sites for recrystallization are predominantly located at the grain boundaries, both the recrystallization time and the recrystallized grain size are affected by initial grain size.

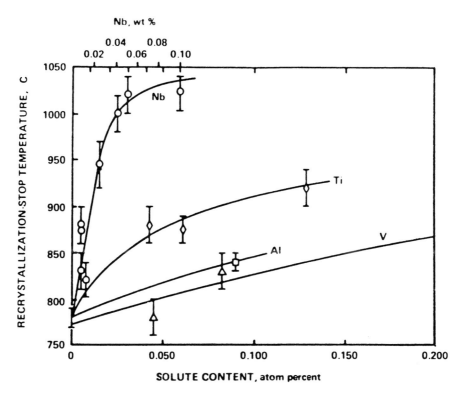

Fig. 25.7 The increase in recrystallization temperature with increase
in the level of microalloy solutes in 0.07% C, 1.40% Mn, 0.25% Si steel.
(From Cuddy. Reprinted with permission from Thermomechanical Processing of
Microalloyed Austenite, edited by A.J. DeArdo, G.A. Ratz, and P.J. Wray,
The Metallurgical Society, 420 Commonwealth Drive, Warrendale, Pennsylvania,
15086, 1982.)

It was found that 50% (f = 0.5) of recrystallization time for the range of C-Mn steels depends on
strain and can be described by the equations [9]:

When $\epsilon < 0.8\epsilon_p$,

$$t_{0.5} = 2.5 \times 10^{-19} d^2 \epsilon^{-4} \exp(3 \times 10^5/RT) \qquad (25\text{-}7)$$

When $\epsilon > 0.8\epsilon_p$,

$$t_{0.5} = 1.06 \times 10^{-5} Z^{-0.6} \exp(3 \times 10^5/RT) \qquad (25\text{-}8)$$

where d_0 = initial grain size, μm

ϵ = strain

R = gas constant, J/mol/K

T = absolute temperature, K.

The recrystallized grain size d_r for stainless steel is given by [12]:

$$d_r = k\epsilon^{-0.5}d_oZ^{-0.06},\qquad\qquad\qquad(25\text{-}9)$$

where k = constant.

Equation (25-9) shows that the recrystallized grain size increases linearly with increase in initial grain size. Similar relationship was found to be also true for HSLA steels [5] as shown in Fig. 25.6.

25.6 EFFECTS OF TEMPERATURE AND MICROALLOYING

The higher rolling temperature the greater number of deformed grains will be recrystallized. The lowest temperature at which austenite recrystallizes completely immediately after deformation is referred to as **recrystallization temperature.** The recrystallization temperature increases with increase in the level of microalloy solutes [13]. This relationship is illustrated in Fig. 25.7.

Columbium, titanium, and, to a lesser degree, vanadium retard both dynamic and static recrystallization [14]. The results obtained on a pearlite-reduced manganese steel are shown in Fig. 25.8. A very pronounced retardation is produced by increasing the columbium content up to 0.06%. This effect increases with decrease in temperature. At temperature below 900°C (1652°F), recrystallization can be retarded by more than two orders of magnitude.

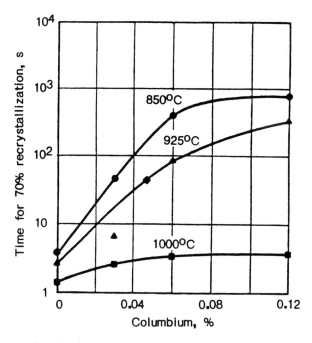

Fig. 25.8 Effect of columbium on the recrystallization of a 0.05% carbon, 1.8% manganese steel. (From Meyer, et al, Microalloying '75, 1977. Copyright STRATCOR, Pittsburgh, Pennsylvania. Reprinted with permission.)

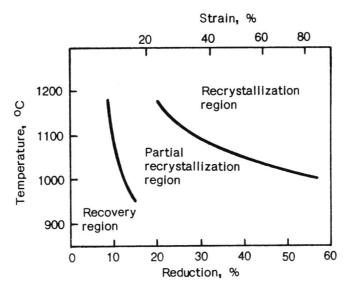

Fig. 25.9 Effect of amount of reduction and rolling temperature on restoration behavior: niobium steel was heated to 1150°C (2102°F) which gave grain size approximately 180 μm (0.007 in.) and rolled in one pass. (From Tanaka, et al. Reprinted with permission from Thermomechanical Processing of Microalloyed Austenite, edited by A.J. DeArdo, G.A. Ratz, and P.J. Wray, The Metallurgical Society, 420 Commonwealth Drive, Warrendale, Pennsylvania 15086, 1982.)

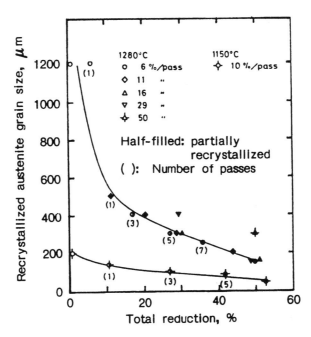

Fig. 25.10 Effect of total reduction on recrystallized austenite grain size in niobium steel which was reheated to 1280°C (2336°F) or 1150°C (2102°F) and subsequently multi-pass rolled. (From Tanaka, et al. Reprinted with permission from Thermomechanical Processing of Microalloyed Austenite, edited by A.J. DeArdo, G.A. Ratz, and P.J. Wray, The Metallurgical Society, 420 Commonwealth Drive, Warrendale, Pennsylvania 15086, 1982.)

25.7 EFFECT OF AMOUNT OF DEFORMATION

Depending on the amount of rolling deformation, the static restoration process may proceed in the following three forms [2] as shown in Fig. 25.9:

1. Recovery - This form of static restoration occurs when rolled with reduction less than critical value for partial recrystallization. In that case, grain coalescence instead of grain refinement occurs due to strain-induced grain boundary migration, producing much larger grains than the initial ones. These large grains formed by slight reduction in recovery region persist even after many passes in the partial recrystallization zone.

2. Partial recrystallization - When rolling reduction is sufficient to initiate partial recrystallization, mixed microstructure of recrystallized grains and recovered grains is produced.

3. Complete recrystallization - The minimum rolling reduction after which austenite recrystallizes completely is often referred to as **critical rolling reduction for recrystallization** [5]. Reduction in the complete recrystallization region produces fine and uniform recrystallized grain structure. The recrystallized austenite grain size markedly decreases with increase in total reduction (Fig. 25.10).

25.8 FACTORS AFFECTING CRITICAL REDUCTION FOR RECRYSTALLIZATION

The critical amount of deformation dividing each form of restoration process increases rapidly with decrease in deformation temperature (Fig. 25.9). It also increases with addition of microalloying elements and specifically niobium (columbium) [3].

Another factor affecting the critical reduction for recrystallization is the initial grain size [3].

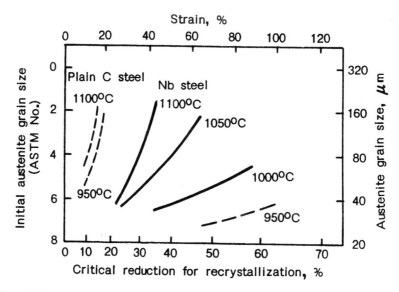

Fig. 25.11 Effects of deformation temperature and initial grain size on critical amount of reduction required for completion of recrystallization in plain carbon and niobium steels. (From Tanaka, et al. Reprinted with permission from Thermomechanical Processing of Microalloyed Austenite, edited by A.J. DeArdo, G.A. Ratz, and P.J. Wray, The Metallurgical Society, 420 Commonwealth Drive, Warrendale, Pennsylvania 15086, 1982.)

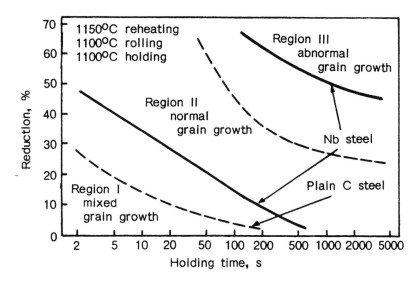

Fig. 25.12 Effects of holding time and amount of reduction on the progress of grain growth in plain-carbon and niobium steels. (From Tanaka, et al. Reprinted with permission from Thermomechanical Processing of Microalloyed Austenite, edited by A.J. DeArdo, G.A. Ratz, and P.J. Wray, The Metallurgical Society, 420 Commonwealth Drive, Warrendale, Pennsylvania 15086, 1982.)

As can be seen from Fig. 25.11 in plain carbon steel, the critical reduction for recrystallization is very small even when initial grain size is very large. However, in niobium-bearing steel, when the initial grain size is large, the critical reduction is extremely high. The influence of rolling temperature is also very strong; the critical reduction becomes very large with decrease in temperature.

25.9 GRAIN GROWTH AFTER DEFORMATION

The grain growth after deformation is markedly affected by both the amount of reduction and holding time as depicted in Fig. 25.12.

The grain growth is divided into three following regions [3]:

Region I - In this region, the grain growth starts from a mixed structure that consists of either recovered grains and giant grains or recrystallized grains and recovered ones, depending on the amount of reduction.

Region II - In this region the grain growth follows to Miller's equation [9, 15]:

$$d^{10} = d_r^{10} + (A't)exp(-Q'/RT) \tag{25-10}$$

where A', Q' = constants

t = time.

Region III - This region corresponds to abnormal grain growth when very large grains suddenly develop among the small grains. The size of grains coalesced from small grain structure during this secondary recrystallization process is much larger than the size of grains coalesced from large grain structure.

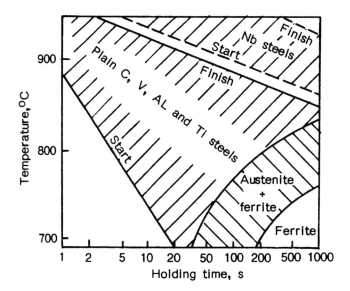

Fig. 25.13 Recrystallization of austenite following single deformation of 50% in various steels. Adapted from Irvine (1970).

As it is shown in Figs. 25.12 and 25.13, the critical holding time required for abnormal grain growth increases with addition of niobium. This differentiates the niobium-bearing steels from plain-carbon steels and also from other steels microalloyed with vanadium, aluminum and titanium [16].

25.10 STRUCTURAL CHANGES IN STEEL DURING COOLING

After hot rolling the workpiece is subjected to a combination of air and water cooling. The ferrite grain size of the rolled steel will be affected by [17, 18]:

1. Finishing rolling temperature
2. Delay time between rolling and an inception of cooling with water
3. Cooling rate.

As can be seen from Fig. 25.14, the grain size increases with increase in delay time. The effect of finishing rolling temperature and cooling rate on the ferrite grain size of mild steel is shown in Fig.

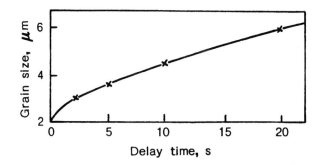

Fig. 25.14 Dependence of ferrite grain size and yield strength on delay time between rolling and accelerated cooling to 600°C (1112°F). (From Dillamore, et al, Metals Technology, July-August 1975. Copyright ASM International. Reprinted with permission.)

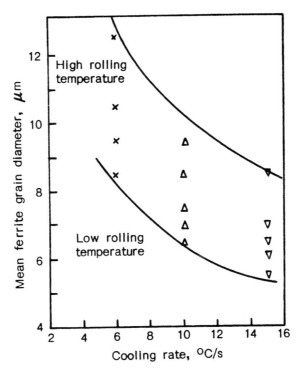

Fig. 25.15 Effect of cooling rate and finishing temperature on ferrite grain size of mild steel. (From Dillamore, et al, Metals Technology, July-August 1975. Copyright ASM International. Reprinted with permission.)

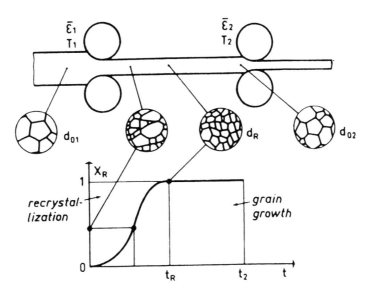

Fig. 25.16 Evolution of structural parameters between passes. (From Choquet, et al, Proceedings of 4-th International Steel Rolling Conference, 1987. Copyright IRSID, France. Reprinted with permission.)

25.15. The lower finishing rolling temperature the smaller the ferrite grain size. The grain size also reduces with increase in cooling rate.

25.11   EFFECT OF STEEL STRUCTURE ON FLOW STRESS

During deformation of steel, energy is stored in the deformed grains in the form of lattice defects (dislocations). Since recrystallization eliminates the lattice defects and reduces energy stored in the deformed grains, the flow stress of a completely recrystallized structure will be smaller than one of a partially recrystallized. Therefore, in order to correctly evaluate the flow stress in hot strip mill, the following two situations shall be considered [19, 20]:

1. Static recrystallization time t_R is less than or equal to interpass time t_I, i.e., $t_R < t_I$.

2. Static recrystallization time t_R is greater than interpass time t_I, i.e., $t_R > t_I$.

The first case is illustrated in Fig. 25.16 where d_{01} is the initial grain size prior to the first pass. The grain structure is completely recrystallized ($X_R = 1$) after time t_R at which the grain size is equal to d_R. Therefore, in that case the flow stress for the second pass may be determined by using the known equation applicable for completely recrystallized grain structure.

If static recrystallization has not been completed prior to the second pass, then the equation for the flow stress may be presented as consisting of two components (Fig. 25.17):

$$\sigma = X_R \sigma_2(\bar{\epsilon}_2) + (1 - X_R)\sigma_1(\bar{\epsilon}_1 + \bar{\epsilon}_2), \qquad (25\text{-}11)$$

where $\sigma_2(\bar{\epsilon}_2)$ = component of flow stress corresponding to recrystallized fraction of the grain structure

$\sigma_1(\bar{\epsilon}_1 + \bar{\epsilon}_2)$ = component of flow stress corresponding to non-recrystallized fraction of the grain structure.

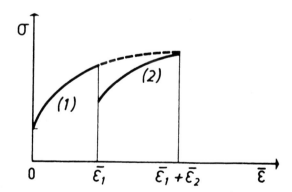

Fig. 25.17 Principle of the Multipass stress model. (From Choquet, et al, Proceedings of 4-th International Steel Rolling Conference, 1987. Copyright IRSID, France. Reprinted with permission.)

REFERENCES

1. G.R. Speich, et al, Proc. Phase Transformation in Ferrous Alloys (Philadelphia), TMS-AIME, Warrendale, 1984, p. 341.

2. R. Kaspar and O. Pawelski, "Austenite Grain in the Process of Thermomechanical Treatment", Steel Research, Vol. 57, No. 5, 1986, pp. 199-206.

3. T. Tanaka, et al, "Formation Mechanism of Mixed Austenite Grain Structure Accompanying Controlled-Rolling of Niobium-Bearing Steel", Thermomechanical Processing of Microalloyed Austenite, Metallurgical Society of AIME, New York, 1981, pp. 195-215.

4. R.A. Grange, "Microstructural Alterations in Iron and Steel During Hot Working", Fundamentals of Deformation Processing, Syracuse Univ. Press, Syracuse, New York, 1964, pp. 299-320.

5. M. Katsumata, et al, "Recrystallization of Austenite in High-Temperature Hot-Rolling of Niobium Bearing Steels", Thermomechanical Processing of Microalloyed Austenite, Metallurgical Society of AIME, New York, 1981, pp. 101-119.

6. E.L. Brown and A.J. DeArdo, "Influence of Hot Rolling on Microstructure of Austenite", Hot Working and Forming Processes, Metals Society, London, 1980, pp. 21-26.

7. T. Tanaka, "Controlled Rolling of Steel Plate and Strip", International Metals Reviews, No. 4, 1981, pp. 185-212.

8. R.A.P. Djaic and J.J. Jonas: Metall. Trans., 1973, 4, pp. 621-624.

9. C.M. Sellars, "The Physical Metallurgy of Hot Working", Hot Working and Forming Processes, Metals Society, London, 1980, pp. 3-15.

10. H.J. McQueen: J. Met., April 1968, 20, pp. 31-38.

11. B.L. Bramfitt and A.R. Marder, "The Influence of Microstructure and Crystallographic Texture on the Strength and Notch Toughness of a Low-Carbon Steel", Processing and Properties of Low Carbon Steel, Metallurgical Society of AIME, New York, pp. 191-224.

12. D.J. Towle and T. Gladman: Met. Sci., 1979, 13, pp. 246-256.

13. L.J. Cuddy, "The effect of Microalloy Concentration on Recrystallization of Austenite During Hot Deformation", Thermomechanical Processing of Microalloyed Austenite, Metallurgical Society of AIME, New York, 1981, pp. 129-140.

14. L. Meyer, et al, "Columbium, Titanium, and Vanadium in Normalized, Thermo-Mechanically Treated and Cold-Rolled Steels", Microalloying '75, Union Carbide Corp., New York, 1977, pp. 153-171.

15. O.O. Miller: Trans. ASM, 1951, 43, p. 260.

16. K.J. Irvine, et al: J. Iron Steel Inst., 1970, 208, pp. 717-726.

17. R.F. Dewsnap, et al: AIME Conf. on "Processing and Properties of Low-Carbon Steel", 369, Nov. 1973, Pittsburgh.

18. I.L. Dillamore, et al, "Metallurgical Aspects of Steel Rolling Technology", Metals Technology, July-Aug. 1975, pp. 294-302.

19. P. Choquet, et al, "FAST: A New Model for Accurate Prediction of Rolling Force Application on the Solmer Hot Strip Mill", Deaville, France, June 1-3, 1987, pp. B5.1-B5.8.

20. S. Licka, et al, "Rolling Load Calculation in Hot Strip Rolling with Respect to Restoration Processes", Proceedings of the International Conference on Steel Rolling, Tokyo, Japan, Sept. 29-Oct. 4, 1980, pp. 840-851.

26

Thermomechanical Treatments Combined with Rolling

26.1 MAJOR PURPOSE OF THERMOMECHANICAL TREATMENTS OF STEEL

Thermomechanical treatments (TMT) is a term used to describe a variety of processes combining controlled thermal and deformation treatments to obtain synergistic effects such as [1, 2]:

1. Higher yield strengths
2. Improved toughness
3. Improved weldability
4. Higher resistance to brittle cleavage
5. Higher resistance to low-energy ductile fractures
6. Lower impact-transition temperature
7. Good cold forming, particularly by bending
8. Lower costs which are possible by using hot-rolled rather than heat-treated sections.

An additional reduction in cost is provided due to the fact that controlled rolling process allows one to achieve the desired properties with less amount of alloying elements than would be required when regular hot rolling process is used.

26.2 U.S. CLASSIFICATION OF TMT

U.S. classification of TMT has been proposed by Radcliffe and Kula [3]. The classification is based on where the deformation occurs relative to the phase transformation. According to this classification the thermomechanical treatments are divided into the following three classes [3-10]:

Class I - Deformation takes place before transformation of austenite. Austenite is deformed in one of the following ranges (Fig. 26.1):

a) in the stable austenite range above the critical temperature (A_1)

b) in the unstable austenite range above the pearlite nose

c) in the unstable austenite range in the bay region between pearlite and bainite noses.

This type of TMT results in formation of martensite in strain-hardened austenite. It has been shown that austenite in hardenable steels can be cold-worked even at temperatures above the critical. In

454

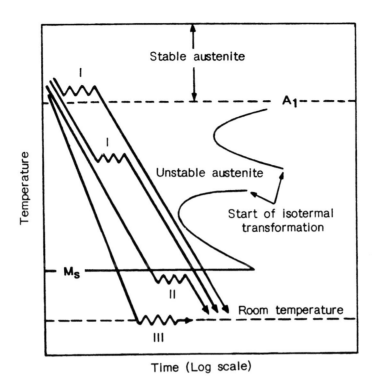

Fig. 26.1 Schematic time-temperature-transformation diagram showing thermomechanical treatments. From John J. Burke, Norman L. Reed, and Volker Weiss, Strengthening Mechanisms: Metals and Ceramics (Syracuse: Syracuse University Press, 1966), p. 85. By permission of the publisher.

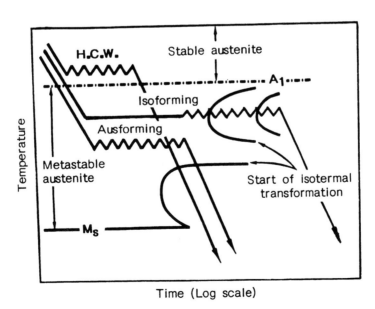

Fig. 26.2 Schematic time-temperature-transformation diagram with the thermal cycles for the hot-cold working and ausforming operations. (From Matas, et al, Mechanical Workings of Steel I, 1964. Copyright Metallurgical Society of AIME. Reprinted with permission.)

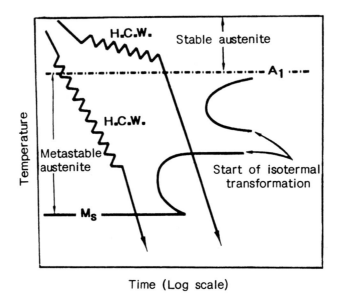

Fig. 26.3 Schematic time-temperature-transformation diagram with the various thermal cycles for the athermal hot-cold working process. (From Matas, et al, Mechanical Workings of Steel I, 1964. Copyright Metallurgical Society of AIME. Reprinted with permission.)

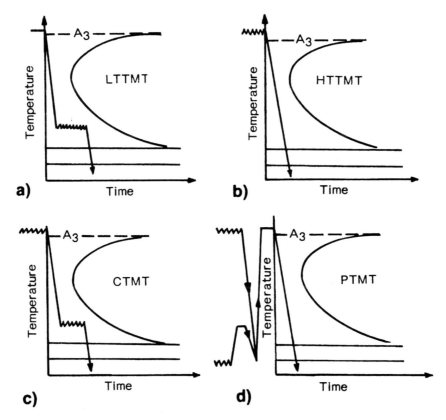

Fig. 26.4 Schematic representation of thermomechanical treatments in relation to a TTT diagram. (From Zackay, Material Science Engineering, Vol. 25, 1976. Copyright Elsevier Sequola S.A., Switzerland. Reprinted with permission.)

that case recrystallization of austenite is suppressed by rapid cooling, so the martensite can be formed from a cold-worked rather than annealed austenite. One of these processes is known as **ausforming** (ausworking, ausrolling) which is shown schematically in Fig. 26.2. Ausforming produces high-strength steel without adverse effects on toughness and ductility. Another process is known as **hot-cold working** (H.C.W.). This process can be either isothermal (Fig. 26.2) or athermal (Fig. 26.3).

      Class II - Deformation takes place during transformation of austenite in the temperature range below the martensite transformation temperature M_S (Fig. 26.1). The transformation products can be either pearlite, bainite or martensite.

      One of the examples of this class TMT is **isoforming** (Fig. 26.2). In this process, the metastable austenite is deformed until the transformation is complete at the deformation temperature. Isoforming does not produce a significant strength improvement, its main contribution is an increase in toughness due to the refined ferrite grain size and the spheroidization of the carbide particles.

      Class III - Deformation takes place after transformation of austenite (Fig. 26.1). This class involves deformation of martensite, tempered martensite, bainite or pearlite, followed possibly by reaging. The most known names for this type of TMT are **strain-aging, flow-tempering, strain-tempering, marstraining,** and **temp-forming.**

26.3 U.S.S.R. CLASSIFICATION OF TMT

According to the U.S.S.R. classification, the thermomechanical treatments are divided into the following seven groups [11]:

      SHT: Standard Heat Treatment - Conventional heat treatment without deformation.

      TMT: Thermomechanical Treatment - A combined thermal and mechanical treatment generally involving a phase transformation.

      LTMT: Low Temperature Thermomechanical Treatment - Deformation below the recrystallization temperature (Fig. 26.4a).

      HTMT: High Temperature Thermomechanical Treatment - Deformation above the recrystallization temperature (Fig. 26.4b).

      CTMT: Combined Thermomechanical Treatment - HTMT followed by LTMT (Fig. 26.4c).

      PTMT: Preliminary Thermomechanical Treatment - Deformation by HTMT or LTMT or cold working followed by rapid reaustenitizing and quenching (Fig. 26.4d).

      MTT: Mechanico-Thermal Treatment - Deformation at room or elevated temperature with or without subsequent annealing or aging applied to a material which does not undergo a phase transformation. Like TMT, deformation can be below (LTMT) or above (HTMT) the recrystallization temperature.

26.4 THERMOMECHANICAL TREATMENTS DURING ROLLING

Hot rolling processes can be classified based on where the deformation occurs relative to the phase transformation. According to this classification, the hot rolling processes are divided into the following four major groups [2, 12, 13] as depicted schematically in Fig. 26.5 and described briefly below.

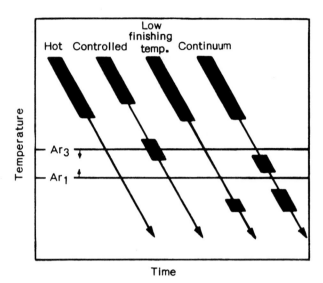

Fig. 26.5 Schematic representation of several common practices used for the TMT production of HSLA steels. (From Melloy and Dennison, The Microstructure and Design of Alloys, Vol. 1, 1973. Copyright Institute of Metals and Iron and Steel Institute. Reprinted with permission.)

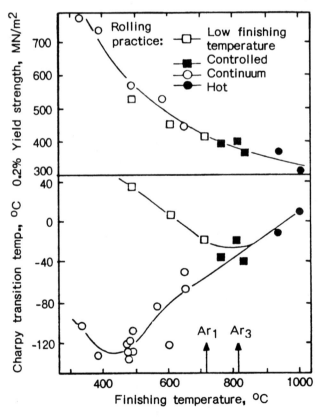

Fig 26.6 Effect of finishing temperature on yield strength and transition temperature of 0.05% C-Mn-Nb steel plate rolled by various procedures. (From Tanaka, International Metals Reviews, No. 4, 1981. Copyright ASM International. Reprinted with permission.)

1. Conventional hot rolling - During this process, the rolling of steel is conducted continuously and is usually finished above the upper cooling transformation temperature A_{r3}. Therefore, the deformation occurs in the gamma-region only.

2. Controlled rolling - In this process the rolling of steel is interrupted by one or two delays which allow one to deform the steel first in the gamma-region and then in the gamma-alpha two-phase region.

3. Low finishing temperature rolling - In this process the finishing rolling passes are conducted below the lower cooling transformation temperature A_{r1} that results in deformation in alpha-region.

4. Continuum rolling - This process provides deformation in the gamma-, (gamma + alpha)-, and alpha-regions.

Figure 26.6 shows the effect of finishing temperature on yield strength and Charpy transition temperature for four different hot rolling processes. The controlled rolling provides an increase in yield strength and improves ductility (transition temperature decreases) in comparison with conventional hot rolling practice. The low finishing temperature rolling allows one to further increase the yield strength. It, however, increases transition temperature. The continuum rolling process allows one to substantially increase the yield strength and to simultaneously decrease the transition temperature.

26.5 TYPES OF CONTROLLED ROLLING PROCESSES
During controlled rolling, the enhanced properties of steel are obtained by refining its structure. Because of the relationship between gamma- and alpha-grain sizes, refinement of the alpha structure is achieved mainly through gamma-grain refinement [2].

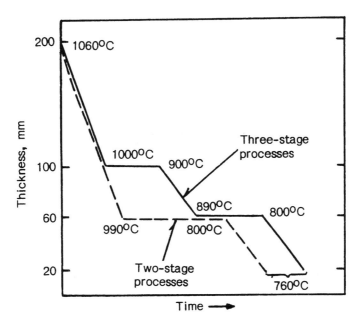

Fig. 26.7 Simplified rolling schedules: conventional two-stage process and new three-stage process. (From Lafrance, et al, Microalloying '75, 1977. Copyright STRATCOR, Pittsburgh, Pennsylvania. Reprinted with permission.)

As was shown in the previous chapter, the grain refinement depends on deformation temperature. The controlled rolling is usually conducted in either two or three stages (Fig. 26.7). The two-stage rolling process involves the following three steps [13, 14]:

Step 1 - Reduction in thickness in the rapid recrystallization region. This region is above 1000°C (1832°F). Deformation in this region produces coarse recrystallized gamma-grains which transform to a relatively coarse alpha and upper bainitic structure.

Step 2 - Delay in rolling in the intermediate temperature range from 1000°C (1832°F) to 900°C (1652°F). The delay is needed to ensure the required amount of deformation in the non-recrystallization region. During the delay, partial recrystallization tends to occur that leads to the formation of a mixed grain structure.

Step 3 - Final reduction in thickness in the non-recrystallization region. Deformation below the recrystallization temperature produces 'worm-worked' gamma-grain structure which leads to a finer alpha-grain structure.

In the three-stage rolling process [13, 15], reduction in the non-recrystallization range is interrupted by a delay. Figure 26.7 gives a comparison between the two- and three-stage rolling processes. During the delay time at high temperature in the two-stage process, recrystallization is rapid,

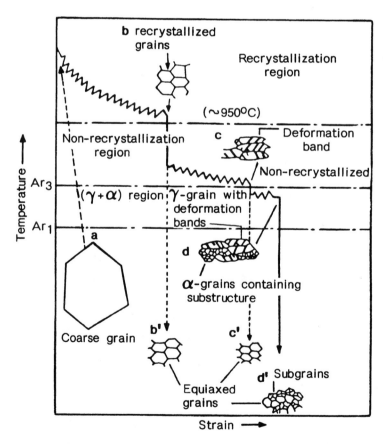

Fig. 26.8 Schematic illustration of change in microstructure with deformation during controlled rolling. (From Tanaka, International Metals Reviews, No. 4, 1981. Copyright ASM International. Reprinted with permission.)

leading to a coarse-grained structure at the end of delay time. In three-stage process, the first delay time at high temperature leads to a coarse-grained structure. During the second delay time, however, recrystallization is sluggish so that the grain size at the end of the three-stage rolling process is finer than that at the end of the two-stage rolling process.

26.6 STRUCTURAL CHANGES IN STEEL DURING CONTROLLED ROLLING

Structural changes in steel during controlled rolling are schematically illustrated in Fig. 26.8. These changes are related to deformation at three following regions [13]:

    1. Deformation in recrystallization region - In this region, coarse austenite grain **a** is refined by repeated deformation and recrystallization producing the recrystallized grains **b**. During cooling these grains would transform into relatively coarse ferrite **b'**.

    2. Deformation in non-recrystallization region - In this region, deformation bands are formed in elongated, unrecrystallized austenite **c**. During cooling ferrite would nucleate on the deformation bands as well as gamma-grain boundaries, giving fine alpha-grain **c'**.

    3. Deformation in the gamma-alpha region - In this region deformation bands continue to be formed and also the deformed ferrite produces a substructure **d**. During cooling after deformation, unrecrystallized austenite transforms into equiaxed alpha-grains, while the deformed ferrite changes into subgrains **d'**.

The formation of the deformation bands is one of the principle features of controlled rolling. In the conventional hot rolling alpha-grains nucleate exclusively at the gamma-grain boundaries, whereas in the controlled rolling the alpha-grain nucleation occurs at both the grain interiors and grain boundaries, Since the deformation band is equivalent to the gamma-grain boundary with regard to the ferrite nucleation, the gamma-grain can be considered as divided into several blocks by deformation bands. This division allows one to produce much more refined grain structure.

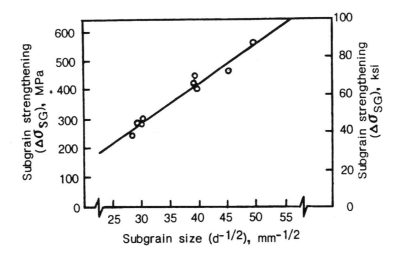

Fig. 26.9 Regression line between subgrain size (d) and its strengthening effect ($\Delta\sigma_{SG}$). (From Cohen and Owen, Microalloying '75, 1977. Copyright STRATCOR, Pittsburgh, Pennsylvania. Reprinted with permission.)

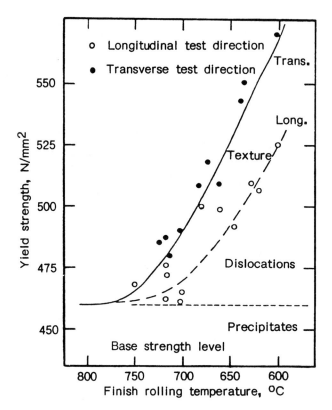

Fig. 26.10 Effect of deformation in two-phase temperature region on yield strength of 0.1% C-1.4% Mn-0.02% Nb steel rolled to 19 mm (0.75 in.). (From Dillamore, et al, Metals Technology, July-August 1975. Copyright ASM International. Reprinted with permission.)

The second important feature of the controlled rolling is a formation of subgrain structure during deformation in two-phase region [16]. As shown in Fig. 26.9, the smaller the subgrain size the stronger its strengthening effect.

Another feature of the controlled rolling is a formation of ferrite-crystallographic texture [17, 18]. Fig. 26.10 shows that for HSLA steel plate rolled at low-finishing temperature there is a marked anisotropy of yield strength attributable to a preferred orientation texture but a substantial strength increment arises from the ferrite dislocation hardening. The effect of precipitation strengthening is rather small; the precipitates are mainly enhancing the retardation of restoration process.

26.7 STRUCTURAL CHANGES DURING CONTINUUM ROLLING

Continuum rolling allows one to obtain a desirable combination of increased strength and toughness in extra-low-carbon steels. This is attributed to [12, 13]:

   a) grain refinement of both gamma and alpha structure by repeated heavy deformations,

   b) dynamic recovery that produces a fine polygonal substructure, and

   c) cube-on-corner crystallographic texturing that results from the deformation process.

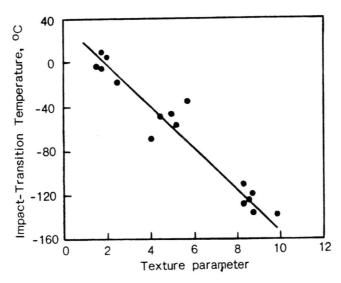

Fig. 26.11 Effect of texture parameter on impact-transition-temperature in continuum rolled steels. (From Pickering, Microalloying '75, 1977. Copyright STRATCOR, Pittsburgh, Pennsylvania. Reprinted with permission.)

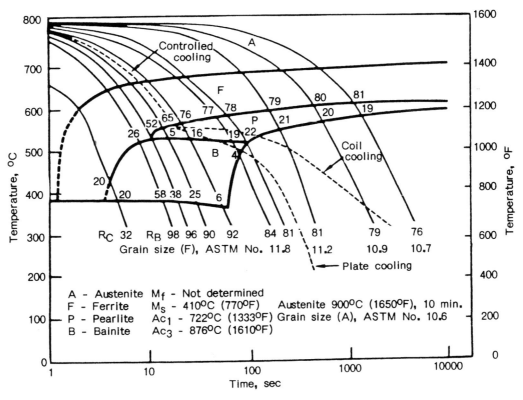

Fig. 26.12 Continuous-cooling transformation diagrams for a vanadium-nitrogen microalloyed steel (0.16% C, 1.4% Mn, 0.004% P, 0.012% S, 0.4% Si, 0.04% Al, 0.11% V, and 0.018% N). Cooling paths for controlled-cooled coils and plates, are superimposed on this diagram.
(From Grozier, Microalloying '75, 1977. Copyright STRATCOR, Pittsburgh, Pennsylvania. Reprinted with permission.)

Texture plays an important role in controlling impact transition temperature (ITT). For continuum rolled steels the transition temperature can be related quantitatively to a **texture parameter** based on the product of the intensities of [111] in the rolling plane and [110] in the transverse plane [2] as shown in Fig. 26.11. The lower finishing temperature in ferrite region, the greater intensity of the [111] < 110 > (cube-on-corner) texture. This results in lower transition temperature.

26.8 STRUCTURAL CHANGES IN STEEL DURING CONTROLLED COOLING

After rolling the plate or strip is usually subjected to the water-quench type cooling. The structure of a steel after quenching will vary with variation of both cooling rate and the temperature at which the water-quenching is ended [19].

The relationship between controlled-cooling path and the resulting microstructure is presented in Fig. 26.12, by superimposing the cooling path for processing of coil and plate on the continuous-cooling diagram for a vanadium-nitrogen steel. It follows from this diagram that bainitic (B) constituents will form in steel of this composition if the water-end temperature is below 550°C (1020°F). If the steel cooling range is between 579 and 635°C (1075 and 1175°F), the mictostructure will consist of fine-grained polygonal ferrite and some pearlite. This temperature also enhances consistency of precipitation strengthening which takes place after coiling.

26.9 EFFECT OF ALLOYING ELEMENTS IN CONTROLLED ROLLING

Alloying elements such as niobium (columbium), vanadium and titanium are frequently added to the controlled rolled steels. Addition of these elements enhances strength of steel [13, 20].

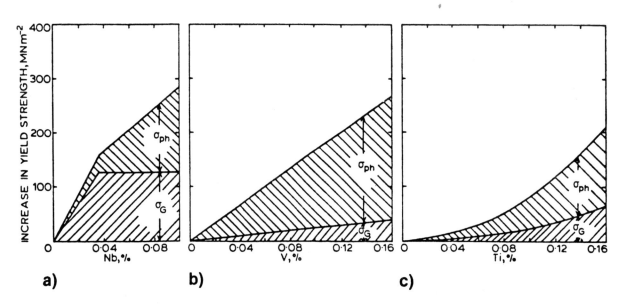

Fig. 26.13 Correlation between alloy content and increase in yield strength as result of grain refinement (σ_G) and precipitation strengthening (σ_{ph}). (From Tanaka, International Metals Reviews, No. 4, 1981. Copyright ASM International. Reprinted with permission.)

The effects of Nb, V, and Ti on strength are shown in Fig. 26.13. Niobium increases strength by promoting both grain refinement and precipitation hardening (Fig. 26.13a). Vanadium causes a large increase in strength mainly through precipitation hardening (Fig. 26.13b). Titanium produces only a slight increase in strength via grain refinement and precipitation strengthening (Fig. 26.13c).

Since niobium promotes grain refinement, both strength and toughness can be simultaneously enhanced in a controlled-rolled niobium steel.

26.10 PRACTICE OF CONTROLLED ROLLING IN HOT STRIP MILLS

Controlled rolling practice in a typical hot strip mill consists of the following six steps [3]:

1. Slab reheating that is accompanying grain growth
2. Rougher rolling that can be regarded as the deformation in the recrystallization region
3. Delay between roughing and finishing rolling
4. Finishing rolling that can be considered as the deformation in non-recrystallization two-phase region
5. Rapid cooling on the runout table
6. Holding at the coiling temperature during which precipitation of Nb and V proceeds causing a large increase in yield strength.

Desired properties of the controlled-rolled steels can be affected by a variety of processing factors. The most important of them are as follows [2, 13, 20]:

a) lowering the slab reheating temperature to obtain small and uniform gamma-grain size, but allowing for a complete solution of alloying elements,

b) selection of a suitable amount of reduction per pass during the initial passes to obtain a fine and uniform recrystallized gamma-grain structure,

c) selection of temperature and time of delay between the recrystallization region and the non-recrystallization region,

d) selection of a suitable amount of reduction and rolling temperature in the two-phase region,

e) selection of appropriate cooling rate, and

f) selection of optimum coiling temperature.

Since all the factors listed above serve the same purpose, a suitable combination of some of them would allow one to obtain the desired properties of the controlled-rolled steels. Diversity of the developed controlled rolling practices is mainly due to the difference in capacity, cooling power and stability of operation in different rolling mills.

26.11 PRECIPITATION STRENGTHENING

The effect of rolling variables on precipitation strengthening is explicitly revealed in high-strength low-alloy (HSLA) steels containing vanadium and nitrogen. It is well known that the precipitation of vanadium carbides and nitrides in ferrite produces precipitation strengthening.

In the experiments conducted by Amin, et al [21], the slab material was reheated for rolling at 1300, 1150, and 1000°C (2372, 2102, and 1832°F) followed by rolling in one pass of 20 or 50% reduction

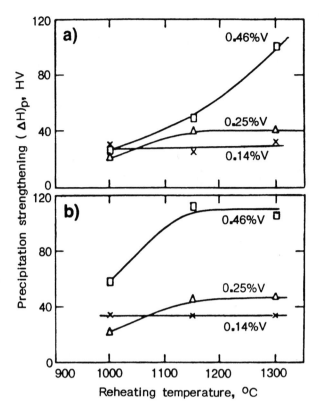

Fig. 26.14 Effect of reheating temperature on precipitation strengthening in high-NV steels rolled to 50% reduction at 950°C and held at 950°C: a) for 100 s; b) for 1000 s. (From Amin, et al, Metals Technology, Vol. 8, 1981. Copyright Institute of Metals. Reprinted with permission.)

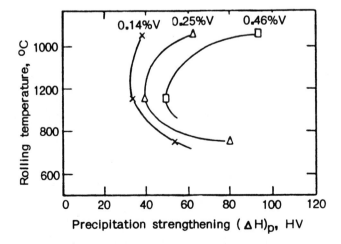

Fig. 26.15 Effect of rolling temperature on precipitation strengthening in high-NV steels reheated at 1300°C and held at 750°C for 100 s. (From Amin, et al, Metals Technology, Vol. 8, 1981. Copyright Institute of Metals. Reprinted with permission.)

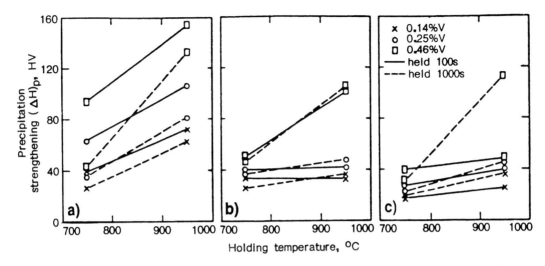

Fig. 26.16 Effect of holding temperature and holding time on precipitation strengthening in high-NV steels rolled to 50% reduction: a) reheated at 1300°C and rolled at 1250°C; b) reheated at 1300°C and rolled at 950°C; c) reheated at 1150°C and rolled at 950°C. (From Amin, et al, Metals Technology, Vol. 8, 1981. Copyright Institute of Metals. Reprinted with permission.)

in thickness. After rolling the rolled workpieces were held for 100 to 1000 seconds at elevated temperatures before being cooled to room temperature at the rate of 400 K/min.

The intensity of the produced precipitation strengthening was evaluated from hardness measurements which were compensated for the variation of the grain size.

Figure 26.14 shows the effect of reheating temperature on the precipitation strengthening of some high-nitrogen vanadium steels. The 0.14% V steel showed no effect of reheating temperature whereas the 0.25% V steel and especially 0.46% V steel show an explicit dependence of precipitation strengthening on the reheating temperature.

Figure 26.15 shows a very distinct dependence of the precipitation strengthening on rolling temperature for all three grades of steels with minimum strengthening when rolling temperature is about 1000°C (1832°F).

Effects of holding temperature and holding time on precipitation strengthening in high-nitrogen vanadium steels are shown in Fig. 26.16. In general, decreasing the holding temperature decreases the precipitation strengthening.

26.12 HOT ROLLING OF ACICULAR FERRITE STEEL

In conventionally rolled low-carbon (less than 0.06%) Mn-Mo-Nb steel, austenite transforms to upper bainite. When transformation is accelerated by controlled rolling, the hot-rolled gamma-grain structure transforms into **acicular ferrite structure [13]**. The acicular ferrite structure consists of fine non-equiaxed ferrite dispersed with cementite and martensite islands.

The process of rolling the acicular ferrite, HSLA steel has been developed by Climax Molybdenum Company. A significant improvement in toughness of this steel is obtained by:

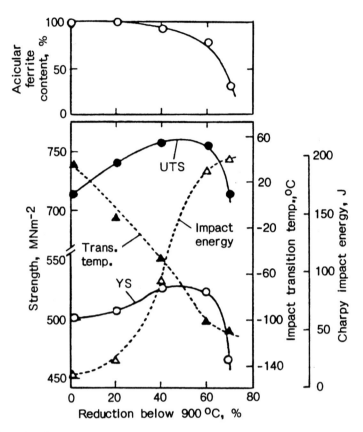

Fig. 26.17 Acicular ferrite volume fraction, strengths, and Charpy impact properties plotted against amount of reduction in non-recrystallization region in 0.07% C-2.0% Mn-0.6% Nb-0.5% Mo acicular ferrite steel. (From Tanaka, International Metals Reviews, No. 4, 1981. Copyright ASM International. Reprinted with permission.)

a) lowering finishing temperature,

b) increasing the amount of deformation in the non-recrystallization region.

An increase in the amount of deformation in the non-recrystallization region increases volume fraction of fine-grained polygonal ferrite at the expense of acicular ferrite (Fig. 26.17). This causes improvement in strength and low-temperature toughness. The optimum combination of strength and toughness corresponds to the structure containing approximately 15% of polygonal ferrite.

26.13 HOT ROLLING OF DUAL-PHASE STEEL

Dual-phase (DP) steel combines high strength with good formability and cost performance.

The hot-rolled dual-phase steel can be classified into the following three types [22]:

1. Si-Mn steel treated after conventional hot rolling

2. Steel with high content of Si, Sr, and Mo produced only by conventional hot rolling

3. Simple Si-Mn steel produced by thermomechanical process utilizing controlled cooling and extra low temperature cooling method.

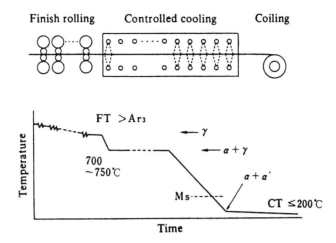

Fig. 26.18 Cooling pattern in manufacturing process of a hot-rolled dual-phase steel. (From Okita, et al, Nippon Kokan Technical Report, Overseas No. 43, 1985. Copyright Nippon Kokan K.K. Reprinted with permission.)

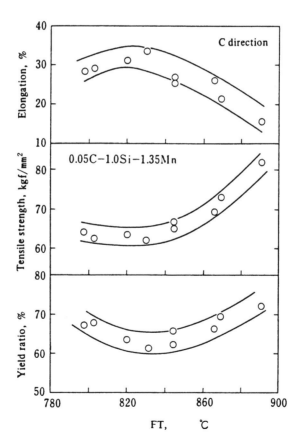

Fig. 26.19 Effect of finish rolling temperature on tensile properties of hot-rolled dual-phase steel. (From Okita, et al, Nippon Kokan Technical Report, Overseas No. 43, 1985. Copyright Nippon Kokan K.K. Reprinted with permission.)

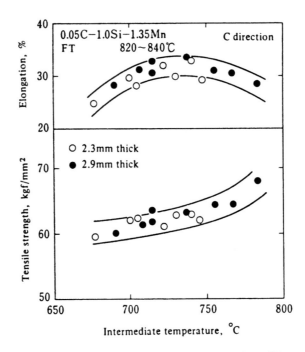

Fig. 26.20 Effect of intermediate temperature on tensile properties of hot rolled dual-phase steel. (From Okita, et al, Nippon Kokan Technical Report, Overseas No. 43, 1985. Copyright Nippon Kokan K.K. Reprinted with permission.)

The thermomechanical process for production of DP steel coil developed by Nippon Kokan is schematically shown in Fig. 26.18. In this process the strip is finished to roll in an austenite region near above A_{r3} critical temperature. The strip is then subjected to cooling in three regulated steps.

The first step is a water cooling of the strip to rapidly reduce its temperature to an intermediate temperature of about 700 to 750°C (1292 to 1382°F). The second step is air cooling during which ferrite (α) of about 80% in volume fraction finely precipitates. The third step is a water cooling to rapidly reduce the strip temperature to below about 200°C (392°F). During this step remaining austenite (γ) is transformed into martensite (α') and the dual-phase ($\alpha + \alpha'$) microstructure is finely formed.

Principal processing factors affecting quality of steel through formation of appropriate microstructure are finish rolling temperature, intermediate temperature, and coiling temperature. These factors are briefly discussed below in application to a low-carbon Si-Mn steel [22].

Finish rolling temperature - As shown in Fig. 26.19, both maximum total elongation and minimum yield ratio (ratio of yield strength to tensile strength) are attained at finish rolling temperature of about 830°C (1526°F). Higher finish rolling temperature leads to higher tensile strength due to fractional increase of the second hardening phase which leads to the deterioration of ductility and higher yield ratio. The finish rolling at temperatures below 800°C (1472°F) is also detrimental to mechanical properties due to the formation of deformed microstructure with high density of dislocation. The most favorable dual-phase structure is obtained at finish temperature just above the A_{r3}.

Intermediate temperature - The intermediate temperature determines the volume fraction of ferrite that will precipitate during air cooling after the first rapid cooling. As shown in Fig. 26.20 high ductility

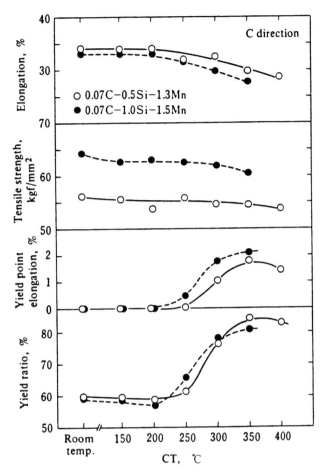

Fig. 26.21 Effect of coiling temperature on tensile properties of hot rolled dual-phase steel. (From Okita, et al, Nippon Kokan Technical Report, Overseas No. 43, 1985. Copyright Nippon Kokan K.K. Reprinted with permission.)

is attained at the intermediate temperature range from 720 to 750°C (1328 to 1382°F). Higher intermediate temperature leads to harder and less ductile steel due to excess of the second phase in volume fraction.

Coiling temperature - As shown in Fig. 26.21, the yield ratio increases with increase of coiling temperature. The yield point elongation appears when coiling temperature exceeds about 200°C (392°F). The reduction in strength and elongation with increasing the coiling temperature up to 350°C (662°F) are relatively small. From the viewpoint of controllability of the coiling process, the coiling temperature is usually set below 100°C (212°F).

26.14 CONTROLLED ROLLING OF ARCTIC GRADE STEEL

The controlled rolling process developed by Sumitomo Metal Industries for rolling of arctic grade steel is another example of practical implementation of the controlled rolling concept [23].

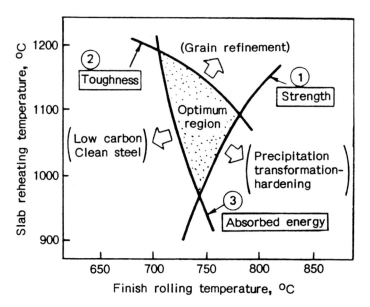

Fig. 26.22 Optimum region of rolling conditions for desired properties.
(From Terasaki, et al, Sumitomo Search, No. 33, Nov. 1986. Copyright
Sumitomo Metals. Reprinted with permission.)

The requirements for arctic grade steel used in large-diameter pipes include high strength, high toughness and high absorbed energy. The controlled rolling allows one to achieve these requirements by appropriate selection of the slab reheating and finishing temperatures and also by application of dynamic accelerated cooling, microalloying, and transformation hardening.

Effect of the slab reheating and finishing temperatures - Figure 26.22 shows the optimum range of the slab reheating temperature and finishing rolling temperature within which the desired mechanical properties can be obtained. Expansion of the strength limit (1) can be obtained by application of precipitation and/or transformation hardening. Grain refinement will expand low temperature toughness limit (2). Higher absorbed energy limit (3) can be achieved by using cleaner steel that has lower sulfur and carbon content and is produced with optimum shape control.

The microstructural changes taking place during controlled rolling of steel are shown schematically in Fig. 26.23. The process provides:

1. Grain refining of heated austenite structure by lowering slab temperature and/or adding of approximately 0.015 Ti as a microalloy.

2. Grain refining of recrystallized austenite grain by increased reduction in the high temperature range.

3. Increase of the amount of fine grain ferrite due to producing of the deformation bands by heavy reduction in the non-recrystallization austenite region just above A_{r3} temperature.

4. Steel strengthening by optimum dual-phase rolling in austenite-ferrite region.

5. Obtaining a mixed structure of fine ferrite and bainite by accelerated cooling process.

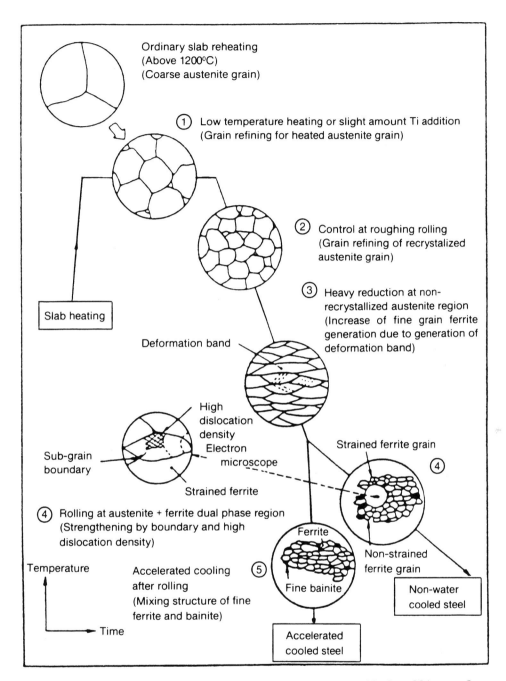

Fig. 26.23 Change in microstructure during controlled rolling and accelerated cooling. (From Terasaki, et al, Sumitomo Search, No. 33, Nov. 1986. Copyright Sumitomo Metals. Reprinted with permission.)

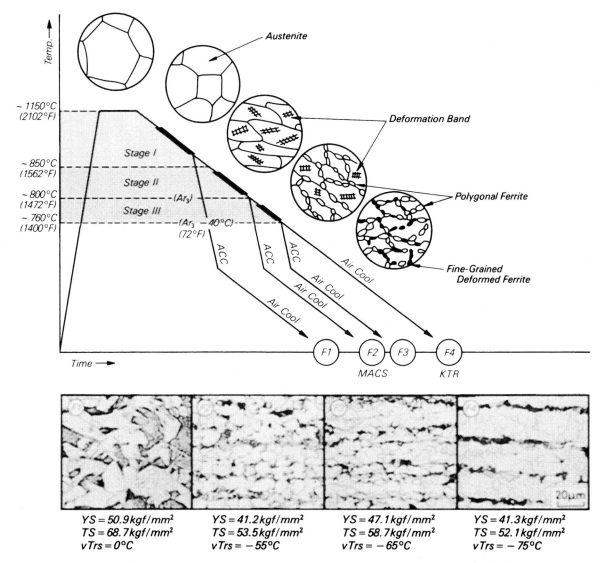

YS = 50.9 kgf/mm²   YS = 41.2 kgf/mm²   YS = 47.1 kgf/mm²   YS = 41.3 kgf/mm²
TS = 68.7 kgf/mm²   TS = 53.5 kgf/mm²   TS = 58.7 kgf/mm²   TS = 52.1 kgf/mm²
vTrs = 0°C   vTrs = − 55°C   vTrs = − 65°C   vTrs = − 75°C

Fig. 26.24 Comparison of the microstructures of Si–Mn steel plates subjected to accelerated cooling from different temperatures. Cooling rate is approximately 10° C/s (18°F/s). (From Kawasaki Steel Corporation Publication YE1A 714 8504 A b, 1985. Reprinted with permission.)

26.15 ACCELERATED COOLING PROCESSES

Accelerated cooling processes are developed in order to enhance tensile strength, toughness, weldability and other properties of the controlled-rolled thick plates.

A number of accelerated cooling processes have been recently developed. The most well-known ones are MACS, OLAC, and DAC which are briefly described below.

MACS - Multipurpose Accelerated Cooling System has been developed by Kawasaki Steel Corporation [24]. The purpose of this process is to enhance tensile strength of Si-Mn, Arctic quality steel plates without sacrificing notch toughness. It is achieved by improving the microstructure.

The desired microstructure can not be generally achieved by the process designated in Fig. 26.24 as F1, a process that subjects the plate to accelerated cooling after uncontrolled hot rolling. Nor can the desired microstructure be attained to any satisfactory extent by the process F4, in which controlled rolling is followed by air cooling. A satisfactory improvement of the microstructure is obtained by both F2 and F3 processes which combine controlled rolling with accelerated cooling.

Tensile strength increases with an increase in cooling rate and decrease in the finish-cooling temperature. However, since the lowering of finish-cooling temperature tends to cause strain in the plate and also adversely affects the cooling effeciency, the finish-cooling temperature range is selected between 500 and 400°C (932 and 752°F).

OLAC - On-Line Accelerated Cooling process has been developed by Nippon Kokan [25]. The process is also applied to a production of Si-Mn steel plates with yield strength of 36 kg/mm^2 (51,192 psi). In this process, the plate is finished-rolled at above A_{r3} temperature (Fig. 26.25). It is then immediately subjected to accelerated cooling at a rate of 3 to 15°C/s (5.4 to 27°F/s) down to 600 or 500°C (1112 or 932°F). The plate is then air-cooled down to room temperature. The self-tempering is being performed at this stage.

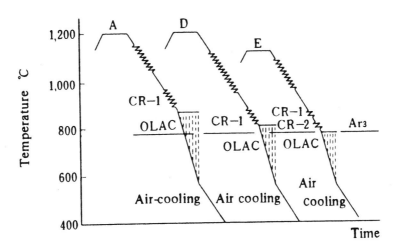

Fig. 26.25 Schematic diagram for production of 25 mm (1 in.)-thick steel plate with yield strength of 36 kg/mm^2 (51,192 psi) by OLAC. (From Tsukada, et al, Nippon Kokan Technical Report, Overseas No. 35, 1982. Copyright Nippon Kokan K.K. Reprinted with permission.)

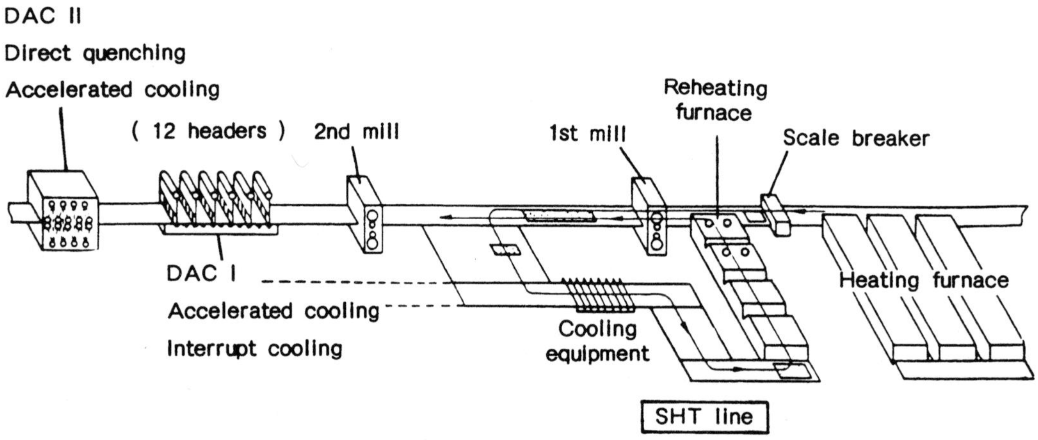

DAC II

Direct quenching

Accelerated cooling

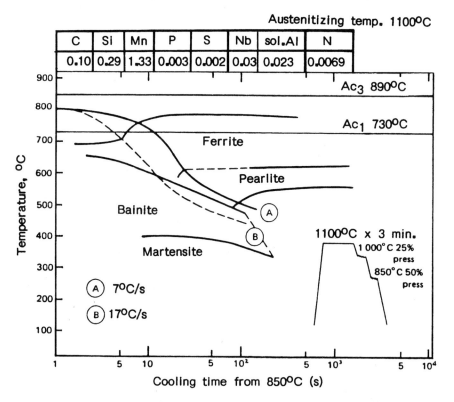

Fig. 26.26 Layout of thermomechanical treatment facilities at plate mill, Kashima Steel Works. (From Takeuchi, et al, Sumitomo Search, No. 32, May 1986. Copyright Sumitomo Metals. Reprinted with permission.)

Fig. 26.27 Relationship between cooling conditions of DAC I process and CCT-diagram after deformation. (From Terasaki, et al, Sumitomo Search, No. 33, Nov. 1986. Copyright Sumitomo Metals. Reprinted with permission.)

DAC - Dynamic Accelerated Cooling process has been developed by Sumitomo Metals Industries [22, 26]. The process consists of two phases. The first phase (DAC-I) is a slow cooling halfway-stop type, and the second phase (DAC-II) is a cooling to room temperature by comparatively fast cooling rate (Fig. 26.26). When cooling rate in DAC-I process is between 7 and $17^{\circ}C/s$, the cooling curve of 0.10% C-0.03% Nb steel goes nearly to the nose of ferrite-pearlite and/or bainite transformation region in the CCT diagram (Fig. 26.27). The resulting structure is clearly affected by both cooling rate and cooling stop temperature.

26.16 COLD ROLLING AND ANNEALING OF STEEL

After hot-rolling, the produced coils are usually uncoiled, pickled, dried, oiled, and recoiled. The oil serves as a protection against rusting and as a lubricant during cold rolling [27].

The prime objective of the cold rolling process is further reduction in thickness of hot-rolled product. The thickness range for the sheet products obtained after hot rolling is usually between 1.24 and 5.0 mm (0.049 and 0.197 in.). The total reduction during cold rolling may run from 45 to 75 percent for most sheet gauges and from 80 to 90 percent for most tin plate.

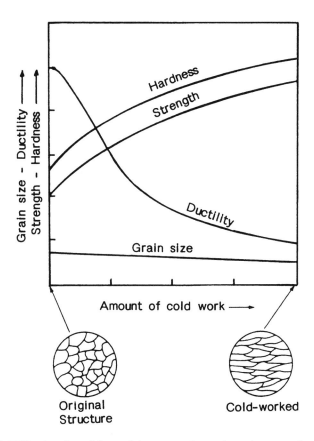

Fig. 26.28 Effect of cold working on microstructure and properties.
(From INTRODUCTION TO PHYSICAL METALLURGY, by S.H. Avner.
Copyright 1974 McGraw-Hill Book Company, Inc. Reprinted with
permission.)

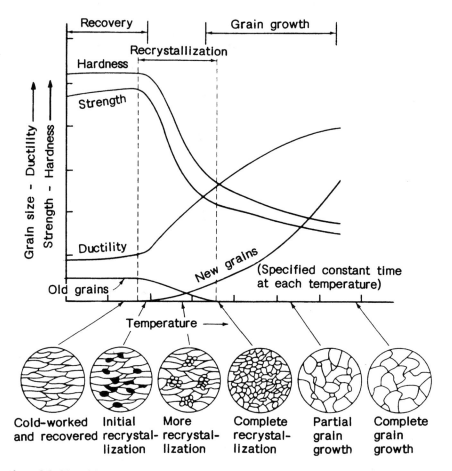

Fig. 26.29 Effect of annealing on microstructure and properties of cold-worked material. (From INTRODUCTION TO PHYSICAL METALLURGY, by Avner, Copyright 1974 McGraw-Hill Book Company, Inc. Reprinted with permission.)

During cold rolling, the material is work hardened. This work-hardening process is due to the jamming of dislocations. As shown in Fig. 26.28, the original equiaxed grain structure is distorted by cold working and the elongated grain structure with smaller grain size is produced [28, 29]. These internal changes are mainly responsible for increasing both hardness and strength as well as for a decrease in ductility of a steel subjected to cold rolling. The increased-energy state of a cold-worked metal makes it more chemically active and consequently less resistant to corrosion [30].

During annealing of cold-rolled steel, the grain restoration process takes place. This process consists of recovery, recrystallization, and grain growth as shown in Fig. 26.29. It was found that recrystallization temperature depends on amount of cold reduction. The greater cold reduction the lower temperature at which steel recrystallizes [31]. The recovery process is accompanied by negligible changes in hardness strength and ductility. During recrystallization both hardness and strength rapidly decrease whereas ductility increase. This trend in change of properties gradually attenuates during the grain growth.

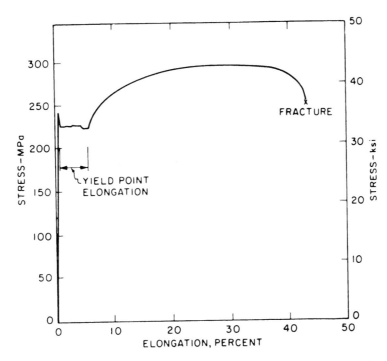

Fig. 26.30 Stress-strain curve for box-annealed rimmed deep-drawing steel. (From THE MAKING, SHAPING AND TREATING OF STEEL, by W.T. Lankford, et al, Tenth Edition, 1985. Copyright AISE, Pittsburgh, Pennsylvania. Reprinted with permision.)

26.17 TEMPER ROLLING OF STEEL

In most sheet products, the principle purpose of temper rolling is to suppress the yield-point elongation (Fig. 26.30) that is present in the as-annealed state for most steels [27]. Suppressing the yield-point elongation allows one to avoid the surface markings during forming which are often referred to as **stretcher-strain markings** or **luder lines**.

Temper rolling is also used to improve the shape, profile, and surface conditions of the flat rolled products. The process features of temper rolling depend on the product requirements as described below.

Tin mill products - Temper rolling is used to develop the proper stiffness or temper by cold working the steel in controlled amount.

High-strength cold-rolled sheets - Heavy temper rolling (1.5 to 3.0 percent elongation) is sometimes used to raise the strength of the as-annealed sheet to a specified minimum yield strength level.

Electrical sheets - Heavy temper rolling (2 to 8 percent elongation) is used, especially for cold-rolled motor laminations, to improve magnetic properties.

Deep-drawing sheets - Light tempering (0.5 to 1.0 percent elongation) is used which is usually sufficient to suppress the formation of stretcher-strain markings during formation without significantly impairing ductility.

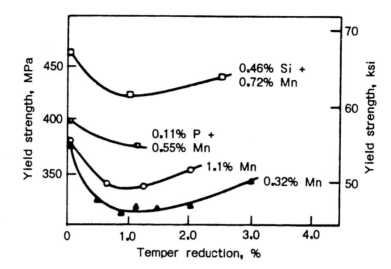

Fig. 26.31 Effect of temper rolling on the yield strength of columbium steels. (From Gupta and Hughes, Microalloying '75, 1977. Copyright STRATCOR, Pittsburgh, Pennsylavania. Reprinted with permission.)

The influence of temper reduction on yield strength and the drop in yield strength from as-annealed values for some columbium steels are shown in Fig. 26.31. Most of the decrease in yield strength for these steels occurs after less than 1 percent temper reduction [32]. Although the initial yield strength is obviously lower at higher annealing temperatures, the decrease in yield strength after tempering is found not to be related to the annealing temperature. However, with similar cementite morphologies, the decrease appears to be larger for finer grain sizes.

REFERENCES

1. ASM Metals Reference Book, American Society for Metals, Metals Park, Ohio, pp. 73, 74 (1981).

2. F.B. Pickering, "High-Strength, Low-Alloy Steels - A Decade of Progress" Microalloying '75, Union Carbide Corp., New York, 1977, pp. 9-31.

3. S.V. Radcliffe and E.B. Kula, "Deformation, Transformation, and Strength", Fundamentals of Deformation Processing, Syracuse University Press, Syracuse (1964), pp. 321-363.

4. E.B. Kula, "Strengthening of Steel by Thermomechanical Treatment of Metals", Journal of Applied Metalworking, Vol. 1, No. 2, 1980, pp. 5-34.

5. G. Krauss, Principles of Heat Treatment of Steel, American Society for Metals, Metals Park, Ohio pp. 235-246 (1980).

6. S.J. Matas, et al, "Ausforming and Hot-Cold Workings - Methods and Properties", Mechanical Working of Steel I, Metallurgical Society of AIME, New York, 1964, pp. 143-178.

7. Encyclopedia of Material Science and Engineering, Vol. 7, pp. 4976, 4977 (1986).

8. V.F. Zackay, "Thermomechanical Processing", Materials Science Engineering, Vol. 25, 1976, pp. 247-261.

9. M.J. May and D.J. Latham, Thermomechanical Treatment of Steels, Part I, Met. Treating (Oct.-Nov. 1972), pp. 3-13.

10. M.J. May and D.J. Latham, Thermomechanical Treatment of Steels, Part II, Met. Treating (Dec.-Jan. 1973), pp. 3-9.

11. M. Azrin, et al, "Soviet Progress in Thermomechanical Treatment of Metals", Journal of Applied Metalworking, Vol. 1, No. 2, 1980, pp. 5-34.

12. G.F. Melloy and J.D. Dennison, "Continuum Rolling - A Unique Thermomechanical Treatment for Plain-Carbon and Low-Alloy Steels", The Microstructure and Design of Alloys, Vol.1, Institute of Metals and Iron and Steel Institute, Cambridge, 1973, pp. 60-64.

13. T. Tanaka, "Controlled Rolling of Steel Plate and Strip", International Metals Reviews, No. 4, 1981, pp. 185-212.

14. J.D. Baird and R.R. Preston, "Processing and Properties of Low-Carbon Steel", Mechanical Working of Steel I, Metallurgical Society of AIME, New York, 1973, pp. 1-46.

15. M. Lafrance, F. Caron, G. Lamant, and J. Leclerc, "Microalloying '75", Union Carbide Corp., New York, 1977, pp. 367-374.

16. M. Cohen and W.S. Owen, "Thermo-Mechanical Processing of Microalloyed Steels", Microalloying '75, Union Carbide Corp., New York, 1977, pp. 2-8.

17. I.L. Dillamore, et al, "Metallurgical Aspects of Steel Rolling Technology", Metals Technology, July-Aug. 1975, pp. 294-302.

18. J.M. Little, et al, Proceedings, 3rd International Conference on the Strength of Metals and Alloys, London, The Institute of Metals/Iron and Steel Institute, Publication No.36, 1973, Vol. 1, pp. 80-84.

19. J.D. Grozier, "Production of Microalloyed Strip and Plate by Controlled Cooling", Microalloying '75, Union Carbide Corp., New York, 1977, pp. 241-250.

20. F. Heisterkamp and L. Meyer, "Thyssenforschung", 1971, 3, pp. 44-65.

21. R.K. Amin, et al, "Effect of Rolling Variables on Precipitation Strengthening in High-Strength Low-Alloy Steels Containing Vanadium and Nitrogen", Metals Technology, Vol. 8, No. 7, July 1981, pp. 250-262.

22. T. Okita, et al, "Production and Quality of Hot Rolled Dual Phase Steel", Nippon Kokan Technical Report, Overseas No. 43, 1985, pp. 25-32.

23. F. Terasaki, et al, "Research and Development on Large-Diameter Line Pipe for Arctic Usage", Sumitomo Search, No. 33, Nov. 1986, pp. 72-85.

24. MACS - Multipurpose Accelerated Cooling System, Kawasaki Steel Corp., Printed in Japan, Apr. 1985, YE1A 714 8504 A b.

25. K. Tsukada, et al, "Development of YS 36kgf/mm^2 Steel with Low-Carbon Equivalent Using On-Line Accelerated Cooling (OLAC) - Development of OLAC, Part 2", Nippon Kokan Technical Report, Overseas No. 35, 1982, pp. 35-46.

26. H. Takeuchi, et al, "Recent Development of Thermomechanical Treatment Technique in Sumitomo Metals", Sumitomo Search, No. 32, May 1986, pp. 8-18.

27. The Making, Shaping and Treating of Steel, 10th Edition, eds. W.T. Lankford, Jr., et al, Association of Iron and Steel Engineers, Pittsburgh, Pennsylvania, pp. 1103-1120 (1985).

28. S.H. Avner, Introduction to Physical Metallurgy, Second Edition, McGraw-Hill Book Company, New York (1974), pp. 129-145.

29. C.O. Smith, The Science of Engineering Materials, Prentice-Hall Inc., Englewood Cliffs, N.J. (1969).

30. R.A. Higgins, Engineering Metallurgy, Part I, Applied Physical Metallurgy, Robert E. Krieger Publishing Company, Melbourne, Florida, pp. 90-99 (1983).

31. A. Sauveur, The Metallography and Heat Treatment of Iron and Steel, Fourth Edition, McGraw-Hill Book Company, Inc., New York and London (1935).

32. I. Gupta and I.F. Hughes, "Metallurgical and Processing Considerations in Producing High-Strength, Cold-Rolled Sheet", Microalloying '75, Union Carbide Corp., New York, 1977, pp. 303-310.

27

Scaling of Steel in Hot Strip Mill

27.1 SCALING OF STEEL DURING REHEATING

Scaling is a process of forming a layer of oxidation products formed on metal at high temperature [1]. The following three types of iron oxides are present in scale at elevated temperatures [2-5]:

    A. Wustite (FeO)

    B. Magnetite (Fe_3O_4)

    C. Hematite (Fe_2O_3)

These chemical compounds are formed by the following reactions:

$$O_2 + 2Fe = 2FeO \tag{27-1}$$

$$FeO + Fe_2O_3 = Fe_3O_4 \tag{27-2}$$

$$O_2 + 4Fe = 2Fe_2O_3 \tag{27-3}$$

Wustite (FeO) is the innermost phase adjacent to metal (Fig. 27.1) with the lowest oxygen content. Below 570°C (1058°F) wustite is not stable. Its content in scale increases with increase in temperature and when the steel temperature is above 700°C (1292°F), wustite occupies about 95% of the scale layer. The wustite phase has a relatively low melting point, 1370 to 1425°C (2498 to 2597°F), compared with that of other phases of scale and of the steel itself. Melting of the wustite layer ('washing') accelerates the scaling rate and further increases grain boundary penetration that produces inferior surface quality, increases fuel consumption and reduces yield.

Magnetite (Fe_3O_4) is the intermediate phase of scale. When the steel temperature is below 500°C (932°F), the scale contains only magnetite (Fig. 27.2). As the temperature increases to about 700°C (1292°F), formation of wustite takes place at the expense of magnetite and, at elevated temperatures, magnetite occupies only 4% of the scale layer. Magnetite is harder and more abrasive than wustite.

Hematite (Fe_2O_3) is the outer phase of scale (Fig. 27.1). Hematite is formed at the temperature above approximately 800°C (1472°F) as shown in Fig. 27.2 and, at elevated temperatures, it occupies about 1% of the scale layer. Similar to magnetite, hematite is hard and abrasive.

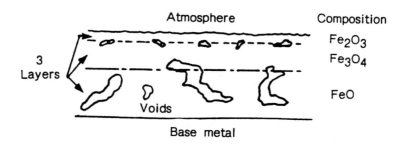

Fig. 27.1 Sketch illustrating the structure of scale. From Roberts, (1983).

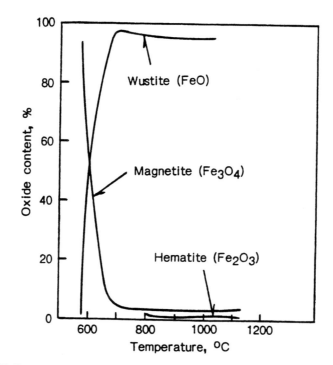

Fig. 27.2 Approximate percentages of FeO, Fe$_2$O$_3$, and Fe$_3$O$_4$ on iron oxidized in oxygen. (From Sacks and Tuck, Iron and Steel Institute Publication 111, 1968. Reprinted with permission.)

27.2 SCALING RATE

The scaling process initiates through the formation of an iron oxide film at the metal surface and progresses by diffusion of oxygen into iron-iron oxide interface. This diffusion process is described by a general relationship which is exponential with temperature and parabolic with time [6, 7]:

$$M = a\sqrt{t} \ \exp[-b/(T + 460)](1/h + 1/w + 1/L)/\rho, \qquad (27-4)$$

where h, w, L = slab thickness, width, and length respectively, in.

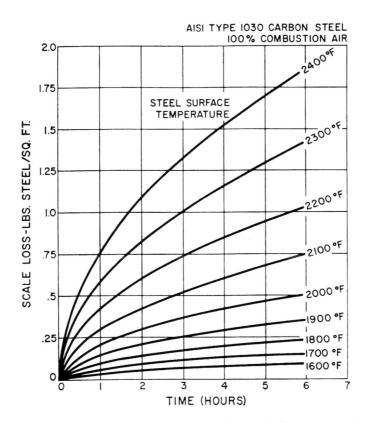

Fig. 27.3 Relationship between time and metal loss at constant temperature. (From Cook and Rasmussen, AISE Year Book, 1970. Copyright AISE, Pittsburgh, Pennsylvania. Reprinted with permission.)

M = weight of scale loss per unit of slab weight, tons/ton

y = scaling time

T = slab temperature, °F

ρ = slab density, lb/in³

a, b = constants depending on type of steel and furnace atmosphere.

For slabs made of carbon steel with 0.30% C, residing in the furnace with 100% combustion air, a = 31.7 and b = 22.8, as derived from Eq. (27-4) and Fig. 27.3.

27.3 EFFECT OF THE ATMOSPHERE ON SCALING

In respect to oxidation of steel, the gases containing in the atmosphere surrounding the steel may be divided into two groups:

　　1. Oxidizing gases: oxygen (O_2), water vapor (H_2O), and carbon dioxide (CO_2).

　　2. Reducing gases: carbon monoxide (CO) and hydrogen (H_2).

The oxidation and reduction processes are reversible and are described by the following formulae [3]:

$$O_2 + 2Fe \rightleftharpoons 2FeO \qquad\qquad\qquad\qquad\qquad (27\text{-}5)$$

$$CO_2 + Fe \rightleftharpoons FeO + CO \qquad\qquad\qquad\qquad (27\text{-}6)$$

$$H_2O + Fe \rightleftharpoons FeO + H \qquad\qquad\qquad\qquad (27\text{-}7)$$

These oxidation-reduction relationship may be best presented by the curves developed by Murphy and Jominy (Fig. 27.4a), and by Marshall (Fig. 27.4b). These curves define the **equilibrium temperature** at which various gas ratios of CO_2/CO and H_2O/H_2 are neutral to iron [6]. Figure 27.4 also depicts the eqiulibrium products of combustion of natural gas related to percentage of theoretical combustion air at 2500°F flue gas temperature. Note that the reduction in percentage of the combustion air produces less oxidizing atmosphere. Quantitative effect of the air-gas ratio on scaling rate is shown in Fig. 27.5 for carbon steel with 0.30% C after one hour of heating. At elevated temperatures the scale-free atmosphere would require the air-gas ratio to be close to 50 percent.

Figure 27.6 illustrates a firing strategy allowing one to reduce furnace scale losses. According to this strategy, the air-gas ratio is gradually reduced from 100 to 50% while the slab surface temperature increases from 760 to 1038°C (1400 to 1900°F).

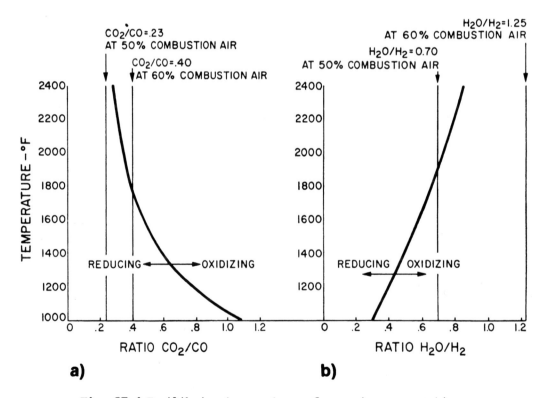

Fig. 27.4 Equilibrium temperatures for various gas ratios:
a) CO_2/Co and b) H_2O/H_2. From Cook and Rasmussen (1970).

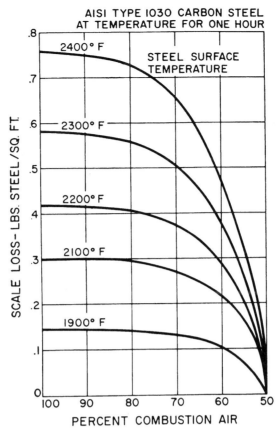

Fig. 27.5 Quantitative effect of air-gas ratio on scaling rate at constant temperatures. From Cook and Rasmussen (1970).

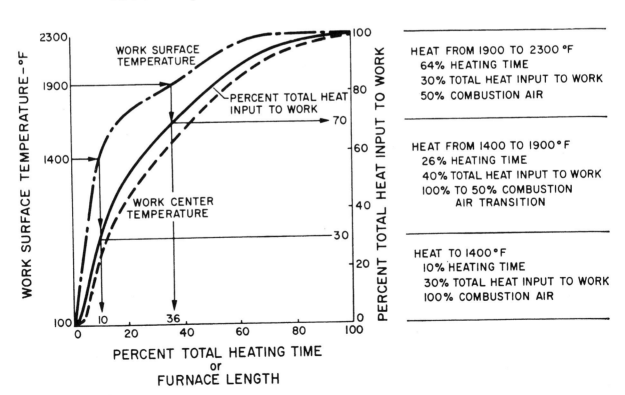

Fig. 27.6 Heating practice utilizing variable air-gas ratio control. From Cook and Rasmussen (1970).

Gas velocity is another factor that affects the scaling rate. It was found [4] that the oxidation rate increases progressively with gas flow rate until critical flow rate was reached. The critical flow rate for both carbon dioxide and air is equal to 1.5 m/min (5 ft/min). For steam it is as high as 7 m/min (23 ft/min).

The other important constituent of furnace gases is **sulfur dioxide** (SO_2). In chemical reaction with steel it forms the liquid sulphides such as FeS in scale. This enhances the scaling process. Additional complication associated with the formation of sulphides is an increase in metal-scale adhesion, which results in difficulty of scale removal [8].

In reducing atmosphere the sulphide phase is formed and grows very rapidly. Therefore, in order to alleviate this problem, the air-fuel ratio is usually increased so that there is at least 4% excess of oxygen in furnace atmoshere.

27.4 EFFECT OF RESIDUAL AND ALLOYING ELEMENTS ON SCALING

Scaling rate may be modified if some residual or alloying elements are present in steel (Fig. 27.7). Effect of some elements is briefly described below.

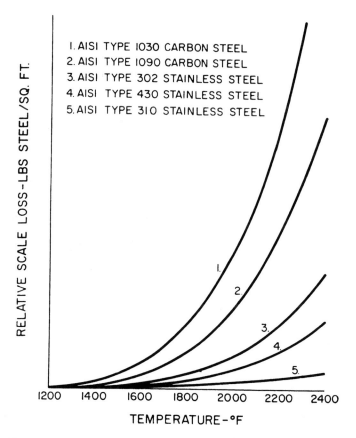

Fig. 27.7 Scale losses of some steels at elevated temperatures. From Cook and Rasmussen (1970).

Carbon - It diffuses to the oxide-metal interface where it forms carbon monoxide CO as a result of chemical reaction with iron oxide FeO. In the presence of the carbon monoxide at the oxide-metal interface, the gap formation between scale and metal enhances and the scale adhesion reduces.

Gap formation without gross cracking of the scale slows the scaling rate. However, at high temperature and high carbon content, gas pressure in the gaps may cause gross cracking of the scale. It increases contact of the metal with the furnace atmosphere and thereby increases the scaling rate.

Manganese - The manganese present in mild steel produces a negligible effect on the oxidation properties [4].

Chromium - Its effect on oxidation rate is negligible when it is present in steel as a residual element. In stainless steels it forms a protective layer of chromium oxide Cr_2O_3 that provides the oxidation resistance. The process of forming this protective layer is often reffered as passivation process [9].

Aluminum - It is also not effective when it is present in steel at a residual level. As an alloying element, it forms a hard protective layer of the aluminum oxide Al_2O_3 that reduces the oxidation rate.

Silicon - It reacts with oxygen diffusing into the steel ahead of the oxide-metal interface and precipitates as a silicon oxide SiO_2 [10]. The particles of SiO_2 form a separate phase, fayalite Fe_2SiO_4, that retards the scaling rate. This phase melts at 1171^oC (2140^oF). With the formation of the molten phase the protective effect disappears and the scaling rate increases sharply [11, 12].

Silicon-killed steels contain only about 0.25%Si. However, this is sufficient to form pools and extended silicate stringers in the scale layer that increase the adhesion of the scale to steel [4].

Nickel - Because the diffusion coefficient for nickel in iron is low, it is concentrated in the thin layer around the alloy core. This layer lowers the scaling rate. However, the resulting nickel-alloy lacework in the oxide makes the scale adhere during rolling giving rise to excessive rolled-in scale [4, 13]. This problem becomes more severe with increase in nickel content.

Copper - Similar to nickel, it is rejected at the oxide-metal interface. However, copper does not create a sticky-scale problem as is the case with nickel. When copper content in the metal oxide interface exceeds 8%, a separate copper-rich phase is precipitated. This phase melts at 1096^oC (2005^oF) and forms a layer between the metal and the oxide that causes hot shortness and surface defects [4].

27.5 SCALING AND DECARBURIZATION

The decarburization process in steels is described by the following formulae [3]:

$$2H_2 + Fe_3C \rightleftharpoons 3Fe + CH_4 \qquad (27\text{-}8)$$
$$O_2 + 2Fe_3C \rightleftharpoons 6Fe + 2CO \qquad (27\text{-}9)$$
$$CO_2 + Fe_3C \rightleftharpoons 3Fe + 2CO \qquad (27\text{-}10)$$

Decarburization below the A_3 and A_1 temperatures is rather slow. However, this process intensifies significantly above these temperatures when austenite is formed. When the steel temperature is about 1200 to 1290^oC (2192 to 2354^oF), decarburization occurs simultaneously with scaling [14]. Both the scaling rate and the decarburization rate increase with temperature.

The carbon-depleted zone is usually known as **decarburization zone.** It is located between the interior of the steel and the oxide-metal interface. Decarburization has a detrimental effect on subsequent processing of certain high-carbon special steels. It usually produces no quality problems in low-carbon steels.

27.6 SCALING OF STEEL DURING ROUGHING PASSES

As was previously discussed, the scale of a reheated slab consists of three distinct layers: wustite (FeO), magnetite (Fe_3O_4), and hematite (Fe_2O_3). Since the bonding between these layers is much stronger than that of the interface bond between the scale and the steel, the descaling and rolling process will tend to break the interface bond rather than bonding between the scale layers.

According to Blazevic [15], the behavior of the scale on the surface of a workpiece during roughing passes may be illustrated as follows. Let us consider a typical slab with 3.2 mm (0.125 in.) thick **primary scale** which was not removed from the slab surface prior to the rolling process. As the slab enters the first pass in the rougher (Fig. 27.8), the scale breaks free from the slab (details X and Y). Some of the scale particles may fly out of the roll bite (detail Z), but most of them will enter the roll bite, become slightly elongated and pressed back into the steel. Since the elongation of the scale particles is not quite equal to the slab elongation, small spaces between the scale particles will be formed.

As the slab enters the second pass the scale becomes less plastic due to lower temperature and breaks into smaller pieces. These pieces will be more deeply embedded into the slab surface with greater gaps between them. The slab surface along those gaps will be exposed to oxidation and a **secondary scale** (oxide) will begin to form. By the time the slab enters the third pass, the scale has lost its plasticity and compressibility due to low temperature and further embeds into the steel.

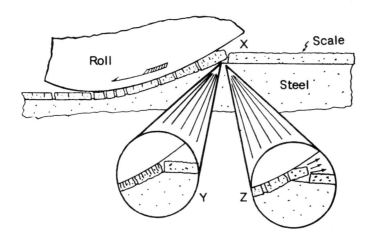

Fig. 27.8 Schematic presentation of the scale behavior during the first roughing pass. (From David T. Blazevic-Hot Rolling Consultants, 1983. Reprinted with permission.)

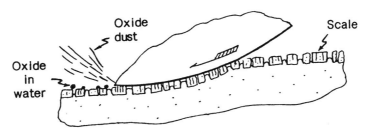

Fig. 27.9 Schematic presentation of the scale behavior during the fourth roughing pass. Adapted from D.T. Blazevic-Hot Rolling Consultants (1983).

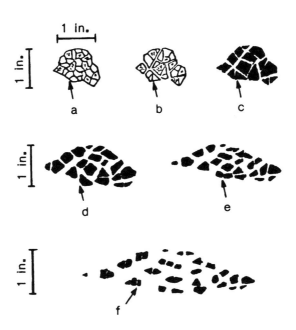

Fig. 27.10 Elongation of a 25 x 25 mm (1 x 1 in.) patch of scale during roughing passes. Adapted from D.T. Blazevic-Hot Rolling Consultants (1983).

During the fourth pass (Fig. 27.9), the scale is totally embedded in the steel. Since the bite angle is reduced, the fracturing of the particles is also reduced. The roll is now sliding across the scale particles resulting in so-called 'sandpapering' the scale off the steel surface.

Elongation of a patch of scale during roughing passes is shown in Fig. 27.10. This particle of scale (a) of about 25 x 25 mm (1 x 1 in.) in size comes out of the first roughing pass being fractured into smaller pieces (b). As the process of fracturing, pinching of the edges, and elongation between the particles continues, the particles take the shapes shown in d, e, and f.

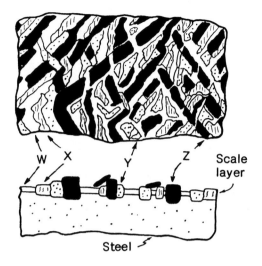

Fig. 27.11 Schematic presentation of scale on the transfer bar prior descaling at the entry of finishing mill. Adapted from D.T. Blazevic-Hot Rolling Consultants (1983).

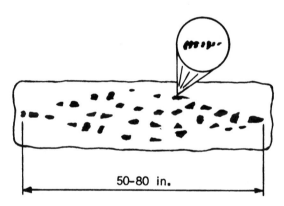

Fig. 27.12 Elongation of a 25 x 25 mm (1 x 1 in.) patch scale after rolling through both roughing and finishing mills. Adapted from D.T. Blazevic-Hot Rolling Consultants (1983).

27.7 SCALING OF STEEL DURING FINISHING PASSES

As the transfer bar approaches the descaling box, located at the entrance of the finishing mill, its surface will be covered with sheets of secondary scale (Fig. 27.11) which vary from 12.7 to 150 mm (0.5 to 6 in.) in length and from 0.08 to 0.40 mm (0.003 to 0.015 in.) in thickness [15].

After descaling at the entry of the finishing mill, the final form of iron oxide, known as **tertiary scale**, starts to appear. During passing through the mill, the iron oxide will elongate while continuing to grow. By the time the strip enters last finishing mill stand its temperature reduces. The bite angle on this stand is usually very small, so little scale fracturing takes place. The rolling action now causes the roll to slide across the oxide and 'sandpaper' the oxide off the strip surface.

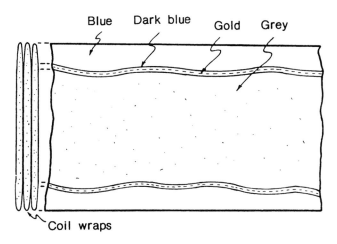

Blue   Dark blue   Gold   Grey

Coil wraps

Fig. 27.13 Schematic presentation of the oxide pattern on the coiled strip. Adapted from D.T. Blazevic-Hot Rolling Consultants (1983).

Figure 27.12 shows the scale pattern after the finishing mill produced by the patch of scale 25 x 25 (1 x 1 in.) in size entering the roughing mill. So the patch of scale is now a series of small jagged 'teardrops' spread in a pattern about 25 mm (1 in.) wide and up to 2030 (80 in.) long.

27.8 SCALING OF STEEL DURING COILING

The oxide growth is slowed during water cooling on the runout table. During coiling, however, the rate of growth increases. The coiled strip shows a so-called 'annealing pattern' [15] as shown in Fig. 27.13. This pattern is due to various cooling rates across the strip width.

The ends and edges of the coil cool off at the fastest rates and are usually covered with blue oxide. This oxide is very dense and is the hardest to pickle. Following the shoulder of the strip cross-sectional profile, there is a narrow band of dark blue and/or gold color. Finally, the center of the strip is varying shades of light grey.

27.9 CLASSIFICATION OF SCALE

Blazevic [16] has identified the following types of scale formed in hot strip mill:

Heavy primary scale	Scale streaks	Contraction gouges
Furnace scale	Plugged nozzle scale	Scratches
Refractory scale	Rebound scale	Gouges
Primary scale	Salt and pepper	Slivers
Secondary scale	Heat pattern	Scabs
Red oxide	Roll wear scale	Pickle line gouges

Table 27.1 Scale color at point of observation; after D.T. Blazevic
- Hot Rolling Consultants [16].

Color coding				
B - Black F - Frosty O - Orange R - Red Y - Yellow BL - Blue G - Grey P - Pits RO - Reddish (Gold or BR - Brown H - Holes PU - Purple Orange Straw) S - Silvery				
Scale descriptions	Hot steel while rolling	Hot rolled surface	Pickled surface	Cold rolled surface	Finished temper rolled surface

Scale descriptions	Hot steel while rolling	Hot rolled surface	Pickled surface	Cold rolled surface	Finished temper rolled surface
Heavy primary	B,R,O	B	B,B&F,F	B,B&F,F,H	B,B&F,F,H
Furnace	B,R,O	B	B,B&F,F	B,B&F,F	B,B&F,F
Refractory	B,BR	BR,R,RO,PU	BR,R,RO,PU	B,BR	B,BR,R,RO
Primary	B,R,O	B	B,B&F,F	B,B&F,F	B,B&F,F,H
Secondary	B,R	B,B&G	B,B&F,F	B,B&F,F	B,B&F,F
Red oxide	-----	R,PU,RO	F,G	-------	----------
Scale streak	B	B,B&G	B,B&F,F	B,B&F,F	B,B&F,F
Plugged nozzle	B	B,B&G	B,B&F,F	B,B&F,F	B,B&F,F
Rebound	B	------	-------	--------	F,F&P
Salt & pepper	---	B	B	-------	--------
Heat pattern	---	B,B&G	B,G	-------	--------
Roll wear	---	B,B&S	B,F	B	B
Scratches	---	B,S,G	B,G,S,B&G	B,B&G	B,F
Gouges	---	B,S	B,B&F,F	B,B&F	B,F&F,F
Contraction gouges	---	B,BL,Y	B,B&G,G	B	B
Slivers	B	B,B&G,G	B,B&G,G	B,B&G,G,S	B,B&G,G,S
Scabs	B	B,P	B,B&P,P	B,P,H	B,H
Pickle line gouges	---	------	B,G,S	B,S	B,S

The scale can be identified by color as well as by shape and patterns. As shown in Table 27.1 the color of the scale or its resulting defect on the strip depends on where the scale is observed. The color varies from black to yellow with many combinations in between.

Table 27.2 identifies the various types of scale by shape and patterns. The probability of occurrence of each type of scale in relation to the location or pattern is shown in Table 27.3. More detailed description of various types of scale is given below according to Blazevic [16].

Table 27.2 Probability of occurance of different types of scale by shape or pattern; after D.T. Blazevic – Hot Rolling Consultants [16].

Probability of occurance H - high M - medium L - low O - none Scale description	Smooth teardrop	Jagged teardrop	Streamlined	Straight line	Random polygon	Free form shape	In a general line	Slightly curving	Curvy line	Repetitious	Very short	1" to 4" long	4" to 12" long	Over 12" long	Arrowhead shapes
Heavy primary	O	H	O	O	M	L	L	O	O	O	O	M	M	L	O
Furnace	H	M	L	O	O	L	L	O	O	O	L	O	M	M	L
Refractory	H	L	M	O	O	M	O	O	L	O	O	O	M	M	L
Primary	L	H	O	O	L	O	L	L	O	O	O	O	H	M	O
Secondary	L	M	L	O	O	O	L	L	O	O	O	H	M	L	O
Red oxide: teardrop	H	O	H	O	O	L	M	O	O	M	H	L	O	O	O
blotchy	O	M	M	O	L	H	O	O	L	M	L	H	L	L	L
scratchy	O	O	O	H	O	O	H	O	O	M	L	M	M	H	O
Scale streak	M	M	L	O	L	O	H	O	O	H	H	M	M	L	O
Plugged nozzle streak	L	H	L	O	L	L	H	O	O	H	M	M	L	L	O
Rebound	M	O	H	O	O	O	O	O	O	L	H	H	O	O	L
Salt & pepper	O	O	O	O	O	O	O	O	O	M	H	H	L	O	M
Heat pattern	O	L	L	O	O	L	O	O	O	M	H	O	O	O	O
Roll wear	O	M	O	O	M	L	M	M	L	H	H	O	O	O	O
Scratches	O	O	H	H	O	O	H	O	O	M	M	M	H	H	O
Gouges	L	M	H	H	O	M	H	M	L	L	H	H	M	L	O
Contraction gouges	M	O	H	M	M	O	M	M	M	H	H	O	O	O	O
Slivers	O	O	M	L	O	O	H	H	M	L	O	L	L	H	L
Scabs	O	O	L	O	L	H	L	L	M	O	O	M	M	L	O

Table 27.3 Probability of occurance of different types of scale by location on the strip; after D.T. Blazevic - Hot Rolling Consultants [16].

Probability of occurance H – high M – medium L – low O – none Scale description	Location or pattern															
	Top and bottom	Mostly top	Mostly bottom	Top only	Bottom only	Top or bottom	Edges – 4" in	Across the strip	Random across strip	Head end only	Heavier – head end	Heavier – tail end	Heavier – middle	By furnace	By coiler	By gage or width
Heavy primary	L	H	O	H	L	L	O	M	M	O	L	M	H	O	O	O
Furnace	L	L	O	L	L	M	L	H	H	O	O	O	M	H	O	O
Refractory	O	H	O	H	O	O	O	H	H	O	O	O	M	H	O	O
Primary	L	H	L	H	L	L	L	H	H	O	O	L	M	L	O	O
Secondary	M	M	L	M	L	H	L	H	H	M	L	M	M	O	O	M
Red oxide: teardrop	L	H	O	H	O	L	L	L	H	L	O	O	M	L	O	L
blotchy	L	H	L	M	L	M	H	M	L	L	L	L	L	L	O	M
scratchy	M	L	M	L	M	M	H	L	M	L	L	L	L	O	O	M
Scale streak	L	L	L	M	M	H	L	O	H	O	O	O	O	O	O	L
Plugged nozzle streak	L	M	L	H	L	M	L	O	H	O	O	O	O	O	O	L
Rebound	O	O	O	H	O	O	L	H	H	O	L	M	H	O	O	L
Salt & pepper	M	L	L	M	M	H	O	H	O	O	L	L	M	O	O	M
Heat pattern	M	L	L	M	M	H	O	H	O	O	L	L	M	O	O	M
Roll wear	M	L	H	M	M	H	L	L	H	O	O	O	O	O	O	L
Scratches	L	O	H	O	H	M	H	M	H	M	M	L	M	O	H	L
Gouges	O	O	H	O	H	L	M	L	H	H	L	L	M	L	H	L
Contraction gouges	H	O	O	O	O	O	O	M	M	O	M	L	H	O	L	H
Slivers	H	O	O	L	L	H	L	L	H	L	L	L	M	O	O	L
Scabs*	H	M	L	L	L	H	L	L	H	O	L	H	L	O	O	L

* For scabs, head end and tail end refers to the top and bottom of the ingot.

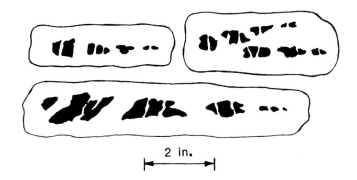

Fig. 27.14 Heavy primary scale. Adapted from D.T. Blazevic-Hot Rolling Consultants (1983).

27.10 HEAVY PRIMARY SCALE

Heavy primary scale is a thick scale, usually 3.2 mm (0.125 in.) thick, that is formed in a soaking pit [16]. It may also be present on the surface of a slab left in a reheating furnace for an extended period of time.

The rolled-in scale appears as a jagged teardrop on the finished hot rolled sheet (Fig. 27.14). Since the particles are embedded in the initial stages of rolling process, the distance between particles on the final sheet can be quite extended. During the rolling operation, the particles can be seen on the top surface of the rolled product as the black elongated teardrops with rounded off corners and smoothed edges. In the pickled sheet, it appears as a black or frosty area which after cold rolling may turn to holes, a rough surface, or a frosty pattern.

The major cause of the resulting rolled-in scale is a failure to break and remove heavy primary scale during initial rolling passes. One of the remedies to the problem is to maintain an excess air practice in the soaking pits so a porous and loose scale is produced.

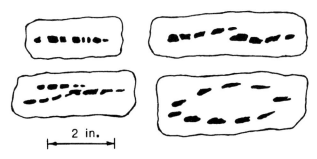

Fig. 27.15 Furnace scale. Adapted from D.T. Blazevic-Hot Rolling Consultants (1983).

27.11 FURNACE SCALE

Furnace scale is a special type of primary scale with a very strong interface bond between the scale and steel [16]. The shape of the rolled-in particles may vary from a slightly jagged to a very smooth teardrop. A cluster of defects may be seen on the finished hot rolled sheet (Fig. 27.15).

The furnace scale is usually a result of overheating the slab, especially in the soak zone. The slab surface temperature becomes so high, that the scale fuses back into the slab. This melting process creates a metallurgical bond at the interface. A furnace reducing atmosphere enhances this process.

The measures for reducing the furnace scale include:

1. Cleaning off all loose scale prior charging the slab into the reheat furnace
2. Maintaining oxidizing atmosphere
3. Firing the bottom zones at a lower temperature than the top zones
4. Setting the burners to produce long soft flames, rather than short hot flames
5. Avoiding overheating the slab surface.

27.12 REFRACTORY SCALE

Refractory scale is a combination of refractory and furnace scale [16]. One of the causes for development of refractory scale is using an improper refractory that has low melting temperature and may drop on the slab. Once the refractory scale gets onto the slab surface, it seals the furnace scale from further oxidation and bonds itself to the steel. The color of the refractory scale on hot rolled surface depends on the type of refractory used and can be red, reddish orange, purple, brown or black (Fig. 27.16).

The measures for reducing the surface defects caused by refractory scale usually involve tighter control on refractory used in the furnace as well as cleaning of the slab surfaces after extensive repairs of the furnace refractory.

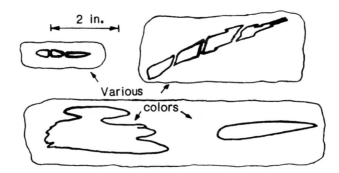

Fig. 27.16 Refractory scale. Adapted from D.T. Blazevic-Hot Rolling Consultants (1983).

27.13 PRIMARY SCALE

Primary scale is similar to heavy primary scale [16]. It is also formed in the reheat furnace and appears on finished hot rolled sheet as a jagged teardrop, or a group of jagged teardrops (Fig. 27.14). The difference is usually in the width of the particles and pattern. The primary scale is more frequently seen on the top surface and away from the edges and coil ends. In the roughing mill it appears as black particles. The scale removed by the descaler may fly back (rebound) on the top surface of the slab and will become rolled into the surface.

Recommendations to reduce the surface defects caused by primary scale include:

1. Establishing the furnace practices to produce a scale that tends to breakup into sheets at the vertical edger and primary descaler.

2. Establishing the width control practice that requires a sufficient edge work to fracture the entire interface bond.

3. Providing a possibility for descaling in front of both the first and second roughing mills (or passes).

27.14 SECONDARY SCALE

Secondary scale is formed after the first descaling of the slab [16]. Its thickness is a function of the chemistry and temperature of the slab, effeciency of previous descaling, and the oxidation time after descaling. Although the individual grain size of the secondary scale is small, they are often grouped.

The surface defects caused by the secondary scale may appear on the finished hot rolled sheet as slightly jagged teardrop (a), wavy pattern (b), smooth teardrop (c), and a streamlined particle, or smooth smeared area (d) as shown in Fig. 27.17. Because of the small grain size the patterns of secondary scale are much more narrow than those of primary scale. Reduction of the surface defects relating to the secondary scale is usually achieved by a proper design and operation of the secondary descaling system.

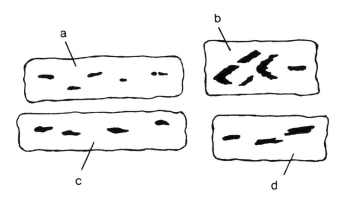

Fig. 27.17 Secondary scale. Adapted from D.T. Blazevic-Hot Rolling Consultants (1983).

27.15   RED OXIDE SCALE

Red oxide scale is a term that identifies various types of scale which are related to iron oxide FeO. The color of the red oxide scale is mostly rusty red, but the color pattern may vary from dusty red to grey black and is probably temperature related [17]. If the red oxide scale is originated in the roughing mill where the temperature is higher, the scale will have a purple or darker color. The red oxide scale originated in the finishing mill has reddish orange or rust red color.

The chemical composition of the rolled product may also affect the color of the red scale. Silicon is found to be a contributing factor to a formation of red oxide. Next element often mentioned is columbium. However, the chemical composition is not considered as the basic cause of red oxide.

Formation of red oxide is originated in the furnace. Under certain conditions, a so-called **sticky scale** is produced. After primary descaling, the scale is getting broken at the intermediate layer between wustite (FeO) and magnetite (Fe_3O_4), thus leaving the former on the slab surface in the form of red oxide. The bounding between red oxide and slab is so strong that the red oxide survives many more descaling operations.

27.16   TEARDROP RED OXIDE SCALE

The teardrop red oxide scale appears on the finished hot rolled sheet as shown in Fig. 27.18. The teardrop shapes are very symmetrical, and the edges of the surface defects are very smooth. The color is usually rust red. The defect can be seen in different parts of the sheet surface but occurs primarily on the top surface.

The teardrop red oxide scale is found to be originated by droplets of water splashing on the rolled sheet surface between finishing mill stands, especially in the F1, F2 and F3 mill areas. The surface defects due to the teardrop red oxide are eliminated by drying up this mill areas. Also, the problem can be alleviated by moving the secondary descaling box closer to F1 mill stand, thus reducing the oxidation time prior to the finishing rolling.

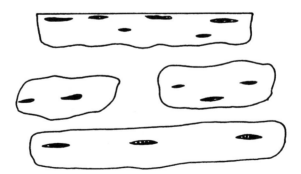

Fig. 27.18 Teardrop red oxide scale. Adapted from D.T. Blazevic-Hot Rolling Consultants (1983).

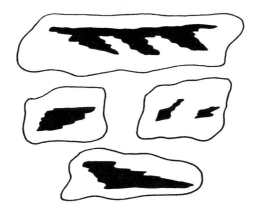

Fig. 27.19 Primary blotchy red oxide. Adapted from D.T. Blazevic-Hot Rolling Consultants (1983).

27.17 BLOTCHY RED OXIDE SCALE

The blotchy red oxide scale is divided into primary and secondary red oxide scale [17]. Primary red oxide is originated in the roughing mills and appears as shown in Fig. 27.19. The pattern often includes geometric corners, jagged straight lines, indentations on the edges within the general free-form or blotchy appearance of the top surface.

The secondary blotchy oxide scale is formed after the secondary descaling. The scale is presumably caused by chilling of the oxide which at this point is mostly iron oxide FeO. The chilling of the oxide may freeze further chemical action. The surface defect appears as a free-form or blotchy pattern. The sides of the pattern are very smooth as shown in Fig. 27.20. The defect occurs mainly on the top surface of the rolled sheet.

27.18 LINE TYPE RED OXIDE SCALE

Line type red oxide scale usually appears on the finished hot rolled sheet as very precise stream-line patterns [17] as shown in Fig. 27.21. It may look like scratches (a) which may be either very fine lines within 6 to 18 mm (0.25 to 0.75 in.) of the edge or much heavier lines extending in further from the

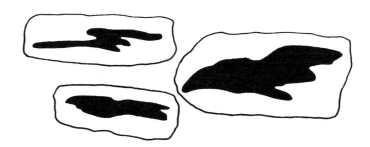

Fig. 27.20 Secondary blotchy red oxide scale. Adapted from D.T. Blazevic-Hot Rolling Consultants (1983).

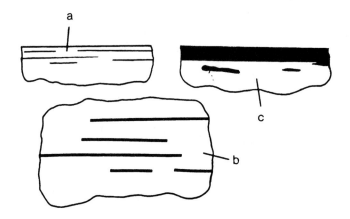

Fig. 27.21 Line type red oxide. Adapted from D.T. Blazevic-Hot Rolling Consultants (1983).

edges. The rust red color can be seen in the scratches. The scratch pattern may also be located in the middle of the strip (b). Another variation of the line type red oxide scale may look like the red oxide painted along the edges (c).

The scratch pattern in the middle of the strip is mostly seen on the top surface while the other types can be seen on both top and bottom surfaces.

27.19 TAIL END RED OXIDE SCALE

Tail end red oxide scale occurs on the tail end of the coil [17]. It appears as a band of red oxide 50 to 125 mm (2 to 5 in.) wide that follows the contour of the tail end. This surface defect is not seen on coils which have the fishtail end but it often occurs on the strip with the tail end curved outwardly as illistrated in Fig. 27.22.

27.20 SCALE STREAKS AND PLUGGED NOZZLE SCALE

Scale streaks is a rolled-in primary, secondary, or even red oxide scale that occurs in a streaky pattern in the rolling direction as shown in Fig. 27.23. The furnace defects related to the scale streaks are

Fig. 27.22 Tail end red oxide scale. Adapted from D.T. Blazevic-Hot Rolling Consultants (1983).

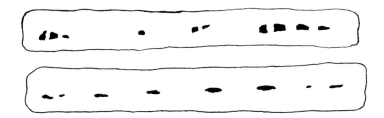

Fig. 27.23 Scale streaks. Adapted from D.T. Blazevic-Hot Rolling Consultants (1983).

Fig. 27.24 Plugged nozzle scale. Adapted from D.T. Blazevic-Hot Rolling Consultants (1983).

usually caused by worn out or leaking descaling nozzles or insufficient overlap between the sprays [16]. The scale streaks can also be caused by a strip bouncing under the sprays.

Plugged nozzle scale (Fig. 27.24) is a more serious type of scale streaks which is caused by a plugged nozzle, or nozzle strainer. This also may be caused by oversized strainer, or by operating without strainers.

27.21 REBOUND SCALE

Rebound scale is a special type of secondary scale. After secondary descaling, the scale is broken into very small particles which will fly toward the F1 mill and land in the water or on the top of the strip [16]. These particles will be then sucked into the mill and rolled into the strip. The surface defect appears on the finished hot rolled sheet as a short streamlined frosty area with a pit in the middle (Fig. 27.25).

The problem is solved by the following measures:

1. Elevating the pass line at the F1 mill stand above the pass line at the descaler, so the water will flow away from the mill.

2. Enclosing the descaling hood.

3. Protecting the entry of the mill from bouncing descaling water by installing soft curtains, damming rolls, etc.

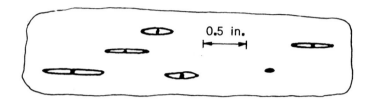

Fig. 27.25 Rebound scale. Adapted from D.T. Blazevic-Hot Rolling Consultants (1983).

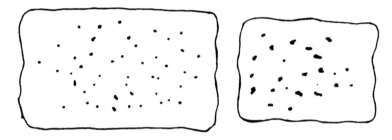

Fig. 27.26 Salt and pepper scale. Adapted from D.T. Blazevic-Hot Rolling Consultants (1983).

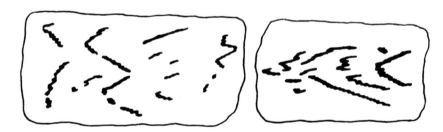

Fig. 27.27 Heat pattern scale. Adapted from D.T. Blazevic-Hot Rolling Consultants (1983).

27.22 SALT AND PEPPER, HEAT PATTERN, AND ROLL WEAR SCALE

Salt and pepper, heat pattern, and roll wear scales are all related, to a certain degree, to the roll wear. Salt and pepper scale owes its name to its appearance on the finished hot rolled sheet (Fig. 27.26). It is believed to come from fracturing the finishing mill scale by worn-out work rolls of the F2 mill stand [16]. The fractured particles are then carried by the rolls and embedded back into the strip on the next revolution.

Heat pattern scale (Fig. 27.27) is caused by the same phenomena as the salt and pepper scale. It usually occurs on heavier thickness sheets. Since these sheets are rolled at higher rolling temperatures

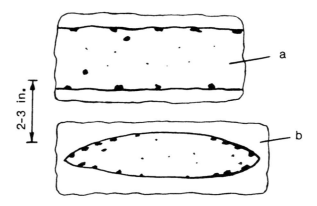

Fig. 27.28 Roll wear scale. Adapted from D.T. Blazevic-Hot Rolling Consultants (1983).

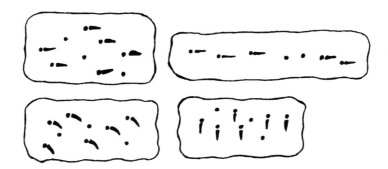

Fig. 27.29 Contraction gouges. Adapted from D.T. Blazevic-Hot Rolling Consultants (1983).

and at slower speeds, the rate of growth of the oxide prior to the entry of the F2 mill increases.

Roll wear scale is related to pealing and banding of the work rolls. Pealing is a result of the breakdown of the roll surface. The broken rough surface of the roll tears the oxide layer off the strip surface and the oxide layer bands around the roll, and embeds back into the strip surface on the next revolution [16, 18]. The heaviest scale particles are usually located near the periphery of the pealing area (Fig. 27.28). The salt and pepper scale can be seen inside the pattern boundary. When a roll is completely pealed in a band (a) around its circumference, the surface begins to polish itself, and the scale may disappear.

The rolled-in scale problem related to the roll wear is minimized by:

1. Improving the roll cooling

2. Introducing the oil lubrication

3. Introducing the width comedown schedule

4. Introducing the roll side-shifting.

27.23 CONTRACTION GOUGES

The contraction gouges are often mistaken for rolled-in scale, because of a presence of small particles of scale in this defect after a sheet is being pickled [16]. The pattern of this defect is always a pit with a comet tail, and small black dots as shown in Fig. 27.29.

The comet tail can be parallel or perpendicular to rolling direction, or have a curved shape. The color of the pits before pickling are usually gold (straw), whereas the comets are dark grey. The small dots are thin layers of oxide and are black in color.

The defect usually occurs during coiling a sheet at high temperature and high tension. This causes the oxide to weld together on the high spots between the coil layers where the pressure is the greatest. The comet effect is created by the spot of oxide pulling free, and scraping along the surface of the opposite wrap. The problem is alleviated by lowering both the coiling temperature and coiling tension.

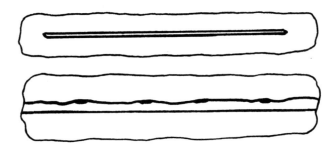

Fig. 27.30 Scratches. Adapted from D.T. Blazevic-Hot Rolling Consultants (1983).

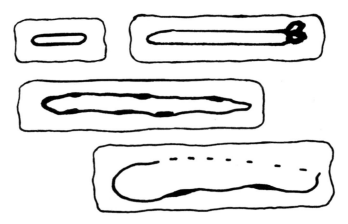

Fig. 27.31 Gouges. Adapted from D.T. Blazevic-Hot Rolling Consultants (1983).

27.24 SCRATCHES AND GOUGES

Scratches (Fig. 27.30) and gouges (Fig. 27.31) are caused by pulling a workpiece across a slower moving roll, a frozen roll, plate, or apron in the mill [16]. The slower moving rolls have usually become firecracked, and the rough surface will cut easier the surface of the rolled piece.

The secondary oxide usually forms in the scratches and near the edges of the gouges. Because of that the gouges are often confused with rolled-in scale. The defects of this nature may be prevented by:

1. Checking the pass line on all mills to eliminate the potential spots for scratching

2. Eliminating the metal stringers, or scablike pieces near the vertical edges

3. Checking the speed matching of the table rolls with the mill

4. Watercooling of all aprons, spacers, plates, etc., to prevent firecracking, and providing a lubrication as the strip passes over.

27.25 SLIVERS AND SCABS

Slivers and scabs are often confused with rolled-in scale. In case of slivers, the confusion comes in two ways: (1) the black edges and surface openings often look like scale, and (2) the slivers are not seen prior to the hot rolling or pickling [16].

The slivers are mainly caused by subsurface defects, or impurities in the steel. During rolling the impurities will be elongated in the rolling direction. This gives them a long stringy appearance. Since the impurities do not elongate the same amount as steel, the surface area often breaks open.

The scabs are originated during casting the ingots and appear as irregular bulges on the ingot skin. These bulges are caused by either the depressions in the mold or by the mold metal adhering to the ingot skin. After being elongated in the mill, the long scabs may appear like slivers. Both surface defects caused by slivers and scabs are usually prevented by a hot scarfing and spot conditioning of the slabs prior to rolling.

REFERENCES

1. ASM Metals Reference Book, American Society for Metals, Metals Park, Ohio, p. 62 (1981).

2. The Making, Shaping and Treating of Steel, 10th Edition, eds. W.T. Lankford, Jr., et al, Association of Iron and Steel Engineers, Pittsburgh, Pennsylvania, pp. 783-860 (1985).

3. A.J. Fisher, "Furnace Atmosphere", AISE Yearly Proceedings, 1937, pp. 131-139.

4. K. Sachs and C.W. Tuck, "Surface Oxidation of Steel in Industrial Furnaces, Reheating for Hot Working", Iron and Steel Institute, Publication No. 111, London, 1968, pp. 1-17.

5. W.L. Roberts, Hot Rolling of Steel, Marcel Dekker, Inc., New York, p. 644 (1983).

6. E.A. Cook and K.E. Rasmussen, "Scale-Free Heating of Slabs and Billets", AISE Yearly Proceedings, 1970, pp. 175-181.

7. V.B. Ginzburg, W.F. Schmiedberg, "Heat Conservation Between Roughing and Finishing Trains of Hot Strip Mills", Iron and Steel Engineer, Apr. 1986, pp. 29-39.

8. N. Birks, "High-Temperature Corrosion in Complex Atmospheres", Chemical Metallurgy of Iron and Steel, Iron and Steel Institute Publication No. 146, London, 1973, pp. 402-411.

9. R.A. Lula, Stainless Steel, American Society for Metals, Metals Park, Ohio, pp. 129,130 (1986).

10. E. Schurman, et al, "On the Scaling of Unalloyed Steel", Archiv fur das Eisenhuttenwesen, Vol. 44, No. 12, December 1973, pp. 927-934.

11. K. Haffe, Oxidation of Metals, Plenum Press, New York (1965).

12. G.C. Wood, "Fundamental Factors Determining the Mode of Scaling of Heat-Resistant Alloys", Werkstoffe und Korrosion, Vol. 22 No.6, June 1971, pp. 491-503.

13. W.E. Boggs, "The Role of Structural and Compositional Factors in the Oxidation of Iron and Iron-Based Alloys, High Temperature Gas-Metal Reactions in Mixed Environments", eds. S.A. Jansson and Z.A. Foroulis, Metallurgical Society of AIME, New York, 1973, pp. 84-128.

14. N. Birks and W. Jackson, "A Quantitative Treatment of Simultaneously Scaling and Decarburization of Steels", Journal of Iron and Steel Institute, Vol. 208, January 1970, pp. 81-85.

15. D.T. Blazevic, "Rolled in Scale: The Continual Problem, Part 1 - Scale Formation and Rolling Characteristics", Hot Rolling Consultants, Ltd., Homewood, Ill., Feb. 1983.

16. D.T. Blazevic, "Rolled in Scale: The Continual Problem, Part II - Rolled in Scale Descriptions, Causes, and Cures", Hot Rolling Consultants, Ltd., Homewood, Ill., Feb. 1983.

17. D.T. Blazevic, "Rolled in Scale: The Continual Problem, Part IV - Red Oxide Scale", Hot Rolling Consultants, Ltd., Olympia Fields, Ill., July 1985.

18. W.H. Betts, "Basic Concepts of Roll Surface Behavior in Stands F1, F2, and F3", AISE Yearbook, 1977, pp. 12-18.

Part IX

Rolling Mills for Flat Products

28

Classification of Rolling Mills

28.1 MAIN COMPONENTS OF A MILL STAND

A typical mill for rolling of flat products includes one or a number of the mill stands which are arranged, usually in line, to produce a sequential reduction in thickness and width of the rolled product.

Although a variety of the mill stand designs are known, it is possible to identify their common components from functional viewpoint.

Figures 28.1 and 28.2 illustrate schematically a mill stand designed for reducing the workpiece thickness. Main functional components of this mill stand are:

1. **Work rolls** between which the rolled product is being squeezed.

2. **Backup rolls** which support the work rolls to reduce their deflection under rolling load.

3. **Roll gap adjustment mechanisms** provide setting of required gap between work rolls and may also allow one to adjust elevation of the pass line.

4. **Housing** is designed to contain the mill stand components and to withstand the rolling load.

5. **Main drive train** provides rotation of the rolls with desired speed and rolling torque.

Generally, the same main components can be found in a typical mill stand designed for reducing the workpiece width as shown in Fig. 28.3.

28.2 CLASSIFICATION OF MILL STANDS

Mill stands for rolling flat products can be classified by the following categories [1-3].

1. Roll arrangement - Depending on the roll arrangement in the mill housing, the mill stands are referred to as

a) Two-high mill stand. Two-high mill stand contains two work rolls (Fig. 28.4a).

b) Three-high mill stand. In three-high mill stands (Fig. 28.4b), the top and bottom rolls revolve in the same direction and the middle roll in the opposite direction. After completing the bottom pass (solid line), the workpiece is lifted to provide the reversing top pass (dotted line).

c) Four-high mill stand. This is the most common roll arrangement that includes two work rolls and two backup rolls (Fig. 28.4c).

511

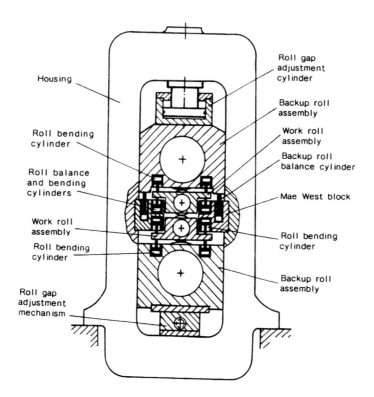

Fig. 28.1 Schematic illustration of a horizontal mill stand (side view).

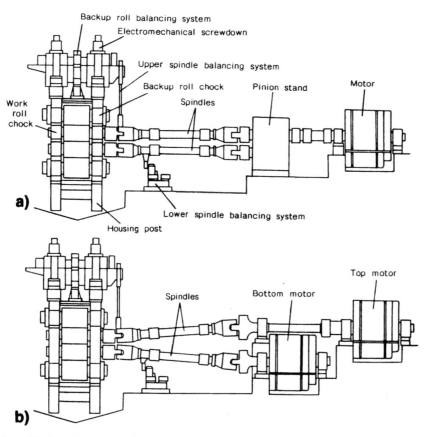

Fig. 28.2 Schematic illustration of two horizontal mill stands (front view) with: a) pinion stand drive and b) independent drive.

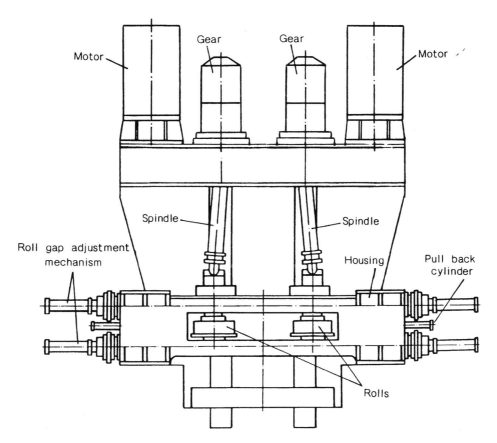

Fig. 28.3 Vertical mill stand (front view). Adapted from IHI Publication 8403-1000.

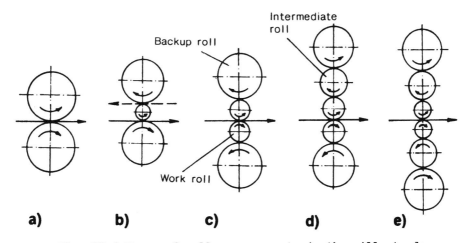

Fig. 28.4 Types of roll arrangements in the mill stands.

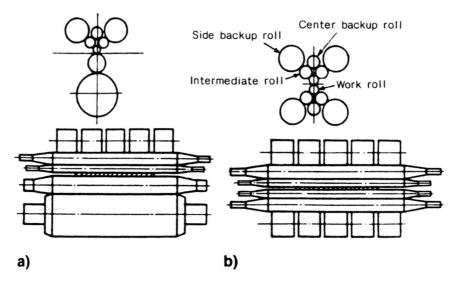

Fig. 28.5 Cluster type mill stand: a) asymmetrical, b) symmetrical. (From Kawamani, et al, Mitsubishi Heavy Industries Technical Review, June 1985. Copyright Mitsubishi Heavy Industries, Ltd. Reprinted with permission.)

d) Five-high mill stand. In the five-high mill stand arrangement, an intermediate roll is added between one of the work rolls and a backup roll (Fig. 28.4d).

e) Six-high mill stand. In addition to two work rolls and two backup rolls, the six-high mill stand has two intermediate rolls (Fig. 28.4e).

f) Cluster type mill stand. Main feature of the cluster type mill stand is that each work roll is

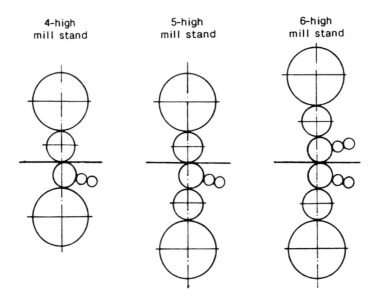

Fig. 28.6 Types of mill stands with offset rolls.

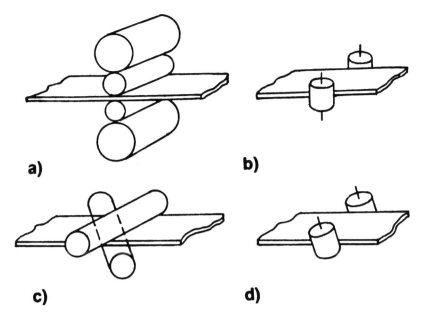

Fig. 28.7 Mill stands with different directions of the roll axes.

surrounded by more than one intermediate roll which in their turn are supported by a number of backup rolls. Both asymmetrical (Fig. 28.5a) and symmetrical (Fig. 28.5b) cluster type rolling mill stands have been developed.

g) Mill stand with off-set rolls. In this mill stand arrangement, at least one of the work rolls is displaced to one side from the vertical centerline of the backup rolls. The displaced roll is laterally supported by intermediate rolls (Fig. 28.6).

2. Direction of roll axes - The mill stands can be identified in association with direction of the roll axes as following:

a) Horizontal mill stand. In the horizontal mill stand all rolls are parallel to the mill floor (Fig. 28.7a).

b) Vertical mill stand. The roll axes of vertical mill stands are perpendicular to the mill floor (Fig. 28.7b).

c) Crossed-roll mill stand. In the crossed-roll mill, the axes of the rolls are tilted in opposite directions in respect to rolling direction (Fig. 28.7c).

d) Mill stand with parallel-tilted rolls. In this type of mill stand, the rolls are tilted at the same angle in respect to rolling direction (Fig. 28.7d).

3. Direction of rolling - Three types of the mill stands are usually considered with regard to rolling direction:

a) Non-reversing mill stand. This mill stand is designed to roll in one direction.

b) Reversing mill stand. This mill stand provides rolling in both directions.

c) Back-pass mill stand. In the back-pass mill stand, the rolls are always rotating in the same direction. After completing the rolling pass, the roll gap is opened and the workpiece is back-passed toward the entry side of the mill. Then the roll gap closes and the next rolling pass proceeds.

4. Main motor type - There are two principal types of main drive motors which are used in rolling mill stands:

a) A-C motors. The alternative current (A-C) motors can be of three modifications: synchronous, squirrel cage and wound-rotor induction motors. They are generally used to provide rotation of the rolls in the same direction at practically constant speed.

Variable speed control have been employed for the wound-rotor induction motors using either Kraemer or Scherbius systems [4]. Another system based on variable frequency speed control has been recently developed for application in rolling mills [5].

b) D-C motors. The direct current (D-C) motors provide reverse rotation of the rolls with wide range of the roll speed control.

5. Drive train arrangement - The principal drive train arrangements can be classified as follows:

a) Direct drive. The direct drive train provides connection of the motor with driven rolls without any change in angular speed.

b) Gear drive. In the gear drive, the angular speed of the motor is either reduced or increased by installing a gear box in the drive train.

c) Pinion stand drive. The drive train with a pinion stand (Fig. 28.2a) allows one to drive both top and bottom rolls from a single motor.

d) Independent drive. In the independent drive train (Fig. 28.2b), top and bottom rolls are driven by independent motors.

The drive train arrangements can also be identified in relation to what type of roll is driven:

a) Train with driven work rolls

b) Train with driven backup rolls

c) Train with driven intermediate roll.

6. Special design mill stands - A number of mill stands of unconventional design have been developed. They include:

a) Planetary mills

b) Rolling-Drawing mills

c) Contact-Bend-Stretch mills

d) Reciprocating mills, etc.

A brief description of these mills will be given in the next chapter.

28.3 GENERAL CLASSIFICATION OF ROLLING MILLS

In general terms, rolling mills can be classified in respect to rolling temperature, type of rolled product, and type of the mill stand arrangement.

1. Rolling temperature - In regard to rolling temperature, the rolling mills are usually identified as:

a) Hot rolling mills. Hot rolling process of steels generally begins when the workpiece temperature is equal or less than 1315°C (2400°F) and completes at the temperatures which are either above or slightly lower than A_3 critical temperature which, for low-carbon steel, is about 900°C (1650°F). Thus, the bulk of the rolling process occurs when the rolled material is in austenitic phase.

b) Cold rolling mills. Cold rolling process is usually conducted with a workpiece that has initial temperature equal to a room temperature. During cold rolling the rolled material temperature may rise to between 50 and 65°C (122 and 150°F).

c) Warm rolling mills. Warm rolling process is generally conducted at elevated temperatures substantially lower than A_1 critical temperature which, for low-carbon steel, is approximately 730°C (1346°F). The desired rolling temperatures are obtained by either preheating of the workpiece or by a controlled lubrication utilizing the heat generated during the rolling process.

2. Rolled product - In relation to a type of rolled product, the rolling mills are identified as:

a) Slabbing mills. The slabbing mills roll ingots into slabs which usually vary in thickness from 150 to 300 mm (6 to 12 in.). The mill stand may have a provision for a large opening of the roll gap ('high lift'), so the width reduction can be made by rolling the slab on its edge.

b) Plate mills. The hot plate mills roll slabs into plates. The rolled product can be either in a flat form or in a coil form with a consequent uncoiling and cutting the product into the plates of desired lengths. The most distinguished characteristic of both hot and cold plate mills is the width of the rolled plates which for some mills may be as wide as 5334 mm (210 in.).

c) Strip mills. Conventional hot strip mills reduce a slab to a strip with thickness as thin as 1.2 mm (0.047 in.). Cold strip mill further reduce the strip to a desired final gauge. The width of the steel coils are usually between 600 and 2000 mm (23 and 79 in.).

3. Mill stand arrangement -Depending on the distance between the two adjacent mill stands in relation to the length of the rolled product, the mill stand arrangements are usually referred to as [6]:

a) Open mill stand arrangement. In open mill stand arrangement, the distance between two adjacent stands is always greater then the length of the rolled bar exiting the upstream mill stand. This allows rolling on these stand with independent speeds.

b) Close-coupled mill stand arrangement. In the close-coupled mill stand arrangement, the distance between adjacent stands is less than the length of the rolled bar exiting the upstream mill stand. So the bar is rolled simultaneously on adjacent stands. This requires synchronization of speed of these mill stands to provide a constant mass flow of the metal.

The close-coupled mill stand arrangements are used in two types of well-known rolling mills.

i) Universal rolling mill. The universal rolling mill includes a horizontal mill stand and, at least, one vertical mill stand. In this mill both edging and flat rolling may be conducted simultaneously.

ii) Tandem rolling mill. In tandem rolling mills, two or more mill stands of the same type (usually horizontal) have a close-coupled arrangement with each other.

28.4 COMPONENTS OF HIGH-PRODUCTION HOT STRIP MILLS

The term 'high-production hot strip mill' is conventionally applied to the mills with yearly production rate equal to or greater than 2,000,000 metric tons. The main components of these mills (Fig. 28.8) are briefly discussed below [7].

1. Reheat furnaces - In the reheat furnaces, cold or warm slabs are heated to a desired

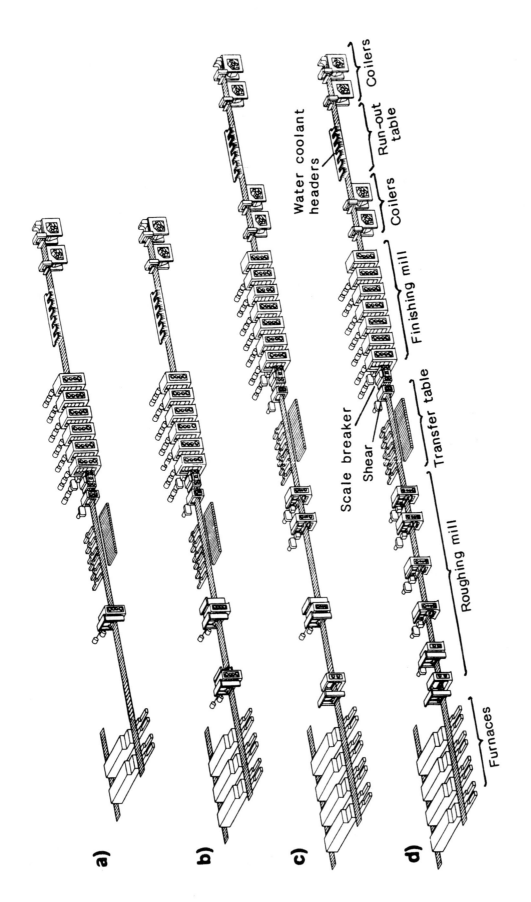

Fig. 28.8 Principal layouts of hot strip mills:
a) semi-continuous, b) with twin reversing roughing train,
c) three-quarter continuous, d) fully continuous.
Adapted from Mitsubishi Brochure HD20-04662.

518

temperature for rolling.

2. **Roughing mill** - The roughing mill, or train, consists of a series of horizontal and vertical mill stands. The main purpose of the roughing mill is to roll a slab into a transfer bar with thickness which is commonly between 19 and 45 mm (0.75 and 1.75 in.).

3. **Transfer table** - The transfer table, or delay table, is located between roughing and finishing mill. The table is usually long enough to accommodate a full length of the transfer bar. This allows one to roll at least two bars independently, one on the roughing mill and another on the finishing mill.

4. **Shear** - The shear is located in front of finishing mill. It is usually designed to cut both the head and tail ends of the transfer bar prior to their entry into the finishing mill.

5. **Finishing mill** - The finishing mill, or train, consists of one or a series of horizontal mill stands. In tandem finishing train, there are **loopers** installed between stands. Each looper maintains a desired interstand strip tension by pushing a free rotating roller against the strip.

6. **Runout table** - The runout table is located between the finishing mill and coilers. A series of water cooling headers are installed above and under runout table. The water coolant system is designed to reduce the strip or plate temperature before the rolled material enters a coiler.

7. **Coilers** - The coilers are usually located at the end of the runout table. In some cases, when cooling of the strip is not required, the coilers may be installed right after the finishing mill.

8. **Descaling system** - Removing the scale from the surfaces of the rolled piece is provided by using a series of high-pressure water spray headers installed at different location of hot strip mills. In some hot strip mills, vertical and horizontal mill stands, known as **scale breakers**, are added to improve efficiency of descaling process.

9. **Roll coolant system** - In the mill stands, the rolls are cooled with the water spray headers located in close vicinity with the rolls. In some finishing mill stands, the roll coolant system is supplemented with the roll lubrication system.

10. **Interstand cooling system** - The interstand cooling system is installed in some high-speed finishing trains to reduce the strip temperature. The strip is cooled by the water sprays located between the mill stands.

28.5 CLASSIFICATION OF HIGH-PRODUCTION HOT STRIP MILLS

Finishing trains of the high-production hot strip mills have customarily from four to seven horizontal mill stands arranged in tandem. The principal difference between hot strip mills is mainly in layouts of the roughing mills. This difference is often used as a base for their classification into four distinct types [6, 7].

1. **Semi-continuous hot strip mill**-The roughing mill of the semi-continuous hot strip mill has one vertical and one horizontal reversing mill stand which are usually combined into one universal roughing mill stand (Fig. 28.8a). These mills may also include both vertical and horizontal scale breakers installed upstream in relation to the reversing rougher.

2. Hot strip mill with **twin reversing roughing train** that contains two universal roughing mill stands (Fig. 28.8b).

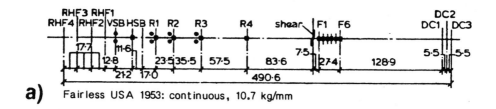

a)   Fairless USA 1953: continuous, 10.7 kg/mm

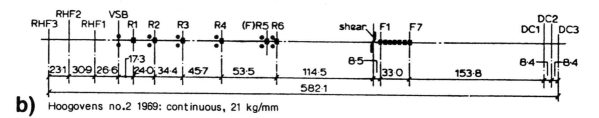

b)   Hoogovens no.2 1969: continuous, 21 kg/mm

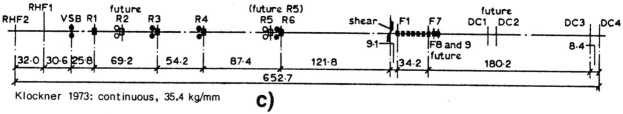

Klockner 1973: continuous, 35.4 kg/mm   **c)**

Fig. 28.9 Evolution in development of high-production hot strip
mills: a) Generation I, b) Generation II, and c) Generation III.
(From Keefe, et al, Ironmaking and Steelmaking, No. 4, 1979.
Copyright Institute of Metals. Reprinted with permission.)

   3. **Three-quarter continuous hot strip mill**—The roughing train of the three-quarter continuous hot
strip mill has one or more horizontal single-pass stands after reversing roughing mill, in open or close-
coupled configuration (Fig. 28.8c).

   4. **Fully continuous hot strip mill**—Fully continuous hot strip mill has four or more horizontal
roughing mill stands, the last two stands being either open or close-coupled.

   Evolution in development of high-production hot strip mills is illustrated in Fig. 28.9. The mills of
different generations can be recognized not only by their layouts but also by their production rate and
the weight of the rolled coils.

Generation I: 1927–1960

Yearly production rate	1.0 to 3.0 Mt
Coil weight per unit width	4.0 to 11.0 kg/mm (224 to 615 PIW)

Generation II: 1961–1969

Yearly production rate	over 3.0 Mt
Coil weight per unit width	18.0 to 22.0 kg/mm (1000 to 1230 PIW)

Generation III: 1970–1978

Yearly production rate	over 5.0 Mt
Coil weight per unit width	up to 36.0 kg/mm (2012 PIW).

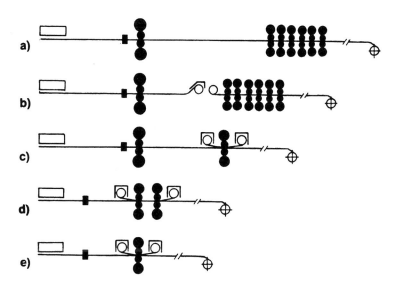

Fig. 28.10 Comparison of layouts of compact hot strip mills (b-e) with a layout of semi-continuous hot strip mill (a).

28.6 COMPACT HOT STRIP MILLS

By definition, the compact hot strip mills are shorter and comprise lesser number of the mill stands. Figure 28.10 gives a diagrammatic comparison of the layouts of some compact hot strip mills with a layout of semi-continuous hot strip mill (Fig. 28.10a).

Coilbox arrangement - The length of the transfer table between roughing and finishing mills can be reduced by installation of a Coilbox [8] in front of the finishing train (Fig. 28.10b). After completing the last roughing pass the transfer bar is coiled in the Coilbox; then the coil is transferred downstream and uncoiled prior to entering the finishing train.

Reversing finishing mill arrangement - This type of mill comprises a reversing roughing mill and a reversing finishing mill known as a **Steckel mill**. In this mill, to conserve heat during intermediate finishing passes, the rolled strip is coiled on the preheated drums adjacent to the mill stand. The length of the hot strip mill with the Steckel mill can be further reduced by using the close-coupled arrangement between roughing and finishing mills as shown in Fig. 28.10c.

Twin reversing hot strip mill - In this arrangement [9], the distance between reversing roughing and finishing mill stands is very short and the coiling furnaces are located as shown in Fig. 28.10d. Roughing passes are intended to be rolled without coiling. Since the mill stands are close-coupled, the rolling passes can be made simultaneously on both stands.

Single-stand reversing hot strip mill - This mill is designed to roll both roughing and finishing passes using the same mill stand (Fig. 28.10e). This is the simplest hot strip mill arrangement which is used for rolling both strip and plate in coil form [10].

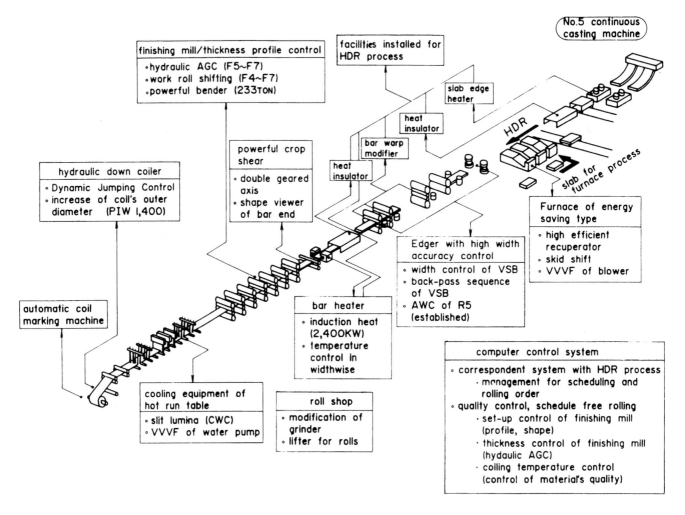

Fig. 28.11 Nippon Kokan Hot Direct charging Rolling (HDR) combined
with continuous casting. (From Emoto, et al, Proceedings of London
Conference on Restructuring Steel Plants for the Nineties, May 1986.
Copyright Institute of Metals. Reprinted with permission.)

28.7 INTEGRATED CONTINUOUS CASTING AND HOT ROLLING PROCESS

Integration of continuous casting process with hot rolling process allows one to eliminate reheating of
slabs and thus to conserve energy, improve quality and yield.

   Integration with thick continuous caster - Figure 28.11 illustrates schematically Hot Direct charging
Rolling (HDR) combined with continuous casting machine developed by Nippon Kokan K.K. [11]. To
achieve desired temperature of the bar prior to its entry into roughing mill, the following equipment is
added:

   a) induction type edge heater that provides uniform bar temperature

   b) heat insulator that reduces heat losses of the bar during its transfer toward hot strip mill

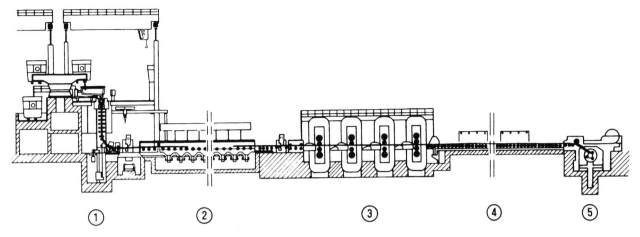

Fig. 28.12 SMS Schloemann-Siemag AG conception of combination of thin caster with hot strip mill: 1 – casting machine, 2 – tunnel furnace, 3 – finishing mill, 4 – run-out table, and 5 – coiler. (From Iron and Steel Engineer, May 1987. Copyright AISE, Pittsburgh, Pennsylvania. Reprinted with permission.)

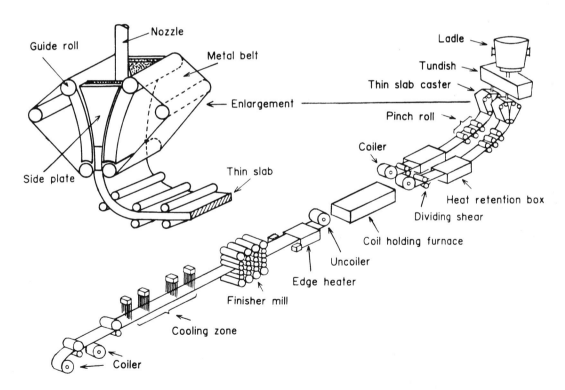

Fig. 28.13 Kawasaki Steel conception of combination of thin caster with hot strip mill. (From Emoto, et al, Proceedings of London Conference on Restructuring Steel Plants for the Nineties, May 1986. Copyright Institute of Metals. Reprinted with permission.)

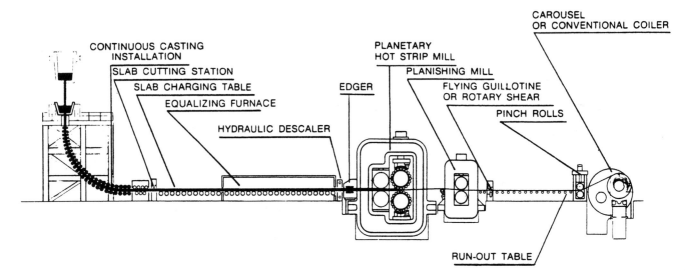

Fig. 28.14 Integration of a planetary mill with thin continuous caster. (From Sendzimir, Iron and Steel Engineer, Oct. 1986. Copyright AISE, Pittsburgh, Pennsylvania. Reprinted with permission.)

c) bar warp modifier that measures the height of the bar warp and adjust the gap between the inductor coils and bar to reduce its warping.

An additional heat insulator is installed on transfer table between roughing and finishing mill.

Integration with thin continuous caster - When thin continuous caster is used, both the slab reheating furnace and the roughing mill are no longer needed for hot strip mill operation.

In the concept developed by SMS Schloemann-Siemag AG, the continuously cast thin slab is fed directly into four-stand finishing mill (Fig. 28.12). Prior to entering the mill, the slab is cut with a shear and then is passed through a reheating tunnel furnace.

Main parameters of the facilities are [12]:

Concast slab dimensions:

thickness	40-50 mm (1.6-2.0 in.)
width	1200-1600 mm (47- 63 in.)
Casting rate	5.0-6.0 m/min (16-20 ft/min)
Rolled strip thickness	1.5-25 mm (0.059-1.0 in.).

In the concept proposed by Kawasaki Steel [13], two twin-belt thin casters are used to feed one finishing hot strip mill with four six-high mill stands (Fig. 28.13). After reheating, the concast slabs are coiled and are then held in a coil holding furnace. When finishing mill is ready to roll, the uncoiled thin slab passes through an edge heater before entering the mill.

The thin continuous caster has also been proposed to be integrated with a planetary mill [14] and with a Steckel mill [15]. Major components of one of the experimental installations of the continuous thin caster is shown in Fig. 28.14. After passing through the equalizing furnace, the continuously cast thin slab is rolled at the planetary mill. The thin strip is then rolled at the planishing mill to smooth the strip surface.

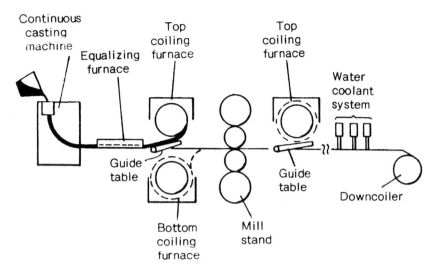

Fig. 28.15 Integration of a Steckel Mill with thin continuous caster. Adapted from Ginzburg and Tippins (1986).

In case of the Steckel mill (Fig. 28.15), the continuously cast thin slab is coiled in one of the two vertically aligned coilers in either side of the pass line. The slab is then payed off into a rolling mill while a subsequent slab is coiled in the other of the two coiler furnaces.

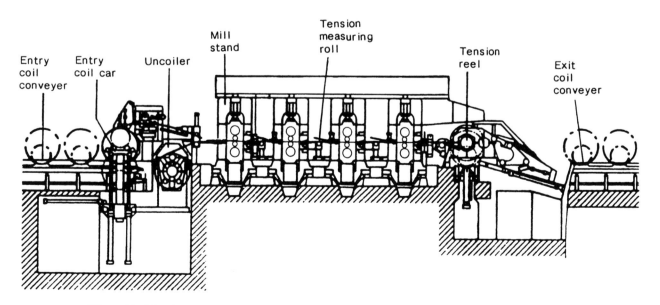

Fig. 28.16 Diagrammatic sketch of a typical tandem cold mill arrangement.

28.8 COLD MILL ARRANGEMENTS

The most common cold mill arrangements are:

 1. **Single-stand cold mill** - Two types of the single-stand cold mills are known:

 a) **high-reduction cold mills** which are capable of taking up to 50% reduction per one pass,

 b) **skin-pass, or temper mills,** which are designed for reductions ranging from 0.5 to 4.0%.

Single-stand mills can be either reversing or non-reversing.

 2. **Twin-stand, or double-reduction cold mill** - This is commonly a non-reversing mill.

 3. **Tandem cold mill** - Three or more close-coupled non-reversing mill stands usually constitute a tandem cold mill. The tandem cold mills can be divided into two groups: stand alone and fully continuous.

 a) **Stand-alone cold mill.**

Figure 28.16 illustrates a typical tandem cold mill arrangement. The coil to be rolled is delivered to the entry side of the mill by the entry coil conveyor and is then lifted to uncoiling position by an entry coil lifting car. Between mill stands and sometimes after last mill stand, there are free-rotating rollers which are used as the parts of the strip tension measuring devices. The strip exiting the last mill stand is coiled on a tension reel. After coiling is completed the exit coil lifting car strips the coil from the tension reel and transfers it onto the exit coil conveyor.

 b) **Fully continuous cold mill.**

In these facilities (Fig. 28.17), the provisions are made for welding the strips head to tail. To provide continuity of rolling during welding, the excess of strips is stored prior to welding operation and is released during welding operation [16].

28.9 PROCESSING LINES INCORPORATING COLD MILLS

The following steps are typically involved in processing of the coils after hot rolling.

 1. **Descaling** - Main purpose of descaling is to remove the scale accumulated on the strip surface during hot rolling and storage. The scale is most commonly removed by pickling the strip while it passes

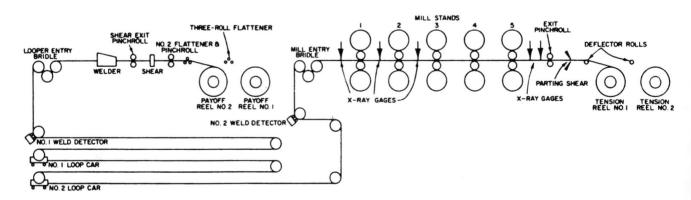

Fig. 28.17 Weirton Steel's fully continuous cold mill. (From FLAT PROCESSING OF STEEL, by W.L. Roberts, 1988. Copyright Marcel Dekker, Inc.)

through the tanks filled with hydrochloric or other types of acids. Another known descaling method utilizes hydroabrasion. The latter method was developed by IHI and Nippon Steel and is known as ISHICLEAN method [17]. In this method, an iron and slurry are mixed onto a high-pressure water jet directed against the surface to be descaled.

Prior to descaling process, the strip is sometimes elongated by a tension leveler by several percent. The purpose is to facilitate the pickling by forming fine cracks in the scale. Skin pass rolling mill is also used in some pickle lines for the same purpose.

2. **Cold rolling** - High-production cold rolling is commonly conducted on tandem cold mills.

3. **Strip cleaning** - During cleaning, the remaining of the rolling solution removed from the strip surface prior to annealing.

4. **Annealing and cooling** - Both batch type and continuous type annealing furnaces and cooling facilities are used to reduce residual stresses in the strip accumulated during cold rolling.

5. **Temper rolling** - Single-stand non-reversing cold mills are customarily used as the temper mills.

6. **Finishing process** - The finishing process involves final leveling, trimming, slitting and cutting the strip to the desired delivery sizes.

There is a strong trend in the steel industry toward integration of the processes described above. The most complete integration has been achieved by introducing a fully integrated processing line [18]. This line (Fig. 28.18) incorporates all cold processing operations.

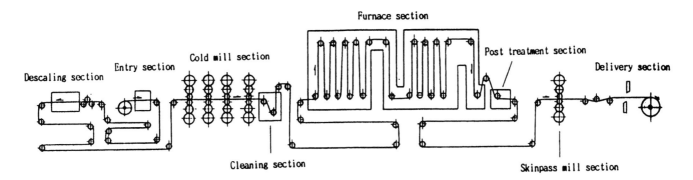

Fig. 28.18 Nippon Steel's fully integrated process line. (From Kawasaki, et al, Proceedings of 4-th International Steel Conference, 1987. Copyright IRSID, France. Reprinted with permission.)

REFERENCES

1. T. Kawamani, et al, "Characteristics of Shape Control in Cluster Type Rolling Mill (CR)", Mitsubishi Heavy Industries Technical Review, June 1985, pp. 171-177.

2. H.G. Hilbert, et al, "MKW Cold Mill-Rolling Silicon Steel Strip", AISE Year Book, 1976, pp. 364-370.

3. Comparison of Rolling Mill Types, Metals Society Conference Proceedings, Cardiff, Wales, November 1984, pp. 142-146.

4. The Making, Shaping and Treating of Steel, 10th Edition, eds. W.T. Lankford, Jr., et al, Association of Iron and Steel Engineers, Pittsburgh, Pennsylvania, pp. 787-840 (1985).

5. K. Sato, et al, "AC Motor Drive Systems for Rolling Mills", Hitachi Review, Vol. 36, No. 1, 1987, pp. 13-20.

6. J.M. Keefe, et al, "Review of Hot Strip Mill Developments", Iron and Steelmaking, No. 4, 1979, pp. 156-172.

7. Mitsubishi Hot Strip Mill Brochure HD 20-04662, (1.0), 85-9,B, pp. 1-2.

8. H.G. Husken and K. Herwig, "Use of the Coilbox in the Wide Hot Strip Mill of Krupp Stahl AG", Metallurgical Plant and Technology, No. 4, 1983, pp. 53-59.

9. H. Wiesinger, et al, "Hot Strip Rolling for Compact Mills: The HSRC Mill", Iron and Steel Engineer, August 1987, pp. 50-55.

10. "Southern Cross Builds Stainless Steel Mill for Half the Cost", Iron and Steel Engineer, Jan. 1984, p. 73.

11. H. Wakatsuki, et al, "The Facilities and Operation of the HDR Process at the No. 2 Hot Strip Mill of Fukuyama Works", Proceedings of the 4th International Steel Rolling Conference, Deauville, France, June 1987, pp. A.3.1-A.3.9.

12. "New SMS Technology for Thin Slab Casting", Iron and Steel Engineer, May 1987, pp. 51, 52.

13. E. Emoto, et al, "Hot Connection Between Steelmaking Shop and Hot Rolling Mill", Proceedings of London Conference on Restructuring Steel Plants for the Nineties, May 1986, pp. 134-139.

14. M.G. Sendzimir, "Hot Strip Mills for Thin Continuous Casting Systems", Iron and Steel Engineer, October 1986, pp. 36-43.

15. V.B. Ginzburg and G.W. Tippins, "Continuous Rolling Method and Apparatus", US Patent No. 4,630,352, Dec. 23, 1986.

16. W. Roberts, Flat Processing of Steel, Marcel Dekker, New York and Basel, pp. 230-234 (1988).

17. S. Yamaguchi, "Recent Improvement of Hydro-Abrasion Type Descaler-ISHICLEAN Method", IHI Engineering Review, Vol. 18, No. 3, July 1985, pp. 134-139.

18. Y. Kawasaki, et al, "Fully Integrated Processing Line at Hirohata Works of Nippon Steel Corporation", Proceedings of the 4th International Steel Rolling Conference, Deauville, France, June 1987, pp. E.34.1-E.34.10.

29

High Reduction Rolling Mills

29.1 LIMITATIONS OF CONVENTIONAL ROLLING MILLS

Reductions which can be taken in a single pass by conventional rolling mills are limited by not only the mill capacity but also by the nature of its design. Indeed, the most obvious way to increase the reduction capability of a rolling mill is to reduce the work roll diameter. However, this leads to higher contact stresses between the rolls, lesser work roll stability in horizontal plane and also produces an increased roll wear.

Although the high reduction rolling mills also utilize small work roll diameters, they have, in addition to that, a number of other features added to their design to avoid the drawbacks typical for conventional rolling mills using small work roll diameters.

These design features include:

1. Offsetting the work rolls in relation to the backup roll center line and incorporating the support rolls for lateral restraining of the work rolls [1-3].

2. Supporting the work rolls by utilizing the cluster roll arrangements [4-6].

3. Combining the rolling deformation process with drawing and bending [7-9].

4. Extending the length of the deformation zone by either reciprocating movement of the work rolls in relation to the workpiece [10-13] or by introducing the planetary arrangement of the rolls [13-16].

The rolling mills utilizing these design features are briefly described below.

29.2 MKW COLD MILL

The MKW mill [1, 2] resembles a 4-high mill (Fig. 29.1). Main difference between these two mills is that in MKW mill the work rolls of substantially smaller diameters are displaced to one side from the vertical center-line of the backup rolls and are laterally restrained by intermediate rolls and a support system.

The torque is transmitted through the backup rolls and the work rolls are driven by frictional forces transmitted from the backup rolls. This design allows one to substantially reduce the work roll diameter that can be as much as six times smaller than the backup roll diameter.

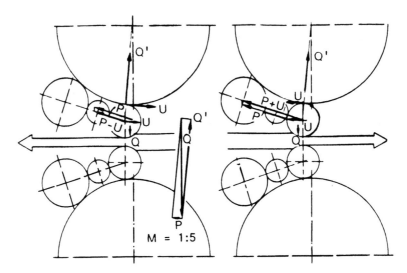

Fig. 29.1 Distribution of forces in a MKW cold mill.
(From Hilbert, et al, AISE Year Book, 1976. Copyright AISE,
Pittsburgh, Pennsylvania. Reprinted with permission.)

Because of the out-of-center arrangement of the work rolls, the vertically acting rolling force Q is divided into two components: a) the vertical component Q' and b) the horizontal component P. The peripheral force U transmitted by friction from the backup rolls to the work rolls in the direction opposite to the direction of rolling. In the reversing mill, the peripheral force U will be either added to or subtracted from the horizontal component P. In the latter case, in order to avoid the work rolls moving out of the support plane, the horizontal component P must always be greater than the peripheral force U.

Table 29.1 presents main data for MKW cold mills designed for production of both nonoriented and oriented silicon steel strip.

Table 29.1 Main data for MKW cold mills. Data from
Hilbert, et al (1976).

Mill parameters	Material: silicon strip	
	Nonoriented	Oriented
Backup roll diameter, mm	1320	1320
Work roll diameter, mm	250	225, 165
Barrel length, mm	1400	1250
Max. rolling speed, mpm	720	900
Total drive, HP	11,000	13,000
Max. mill speed, mpm	720	900
Max. strip width, mm	1250	1100
Min. strip thickness, mm	0.5	0.28

29.3 FLEXIBLE FLATNESS CONTROL (FFC) MILL

In the Flexible Flatness Control (FFC) mill, higher reductions and improved strip profile and shape control are achieved by introduction of the following concepts [3]:

1. A combination of two work rolls with different diameters. The average diameter of the two rolls is usually smaller than in a conventional mill.

2. Driving the two rolls at different peripheral speeds.

3. Horizontal bending of the work rolls with a horizontal roll bending mechanism.

4. Vertical bending of the work rolls with either single-chock or double-chock roll bending device.

There are many types of FFC mills. Some of them are shown schematically in Fig. 28.6. The horizontal bending mechanism consists of a support roll and a sectional backup roll. Each section of the backup roll is supported by a position regulated hydraulic cylinder. These cylinders allow one to establish the desired pressure distribution on the work roll. This pressure pattern in horizontal plane is correlated with the pressure pattern in vertical plane produced by the horizontal bending mechanism striving to obtain the desired strip profile and flatness.

Technical data of a typical 5-high 1425 mm (56 in.)-wide FFC mill are the following:

Top work roll diameter:

        minimum   460 mm (18.1 in.)

        maximum   520 mm (20.5 in.)

Bottom work roll diameter:

        minimum   290 mm (11.4 in.)

        maximum   320 mm (12.6 in.)

Intermediate roll diameter:

        minimum   460 mm (18.1 in.)

        maximum   520 mm (20.5 in.)

Horizontal support roll diameter:

        minimum   180 mm (7.1 in.)

        maximum   200 mm (7.9 in.)

Sectional backup roll diameter:

        minimum   390 mm (15.4 in.)

        maximum   400 mm (15.8 in.)

Number of sectional backup rolls   6

Maximum work roll speed   756 rpm.

29.4 CLUSTER ROLLING MILLS

Main purpose of the multi-high, or cluster, roll arrangement in rolling mills is to provide an adequate support for the work rolls with a small roll diameter. This would allow one to roll thinner gauges and achieve high reductions on the materials which quickly work-hardened.

Figure 29.2 shows the predominant layout of the Sendzimir mill [4, 5] that is known as the 1-2-3-4 arrangement. In this mill, there are eight backing shafts A through H. Shafts B and C have roller

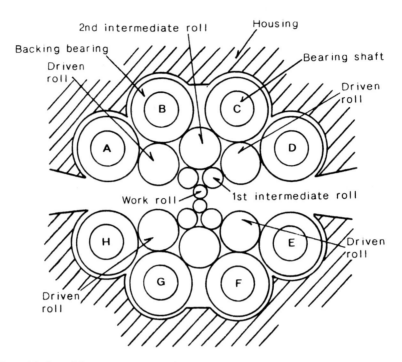

Fig. 29.2 Roll arrangement in type 1-2-3-4 Sendzimir cold mill. (From T. Sendzimir, Inc. Publication 3/80/5M, 1980. Copyright T. Sendzimir, Inc. Reprinted with permission.)

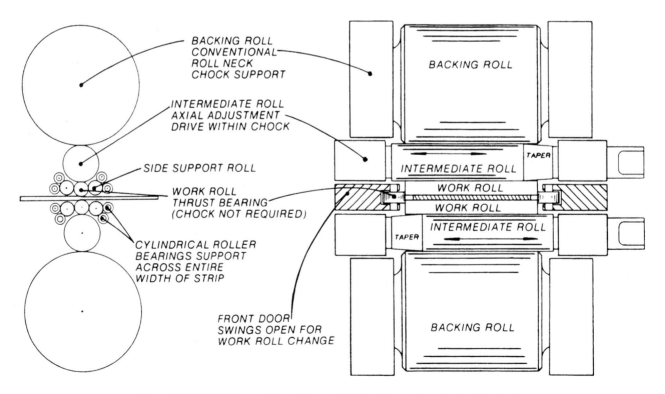

Fig. 29.3 Z-high cold rolling mill. (From T. Sendzimir, Inc. Publication 3/80/5M, 1980. Copyright T. Sendzimir, Inc. Reprinted with permission.)

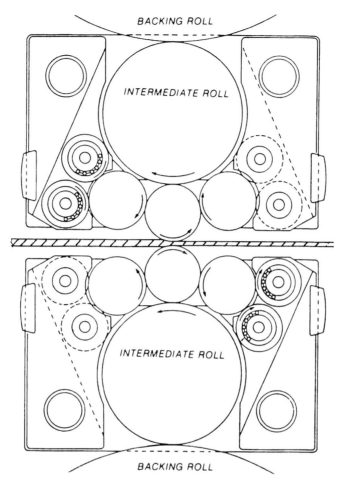

Fig. 29.4 Typical Sendzimir Z-high insert. Adapted from 33 Metal Producing Publication (1981).

bearings. Vertical position of these shafts is controlled by hydraulic cylinders installed on the top of the mill. This control allows one to adjust the roll gap. Vertical position of shafts F and G is controlled by hydraulic cylinder installed in front of the mill. This control serves two purposes: first, it adjusts the passline in the mill and second, it allows one to take out all the slack between the rolls.

There are two top and two bottom driven rolls with the nondriven intermediate rolls located between the driven rolls. The first intermediate rolls can be axially adjusted. This adjustment is similar to that for Z-high cold rolling mill shown in Fig. 29.3. The shifted top and bottom rolls are grounded with a taper that is located at the opposite ends of the rolls. The roll shifting is used for strip profile and flatness control.

Another development is the Sendzimir Z-high insert (Fig. 29.4) that is designed to be installed in place of the two work rolls of a conventional 4-high mill. Typical roll diameters for the 2-high insert designed for rolling of the 660 mm (26 in.)-wide strip are following [6]:

Work roll:
    minimum   66.7 mm (2.625 in.)
    maximum   165 mm (6.50 in.)
Intermediate roll   216 mm (8.5 in.)
Backing roll:
    minimum   660 mm (26.0 in.)
    maximum   711 mm (28.0 in.).

29.5 ROLLING-DRAWING (PV) MILL

PV are the initial letters of the Russian term "Prokatka Volocheniem" that translates as the rolling-drawing process. In this process [7], the work rolls (Fig. 29.5) are driven at different peripheral speeds so the peripheral speed of the slower speed roll V_{r1} is equal to the strip entry speed V_1 and the speed of the other work roll V_{r2} is equal to the strip exit speed V_2, i.e.,

$$V_1 = V_{r1} ; \quad V_2 = V_{r2} \tag{29-1}$$

Therefore, the elongation of the strip is equal to

$$e = h_1/h_2 = V_{r2}/V_{r1} \tag{29-2}$$

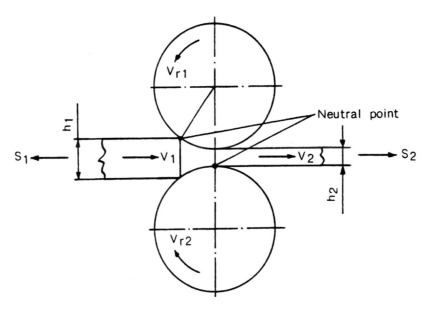

Fig. 29.5 Schematic presentation of the rolling-drawing (PV) process.

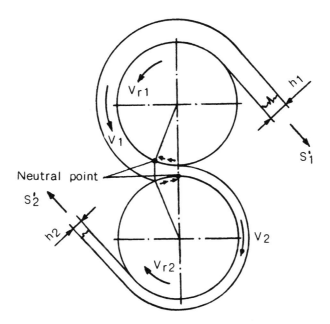

Fig. 29.6 Schematic presentation of the strip encompassing rolling-drawing (PV-E) process.

By applying the proper amount of entry tension S_1 and exit tension S_2 the neutral plane of the slower speed roll contact zone is shifted toward the extreme entry side position whereas the neutral plane of the higher speed roll contact zone is shifted toward the extreme exit side position.

The desired shifting of the neutral points occurs when the strip tension is of the same order of magnitude as the resistance to deformation of the rolled material. To achieve this condition more effectively, either one of the rolls or both rolls are encompassed by the strip as in the PV-E mill shown in Fig. 29.6. In that case the work rolls behave like bridle rolls allowing one to achieve the same magnitude of the strip tensions S_1 and S_2 at the roll bite as in the first case by applying the strip tensions S_1' and S_2' of much lesser magnitudes.

The rolling characteristics of the PV process are modified in the following way:

1. Roll separating force is reduced in comparison with a conventional rolling process. This is due to the fact that frictional forces at the deformation zone of PV process act in opposite direction to each other and therefore the friction hill is eliminated.

2. The roll torque is input through the higher speed roll and is dragged out from the slower speed roll. This occurs because the frictional forces acting on the opposing contact surfaces at the deformation zone are equal and have opposite directions.

3. The energy savings are most significant when rolling hard materials that produce a high friction hill in a conventional rolling process.

The performance characteristics of the PV process have the following features:

1. Improved rollability of the minimum gauge

2. Possibility of heavy reductions

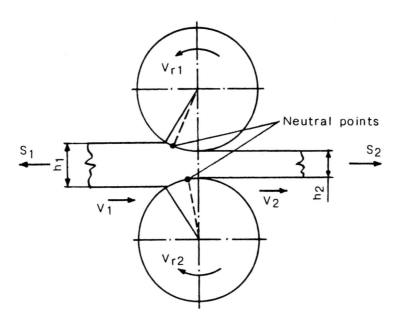

Fig. 29.7 Schematic presentation of IPV mill. Adapted from IHI Engineering Review (1978).

3. Reduced edge drop and edge cracking

4. Improved gauge performance.

Major shortcomings of the PV-E mills include the difficulty to apply roll coolant and lubricant and also bigger roll deflection caused by the 2-high mill configuration. These problems are alleviated in IPV-S mill.

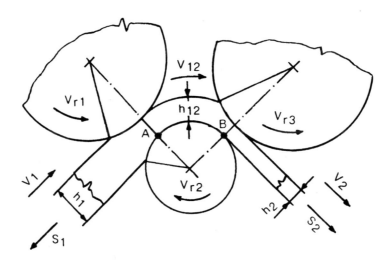

Fig. 29.8 Double rolling-drawing process. (From Vydrin, et al, Steel in USSR, Vol. 14, Nov. 1984. Copyright Institute of Metals. Reprinted with permission.)

The IPV-S mill is a conventional 4-high mill which adopts the PV characteristics of differential roll speed between two rolls. In this mill (Fig. 29.7) strip passes straight through the roll bite. The magnitudes of the entry and exit strip tensions S_1 and S_2 are generally smaller than in the PV mill, so the neutral points, NP, related to the top and bottom roll contact surfaces, are not shifted to the extreme entry and exit planes. Therefore, the frictional hill is being reduced in comparison with a conventional rolling process, nevertheless does not disappear completely as in the PV process.

These specifics of the IPV-S mill allow one to incorporate some advantages of the PV mill such as improved gage and strip profile, increased reduction and improved minimum gauge rollability with the advantages of the conventional 4-high mill which include higher rolling speeds, possibility to roll wider product, simplicity in operation, etc.

29.6 DOUBLE ROLLING-DRAWING MILL

In the double rolling-drawing mill [8], a workpiece is deformed by three work rolls (Fig. 29.8). Thus, there are two deformation zones; each one is formed by the center work roll and one of the outer work rolls. The distance between the successive deformation zones AB, along the outer surface of the center work roll, is less than half of the perimeter of the center roll.

The speed of the workpiece V_{12} in the zone AB is equal to the peripheral speed of the center roll, V_{r2}, whereas the speed of the workpiece after the second deformation zone V_2 is equal to the peripheral speed of the second outer work roll V_{r3}, i.e.,

$$V_{12} = V_{r2} ; \quad V_2 = V_{r3} \tag{29-3}$$

Thus, according to the kinematic conditions, a rolling-drawing process always takes place in the second deformation zone.

The mismatch in the peripheral speeds of the first outer roll and the center roll may vary in the range $V_{r2}/V_{r1} > 1$. In the case when the workpiece entry speed V_1 is equal to the peripheral speed of the first outer roll, i.e., when

$$V_1 = V_{r1} ; \quad V_{12} = V_{r2}, \tag{29-4}$$

the rolling-drawing process occurs in both first and second deformation zones.

The elongations in these zones are respectively equal to

$$e_1 = h_1/h_{12} = V_{r2}/V_{r1} \tag{29-5}$$

$$e_2 = h_{12}/h_2 = V_{r3}/V_{r2} \tag{29-6}$$

29.7 CONTACT-BEND-STRETCH ROLLING PROCESS

The Contact-Bend-Stretch (C-B-S) rolling process [9] incorporates plastic bending in conjunction with longitudinal tension and rolling pressure. It is applicable for reduction of strip and foil.

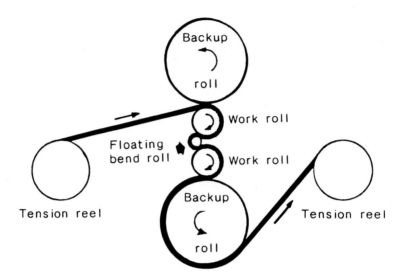

Fig. 29.9 4-high C-B-S rolling mill. (From Coffin. Reprinted with permission from JOURNAL OF METALS, Vol. 19, No. 8, 1967, a publication of The Metallurgical Society, Warrendale, Pennsylvania, 15086, USA.)

In the basic configuration of the C-B-S single stand rolling mill (Fig. 29.9), the strip enters the mill from the left, and is threaded around a large entry contact roll, then around a small bend roll and finally passes around a large exit contact roll and emerges from the mill. The small bending roll is cradled in the gap between the large contact rolls. In order to prevent slipping between the strip and the two contact rolls, the strip is maintained under tension by means of entry and exit tension reels.

Reduction takes place at two points of contact between the bend roll and the two contact rolls. The reduction is determined by a ratio V_{r1}/V_{r2} of the peripheral speeds of the contact rolls (Fig. 29.10). The rolling mill is insensitive to the roll gap setting. Indeed, the roll gap, or distance between the two contact rolls, defines the relative position of the bend roll, as given by the toggle angle α. The smaller toggle angles the greater roll separating force and a smaller strip tension is necessary to satisfy the established speed ratio.

Another feature of this rolling process is a constant reduction across the width of the strip. This is due to an absence of the roll bending forces which are usually present in the conventional rolling mills.

Figure 29.9 illustrates 4-high reversing cold mill with driven work rolls. The mill rolling characteristics are the following:

Rolled material	stainless steel, grade 304
Material width	305 mm (12 in.)
Entry thickness	1.27 mm (0.050 in.)
Final thickness	0.1 mm (0.004 in.)
Number of passes	8 (no intermediate anneal)
Max. reduction per pass	55 %.

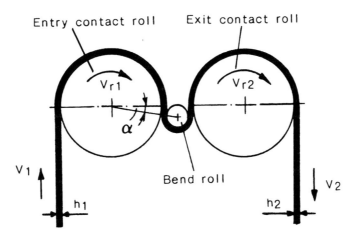

Fig. 29.10 Basic configuration of C-B-S single stand rolling mill. (From Coffin. Reprinted with permission from JOURNAL OF METALS, Vol. 19, No. 8, 1967, a publication of The Metallurgical Society, Warrendale, Pennsylvania, 15086, USA)

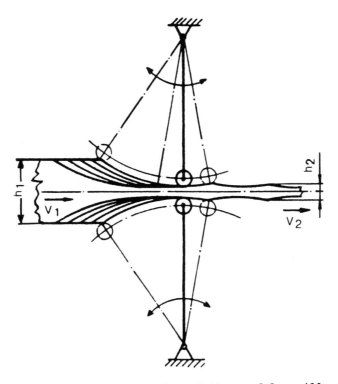

Fig. 29.11 Schematic presentation of the pendulum mill. (From Frohling, Metals Society Conference Proceedings, Cardiff, Wales, 1978. Copyright Institute of Metals. Reprinted with permission.)

29.8 PENDULUM COLD MILL

The pendulum mill [10] has two small work rolls, each being backed up by two or five rolls. All rolls are mounted in upper and lower saddles which are reciprocated via rods by driven eccentrics.

The workpiece is fed into the mill by a pair of driven pinch rolls. The work rolls remain in contact with the material during the complete cycle. Because they move around a fixed radius (Fig. 29.11), the exit strip shows a wave or scallop across its width. This wave, however, disappears completely after 10-15% reduction by conventional rolling.

A production unit was reportedly able to reduce bronze and copper-nickel alloys being in the order of 200 mm (8 in.) wide with 12.7 mm (0.5 in.) entry thickness and 1.4 mm (0.055 in.) exit thickness.

29.9 KRAUSE RECIPROCATING MILL

The principle of the reducing operation of the Krause reciprocating mill is essentially a combined drawing and pressing operation [11].

The main parts of the mill are (Fig. 29.12): oscillating backup frame with double wedge-shaped paths and set of work rolls installed inside the backup frame. The backup frame is moved in bed back and forth by a crank drive. The rolls are held in alignment by a cage which is manipulated by an air cylinder. When there is no material between the rolls they are held against the wedge-shaped paths by springs.

During a rolling cycle the workpiece is fed forward and then it is held by a hydraulic clamp. At that time the cage with the rolls is held by an air cylinder in the starting position. As the crank rotates counterclockwise, the backup frame moves and forces the rolls to close the roll gap while they are rolled over the workpiece toward the exit side of the mill. As the roll gap closes, the increment of the workpiece length that was fed forward prior to the beginning of the cycle is being rolled out to the finish gauge. On the neutral stroke of the crank both backup frame and the rolls are returned to the starting position.

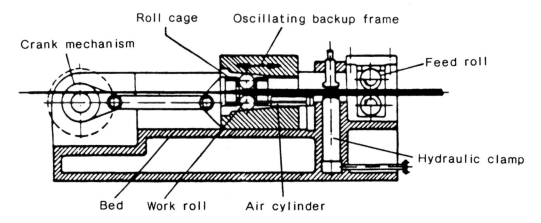

Fig. 29.12 Schematic presentation of Krause reciprocating rolling mill. (From Fink and Jungmann, AISE Year Book, 1971. Copyright AISE, Pittsburgh, Pennsylvania. Reprinted with permission.)

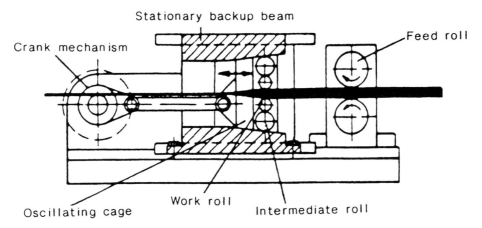

Fig. 29.13 Schematic presentation of Platzer reciprocating rolling mill. (From Fink and Jungmann, AISE Year Book, 1971. Copyright AISE, Pittsburgh, Pennsylvania. Reprinted with permission.)

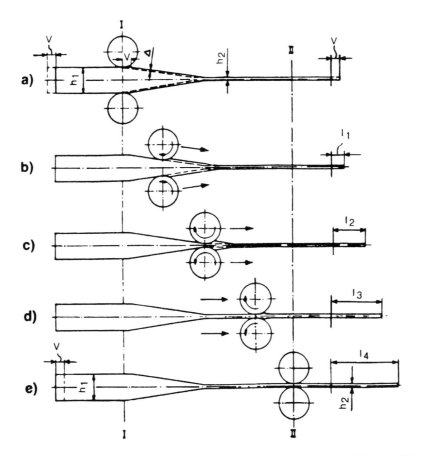

Fig. 29.14 Rolling cycle of Platzer reciprocating rolling mill: a) starting, b) wedge rolling, c) maximum load position, d) parallel rolling, e) finish. (From Buch and Fink, Metals Society Conference Proceedings, Cardiff, Wales, 1978. Copyright Institute of Metals. Reprinted with permission.)

Table 29.2 Data of Platzer reciprocating mills of various sizes.
From Fink and Buch (1978).

Parameters	Mill size (strip width), mm		
	200	400	1200
Rolled stock initial thickness, mm....3-15		3-30	5-35
Rolled stock final thickness, mm..... 0.3-1.5		0.3-2.0	0.5-2.0
Production output for non-ferrous heavy metal, t/h 0.3-1.0		0.5-4.0	1-12
Work roll diameter, mm............... 70		120	220
Drive rating (main motor), kW........ 160		280	650
Number of roll strokes per unit of time, sec............................. 10		8	5

29.10 PLATZER RECIPROCATING MILL

In the Platzer reciprocating mill [12, 13], the workpiece is continuously fed at low speed into the rolling gap by a pair of driven feed rolls (Fig. 29.13). Within the mill the workpiece is reduced to a final thickness by the reciprocating rolling mechanism. This mechanism consists of an oscillating cage that contains two work rolls and two intermediate rolls. The oscillatory movement of the cage in the horizontal direction is provided by a crank drive. The intermediate rolls are restrained in the vertical direction by stationary backup beams which have wedge-type paths verging into horizontal parallel paths. Under practical conditions, the frequency of oscillations can be as high as 600 strokes/min.

As the material is advanced to the roll gap, the work rolls are forced to roll down thin key-shaped material layers (Fig. 29.14a) from top and bottom sides of the slab during each stroke of the cage. The rolled material is pushed in front of the rolls as a bulge (Fig. 29.14b-d) which is flattened to a completely flat strip after the cage reaches the parallel path (Fig. 29.14e).

Reduction of 95% can be achieved in cold rolling. Table 29.2 shows data of Platzer reciprocating mills.

29.11 CYCLOIDAL MILL

In the cycloidal mill, the workpiece is fed into the roll gap by a pair of hydraulically operated pincers (Fig. 29.15). These pincers are used intermittently. As one pincer pushes the material in the direction of the rolling, the second pincer is moving in opposite direction in preparation for taking over the function of the other pincer [10].

The roll assembly consists of two work rolls which are supported by a cluster arrangement of two intermediate rolls and three axes supporting a series of backup roll bearings. The whole assembly is contained in two saddles located in the housing symmetrical in relation to the pass line. The rolling load is transferred through the cycloidal shape surfaces of the saddles contacting the flat surfaces of the inserts installed in the mill housing. The top and bottom saddles reciprocate synchronously through a crank gear. The roll gap adjustment is provided by a hydromechanically operated screwdown mechanism.

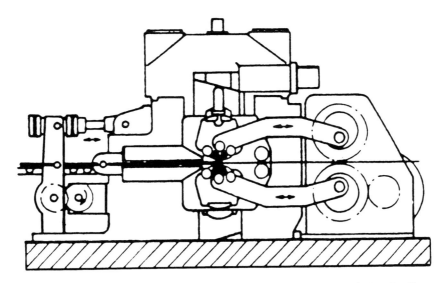

Fig. 29.15 Cycloidal mill. (From Frohling, Metals Society Conference Proceedings, Cardiff, Wales, 1978. Copyright Institute of Metals. Reprinted with permission.)

When the saddles are positioned at the left side of the stroke (Fig. 29.16), the work roll moves along the path with radius r. In that time, the work roll bite opens sufficiently to grip the incoming material. As the material is fed into the mill, the reduction takes place determined by radius r and feeding speed. As the saddles move from left to right, the work rolls now begin to follow the path determined by radius R. This would provide a constant roll gap, thus producing a finished material with a flat service.

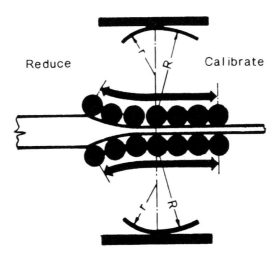

Fig. 29.16 Principle of the cycloidal mill. (From Frohling, Metals Society Conference Proceedings, Cardiff, Wales, 1978. Copyright Institute of Metals. Reprinted with permision.)

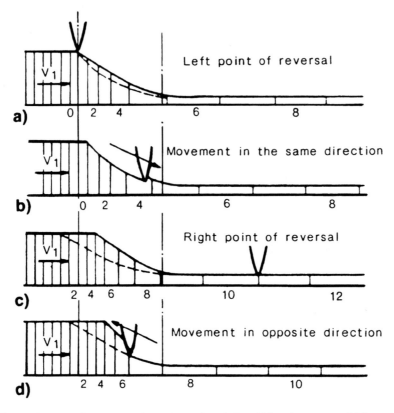

Fig. 29.17 Rolling cycle of cycloidal mill. (From Frohling, Metals Society Conference Proceedings, Cardiff, Wales, 1978. Copyright Institute of Metals. Reprinted with permission.)

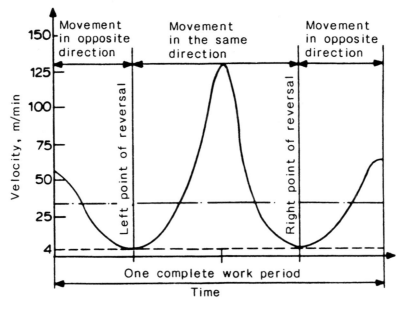

Fig. 29.18 Speed of the strip leaving the cycloidal mill. (From Frohling, Metals Society Conference Proceedings, Cardiff, Wales, 1978. Copyright Institute of Metals. Reprinted with permission.)

544

Table 29.3 Material rolled by Cycloidal mill. Data from Frohling (1978).

Rolled material	Width mm	Thickness, mm	
		entry	exit
Brass 70/30.........185		16.0	1.55
Bronze 6............200		19.0	1.98
Bronze 8............190		11.0	0.80
CuNi 75/25.........196		12.0	1.50
Mild Steel.........187		6.0	0.65
CrNi 18/8..........195		20.0	2.14
NiFe 75/25.........175		8.5	0.98
Titanium...........158		15.0	1.36

Figure 29.17a shows the extreme left position of the work rolls which are gripping the ingoing material. The rolls then move to the right in the same direction as the metal flow (Fig. 29.17b) until the extreme right position is reached (Fig. 29.17c) and the direction of movement is reversed (Fig. 29.17d). The speed of the outgoing strip varies within each cycle as shown in Fig. 29.18. The cycloidal mill is used for cold rolling of both ferrous and nonferrous metals. Table 29.3 illustrates the product rolled by the cycloidal mill with the following technical data:

Work roll diameter	58 mm (2.28 in.)
Intermediate roll diameter	100 mm (3.94 in.)
Barrel length	540 mm (21.3 in.)

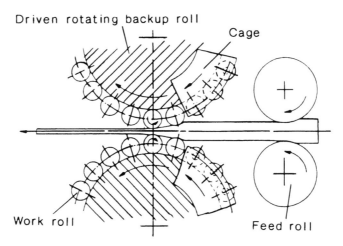

Fig. 29.19 Schematic presentation of Sendzimir planetary rolling mill. (From Fink and Jungmann, AISE Year Book, 1971. Copyright AISE, Pittsburgh, Pennsylvania. Reprinted with permission.)

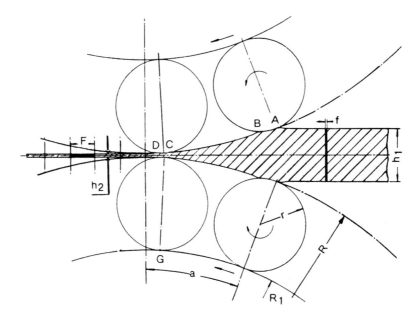

Fig. 29.20 Schematic sketch of planetary roll bite shows one pair of work rolls at the start of the bite, a second pair at the end. (From Sendzimir, AISE Year Book, 1958. Copyright AISE, Pittsburgh, Pennsylvania. Reprinted with permission.)

29.12 SENDZIMIR PLANETARY MILL

In the Sendzimir planetary rolling mill [13–15] two driven backup rolls rotate by friction the work rolls installed in a cage around the backup roll as shown in Fig. 29.19. A pair of feed rolls take a light reduction on a workpiece. Their prime purpose is to provide the necessary feeding force to be extended in order to secure a reliable forward movement of the workpiece.

Figure 29.20 illustrates a principle of operation of the mill. One pair of the work rolls is shown in the position where the workpiece is reduced to almost its final gauge, whereas the next pair has just started to squeeze the material. In this figure, R_1 and r are the radiuses of the backup rolls and work rolls respectively. For each pair of work rolls passing through the bite, the workpiece is advanced into the planetary rolls a small distance f. After reduction this workpiece length is elongated to a length of strip F. Since work rolls move on circular paths (AB, CD), there is some thickness variation along the strip length at the exit side of the mill so the sections F appear concave. This thickness variation, however, is much less than would be expected from geometrical relation because of the elasticity of the mill housing. The reduction from hot slab to strip in one single pass is reported to be over 97% and the production rate is as high as 200 t/h.

In the planetary mill, the metal is subjected to a quasi-rolling quasi-forging deformation and most of the energy required for the deformation is absorbed by the workpiece. Since the workpiece heats up continuously, less energy is needed than in a tandem finishing mill where it gets cooler while passing

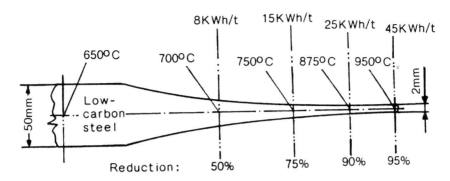

Fig. 29.21 Energy input and reduction in the planetary mill. (From Sendzimir, Metals Society Conference Proceedings, Cardiff, Wales, 1978. Copyright Institute of Metals. Reprinted with permission.)

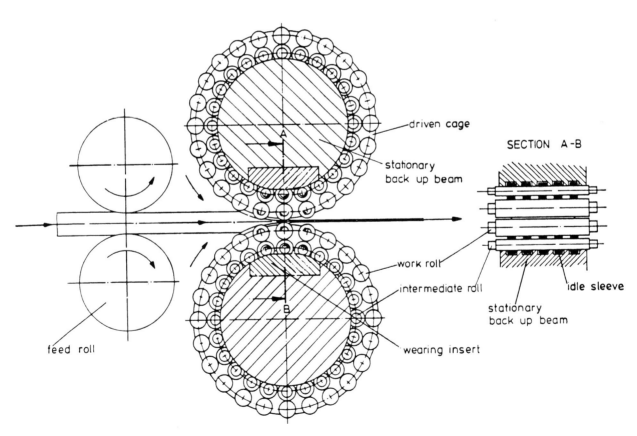

Fig. 29.22 Schematic presentation of Platzer planetary rolling mill. (From Fink and Jungmann, AISE Year Book, 1971. Copyright AISE, Pittsburgh, Pennsylvania. Reprinted with permission.)

Table 29.4 Data for Platzer planetary mills for 200, 400 and 1200 mm strip width. From Fink and Buch (1978).

	Mill size (strip width), mm		
Parameters	200	400	1200
Rolled stock initial thickness, mm...up to 75		up to 100	up to 180
Rolled stock final thickness, mm.....1.5 min		1.8 min	2.5 min
Production output referred to:			
steel, t/h...................... 8		20	100
copper, t/h.....................10		25	150
aluminum, t/h.................. 5		10	60
Work roll diameter, mm.................75		105	225
Drive rating, kW (main motor)........450		800	4500
Number of roll contacts per unit of time (per second)..................60		50	40

from one mill stand to the next one. Figure 29.21 shows, as an example, the calculated energy input into a workpiece with the initial thickness 50 mm (2 in.) and exit thickness 2 mm (0.080 in.) and corresponding theoretical workpiece temperature at the different planes of the roll bite.

29.13 PLATZER PLANETARY MILL

In the Krupp-Platzer planetary mill [13, 16], groups of work rolls and intermediate rolls are mounted in ring-shaped cages which revolve around two rigid stationary backup beams (Fig. 29.22).

The work rolls and intermediate rolls can be radially shifted in the driven cages. Feed rolls guide the preheated slab to the roll gap. As the material is pushed on in front of the work rolls, a resulting small bulge is flattened to a completely even piece of strip. This is achieved by installing two flattened wearing inserts in the roll gap. The work rolls can rotate in the cage freely. The intermediate rolls consist of a central shaft and idle rotating sleeves. The rolling pressure is transferred from the work rolls to these sleeves and then via the central shaft to the comb-type runways.

Reduction of 98% can be achieved during hot rolling in this type mill. Table 29.4 shows data of Platzer planetary mills.

REFERENCES

1. H.G. Hilbert, et al. "MKW Cold Mill - Rolling Silicon Steel Strip", AISE Year Book, 1976, pp. 364-370.

2. G. Giermann and K.A. Kennepol, "MKW Mills for Rolling Silicon Strip and Non-Ferrous Metals", Flat Rolling: A Comparison of Rolling Mill Types, Metals Society Conference Proceedings, Cardiff, Wales, September 1978, pp. 142-146.

3. F. Fujita, et al, "Development of a New Type of Cold Rolling Mill for Sheet Products", Iron and Steel Engineer, June 1985, pp. 41-48.

4. T. Ohama, et al, "World's First Sendzimir Tandem Mill", AISE Yearly Proceedings, 1973, pp. 173–179.

5. Sendzimir Cold Rolling Mills, Sendzimir Publication 3/80/5M, Waterbury, Connecticut.

6. "Universal-Cyclops Gives Z-Hi Retrofit Scheme its First Break in the US", 33 Metal Producing, October 1981.

7. "What are PV and IPV Mill?", IHI heavy Industries Co. Publication, Tokyo, Japan, January 1978.

8. V.N. Vydrin, et al, "Double Rolling-Drawing Process', Steel in the USSR, Vol. 14, November 1984, pp. 542, 543.

9. L.F. Coffin, "Status of Contact-Bend-Stretch Rolling", Journal of Metals, August 1967. pp. 14-22.

10. P. Frohling, "Cycloidal High Reduction Mill", Flat Rolling: A Comparison of Rolling Mill Types, Metal Society Conference Proceedings, Cardiff, Wales, September 1978, pp. 131-135.

11. F.R. Krause, " A New Metal Rolling Process - Its Theory and Operation", AISE Yearly Proceedings, 1938, pp. 414-427.

12. E. Buch and P. Fink, "Platzer Reciprocating Mill for Cold High Reduction of Metal Strip", Flat Rolling: A Comparison of Rolling Mill Types, Metal Society Conference Proceedings, Cardiff, Wales, September 1978, pp. 136-141.

13. P. Fink and H.D. Jungmann, "Economic Application of the Krupp-Platzer Planetary Mill for the Production of Hot Rolled Strip", AISE Yearly Proceedings, 1971, pp. 81-90.

14. T. Sendzimir, "The Planetary Mill and Its Uses", AISE Yearly Proceedings, 1958, pp. 49-55.

15. M.G. Sendzimir, "Planetary Hot Strip Mill Development, Operation, and Potential", Flat Rolling: A Comparison of Rolling Mill Types, Metals Society Conference Proceedings, Cardiff, Wales, September 1978, pp. 69-78.

16. P. Fink and E. Buch, "Platzer Planetary Mill for Hot High Reduction of Metal Strip", Metals Society Conference Proceedings, Cardiff, Wales, September 1978, pp. 73-81.

30

Optimization and Modernization of Hot Strip Mills

30.1 MAIN STRATEGY IN OPTIMIZATION OF ROLLING PROCESS

In the process of rolling a uniformly preheated slab in hot strip mill, its temperature changes due to the various types of the heat transfer have been described earlier. The following three temperature profiles are usually used for evaluating the temperature rundown of the workpiece as well as a degree of uniformity of the temperature along its length and width:

1. Temperature rundown of a selected portion (for example, a head end, tail end, or a middle portion of the workpiece expressed) in relation to each rolling pass.

2. Temperature variation along the workpiece length after the same rolling pass.

3. Temperature variation across the workpiece width.

The temperature rundown in hot strip mill is shown in general form in Fig. 30.1. The main parameters of the temperature rundown include [1]:

T_0 = slab reheat furnace dropout temperature

T_R = bar temperature after leaving roughing train

T_F = bar temperature at finishing train entry

T_E = strip temperature at finishing train exit.

The temperature variation along the workpiece length is usually defined by the following parameters:

ΔT_0 = absolute value of temperature differential between slab head and tail ends

ΔT_F = absolute value of temperature differential between transfer bar head and tail ends at finishing train entry

ΔT_E = absolute value of temperature differential between strip head and tail ends at finishing train exit.

The temperature variation across the workpiece length can be defined as a difference between the temperatures measured at the middle and near the edge of the workpiece ΔT_W.

The strategy of controlling the workpiece temperature during hot rolling is twofold. Firstly, it is necessary to maintain the optimum temperature of the rolled piece which allows one to obtain the

550

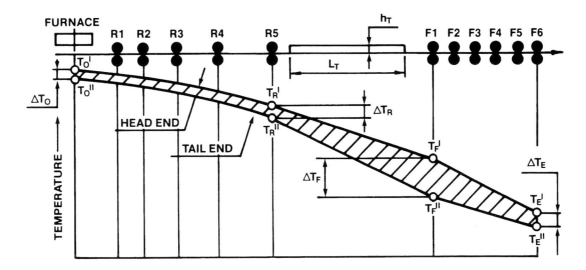

Fig. 30.1 Main parameters of temperature rundown in hot strip mill. (From Ginzburg and Schmiedberg, Iron and Steel Engineer, 1986. Copyright AISE, Pittsburgh, Pennsylvania. Reprinted with permission.)

desired properties of the rolled product with minimum energy consumption, required production rate, and maximum yield. Secondly, it is desirable to achieve a uniform workpiece temperature in both longitudinal and transverse direction during each rolling pass which helps to improve quality of the rolled product.

30.2 METALLURGICAL REQUIREMENTS

The boundary conditions for material temperature during the rolling deformation process are defined by metallurgical requirements [1].

   To ensure the homogeneity of the rolled product, all deformations in conventional hot rolling process are usually made in the austenitic phase. For low-carbon steel, this implies that the last deformation must occur at a strip temperature T_E above the phase transformation point between austenite and ferrite; for low-carbon steel the optimum range for T_E is 1550 to 1650°F.

   The second important metallurgical requirement is that the slab temperature T_0 be high enough to ensure dissolution of intermetallic phases or compounds resulting from the addition of alloying elements. From this point of view, the minimum value of T_0 for low-carbon steel is approximately 2000°F.

   The maximum value of T_0 is usually limited because of another metallurgical phenomenon related to excessive grain coarsening, which can have a detrimental effect on the final product. The maximum value of T_0 for low-carbon steel is approximately 2400°F [2].

   More detailed description of the metallurgical requirements for rolling of different types of steels is given in the following chapters.

30.3 ENERGY CONSUMPTION REQUIREMENTS

Energy consumption directly related to the hot rolling process can be divided into three components:

a) energy required for heating the slab in the reheat furnace

b) energy required for maintaining heat during transfer of the workpiece between rolling mill stands

c) energy required for hot rolling of the workpiece [1].

Savings in fuel energy consumption resulting from changing the slab dropout temperature from T_{01} to T_{02} can be defined as follows:

$$E_f = \frac{2}{\eta_m} c_s (T_{01} - T_{02}) 10^{-3}, \tag{30-1}$$

where E_f = fuel energy savings, MBtu/ton

c_s = slab specific heat, Btu/lb/°F

η_m = reheat furnace efficiency.

Savings in electrical energy consumption by the main mill drive motors can be achieved by increasing the slab dropout temperature, redistributing reductions between stands and changing mill speeds. These savings can be calculated as follows:

$$E_e = \frac{0.746}{\eta_r} (H_1 - H_2) \tag{30-2}$$

where H_1, H_2 = total specific power for rolling slab to coil for cases 1 and 2, hp-hr/ton

E_e = savings in electrical energy consumption, kwhr/ton

η_r = average mill drive efficiency.

30.4 YIELD REQUIREMENTS

Scaling of steel during reheating and rolling is a principal reason for reduction in yield of the hot rolled product. One of the most effective ways to reduce formation of scale is by lowering the slab dropout temperature.

Reduction in scale weight per unit of slab weight due to reduction in slab dropout temperature from T_{01} to T_{02} can be calculated from the following equation [1]:

$$m = \frac{a \sqrt{t_s}}{\rho_s} \left[\exp \left(\frac{-b}{T_{01} + 460} \right) - \exp \left(\frac{-b}{T_{02} + 460} \right) \right] \left(\frac{1}{h_s} + \frac{1}{w_s} + \frac{1}{L_s} \right) \tag{30-3}$$

where T_{01}, T_{02} = slab dropout temperatures, °F

m = weight of scale loss per unit of slab weight, tons/ton

t_s = scaling time, hr

w_s = slab width, in.

L_s = slab length, in.

h_s = slab thickness, in.

ρ_s = slab density, lb/cu in.

a, b = constants depending on type of steel. For low-carbon steel a = 31.7 and b = 22.83.

The slab dropout temperature is usually determined by metallurgical requirements and by power limitations of rolling mill.

30.5 PRODUCT QUALITY REQUIREMENTS

Temperature variation of the rolled material in both the longitudinal and the transverse direction is a major obstacle in maintaining the required strip gage, profile and shape tolerance.

The most drastic variation in the longitudinal direction occurs when the transfer bar enters the first finishing stand. Because the head end of the bar is usually transferred from the last roughing stand to the first finishing stand in less time than the tail end of the bar, the tail end is subjected to heat radiation loss for a longer time than the head end. The resulting temperature rundown increases with increasing slab weight. As will be shown later, if no preventive measures are taken to reduce this rundown, the temperature differential between head and tail end of the bar at the entry of the finishing train ΔT_F can be as much as 300°F for a 1000-PIW coil.

The adverse effect of this temperature differential on strip shape is inversely proportional to the rolled material thickness. The mean temperature differential (MTD) of the entire hot strip mill can be therefore calculated as follows [1]:

$$(MTD)^2 = \frac{\displaystyle\sum_{i=1}^{n} (\Delta T_i^2 / h_i^c)}{\displaystyle\sum_{i=1}^{n} h_i^{-c}} \tag{30-4}$$

where ΔT_i = temperature differential at exit of each pass

h_i = exit thickness for each pass

n = total number of rolling passes

c = constant depending on adverse effect of thickness on strip shape. For a conventional hot strip mill it is assumed that $c = 0.5$.

Rolled material temperature variation in the transverse direction is mainly due to excessive radiation near the edges where the surface-to-volume ratio increases substantially. If no measures are taken to reduce edge cooling, the transverse temperature variation ΔT_F can be as much as 180°F [3].

30.6 ANALYSIS OF TEMPERATURE CONDITIONS IN HOT STRIP MILL

Review of the foregoing requirements shows that there is no universal definition of optimum temperature conditions for hot strip mills. For example, a possible reduction in reheat furnace temperature due to heat conservation on the transfer table might not be fully utilized because of power limitations of the roughing train or, in another case, because of poor surface quality of the slabs loaded into the furnace, which requires maintaining the higher reheat furnace temperature needed to enhance the scaling process that helps to improve the slab surface.

These facts suggest that the optimum temperature conditions must be found for each hot strip mill on an individual basis. However, the following common criteria can be applied for objective evaluation of different solutions [1]:

a) reduction in mean temperature differential, MTD

b) reduction in primary scale, m

c) savings in fuel energy, E_f

d) savings in electrical energy consumption, E_e

e) total annual cost savings due to reheating and rolling optimization, S_t

f) total additional capital cost, C_t

g) payback time, PBT.

On the basis of equations (30-1) through (30-3), the following equation for total annual cost savings due to optimization of the reheating and rolling processes can be derived:

$$S_t = (C_m M + C_f E_f + C_e E_e)G, \tag{30-5}$$

where S_t = annual cost savings, \$/year

C_m = rolled material cost, \$/ton

C_f = fuel energy cost, \$/MBtu

C_e = electrical energy cost, \$/kwhr

G = annual production, tons.

The payback time can be defined as:

$$PBT = C_t/S_t, \tag{30-6}$$

where PBT = payback time, years

C_t = total additional capital cost, \$.

30.7 METHODS OF OPTIMIZING TEMPERATURE CONDITIONS

Because the temperature rundown in a hot strip mill is primarily due to heat radiation losses, factors affecting the heat radiation process must be considered. For the case when the workpiece thickness is much smaller than its width, Eq. (24-3) may be rewritten as:

$$\alpha_r = \frac{S\xi}{\rho ch} [(T - 460)^4 - (T_a + 460)^4] \tag{30-7}$$

Thus, the temperature loss rate due to radiation α_r can be reduced by the following methods:

a) decreasing the radiation time, t_r

b) increasing the rolled material thickness, h

c) increasing the ambient temperature, T_a

d) decreasing the initial transfer bar temperature, T.

These methods are utilized in the technology for optimizing temperature condition in hot strip mills listed below [1].

1. **Optimizing operating parameters** - The most effective methods for optimizing operating parameters include:

Lowering slab heating temperature

Uneven slab heating

Increasing the transfer bar thickness, and

Zoom rolling, which is a gradual acceleration of rolling speed in the finishing train.

2. **Optimizing mill configuration** - The main approaches to optimizing mill configuration may include:

Close coupling of continuous rougher with finishing mill

Close coupling of a reversing rougher with finishing mill

Combining a reversing rougher with a continuous finishing mill.

3. **Conserving and adding heat** - Typical systems for conserving and adding heat in hot strip mills are:

Coilbox

Intermediate Steckel mill

Intermediate reheat facilities

Thermal covers.

A more detailed description of the technology for optimizing temperature conditions in hot strip mill is given in the following text.

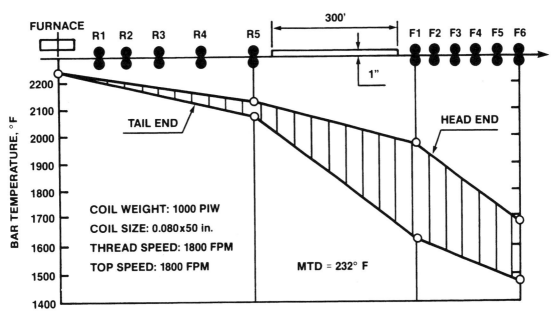

Fig. 30.2 Simulated temperature rundown in hot strip mill prior to optimization of operating parameters. (From Ginzburg and Schmiedberg, Iron and Steel Engineer, 1986. Copyright AISE, Pittsburgh, Pennsylvania. Reprinted with permission.)

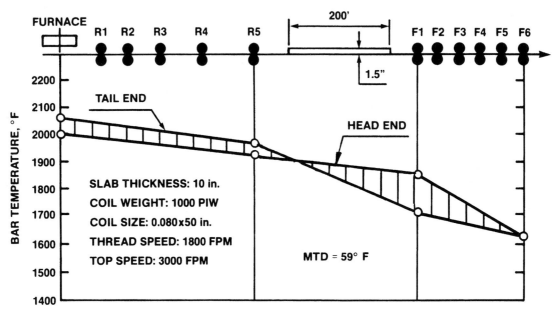

Fig. 30.3 Simulated temperature rundown in hot strip mills after optimization of operating parameters. (From Ginzburg and Schmiedberg, Iron and Steel Engineer, 1986. Copyright AISE, Pittsburgh, Pennsylvania. Reprinted with permission.)

30.8 OPTIMIZING OPERATING PARAMETERS

For quantitative evaluation of some methods for optimizing operating parameters, an off-line computer model has been utilized [1]. Rolling parameters of a continuous hot strip mill have been calculated for the following conditions: slab thickness, 10.0 in.; slab width, 50.0 in.; slab weight, 1000 PIW; slab material, low-carbon steel; and strip thickness, 0.080 in.

The simulated temperature rundown for both head and tail ends of the product rolled in a continuous hot strip mill prior to the optimization of operating parameters is shown in Fig. 30.2. In this simulation, the slab was assumed to be evenly heated. It was rolled in roughing train to a transfer bar thickness of 25.4 mm (1.0 in.) and then rolled in finishing mill at constant speed (no zoom). After optimizing operating parameters by lowering slab heating temperature, uneven slab heating, increasing the transfer bar thickness to 38.1 mm (1.5 in.) and introducing the zoom rolling, the mean temperature differential, MTD, was reduced from 129°C (232°F) to 33°C (59°F) as shown in Fig. 30.3. In addition, this optimization produces substantial savings in both fuel energy and scale losses with a relatively small increase in electrical energy that is due to higher loading of the roughing mill stands.

It is worth mentioning that uneven slab heating leads to cost savings only if the average slab dropout temperature is lowered as a result of the optimization. Also, in zoom rolling, it is difficult to achieve strip temperature uniformity in the longitudinal direction throughout the entire length of the strip leaving the finishing train [3]. This is mainly because the finishing train is usually accelerated only after the head end of the strip is engaged in the coiler. As a result, the strip temperature profile in the

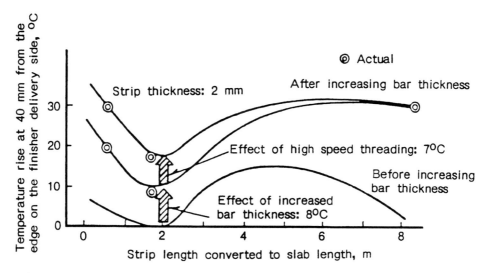

Fig. 30.4 Temperature distribution in the longitudinal direction of coil. Adapted from Sugita and Oi (1984).

longitudinal direction has a pronounced dip. This problem may be alleviated by increasing transfer bar thickness and by using high speed threading as shown in Fig. 30.4.

Another drawback of zoom rolling is that the strip temperature rundown at the entry of the finishing train remains large. In addition, zoom rolling requires a substantial increase in available rolling power, which calls for a substantial increase in capital costs. This investment, however, can be justified if an increase in the production rate is also required.

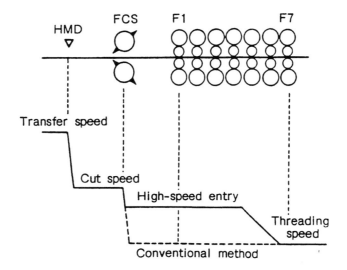

Fig. 30.5 Principle of high-speed entry rolling method. Adapted from Sasada, et al (1981).

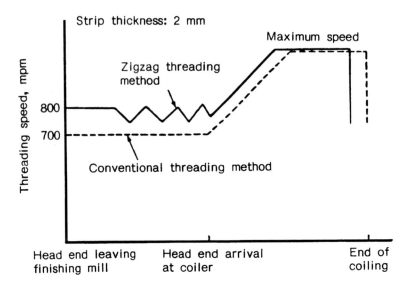

Fig. 30.6 Principle of zigzag threading method. Adapted from Sasada, et al (1981).

Some improvements in zoom rolling practice may be achieved by applying two methods developed by Nippon Steel [4]:

1. High-speed entry rolling method
2. Zigzag threading method.

The high-speed entry rolling method involves threading the strip through the finishing mill at higher speed until its head end reaches the last mill stand as shown schematically in Fig. 30.5.

Under the zigzag method, the strip exits finishing mill at the speed higher than that in conventional rolling. The strip loop caused by high traveling speed is eliminated by short-time deceleration. The strip is repeatedly accelerated and decelerated in a zigzag manner to increase the average threading speed as illustrated in Fig. 30.6.

30.9. CLOSE COUPLING OF CONTINUOUS ROUGHER WITH FINISHING MILL

Close coupling of continuous roughing mill stands with the finishing train allows an increase in transfer bar thickness. According to Eq. (30-7), since rolling cycle time remains practically the same, temperature losses of the transfer bar decrease as its thickness increases. The effect of increasing the transfer bar thickness on heat conservation was recognized in 1979 when a heavy reduction intermediate stand (an M stand) was installed in front of the finishing train at Nippon Steel's Muroran works [5].

This method of heat conservation has also been utilized in the concept of a continuous tandem hot strip mill [6, 7]. An example is shown in Fig. 30.7 which illustrates steps by which a typical continuous hot strip mill with five roughing mill stands is gradually converted to a continuous tandem hot strip mill by relocating the roughing mill stands closer to the finishing train. As a result, the transfer bar

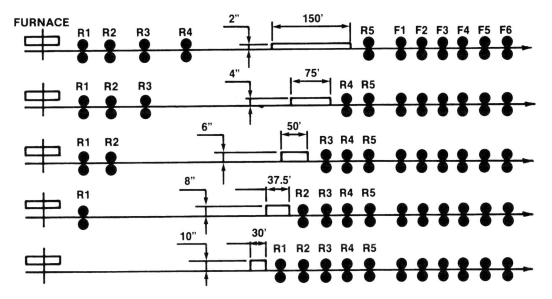

Fig. 30.7 Conversion of continuous hot strip mill to continuous tandem hot strip mill. (From Ginzburg and Schmiedberg, Iron and Steel Engineer, 1986. Copyright AISE, Pittsburgh, Pennsylvania. Reprinted with permission.)

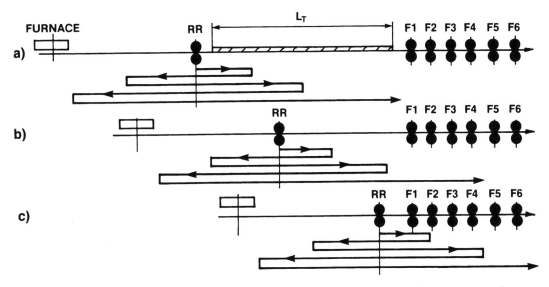

Fig. 30.8 Close coupling of reversing roughing mill with finishing train. (From Ginzburg and Schmiedberg, Iron and Steel Engineer, 1986. Copyright AISE, Pittsburgh, Pennsylvania. Reprinted with permission.)

thickness increases each time another roughing mill stand is relocated. After relocating the last mill stand, the transfer bar thickness will be equal to the slab thickness. The computer study shows [1] that the final mill configuration will reduce the mean temperature differential to 14°C (26°F).

In this configuration, however, the roughing passes are rolled at much lower speed than originally, which may produce two negative effects: overheating of rolls and excessive formation of secondary scale. A possible solution is the use of high-pressure descaling water at each roughing pass, which would require a higher slab dropout temperature.

30.10 CLOSE COUPLING OF A REVERSING ROUGHER WITH FINISHING MILL

Close coupling of a reversing roughing mill with the finishing train provides results that are similar to those obtained with the previous approach, although with less efficiency. Three possible arrangements for close coupling of a reversing roughing mill are shown in Fig. 30.8:

1. A conventional semi-continuous hot strip mill with the distance between the reversing roughing mill and finishing train being greater than the length of the transfer bar L_T (arrangement a).

2. A semi-continuous hot strip mill with the distance between the reversing roughing mill and finishing train being great enough to accommodate the rolled bar during all intermediate roughing passes but less than the length of the transfer bar (arrangement b). As a result, the roughing mill will be close-coupled with the finishing train during the last roughing pass [8].

3. A semi-continuous hot strip mill with the distance between the reversing roughing mill and finishing train being less than the length of the rolled bar during intermediate roughing passes (arrangement c). In this case, the roll gaps of the finishing mill stands are kept open until all intermediate roughing passes are completed [9].

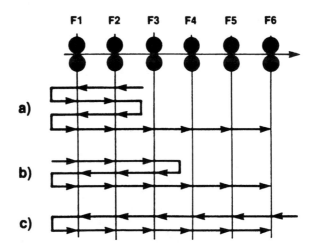

Fig. 30.9 Combined reversing roughing mill and continuous tandem finishing train. (From Ginzburg and Schmiedberg, Iron and Steel Engineer, 1986. Copyright AISE, Pittsburgh, Pennsylvania. Reprinted with permission.)

30.11 COMBINING A REVERSING ROUGHER WITH FINISHING MILL

Combining reversing roughing mill with continuous finishing train is a logical extension of the two previous approaches. This idea is illustrated in Fig. 30.9, which shows three pass schedules (a, b and c). In all three schedules, the last six finishing passes are made in a conventional way by continuous rolling on all six stands. The difference between the three schedules is in the sequencing of the roughing passes [1].

In schedule a, six roughing passes are made on stands F1 and F2. In schedule b, the roughing passes are made on stands F1, F2 and F3. In schedule c, an ultimate utilization, all stands are used for roughing passes. Although the use of schedule c will result in maximum production, it is the least desirable when strip surface quality is critical, since the last mill stands are used for both roughing and finishing passes.

A combination of reversing roughing mill with continuous finishing train is the shortest hot strip mill arrangement which provides finishing passes with close-coupled mill stands. Although its production rate is lower than that of the continuous tandem hot strip mill, its capital investment may be less. Also, it can roll the roughing passes at much higher speeds, which reduces roll firecracking as well as secondary scale formation.

30.12 COILBOX

The Stelco Coilbox [10] is usually installed at the entry side of the finishing train. Upon leaving the mill after the last roughing pass, the transfer bar is directed into the Coilbox entry chute and through the bending rolls to form the coil (Fig. 30.10a). This coil is then transferred to the uncoiling position for threading through the finishing train while the next transfer bar is being coiled.

Heat conservation in the Coilbox is due to the increase in equivalent thickness of the body subjected to radiation, and to the increase in ambient temperature differential inside the eye of the coil.

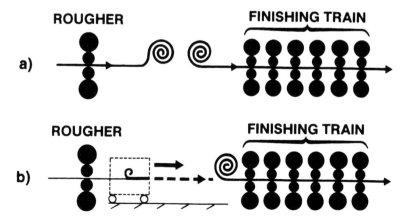

Fig. 30.10 Basic concept of Coilbox (a) and inverse arrangement (b). (From Ginzburg and Schmiedberg, Iron and Steel Engineer, 1986. Copyright AISE, Pittsburgh, Pennsylvania. Reprinted with permission.)

Also, the change in direction of the transfer bar helps reduce the temperature differential of the bar in the longitudinal direction. In some cases, however, special measures must be taken to prevent an excessive temperature rise, especially when the finishing train is being accelerated to increase production rate.

Use of the Coilbox allows a substantial decrease in transfer table length and reduces installed rolling mill power. These features may partly offset the Coilbox initial cost. Additional maintenance and operating costs may be offset by savings due to lower temperature of the slab leaving the reheat furnace, improved yield and better gage control. However, with the Coilbox, it is difficult to automatically control the longitudinal temperature gradient. The most convenient method of control, by varying the finishing train speed, is not practical because a speed increase would result in an undesirable temperature rise, whereas speed decrease would lower production rate.

This problem has been solved in the inverse design of the Coilbox (Fig. 30.10b), which provides coiling of the transfer bar beginning from its tail end rather than head end [11]. In the inverse arrangement, the coiling process begins after the transfer bar head end enters the finishing train and the tail end leaves the roughing train. This method of control allows the required temperature rundown to be maintained by adjusting the coiling rate. It can also be applied when mill acceleration is desirable.

30.13 INTERMEDIATE STECKEL MILL

Installation of Steckel mill in front of the finishing train [12] can be expected to have effects combining those results from installation of the M stand and the Coilbox. In this arrangement, a thicker transfer bar can enter the Steckel mill (a feature similar to that with the M stand arrangement), and then after passing through the mill stand, the material is coiled (a feature similar to that with the Coilbox).

As an example, the case when three reversing passes are being made at the Steckel mill before material enters the finishing train is illustrated in Fig. 30.11a. A major drawback of this arrangement is

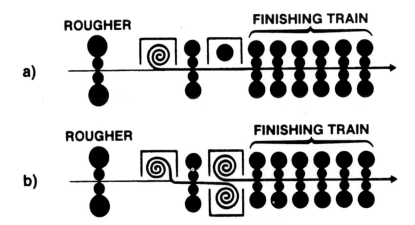

Fig. 30.11 Steckel mill as intermediate stand between roughing and finishing trains. (From Ginzburg and Schmiedberg, Iron and Steel Engineer, 1986. Copyright AISE, Pittsburgh, Pennsylvania. Reprinted with permission.)

that it does not allow use of the Steckel mill to roll the next bar during the final finishing pass of the previous bar; consequently, it results in a lower production rate. This problem is eliminated by introduction of a third coiling furnace [13] (Fig. 30.11b).

30.14 RERADIATING THERMAL COVER SYSTEM

The concept of a complete reradiating thermal cover system for a hot strip mill [1] is illustrated in Fig. 30.12. Each top cover can have three positions over the transfer table: top (maintenance) position, intermediate (threading) position, in which the covers are elevated approximately 1250 mm (50 in.) above the table, and rolling position, in which the gap between the top cover and the transfer table can be as low as 300 mm (12 in.). The ability to position the covers allows for reduction of the air gap between the transfer bar and the reradiating shield during rolling on the finishing train. Also, it insures maximum system reliability, thus reducing maintenance cost.

For example, during transfer of the head end of the bar from the roughing train to the shear, the thermal covers can be kept in threading position, which would protect them from possible damage due to abnormal rolling conditions such as bar turn-up, cobble, etc. Then, as soon as the head end of the bar reaches the shear, the covers can be lowered to rolling position. The required temperature rundown can be achieved by regulating the elevation of the thermal covers during threading as suggested by Gray [14].

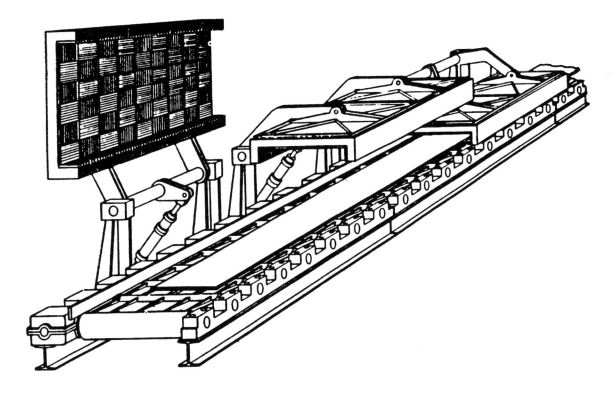

Fig. 30.12 Schematic presentation of United Engineering reradiating thermal cover system. (From Ginzburg and Schmiedberg, Iron and Steel Engineer, 1986. Copyright AISE, Pittsburgh, Pennsylvania. Reprinted with permission.)

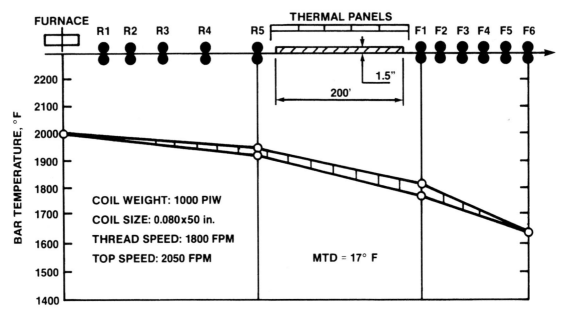

Fig. 30.13 Simulated temperature rundown in hot strip mill with the thermal cover system installed between the roughing and finishing mills. (From Ginzburg and Schmiedberg, Iron and Steel Engineer, 1986. Copyright AISE, Pittsburgh, Pennsylvania. Reprinted with permission.)

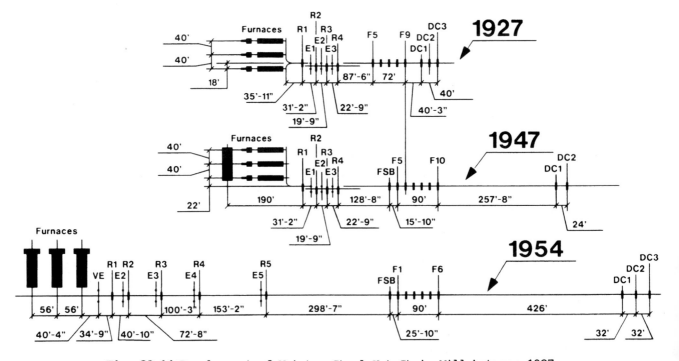

Fig. 30.14 Development of Weirton Steel Hot Strip Mill between 1927 and 1954. (From Ginzburg, et al, Iron and Steel Engineer, 1988. Copyright AISE, Pittsburgh, Pennsylvania. Reprinted with permission.)

The effect of the thermal cover system with reradiating insulating blocks installed above and underneath the transfer table on the entire temperature rundown of the hot strip mill is illustrated in Fig. 30.13, showing a mean temperature differential of 9.5°C (17°F).

30.15 MAIN OBJECTIVES IN MODERNIZATION OF HOT STRIP MILL

The first continuous hot strip mills were built more than sixty years ago. Some of them are still in operation in spite of a tremendous change in the production requirements. The survival of these mills was possible only due to their periodic modernization. As an example, Fig. 30.14 illustrates the changes in the layout of Weirton Steel Hot Strip mill which took place between 1927 and 1954. The main parameters of the mill have been changed as shown below [15]:

Year	1927	1947	1954
Mill design capacity, short million/year	0.38	1.62	2.4
Total main drive horse power, HP	16,600	28,000	57,000
Maximum mill speed, fpm	1,100	1,590	2,400
Maximum unit coil weight, PIW	115	255	830

The main objectives for modernization of the existing hot strip mills may vary from one mill to another. At Weirton Steel those objectives in the recent modernization program have included:

1. Increase in the mill design capacity
2. Improving quality of the rolled product

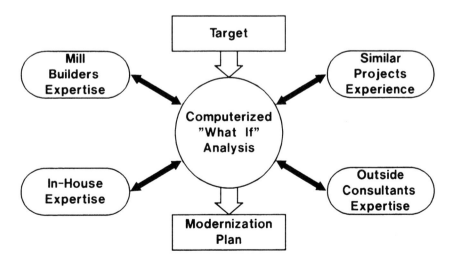

Fig. 30.15 Strategy in determining the design for modernized rolling mill.

3. Modification of the rolled product mix

4. Increase in weight of the rolled coils

5. Increasing the edging capabilities

6. Providing high degree of automation of the rolling process.

Once the main objectives for modernization were established, Weirton Steel has utilized its own experience as well as the experience of other experts to limit a number of possible solutions viable for a further detailed analysis (Fig. 30.15). The comparative analysis of the proposed solutions has been made by utilizing the off-line computer model MILLMAX^R, developed by United Engineering, Inc. and International Rolling Mill Consultants, Inc. (Pittsburgh, PA). The model is designed for an objective analysis of the capabilities of hot strip mills with different layout.

30.16 REQUIREMENTS FOR THE EVALUATION MODELS

In order to accurately evaluate and also to objectively compare different proposed solutions for modernization of the mill, the off-line computer model has to be provided with the following properties [16]:

1. Applicability to wide variety of existing or perspective mill configurations.

2. Accuracy in simulation of existing rolling conditions.

3. Extrapolative accuracy, or the capability to be accurate for conditions different than those that existed during calibration of the model.

4. Efficiency in utilization, or the capability to be accurate when a limited number of empirical constants is used for wide variety of rolled products.

5. Efficiency in calibration, or the capability to utilize standard test equipment and on-line instrumentation available in the mill.

The best known method to achieve the accuracy in simulation of existing rolling conditions is by calibrating the model using actual rolling data. This method allows one to take into account specific features of each particular mill. The process of calibration involves statistical analysis of the data obtained during rolling of the preselected products.

Another vital aspect is a provision of the extrapolative accuracy of the model. This is usually achieved by using in the model the algorithms, which are based on the physics of the processes taking place during rolling of metals.

30.17 EVALUATION OF THE SOLUTIONS FOR MILL MODERNIZATION

Figure 30.16 presents schematically the hot strip mill arrangements used in sensitivity analysis at Weirton Steel [15]. The following criteria have been used in this analysis:

1. Mill production rate has to be 3,000,000 tons/year.

2. Mill has to provide rolling of both carbon steel and HSLA products.

3. Mill has to roll the required strip width with the slab width increments up to 4 in.

4. Mill has to provide optimum temperature condition for rolled materials.

Table 30.1 Comparative performances of different mill arrangement shown in Fig. 30.16. Rolled product: low-carbon steel 2.92 mm (0.115 in.)-thick and 1024 mm (40.3 in.)-wide [15].

Parameter	Units	Mill arrangement 1	2	3&4	5
Slab thickness	in.	8.5	10.0	10.0	10.0
Unit coil weight	PIW	712	1144	1144	1144
Thread speed	fpm	1800	1800	1800	1800
Top speed	fpm	1800	2170	2200	2750
Production rate:	tons/hr				
furnace		825	800	800	800
roughing train		620	1300	1070	915
finishing train		550	745	775	830
Material temperature differential:	°F				
mean		65	47	44	37
exit		55	0	0	0
Savings:	$/ton				
fuel		base case	0.096	0.0479	0.1466
yield		base case	0.233	0.121	0.334
electric power		base case	-0.106	-0.0452	-0.165
Total savings	&/year	base case	667,980	371,876	948,693

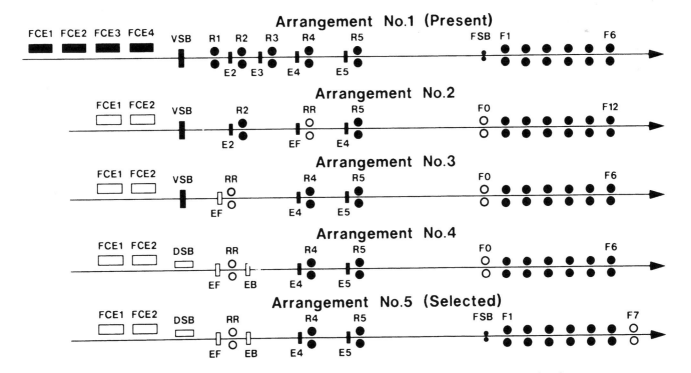

Fig. 30.16 Hot Strip Mill arrangements used in sensitivity analysis at Weirton Steel. (From Ginzburg, et al, Iron and Steel Engineer, 1988. Copyright AISE, Pittsburgh, Pennsylvania. Reprinted with permission.)

In addition to the above criteria, the following factors have been taken into account:

1. Adverse effect of temperature differential between head and tail ends of the strip on its shape.

2. Scale losses in reheat furnace.

3. Conservation of energy.

The results of the comparative analysis are summarized in Table 30.1. The mill arrangement No. 5 was found superior to the others in meeting the main objectives and, therefore, was selected to be used for the modernized hot strip mill. As follows from Fig. 30.16, the selected mill arrangement calls for a replacement of the existing vertical scalebreaker VSB, the roughing mill stands R1, R2 and R3 and the attached edgers E2 and E3 with a new equipment that included a descaling box DSB, a reversing rougher RR with front and back edgers EF and EB. It also has a provision for an additional finishing mill stand F7.

Later development of the modernization program at the Weirton Steel had led to an installation of the reradiating heat covers [1] on the transfer table between roughing and finishing mills.

REFERENCES

1. V.B. Ginzburg and W.F. Schmiedberg, "Heat Conservation Between Roughing and Finishing Trains of Hot Strip Mills", Iron and Steel Engineer, April 1986, pp. 29–39.

2. The Making, Shaping and Treating of Steel, 10th Edition, eds. W.T. Lankford, Jr., et al, Association of Iron and Steel Engineers, Pittsburgh, Pennsylvania, p. 851 (1985).

3. K. Sugita and J. Oi, "Maintenance of Temperature in Hot Strip Mill Under Continuous Casting - Direct Rolling Process", Nippon Steel Technical Report No. 23, June 1984.

4. T. Sasada, et al, "Modernization Technology of Conventional Hot Strip Mills", Nippon Steel Technical Report, No. 18, Dec. 1981, pp. 1-21.

5. N. Kamii and R. Terakado, "Modernization of 56-Inch Hot Strip Mill at Nippon Steel's Muroran Works", AISE Year Book, 1981, pp. 66-70.

6. V.B. Ginzburg, US Patent No. 4,444,038, Apr. 24, 1984.

7. V.B. Ginzburg, US Patent No. 4,430,876, Feb. 14, 1984.

8. G.W. Tippins and V.B. Ginzburg, US Patent No. 4,433,566, Feb. 28, 1984.

9. G.W. Tippins and V.B. Ginzburg, US Patent No. 4,503,697, March 12, 1985.

10. W. Smith and A.G. Watson, "The Coilbox - A New Approach to Hot Strip Rolling", AISE Year Book, 1981, pp. 342-436.

11. V.B. Ginzburg and J.E. Thomas, US Patent No. 4,491,006, Jan 1, 1985.

12. G.W. Tippins, US Patent No. 4,348,882, Sept. 14, 1982.

13. G.W. Tippins, V.B. Ginzburg and W.G. Pottmeyer, US Patent No. 4,430,874, Feb. 14, 1984.

14. R. Gray, US Patent No. 3,344,648, Oct. 3, 1967.

15. V.B. Ginzburg, et al, "Application of the Off-line Computer Model MILLMAX at Weirton Steel's Hot Strip Mill", Iron and Steel Engineer, June 1988, pp. 24-33.

16. V.B. Ginzburg, "Basic Principles of Customized Computer Models for Cold and Hot Strip Mills", Iron and Steel Engineer, Sept. 1985, pp. 21-35.

Part X

Geometry of Flat Products

31

Geometrical Characteristics of Flat Products

31.1 FACTORS AFFECTING WORKPIECE PROFILE

Profile of a flat product may be described as 'an outline of a section across the width of the product (Fig. 31.1), as defined by a set of thickness measurements at assigned points, or at assigned increments of width' [1].

When the workpiece is reduced by plastic deformation only, there is no elastic recovery after rolling. In that case, the workpiece profile is completely determined by the roll gap profile.

There are four principal factors affecting the roll gap profile:

1. Vertical plane displacement of the rolls
2. Horizontal plane displacement of the rolls
3. Roll thermal crown
4. Roll wear.

Vertical plane displacement - Displacement of the rolls in vertical plane is produced by:

a) mill stretch that is a result of extension and contraction of the mill parts, retaining the rolls, due to both rolling load and heat,

b) roll bending that is caused by both rolling load and force generated by vertical plane roll bending cylinders,

c) change in thickness of the hydrodynamic lubricant film in the roll bite,

d) change in thickness of the hydrodynamic lubricant film in the backup roll bearings.

Horizontal plane displacement - Displacement of the rolls in horizontal plane may be caused by:

a) horizontal component of rolling load acting on a work roll which has a centerline offset from the centerline of an adjacent backup roll,

b) roll bending due to force generated by horizontal plane roll bending mechanism,

c) roll displacement and bending due to unequal strip tension forces acting at the entry and exit sides of the roll bite.

Roll thermal crown - The roll thermal crown is defined as an increment in roll diameter caused by roll heating and cooling during rolling. In some instances, the roll thermal crown is purposely introduced by roll preheating.

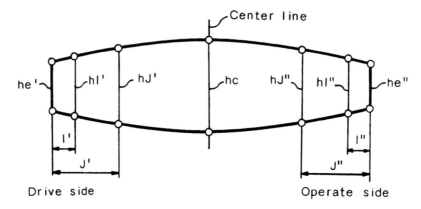

Fig. 31.1 Outline of a section across the width of the strip.

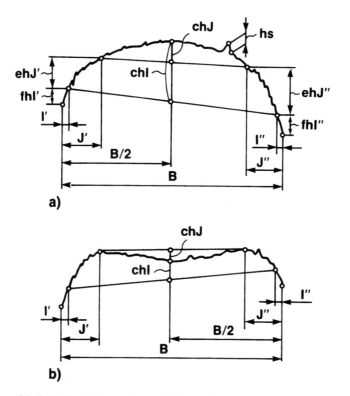

Fig. 31.2 Definition of profile: a) overall and partial center crowns with the same signs and b) overall and partial center crowns with the opposite signs. (From Ginzburg, Iron and Steel Engineer, 1987. Copyright AISE, Pittsburgh, Pennsylvania. Reprinted with permission.)

Roll wear - This is a gradual deterioration of the roll contour due to abrasive, corrosive and adhesive wear.

31.2 DEFINITION OF ELEMENTS OF PROFILE

Profile of flat products is usually described in terms of center gauge, edge thickness, level, wedge, crown, edge drop, feather, ridge, and valley [1, 2], as shown in Fig. 31.1 and 31.2.

Center gauge hc - The center gauge is the thickness of a workpiece at its centerline.

Edge thicknesses hI and hJ - The edge thicknesses hI and hJ are measured at distances I and J from extreme edges of the workpiece respectively:

a) Drive side hI' and hJ'

b) Operate side hI'' and hJ''.

Selected distances for I vary from 9.5 to 19 mm (0.375 to 0.75 in.) and for J from 50 to 75 mm (2 to 3 in.).

Level δhI - The profile level is determined as the difference in edge thickness between drive and operating sides

$$\delta hI = hI' - hI'' \tag{31-1}$$

Wedge - Drive side and operate side are considered:

a) Drive side wedge is when

$$hI' > hc > hI'' \tag{31-2}$$

b) Operate side wedge is when

$$hI'' > hc > hI' \tag{31-3}$$

Crown - The crown is defined as the difference between the center gauge hc and specified edge thickness. Four types of crown are identified below.

a) Center crown, or overall center crown chI, is given by

$$chI = hc - \frac{(hI' + hI'')}{2} \tag{31-4}$$

b) Partial center crown chJ is equal to

$$chJ = hc - \frac{(hJ' + hJ'')}{2} \tag{31-5}$$

c) Drive side crown chI' is equal to

$$chI' = hc - hI' \tag{31-6}$$

d) Operate side crown is equal to

$$chI'' = hc - hI'' \qquad (31\text{-}7)$$

Edge drop - The edge drop is defined as the difference between the edge thicknesses hI and hJ. Three types of edge drop are described below:

a) Average edge drop is given by

$$eh = \frac{(hJ' + hJ'' - hI' - hI'')}{2} \qquad (31\text{-}8)$$

b) Drive side edge drop is given by

$$eh' = hJ' - hI' \qquad (31\text{-}9)$$

c) Operate side edge drop is given by

$$eh'' = hJ'' - hI'' \qquad (31\text{-}10)$$

Feather - The feather defines the change in thickness near extreme edges along distances I. Three types of feather may be considered:

a) Average feather is expressed as

$$fh = \frac{(hI' + hI'' - he' - he'')}{2} \qquad (31\text{-}11)$$

where he', he'' = thicknesses at the extreme edges of drive and operate side respectively.

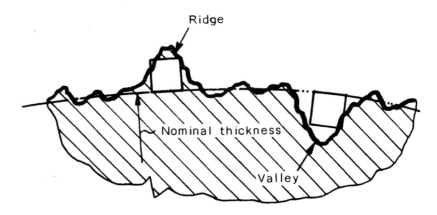

Fig. 31.3 Schematic presentation of ridge and valley.

Thicknesses he' and he'' may be measured from 2 to 3 mm (0.08 to 0.12 in.) from the extreme edges.

b) Drive side feather is expressed as:

$$fh' = hI' - he'$$ (31-12)

c) Operate side feather is expressed as:

$$fh'' = hI'' - he''$$ (31-13)

Ridge/Valley - The ridge and valley are defined as the deviations in thickness either above or below the nominal thickness in the local strip profile that projects beyond the rectangular window of predefined size (Fig. 31.3). Recommended minimum dimensions of the window: width = 10 mm (0.4 in.) and height = 10 μm (0.0004 in.). The nominal thickness can be determined by a non-linear curve-fitting method utilizing regression analysis.

31.3 ORIGIN OF BAD FLATNESS

Strip flatness, or shape, may be described as a 'measurable parameter describing the presence or absence of waves or buckles and their size and location' on the workpiece [1].

Bad flatness, or shape, of both hot and cold rolled strip is caused by differential elongation across its width [3-5]. These differences produce corresponding internal stresses within the strip. These stresses, giving rise to shape deficiencies, can be compressive or shear in nature. The compressive

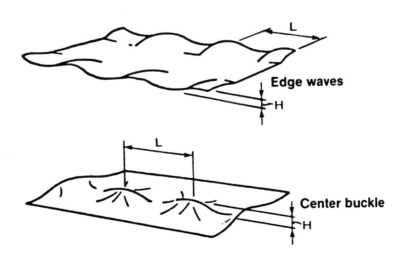

Fig. 31.4 Strip shape parameters. (From Sivilotti, et al, AISE Year Book, 1973. Copyright AISE, Pittsburgh, Pennsylvania. Reprinted with permission.)

stresses produce buckles parallel to the strip width (Fig. 31.4). According to Wistreich [6], these strip forms were defined as long edge (edge waves), long middle (center buckle) and quarter buckle. The shear stresses result in buckles diagonally disposed across the strip, known as herringbone.

When external tension is applied to the strip, the external tensile stresses will be combined with internal stresses. As a result, the compressive stresses in the strip will be reduced, giving the appearance of better shape. The differential elongation of the strip is caused by local mismatch between strip and roll-gap profiles under load. Long edge and long middle are due to differences in crown between strip and roll, whereas the more complex irregularities in shape are caused by roll wear, uneven temperature profiles, ridges of incoming strip, inhomogeneous metallurgy of strip, nonuniform lubrication, etc. [4].

31.4 FORMS OF STRIP SHAPE

With respect to shape control, the following four major forms of strip shape (Fig. 31.5) are usually considered [4].

Ideal - Ideal shape relates to a purely theoretical case when the internal stresses are equal across the width of the strip. This ideal flat shape will be maintained after external tension is removed and also after the strip has been slit.

Latent - Latent shape corresponds to the case when the internal stresses are not equal across the width of the strip but the section modulus of the strip is sufficiently large to resist formation of buckles. Strip with latent shape will appear flat without external tension. However, slitting of the strip will, by releasing the latent forces, cause irregularities in shape.

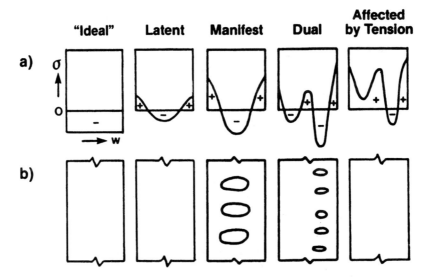

Fig. 31.5 Form of strip shape: a) stress distribution and b) strip appearance. (From Kelk, et al, Iron and Steel Engineer, 1986. Copyright AISE, Pittsburgh, Pennsylvania. Reprinted with permission.)

Manifest - Manifest shape appears when the internal stresses are not equal across the width of the strip and, at the same time, the section modulus of the strip is not sufficient to resist formation of buckles. It results in local elastic buckling. Under appropriate external tension, the overall compressive stresses may be reduced to such a level that it results in transformation of manifest shape into latent shape. On the other hand, removing the external tension and slitting emphasizes the manifest shape.

Dual - Dual shape covers such cases as when one portion of the strip has latent shape while another portion has manifest shape [7]. Long edge or quarter buckle on one side of the strip would be typical examples of this form of strip shape.

31.5 DEFINITION OF STRIP FLATNESS

Strip flatness can qualitatively be defined as a deviation of latent, manifest and dual shapes from ideal shape. Practically, this deviation can be found by slitting the strip into separate narrow ribbons and subsequently measuring the lengths. Using this concept, Pearson [8] has proposed the following definition of strip shape:

$$\Sigma = \frac{(L_m - L_n)}{L_n b} \qquad (31\text{-}14)$$

where $\quad \Sigma$ = strip shape, mons/cm

L_m, L_n = lengths of two ribbons of the strip, cm

$\quad b$ = distance between two ribbons, cm.

Wilmotte, et al [9] have defined strip shape as:

$$\rho = \frac{L - L_{40}}{L} \qquad (31\text{-}15)$$

where ρ = shape index (nondimensional)

$\quad L$ = length of the middle ribbon of the strip

L_{40} = length of the ribbon 40 mm (1.575 in.) from the strip edge.

However, the following three flatness unit formulae have mostly been adopted in steel industry [1, 10]:

I-unit - The I-unit is expressed in the form (Fig.31.4)

$$I = \frac{\Delta L}{L} 10^5 \qquad (31\text{-}16)$$

where $\quad I$ = strip shape, I-units

$\Delta L/L$ = strip waviness

$\quad \Delta L$ = difference between the longest and shortest ribbon across the width of the strip

$\quad L$ = wavelength.

Peak-to-peak height - The peak-to-peak height H is defined as a distance between peaks of two adjacent buckles or waves.

% Steepness - The percent steepness is equal to

$$S = \frac{H}{L} 10^2 \tag{31-17}$$

The relationships between the I-unit, peak-to-peak height H, and steepness S are [1]

$$I = \left(\frac{\pi H}{2L}\right)^2 10^5 = 2.5(\pi S)^2 \tag{31-18}$$

$$H = \frac{2L}{\pi} \sqrt{I \times 10^{-5}} = \frac{LS}{100} \tag{31-19}$$

$$S = \frac{2}{\pi} \sqrt{I \times 10^{-1}} = \frac{H}{L} 10^2 \tag{31-20}$$

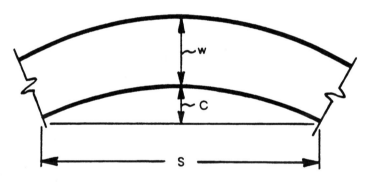

Fig. 31.6 Camber.

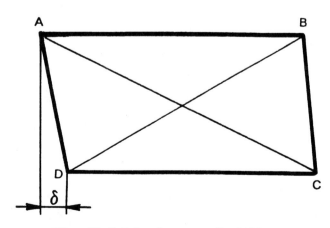

Fig. 31.7 Out-of-square deviation.

31.6 PLAN VIEW GEOMETRY

Two parameters are conventionally used to evaluate workpiece plan view geometry, camber and out-of-square deviation.

Camber - As shown in Fig. 31.6, 'camber is the deviation of a side edge from a straight line, the measurement being taken on the concave side with a straight edge' [11].

Out-of-square deviation - The out-of-square deviation (Fig. 31.7) is defined as 'the greatest deviation of an end edge from a straight line at right angle to a side and touching one corner. It is also obtained by measuring the difference between the diagonals (AC and BD) of the cut length sheet. The out-of-square deviation is one-half of that difference' [11].

31.7 TYPES OF TOLERANCE FOR WORKPIECE GEOMETRY

There is presently no unified system of presentation of tolerances for geometry of rolled product. The known types of tolerances can be classified as follows.

Absolute tolerances - The absolute tolerances are expressed in the same units as the measured parameter.

Relative tolerances - The relative tolerances are usually given as percentage of nominal value of the measured parameter.

It is also customary to apply the tightest tolerances to a limited length, say 98%, of the product length while having less tight tolerances for the remainder of the length. In case of the profile measurements, the data taken near the edge, say as close as 19 mm (0.75 in.) from the edge, are often excluded.

One of the most significant trends in quality control of rolled products is a recognition of the inherent variability of the rolling process and application of statistical terms, such as standard deviation, for evaluation of the geometrical parameters of the products.

31.8 STANDARD DEVIATION

Data obtained from measurement of geometrical parameters of the rolled product may be presented as a set of numerical values measured at the predetermined length intervals. This set of data is called **population**.

If in the population, there are N observations, labeled $x_1, x_2, \ldots x_n$, the **population mean** is [12]:

$$\mu = \frac{\sum_{i=1}^{N} x_i}{N} \tag{31-21}$$

The dispersion of the measured values from the mean is defined by the **population variance**

$$\sigma^2 = \frac{\sum_{i=1}^{N} (x_i - \mu)^2}{N} \tag{31-22}$$

The population variance can be used to compare dispersion of two or more population distributions.

The variance of a single population is analyzed by using the square root of the variance. The resulting quantity is called **standard deviation**

$$\sigma = \sqrt{\sigma^2}$$

(31-23)

31.9 HISTOGRAM

The information obtained from the measurements can be presented pictorially, using a **histogram.** To draw a histogram (Fig. 31.8), the following steps are taken:

1. The **range** of a set data is determined as a difference between the largest and smallest measured values and is defined as

$$R = x_{max} - x_{min}$$

(31-24)

2. The range is divided into m equal subranges known as **classes** and equal to:

$$r = R/m$$

(31-25)

3. The number of measurements in each class, referred to as **frequencies,** is determined

4. The class boundaries are marked along a horizontal scale of the histogram

5. On the top of each class interval is drawn a rectangle, whose area is proportional to the frequency of that class. Since the class intervals are all of the same width, the heights of the rectangles are also proportional to the frequencies.

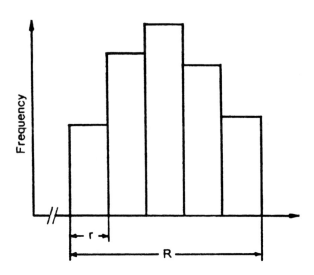

Fig. 31.8 Histogram.

31.10 NORMAL DISTRIBUTION CURVE

When a number of measurement is very large, the histogram can be mathematically expressed by the equation for normal distribution:

$$f(x) = \frac{1}{\sigma\sqrt{2\pi}} \exp\left[-\frac{1}{2}\left(\frac{x - \mu}{\sigma}\right)^2\right] \tag{31-26}$$

where $-\infty < x < \infty$

This curve (Fig. 31.9) has its maximum at the point $x = \mu$ and tapers off toward zero as x becomes either very large or very small.

The following significant conclusions can be derived from the normal distribution curve representing a very large populations:

1. Approximately **68%** of the population members lie within **one standard deviation** of the mean
2. Approximately **95%** of the population members lie within **two standard deviations** of the mean
3. Approximately **99.7%** of the population members lie within **three standard deviations** of the mean.

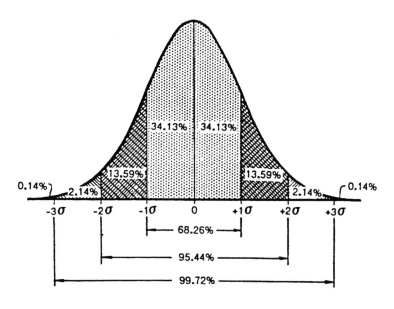

Fig. 31.9 Normal distribution curve.

31.11 GEOMETRICAL TOLERANCES

Geometrical tolerances of the flat-rolled product are generally specified as a function of:

   a) type of the product
   b) type of the metal
   c) width range
   d) thickness range.

Three types of flat-rolled products are commonly known: plates, sheet, and strips. American Society for Metals (ASM) identifies these products as follows [13]:

'**Plate** is a flat-rolled product of some minimum thickness and width arbitrarily dependent on the type of metal'.

'**Sheet** is a flat-rolled product of some maximum thickness and minimum width arbitrarily dependent on the type of metal. It is thinner than plate and has width-to-thickness ratio greater than about 50'.

'**Strip** is a flat-rolled product of some maximum thickness and width arbitrarily dependent on the type of metal. It is narrower than sheet'.

For example, the following thickness and width ranges are used for hot-worked, hot-cold-worked, and cold-worked alloy plate, sheet, and strip for high strength at elevated temperatures [11].

Plate:

thickness range	4.76 to 152 mm (0.1875 to 6.0 in.)
width range	not specified

Sheet:

thickness range	0.13 to 4.76 mm (0.005 to 0.1875 in.)
minimum width	610 mm (24.0 in.)

Strip:

thickness range	0.13 to 4.76 mm (0.005 to 0.1875 in.)
maximum width	less than 610 mm (24.0 in.)

Tables 31.1-31.8 show some geometrical tolerances for carbon and high-strength low-alloy sheets as specified by American Society for Testing and Materials (ASTM). The latest trend in the steel industry is to tighten the tolerances for the rolled flat products by 50% or even 75% in comparison with the ASTM tolerances.

Table 31.1 ASTM thickness tolerances of hot-rolled sheet (carbon steel). Coils and cut lengths, including pickled. Data from Annual Book of ASTM Standards (1987).

	Thickness tolerances over, in. No tolerance under					
	Specified minimum thickness, in.					
Specified Width, in.	0.180 to 0.230 excl	Over 0.098 to 0.180 excl	Over 0.071 to 0.098 incl	Over 0.057 to 0.071 incl	Over 0.051 to 0.057 incl	0.044 to 0.051 incl
12 to 20 incl	0.014	0.014	0.012	0.012	0.010	0.010
Over 20 to 40 incl	0.016	0.014	0.014	0.012	0.010	0.010
Over 40 to 48 incl	0.018	0.016	0.014	0.012	0.012	0.010
Over 48 to 60 incl	.....	0.016	0.014	0.014	0.012	.....
Over 60 to 72 incl	.....	0.016	0.016	0.014	0.014	.....
Over 72	.....	0.016	0.016	.....	.....	.....

See notes in Table 31.2.

Table 31.2 ASTM thickness tolerances of hot-rolled sheet (high-strength, low-alloy steel). Coils and cut lengths, including pickled. Data from Annual Book of ASTM Standards (1987).

| | Thickness tolerances over, in. No tolerance under | | | |
| | Specified minimum thickness, in. | | | |
Specified Width, in.	0.180 to 0.230 excl	Over 0.098 to 0.180 excl	Over 0.082 to 0.098 incl	0.071 to 0.082 incl
12 to 15 incl	0.014	0.014	0.012	0.012
Over 15 to 20 incl	0.016	0.016	0.014	0.014
Over 20 to 32 incl	0.018	0.016	0.014	0.014
Over 32 to 40 incl	0.018	0.018	0.016	0.014
Over 40 to 48 incl	0.020	0.020	0.016	0.014
Over 48 to 60 incl	.....	0.020	0.016	0.014
Over 60 to 72 incl	.....	0.022	0.018	0.016
Over 72 to 80 incl	.....	0.024	0.018	0.016
Over 80	.....	0.024	0.020	.....

Note 1 - Thickness is measured at any point across the width not less than 3/8 in. from a cut edge and not less than 3/4 in. from a mill edge. This table does not apply to the uncropped ends of mill edge coils.

Note 2 - The specified thickness range captions also apply when sheet is specified to a nominal thickness, and the tolerances are divided equally, over and under.

Table 31.3 ASTM width tolerances of hot-rolled mill
edge sheet (carbon and high-strength low-alloy steel).
Coils and cut lengths, including pickled. Data from
Annual Book of ASTM Standards (1987).

Carbon steel	
Specified width, in.	Tolerances over specified width, in. No tolerance under
12 to 14	7/16
Over 14 to 17	1/2
Over 17 to 19	9/16
Over 19 to 21	5/8
Over 21 to 24	11/16
Over 24 to 26	13/16
Over 26 to 30	15/16
Over 30 to 50	1 1/8
Over 50 to 78	1 1/2
Over 78	1 7/8
High-strength low-alloy steel	
12 to 14	7/16
Over 14 to 17	1/2
Over 17 to 19	9/16
Over 19 to 21	5/8
Over 21 to 24	11/16
Over 24 to 26	13/16
Over 26 to 28	15/15
Over 28 to 35	1 1/8
Over 35 to 50	1 1/4
Over 50 to 60	1 1/2
Over 60 to 65	1 5/8
Over 65 to 70	1 3/4
Over 70 to 80	1 7/8
Over 80	2

The above tolerances do not apply to the
uncropped ends of mill edge coils.

Table 31.4 ASTM camber tolerances for hot-rolled, including pickled, and cold-rolled sheet over 12 in. width (carbon and high-strength low-alloy steel). Cut lengths, not resquared. Data from Annual Book of ASTM Standards (1987).

Cut length, ft	Camber tolerances, in.
To 4 incl	1/8
Over 4 to 6 incl	3/16
Over 6 to 8 incl	1/4
Over 8 to 10 incl	5/16
Over 10 to 12 incl	3/8
Over 12 to 14 incl	1/2
Over 14 to 16 incl	5/8
Over 16 to 18 incl	3/4
Over 18 to 20 incl	7/8
Over 20 to 30 incl	1 1/4
Over 30 to 40 incl	1 1/2

Note 1 - Camber is the greatest deviation of a side edge from a straight line, the measurement being taken on the concave side with a straightedge.

Note 2 - The camber tolerance for coils is 1 in. in any 20 ft.

Table 31.5 ASTM out-of-square tolerances of hot-rolled cut-edge, including pickled, and cold-rolled sheet over 12 in. wide (carbon and high-strength, low-alloy steel). Cut lengths, not resquared. Data from Annual Book of ASTM Standards (1987).

Out-of square is the greatest deviation of an end edge from a straight line at right angle to a side and touching one corner.
It is also obtained by measuring the difference between the diagonals of the cut length. The out-of-square deviation is one half of that difference. The tolerance for all thicknesses and all sizes is 1/16 in. per 6 in. of width or fraction thereof.

Table 31.6 ASTM flatness tolerances[a] of hot-rolled sheet, including pickled cut lengths not specified to stretcher-leveled standard of flatness (carbon and high-strength low-alloy steel). Data from Annual Book of ASTM Standards (1987).

Specified minimum thickness, in.	Specified width in.	Flatness tolerances[b], in. Specified yield point, min, ksi	
		Under 45	45 to 50[cd]
0.044 to 0.057 incl	over 12 to 36 incl	1/2	.....
	over 36 to 60 incl	3/4	.....
	over 60	1	.....
0.057 to 0.180 excl	over 12 to 60 incl	1/2	3/4
	over 60 to 72 incl	3/4	1 1/8
	over 72	1	1 1/2
0.180 to 0.230 excl	over 12 to 48 incl	1/2	3/4

[a] The above table also applies to lengths cut from coils by the consumer when adequate flattening operations are performed.

[b] Maximum deviation from a horizontal flat surface.

[c] Tolerances for high-strength low-alloy steels with specified minimum yield point in excess of 50 ksi are subject to negotiation.

[d] 0.071 minimum thickness of HSLA.

Table 31.7 ASTM thickness tolerances of cold-rolled sheet (carbon and high-strength, low-alloy steel)[a].
Coils and cut lengths over 12 in. in width. Data from Annual Book of ASTM Standards (1987).

Specified Width, in.	Thickness tolerances over, in. No tolerance under Specified minimum thickness, in.					
	Over 0.098 to 0.142 incl	Over 0.071 to 0.098 incl	Over 0.057 to 0.071 incl	Over 0.039 to 0.057 incl	Over 0.019a to 0.039 incl	0.014 to 0.019 incl
12 to 15 incl	0.010	0.010	0.010	0.008	0.006	0.004
Over 15 to 72 incl	0.012	0.010	0.010	0.008	0.006	0.004
Over 72	0.014	0.012	0.010	0.008	0.006	.....

[a] 0.020 in. minimum thickness for high-strength low-alloy steel.

Table 31.8 ASTM flatness tolerances of cold-rolled sheet, (carbon and high-strength, low-alloy steel). Cut lengths over 12 in. in width, not specified to stretcher-leveled standard of flatness. Data from Annual Book of ASTM Standards (1987).

Specified thickness, in.	Specified width in.	Flatness tolerances[a], in. Specified yield point, min, ksi	
		To 40 incl	45 to 50[b]
To 0.044 incl	to 36 incl	3/8	3/4
	over 36 to 60 incl	5/8	1 1/8
	over 60	7/8	1 1/2
Over 0.044	to 36 incl	1/4	3/4
	over 36 to 60 incl	3/8	3/4
	over 60 to 72 incl	5/8	1 1/8
	over 72	7/8	1 1/2

[a] Maximum deviation from a horizontal flat surface.

[b] Tolerances for high-strength low-alloy steels with specified minimum yield point in excess of 50 ksi are subject to negotiation.

Note 1 – This table does not apply when product is ordered full hard to a hardness range, or 'annealed last' (dead soft).

Note 2 – This table also applies to lengths cut from coils by the consumer when adequate flattening measures are performed.

REFERENCES

1. Hot Strip Mill Profile and Flatness Study, Phase 1, Association of Iron and Steel Engineers, 1986, pp. 20-22.

2. V.B. Ginzburg, "Strip Profile Control with Flexible Edge Backup Rolls", Iron and Steel Engineer, July 1987, p. 23.

3. J.P. Barreto and M.J. Hillier, "Shape Control in Cold Strip Rolling", Sheet Metal Industries, Vol. 45, Oct. 1986, pp. 707-709.

4. T. Sheppard, "Shape in Metal Strip: The State of the Art", Proceedings of the Metals Society Conference on Shape Control, Chester, England, April 1976, pp. 11-18.

5. G.F. Kelk, et al, "New Development Improve Hot Strip Shape: Shapemeter-Looper and Shape Actimeter", Iron and Steel Engineer, Aug. 1986, pp. 50-56.

6. J.G. Wistreich, "Control of Strip Shape During Cold Rolling", Journal of the Iron and Steel Institute, Vol. 206, Dec. 1986, pp. 1203-1206.

7. W.K.J. Pearson, "Shapemeter II", Proceedings of the Metals Society Conference on Shape Control, Chester, England, April 1976, pp. 46-54.

8. W.K.J. Pearson, "Shape Measurement and Control", Journal of the Institute of Metals, Vol. 93, 1964-1965, pp. 169-178.

9. S. Wilmotte, et al, "The SIGMA-RO Process: A New Approach to the Hot Strip Mill Computer Control", Centre de Recherches Metallurgigues (C.R.M.) Report, No. 52, May 1978, pp. 7-15.

10. O.G. Sivilotti, et al, "ASEA-ALCAN AFC System for Cold Rolling of Flat Strip", AISE Yearly Proceedings, 1973, pp. 263-270.

11. Annual Book of ASTM Standards, American Society for Testing and Materials, pp. 249-357 (1987).

12. P. Newbold, Statistics for Business and Economics, Prentice-Hall, Englewood Cliffs, New Jersey, pp. 1-23 (1984).

13. ASM Metals Reference Book, American Society for Metals, Metals Park, Ohio, pp. 54-71 (1981).

32

Measurement of Geometrical Parameters of Flat Products

32.1 PRECISION OF INSTRUMENTS

Measurement of geometrical parameters of flat products requires an application of the instruments which have certain standard performance characteristics. Below are the basic terms used to evaluate the precision of different instruments as defined by the Dictionary of Instrumentation Terminology [1].

Accuracy - The accuracy of an instrument is 'the capability of the instrument to follow the true value of a given phenomenon'.

Combined error - The combined error is 'the largest possible error in an instrument resulting from a combination of adverse conditions'.

Full scale (FS) - This is 'the total interval over which an instrument is intended to operate'.

Hysteresis - The hysteresis may be defined as 'the summation of all effects, under constant environmental conditions, which cause the output of an instrument to assume different values at a given stimulus point when that point is approached first with an increasing stimulus and then with a decreasing stimulus' (Fig. 32.1).

Noise - The noise is any unwanted electrical disturbance or spurious signal, which modifies the transmission, measurement, or recording of desired data'.

Nonlinearity - The nonlinearity is a deviation from a straight line of the relationship between a measured parameter and a corresponding output signal provided by an instrument.

Nonlinearity is measured in several ways:

1. Maximum-deviation-based nonlinearity is 'the maximum departure between the calibration curve and a straight line drawn to give the most favorable accuracy'. It is usually expressed as a percentage of full-scale deflection.

2. Slope-based nonlinearity is 'the ratio of maximum slope error anywhere on the calibration curve to the slope on the nominal sensitivity line'. It is normally given as a percentage of nominal slope.

Operating temperature - This is 'the temperature or range of temperature over which an instrument is expected to operate within specified limits of error'.

Precision - Precision is 'the smallest part that a system or device can distinguish. This term is

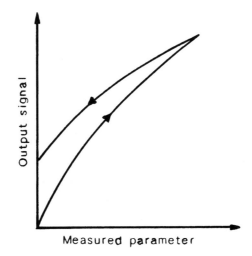

Fig. 32.1 Hysteresis.

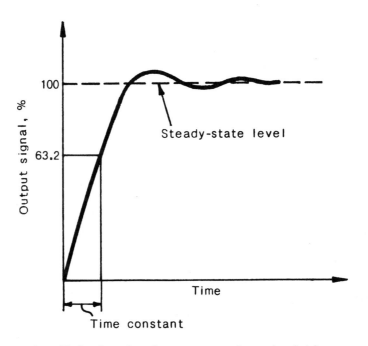

Fig. 32.2 Time domain response characteristics.

closely related to resolution and is associated with intended or designed performance'.

Repeatability - Two following definitions of this term are in use:

1. 'The maximum deviation from the mean of corresponding data points taken from repeated tests under identical conditions'.

2. 'The maximum difference in output for any given identically repeated stimulus with no change in other test conditions'.

Resolution - The resolution is described as 'the smallest change in applied stimulus that will produce a detectable change in the instrument output'.

Scale factor - The scale factor is 'the amount by which a quantity being measured must change in order to produce unit per deflection'.

Sensitivity - This term shall more correctly be used as 'a maximum sensitivity that is the maximum scale factor for which an instrument is capable of responding'.

Stability - The stability may be defined in two ways:

1. 'Independence or freedom from changes in one quantity as a result of change in another'.
2. 'The absence of drift'.

Temperature effect - The temperature effect is 'the change in performance because of temperature changes'. This includes 'a change in sensitivity and a change or shift in zero indication'.

Time constant - When a measurement has taken place, it requires a certain time for an instrument to develop an output signal corresponding to a measured quantity. The time constant is 'the time required for exponential quantity to change by an amount equal to 63.2% of the change required to reach steady-state' (Fig. 32.2).

32.2 MEASUREMENT OF WORKPIECE THICKNESS

Measurement of thickness of flat products can be accomplished by using both contact and non-contact systems. The non-contact systems are usually utilized in high-speed rolling mills. The basic principles of these systems are as follows. A high intensity beam of radiation is generated from an x-ray tube, or from other source of radiation, and directed at the workpiece which thickness is measured. Some of the radiation is attenuated by the workpiece, the remainder passes through. The attenuation is predictable and follows the formula:

$$I = I_0 exp(-\mu h), \tag{32-1}$$

where I_0 = intensity of radiation before attenuation

I = intensity of radiation after attenuation

μ = linear absorption coefficient

h = workpiece thickness.

The radiation is measured by a radiation detector that is generally a combination of a scintillator crystal with photomultiplier.

Typical performance characteristics of the x-ray thickness gauges are [2, 3]:

Thickness range:

hot strip mill	1 - 16 mm (0.040 - 0.650 in.)
cold strip mill	1 - 8 mm (0.040 - 0.325 in.)

Setting accuracy: 0.1% of setting thickness

Time constant:

hot strip mill	0.030 sec
cold strip mill	0.010 sec

Noise as a percentage of measured thickness:

    hot strip mill   (0.05 - 0.15)%

    cold strip mill   (0.05 - 0.10)%.

32.3 MEASUREMENT OF STRIP PROFILE

The radiation principle has been successfully applied to development of the strip profile gages.

The profile meter developed by Kawasaki Steel [4] consists of three main components (Fig. 32.3): a stationary thickness gage that measures the center thickness of the strip, a scanning gage that measures strip thickness while moving at the right angle to the rolling direction, and a profile calculator. The calculator computes the strip profile as the difference between the two measured values obtained by the fixed gauge and by the scanning gage. The scanning gage is driven through rack and pinion by DC motor with variable speed that ranges from 1.8 to 18 m/min. In alternate designs of profile meters, an increased scanning rate is achieved by adding the second scanning gage that moves in opposite direction in relation to the movement of the first scanning gage.

In the profile meter developed by Isotope Measuring Systems [5], two cesium 137 radioactive sources, mounted in the top yoke of a C-frame (Fig. 32.4), irradiate a total of 22 ionization chamber detectors mounted in the bottom yoke of the C-frame. In order to measure the strip thickness at 10 mm (0.394 in.) width intervals the C-frame is made to oscillate at right angles to the direction of rolling, so that each channel covers a total strip width of 100 mm (3.94 in.) over a period of 1 sec (Fig. 32.5). During one cycle of one second duration the strip thickness is measured twice per 10 mm (0.394 in.) intervals across the strip.

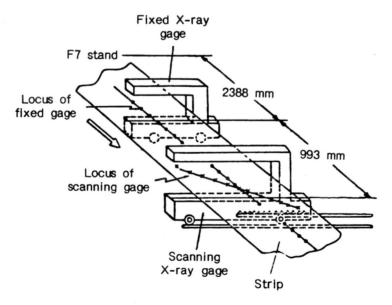

Fig. 32.3 Outline of Kawasaki Steel Profilemeter. (From Tamiya, Kawasaki Steel Technical Report No. 5, 1982. Copyright Kawasaki Steel Corporation. Reprinted with permission.)

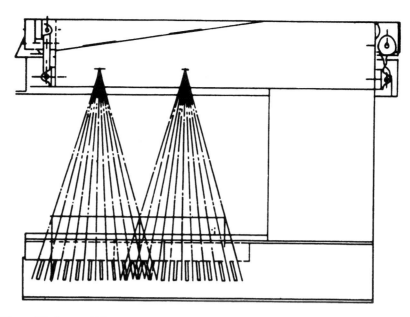

Fig. 32.4 Profile Thickness Gage designed by Isotope Measuring Systems, Ltd. (From Isotope Measuring Systems Publication. Copyright Isotope Measuring Systems, Ltd., England. Reprinted with permission.)

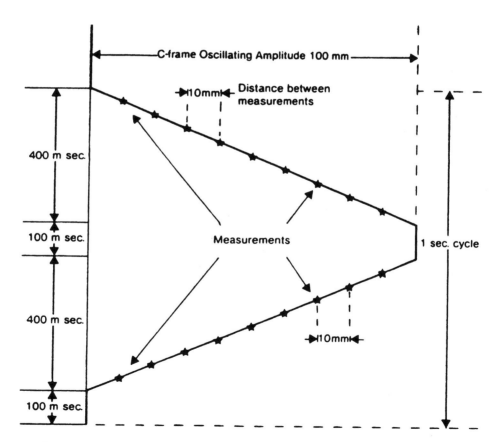

Fig. 32.5 Profile measuring period (1 second cycle) of Profile Thickness Gage shown in Fig. 32.4. (From Isotope Measuring Systems Publication. Copyright Isotope Measuring Systems, Ltd., England. Reprinted with permission.)

Main characteristics of the Isotope Profile Thickness Gage are:

Accuracy	0.2%, not better than + 5 μm
Nonlinearity	0.1%, not better than + 2 μm
Measuring time constant	0.050 sec
Measured value acquisition period	0.010 sec
Measured value processing time	0.040 sec
Statistical noise due to radioactive decay	(5-15) μm.

32.4 METHODS OF STRIP FLATNESS MEASUREMENT

On-line measurement of strip flatness must be based on non-destructive methods. This becomes possible due to the direct relationship between strip waviness determined by slitting of the strip and the variation of stresses and geometry along the width of uncut strip. Depending on which of these indirect parameters is used for shape measurement, the following three methods can be identified [6]:

1. Internal stress method
2. External stress method
3. Geometrical method.

A brief description of these methods follows.

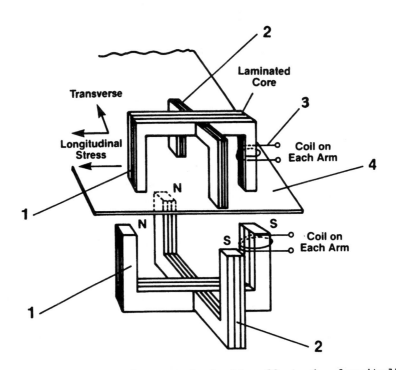

Fig. 32.6 Shapemeter using magnetoelastic effect: 1 - longitudinal laminated core; 2 - transverse laminated core; 3 - coil; 4 - strip. (From Sheppard, Proceedings of the Metals Society Conference on Shape Control, Chester, England, 1976. Copyright Institute of Metals. Reprinted with permission.)

32.5 INTERNAL STRESS METHOD OF FLATNESS MEASUREMENT

Internal stress method utilizes the fact that the variation in lengths of the ribbons of the strip after slitting is proportional to the variation in internal stresses across the width of uncut strip. Therefore, the strip waviness is equal to:

$$\frac{\Delta L}{L} = \frac{\Delta \sigma_i}{E_s} \tag{32-2}$$

where $\Delta \sigma_i$ = variation of internal stresses across the width of the strip

E_s = strip modulus of elasticity

$\Delta L/L$ = strip waviness

ΔL = difference between the longest and shortest ribbon across the width of the strip

L = wavelength.

Typical examples of the shapemeters which are sensitive to the internal stresses would be those (Fig. 32.6) that utilize the magnetoelastic effect, which is the variation of magnetic properties of ferromagnetic materials with variation in stresses [7]. Magnetoelastic shapemeters have been developed in BISRA (England), by Jones & Laughlin (U.S.), and in the USSR [8]. Due to their nature, magnetoelastic shapemeters are applicable only for magnetic materials. Since during hot rolling the rolled material is normally in non-magnetic state, these shapemeters are excluded from application to hot strip mills.

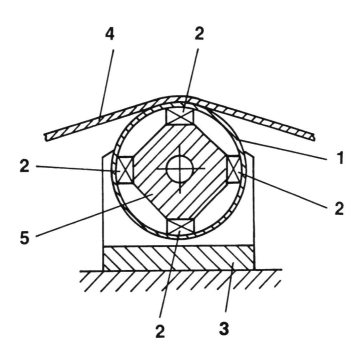

Fig. 32.7 Shapemeter with built-in roll sensors: 1 - segmented roll; 2 - sensors; 3 - support; 4 - strip; 5 - arbor. (From Kelk, et al, Iron and Steel Engineer, 1976. Copyright AISE, Pittsburgh, Pennsylvania. Reprinted with permission.)

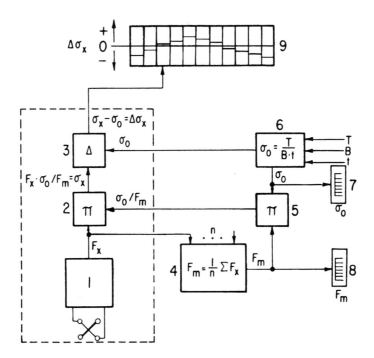

Fig. 32.8 Block diagram of measuring system for ASEA – ALCAN Shapemeter-Stressometer. (From Sivilotti, et al, AISE Year Book, 1973. Copyright AISE, Pittsburgh, Pennsylvania. Reprinted with permission.)

32.6 EXTERNAL STRESS METHOD OF FLATNESS MEASUREMENT

External stress method is applicable when uniform external tension is applied to the strip. When all parts of the strip are under tension, the strip waviness can be expressed as:

$$\frac{\Delta L}{L} = \frac{\Delta \sigma_e}{E_s} \quad , \tag{32-3}$$

where $\Delta\sigma_e$ = variation in tensile stresses across the width of the strip.

One example of the shapemeters based on this method would be the stressometers with segmented rolls (Fig. 32.7) contacting the strip and measuring the tensile stress distribution across the strip width [9]. The measurement process may be described with reference to the block diagram in Fig. 32.8, where the following notations are used [10]:

F_x = force on measuring zone

σ_x = mechanical stress in the strip across zone x

F_m = mean value of all forces acting on zones covered by the strip

σ_0 = mean value of mechanical stress

T = strip tension

B = strip width

t = strip thickness

n = number of zones.

The figure shows the signal processing for one measuring zone (the blocks within the broken line) as well as the common part for the entire equipment. The transducer signal from zone x is rectified and filtered in block 1. The output signal is a direct voltage proportional to the force F_x. In block 4 the mean value F_m from all zones covered by the strip is formed. The mechanical mean stress σ_0 is calculated in block 6 with the aid of data for the strip. The quotient σ_0/F_m is formed in block 5 and multiplied by F_x in block 2. The output signal $F_x\sigma_0/F_m$ is equal to the local mechanical stress σ_x in the strip. Finally, σ_0 is subtracted in block 3 and the difference $\Delta\sigma_x = \sigma_x - \sigma_0$ constitutes the output signal and is displayed on a separate instrument for each measuring zone.

32.7 GEOMETRICAL METHOD OF FLATNESS MEASUREMENT

Geometrical method is based on measurement of manifest shape of the strip (Fig. 31.4). In this case, the latent component of shape will be left undetected.

The strip waviness corresponding to the manifest component of the shape is equal to [10]:

$$\frac{\Delta L}{L} = \left(k\,\frac{R}{L}\right)^2 , \tag{32-4}$$

where R = wave amplitude

k = wave shape factor, for sinusoidal wave $k = \pi/2$.

One of the first shapemeters developed by BISRA [11] realized this method by measuring the differential speed of the strip across its width. The shapemeter developed by Vollmer America [12], measures the strip deflection with contact linear position transducers (Fig. 32.9). The noncontact optical sensors in combination with pattern recognition technology are used in the shapemeters developed by CRM [13, 14] and Nippon Steel [15].

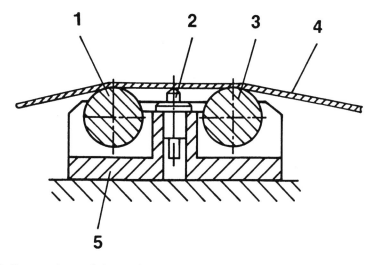

Fig. 32.9 Shapemeter with strip contact sensors using geometrical method: 1, 3 - supporting rolls; 2 - linear position transducer; 4 - strip; 5 - support. (From Kelk, et al, Iron and Steel Engineer, 1986. Copyright AISE, Pittsburgh, Pennsylvania. Reprinted with permission.)

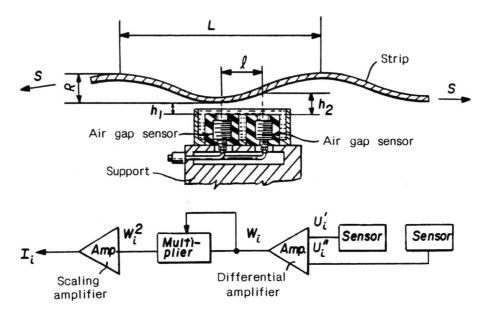

Fig. 32.10 Shapemeter with non-contact sensors using geometrical method. Adapted from Ginzburg, U.S. Patent No. 4,771,622 (1988).

Figure 32.10 illustrates the principle of operation of the shapemeter proposed by United Engineering and International Rolling Mill Consultants [16]. The shapemeter has a number of air gap sensors which are grouped in two parallel arrays stretched in transverse direction of the strip. Each pair of two sensors adjacent in longitudinal direction measures the flatness of a section of the strip which passes these sensors.

The output signals from each upstream and downstream sensor that belongs to the i-th pair may be expressed as follows:

$$U_i' = \frac{R_i}{2}\sin(2\pi f_i t) \tag{32-5}$$

$$U_i'' = \frac{R_i}{2}\sin(2\pi f_i t - \alpha_i)\ , \tag{35-6}$$

where R_i = wave amplitude

   f_i = wave frequency

   t = time

   α_i = phase shift between the signals U_i' and U_i''.

The amplitude of the differential signal between U_i' and U_i'' is equal to

$$U_i' - U_i'' = R_i \sin\frac{kl}{L_i}\ , \tag{32-7}$$

where l = distance between upstream and downstream sensors

   L_i = wave length.

When l is substantially less than L_i, Eq. (32-7) reduces to

$$U_i' - U_i'' = k \frac{R_i}{L_i} \qquad (32\text{-}8)$$

The differential multiplier shown in Fig. 32.10 performs the calculations according to Eq. (32-8). The multiplier calculates the strip waviness according to Eq. (32-4). The strip waviness is then converted into I-units by the scaling amplifier as dictated by Eq. (31-16).

One of the advantages of the double row of sensors is that it minimizes error due to temperature and material hardness variation in the longitudinal directions as the differential signals are used from each pair of adjacent sensors located in two parallel arrays.

The shapemeter developed by Kawasaki Steel Corporation employs a so-called 'water column method' [17]. The water column method is based on two kinds of principles. One of the principles is the ultrasonic measurement. Ultrasonic pulses are transmitted and received via water column to and from the object. The distance to the object and its thickness can be obtained from time intervals between echo pulses. The second principle is the electric resistance measurement. A constant electric current is fed to a water column which connects an electrically insulated fixed nozzle with an electrically grounded object. By measuring the electric potential of the nozzle, the length of the water column and hence one of the factors of the object's shape can be known.

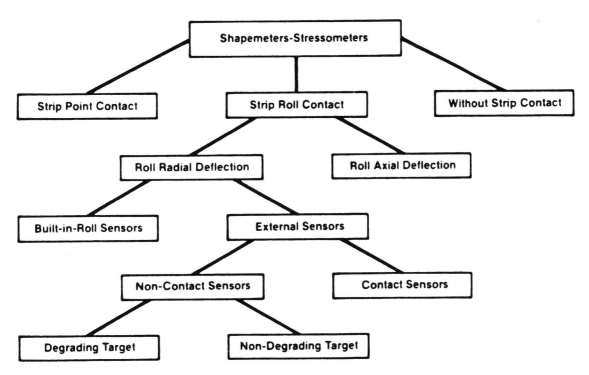

Fig. 32.11 Classification of shapemeters based on external stress method. (From Kelk, et al, Iron and Steel Engineer, 1986. Copyright AISE, Pittsburgh, Pennsylvania. Reprinted with permission.)

The geometrical method has found its best application for measurement of the strip shape at the exit of a hot strip mill where the strip remains without tension during threading before engaging the downcoiler. After the strip tension between the last finishing stand and the downcoiler is established the manifest component of the strip shape is accordingly reduced. This problem is reportedly solved by using the sensors with increased resolution.

32.8 CLASSIFICATION OF SHAPEMETERS-STRESSOMETERS

A review of the methods for measuring strip shape indicates that the external stress method may be attractive in application to the hot rolled strip. Depending on the type of contact with the strip, the shapemeters can be divided into three groups (Fig. 32.11):

1. Without strip contact
2. With strip point contact
3. With strip roll contact.

The shapemeter developed by Mitsubishi Electric [18] is an example of a device without strip contact. This shapemeter (Fig. 32.12) measures the deflection of the strip across its width after this deflection is forced by a periodic magnetic field. Another shapemeter of this type was developed by

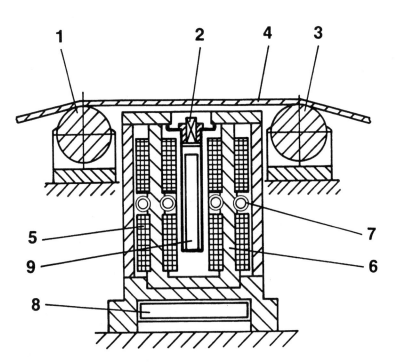

Fig. 32.12 Shapemeter without strip contact: 1, 3 - supporting rolls; 2 - air gap sensor; 4 - strip, 5 - electromagnetic force generator; 6 - magnetic core; 7 - cooling water pipes; 8 - cooling water tank; 9 - signal processing unit. (From Kelk, et al, Iron and Steel Engineer, 1986. Copyright AISE, Pittsburgh, Pennsylvania. Reprinted with permission.)

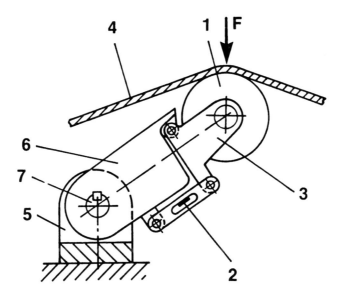

Fig. 32.13 Shapemeter based on measurement of roll radial deflection with external contact sensors: 1 - segment rolls; 2 - strain gage; 3, 6 - support arms; 4 - strip; 5 - support; 7 - looper shaft. (From Kelk, et al, Iron and Steel Engineer, 1986. Copyright AISE, Pittsburgh, Pennsylvania. Reprinted with permission.)

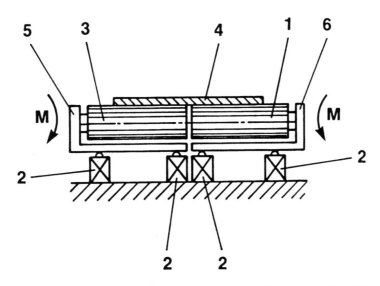

Fig. 32.14 Shapemeter based on measurement of deflection of roll axis: 1, 3 - deflecting rolls; 2 - load cells; 4 - strip; 5, 6 - roll supporting frames. (From Kelk, et al, Iron and Steel Engineer, 1986. Copyright AISE, Pittsburgh, Pennsylvania. Reprinted with permission.)

Table 32.1 Stressometers with built-in roll sensors.

No.	Company	Country	Type of transducers
1	ASEA	Sweden	Magnetoelastic Load Cells
2	BBC Brown Boveri	W. Germany	Piezo-electric Load Cells
3	Davy McKee	England	Air Pressure Transducers
4	Clecim	France	Inductive Position Transducers
5	E.W. Bliss	U.S.A.	Magnetic Bearings

Nippon Steel [19], which uses indirect measurement of the stresses by detecting the variation in amplitude and frequency of natural or forced oscillations of the strip across width. Barreto and Hillier [20] have proposed a shapemeter with one point contact in the middle of the strip. However, this idea has never been practically realized.

Shapemeters with rolls contacting the strip can be designed either for measurement of radial deflection of the rolls or measurement of deflection of the axis of the rolls. As shown in Fig. 32.13, in the shapemeter proposed by Fabian et al [21], the radial deflection of the rolls is measured in the plane coinciding with the longitudinal direction of the strip. In the shapemeter proposed by Kajiwara [22], the

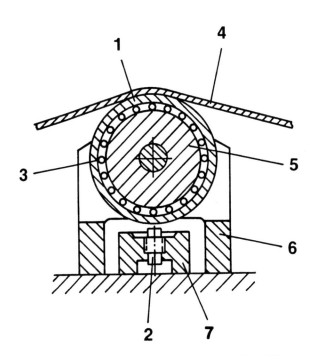

Fig. 32.15 Shapemeter with noncontact sensors and roll segments used as measuring targets; 1 - segmented shells; 2 - noncontact sensors; 3 - cylindrical springs; 4 - strip; 5 - stationary shaft; 6 - support for stationary shaft; 7 - sensor bar. (From Kelk, et al, Iron and Steel Engineer, 1986. Copyright AISE, Pittsburgh, Pennsylvania. Reprinted with permission.)

deflection of the roll axes is measured in the plane coinciding with the transverse direction of the strip (Fig. 32.14). It is easier to achieve better resolution in measurement of the stress distribution with the shapemeters using radial deflection of the rolls rather than using the deflection of the roll axis. This may be the primary reason why the latter design has not found wide recognition.

Shapemeters with radial deflection of the rolls can be either with built-in roll sensors or external sensors. Devices with built-in roll sensors (Fig. 32.7) have been developed by several companies (Table 32.1) and have been successfully implemented in cold rolling mills. Because of the proximity of their transducers to the rolled strip, these devices have not been adopted for application in hot strip mills. From that standpoint, it is more advantageous to use shapemeters with external sensors. That was confirmed by Hoesch Stahl AG [21], which developed and successfully implemented this design for a hot strip mill. In this shapemeter (Fig. 32.13), the strain gages are installed on the bar attached to the arms and holding the segmented rolls. The lower part of the arms is pivotally attached to the looper shaft. It allows use of the unit as a part of the looper.

Shapemeters with external sensors can be further divided into two types: with contact sensors, and with noncontact sensors. The shapemeter shown in Fig. 32.13 and described previously is a typical example of a design with contact sensors. Its major disadvantage would be the necessity of shutting down the mill to repair or replace any of the transducers. From this viewpoint, units with noncontact sensors have an obvious advantage. In the shapemeter proposed by Ishimoto et al [23], the outer rotating segmented shells are supported by cylindrical springs on the stationary shaft (Fig. 32.15). The noncontact sensors are installed inside the support which can be easily removed for maintenance purposes.

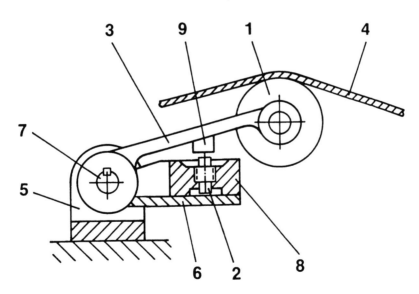

Fig. 32.16 Shapemeter based on measurement of roll radial deflection with external sensors and nondegrading targets: 1 - segmented roll; 2 - sensor; 3 - cantilever beam; 4 - strip; 5 - support for shaft; 6 - sensor bar; 7 - shaft; 8 - subframe; 9 - target. (From Kelk, et al, Iron and Steel Engineer, 1986. Copyright AISE, Pittsburgh, Pennsylvania. Reprinted with permission.)

Shapemeters with noncontact sensors can be further divided into two types: with degrading target, and with nondegrading target. The latter classification is important because noncontact sensors are usually sensitive to deterioration of the target surface. In the unit shown in Fig. 32.15, the outer segmented shells serve as targets for noncontact sensors. Continuous contact of these shells with rolled strip would result in degrading its performance. This problem has been eliminated in the shapemeter shown in Fig. 32.16, which has nondegrading targets in conjunction with noncontact sensors [24].

32.9 SHAPEMETER-LOOPER

The Shapemeter-Looper has a dual purpose. It must provide a continuous measurement of strip shape and strip tension and at the same time perform the functions of a conventional strip tension looper between two stands of finishing train. Therefore, it must meet the following requirements:

1. Design must provide high reliability of operation in the hot strip mill environment.

2. Whenever it is designed to replace an existing looper, it must be compatible with the existing drive and existing mill configuration.

3. Shapemeter must have adequate resolution in measuring both strip shape and strip tension. Its control range must be adequate for the rolled product mix.

4. Inertia of the segmented rolls and lifting system as well as the natural frequency of its mass-spring system must be within justifiable limits.

5. Shapemeter-Looper must provide all appropriate signals to the closed-loop shape control system as well as to the interstand tension regulator.

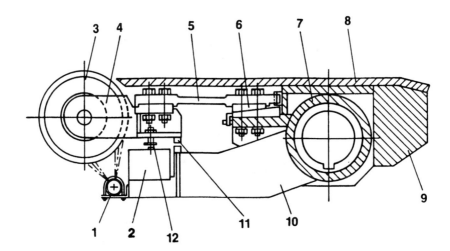

Fig. 32.17 Cross-section of Shapemeter-Looper: 1 – sensor bar; 2 – coolant water header; 3 – segmented roll; 4 – roll segmented assembly; 5 – cantilever beam; 6 – wedge assembly; 7 – looper shaft; 8 – apron; 9 – counter-balance weight; 10 – subframe; 11 – adjustable stop; 12 – target. (From Kelk, et al, Iron and Steel Engineer, 1986. Copyright AISE, Pittsburgh, Pennsylvania. Reprinted with permission.)

A plan view of the Shapemeter-Looper (Fig. 32.17), proposed by United Engineering, International Rolling Mill Consultants and Kelk, Ltd [6], shows the similarity to traditional loopers used between finishing stands of hot strip mills. The shapemeter can be tailored to replace the conventional looper arm and retain the original drive mechanism, whether it be electromechanical, pneumatic or hydraulically actuated. The new unit would continue to perform the constant mass flow function as well as maintain constant strip tension between mill stands.

To perform the second function of measuring variations in tension across the strip width, the traditional full length roll is replaced by a row of roll segments each with its own bearings and arbor. The rolls are 10-in. diameter on 6-in. centers. For a 66-in. wide mill, 11 rolls would be used. Each roll segment assembly is attached to the main looper shaft. A wedge assembly is supplied to adjust the height of any roll on dressing. A subframe is also fastened to the main looper shaft, rotating with it and extending outward to the roll assemblies. This frame supports the sensor bar which houses the electromagnetic pick-up heads, one for each roll segment. Mounted on each roll segment arbor is a target, which will be deflected to a greater or lesser degree toward the pick-up head due to the deflection of the beam when various tensions are imposed on the roll segment. Variations in the air gap will be sensed by the pick-up and further processed.

32.10 SHAPE ACTIMETER

The basic difference between a conventional Shapemeter-Stressometer and Shape Actimeter is that in the former the roll segments are generally aligned in the same plane whereas in the latter the elevation of the roll segments can be individually adjusted. This feature may provide several operating advantages.

Increased range of shape control - The following two cases should be considered:

1. Strip tension is high enough to transform the manifest shape into latent shape. In this case, it is possible to use the roll segments aligned in the same plane (Fig. 32.18a).

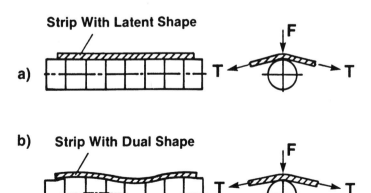

Fig. 32.18 Application of Shape Actimeter. (From Kelk, et al, Iron and Steel Engineer, 1986. Copyright AISE, Pittsburgh, Pennsylvania. Reprinted with permission.)

2. The strip tension is not sufficient to completely eliminate the manifest shape. In this case, to measure both latent and manifest components of the strip shape, it is more appropriate to use a shapemeter that provides adjustable roll contour (Fig. 32.18b).

Improved tracking capability - By introducing the desired crown between the segmented roll and the strip, improved tracking capability can be achieved.

On-line shape correction - Tension distribution in the strip may affect the stress distribution in the roll bite and consequently the strip profile and shape. This concept has been utilized in the tension distribution control as shown in Figs. 32.19 and 32.20 [25-28]. However, for effective use in cold rolling, it is necessary to locate the tension distribution roll in close vicinity to the roll bite. In hot rolling, the yield stresses and the modulus of elasticity of the strip are relatively low. It is, therefore, possible for the contoured roll itself to produce local yielding, resulting in the desired shape correction.

In the Shape Actimeter proposed by United Engineering and International Rolling Mill Consultants [28], each roll segment is mounted on a pivoted radius arm, the position of which is determined, in part, by a hydraulic cylinder (Fig. 32.21). The inner races of roll segments are coupled to the radius arms. The result is that the segmented roll can both rise and fall as in a conventional looper, but can also take any shape (convex to concave) within a certain degree of curvature. Each hydraulic cylinder is pivoted at its upper end. At the lower end, a servovalve and a differential pressure transducer are fitted. An axial position transducer is installed inside the cylinder.

Two basic modes of operation are available: position mode and tension mode. In the position mode the elevation of individual rolls H_A is maintained based on magnitudes of the position reference signals H_R. The strip tension signals S_A are calculated by using data from both pressure and position transducers. In the tension mode, the strip tension maintained by individual rolls follows the reference signals S_R. The strip flatness is calculated based on actual elevations of the individual rolls.

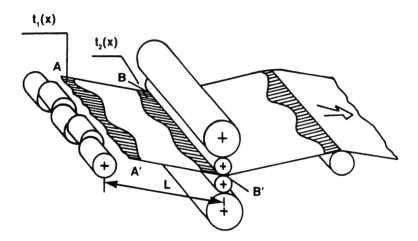

Fig. 32.19 Schematic diagram of tension distribution controller for cold strip mill. From Kelk, et al, Iron and Steel Engineer, 1986. Copyright AISE, Pittsburgh, Pennsylvania. Reprinted with permission.)

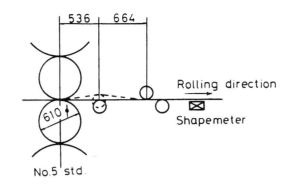

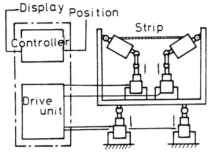

Fig. 32.20 Schematic diagram of prototype tension distribution controller for cold mill. (From Okado, et al, Iron and Steel Engineer, 1982. Copyright AISE, Pittsburgh, Pennsylvania. Reprinted with permission.)

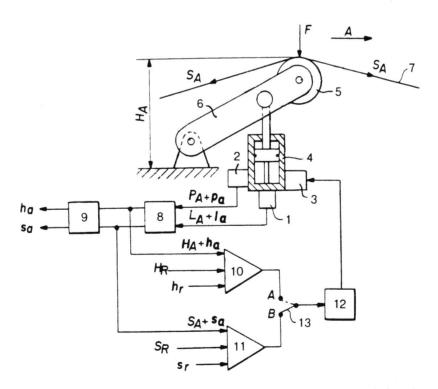

Fig. 32.21 Schematic presentation of Shape Actimeter: 1 – position transducer, 2 – pressure transducer, 3 – servovalve, 4 – hydraulic cylinder, 5 – segmented roll, 6 – lever arm, 7 – strip, 8 – calculating device, 9 – output signal conditioner, 10 – position regulator, 11 – tension regulator, 12 – servovalve controller, 13 – mode of operation selector switch. Adapted from Ginzburg, U.S. Patent No. 4,674,310 (1987).

607

Friction in the hydraulic cylinders is negated by using a dither signal, a low-amplitude, rapid oscillation sufficient to move the piston rod on the seals. In another mode of operation, to increase resolution of the actimeter, periodic roll position references h_r or strip tension references s_r are introduced in addition to the steady-state references H_R or S_R and the resulting periodic variations of the strip tension s_a or displacement h_a is sensed. The resulting display is a plot of stress variations or strip shape against width.

32.11 NONCONTACT STRIP FLATNESS MEASUREMENT SYSTEM

A system for on-line measurement of hot strip flatness developed by IRSID [29, 30], is a typical example of application of a noncontact method which is known as an 'optical triangulation method'. The method is based on the principle of shifting images illustrated in Fig. 32.22. Lasers are used as emitters with line-scan cameras and photosensitive diode arrays as receivers. The laser beam axis SBA and the camera axis AA' are fixed with respect to the roller table. The image A' of the laser spot A on the receiver shifts to B' when the laser spot moves to B, due to change in strip level from y_0 to y_1 above the roller table.

The operating principle, illustrated in Fig. 32.23, shows a longitudinal section of the strip, i.e., the fiber whose length along waves L has to be calculated over selected time intervals, e.g., every 3 to 6 s. The strip levels above a reference plane (the runout table) y_0, y_1, y_2.....y_n are measured periodically at the moments t_0, t_1, t_2......t_n. The fiber length is then obtained from the following relationship:

$$L = \sum_i^n \sqrt{(y_i - y_{i-1})^2 + V_i^2 (t_i - t_{i-1})^2} \qquad (32\text{-}9)$$

where V_i = strip speed when measuring y_i

n = number of measurements during the integration time.

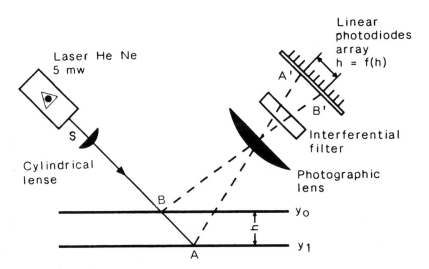

Fig. 32.22 Schematic presentation of Optical triangulation method used in IRSID - SPIE - TRINDEL Lasershapemeter. Adapted from Spie-Trindel Publication.

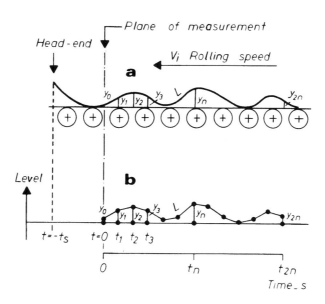

Fig. 32.23 Principle of flatness measurement of Lasershapemeter.
(From Pirlet, et al, AISE Year Book, 1983. Copyright AISE, Pittsburgh,
Pennsylvania. Reprinted with permission.)

Successive y_i values are measured simultaneously and corresponding L values computed for the chosen fibers across the strip: e.g., in the middle (L_{mi}), on the operator side (L_{op}), and motor side (L_{mo}) of the strip. The x-image position along the array is detected by the cameras. The x-signals delivered by the control units of the cameras are processed by a minicomputer which calculates L and RO indices. Typical RO indices define flatness and level as follows:

$$RO = \frac{L_{op} + L_{mo}}{2L_{mi}} - 1 \qquad (32\text{-}10)$$

$$RO' = 2\frac{L_{op} - L_{mo}}{L_{op} + L_{mo}} \qquad (32\text{-}11)$$

32.12 MEASUREMENT OF WORKPIECE WIDTH

Both contact and noncontact method can be applied for measurement of the product width. The non-contact method is commonly used in the hot strip and plate mills.

The noncontact width measurement system usually incorporates an optical device. This device operates on the principle that the position of a slab edge may be optically located by virtue of the light contrast that occurs between the slab and its background. During measurement the slab image is detected by a light sensitive transducer which transmits information to the measuring electronics.

Various width measuring systems have been proposed. These systems differ mainly by arrangements of optical cameras in respect to the object of measurement.

Figure 32.24 illustrates a geometry in the slab width and thickness measurement which was utilized in the system developed by Broken Hill Proprietary Co. Ltd. (BHP) which uses two cameras. The measured slab width and thickness are given by [31]:

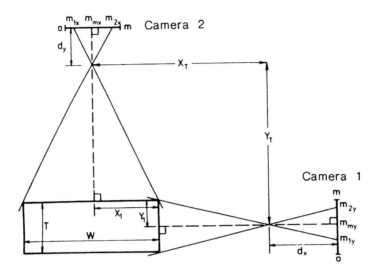

Fig. 32.24 Principle of width and thickness measurement utilized in the BHP system. (From Kenyon, Iron and Steel Engineer, 1985. Copyright AISE, Pittsburgh, Pennsylvania. Reprinted with permission.)

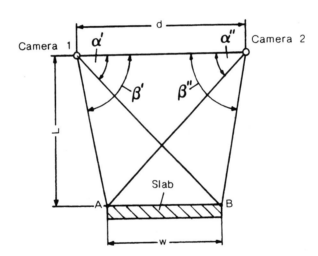

Fig. 32.25 Principle of width measurement utilized in the system developed by George Kelk, Ltd.

$$w = (Y_T - Y_1)(m_{2x} - m_{1x})/d_y \tag{32-12}$$

$$h = (X_T - X_1)(m_{2y} - m_{1y})/d_x \tag{32-13}$$

In the width measuring system developed by George Kelk, Ltd., the stereoscopic principle of measurement is utilized. As shown in Fig. 32.25, the positions A and B of each edge of a slab are scanned with two cameras. Since the distances d and L are known, the locations of the points A and B can readily be calculated by using the triangulation method.

The UPL-Scanex system uses a flying image technique. In this system, the image of the product to be measured is scanned across a detector. This produces a scan signal which is a variation of voltage

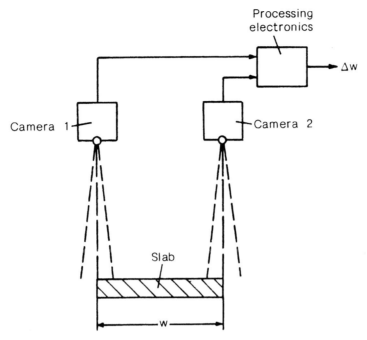

Fig. 32.26 Principle of width measurement utilized in the system developed by Exatest Schlumberger. Adapted from Mall and Meier-Engelen (1974).

with time. After further processing, the scan signal is converted into a square pulse proportional to the product dimension. To reduce distance errors, two cameras are used, one for each edge of the material being measured (Fig. 32.26). The cameras are set to the nominal width of the material. The deviations of the material width from the set nominal width are detected with the scanning cameras [32].

32.13 MEASUREMENT OF WORKPIECE CAMBER

The measurement of the workpiece camber (Fig. 31.6) can be accomplished by measuring the workpiece lengths along both edges as proposed by Ichihara [33]. Indeed, the relationship between the average radius of curvature of the plate center line and the plate lengths along its edges is given by

$$R_c = 0.5w(s_o + s_i)/(s_o - s_i), \qquad (32\text{-}14)$$

where w = plate width

s_i = plate length along inner edge

s_o = plate length along outer edge.

In the proposed measuring device (Fig. 32. 27) the plate lengths are measured with two pulse counters connected with axes of two measuring rolls. These rolls are pivotally mounted on movable blocks to accommodate different plate widths. The measuring rolls are pressed against the plate edges with two independent pressure cylinders and the plate width is measured with the position transducers attached to the pistons of the pressure cylinders. Once the plate width w and the plate lengths s_i and s_o are measured, the camber curvature radius R_c can be calculated by a computer according to Eq. (32-14).

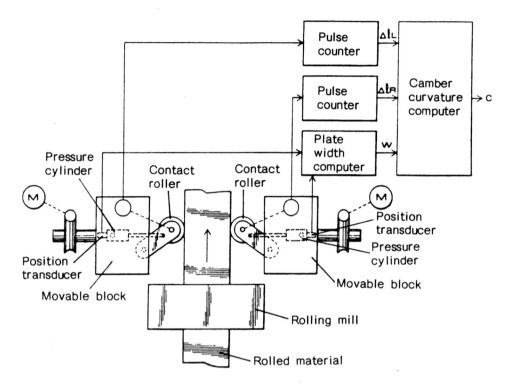

Fig. 32.27 System for detecting camber of rolled material. Adapted from Ichihara, U.S. Patent No. 4,528,756 (1985).

REFERENCES

1. "Dictionary of Instrumentation Technology", Machine Design, November 23 (1967).

2. "Xactray Series 2000 Gage", Fairchild Weston Schlumberger, Weston Controls, Inc., Archbald, Pa, 1983.

3. "Tosgage 5000 Series X-ray Thickness Gage ", Toshiba Corporation, Printed in Japan.

4. T. Tamiya, "An On-Line Measurement of Hot Strip Profile", Kawasaki Steel Technical Report No. 5, May 1982, pp. 55-67.

5. "Profile Thickness Gage", Isotope Measuring Systems, Ltd., Beckett House, England.

6. G.F. Kelk, R.H. Ellis, and V.B. Ginzburg, "New Developments Improve Hot Strip Shape: Shapemeter-Looper and Shape Actimeter", Iron and Steel Engineer, August 1986, pp. 48-56.

7. V.B. Ginzburg, "Magnetoelastic Properties of a Simplified Model of a Ferromagnetic Body in a Low Magnetic Field", IEEE Transactions on Magnetics, Vol. Mag-13, No. 5, Sept. 1977, pp. 1657-1663.

8. T. Sheppard, "Shape in Metal Strip: The State of the Art", Proceedings of the Metals Society Conference on Shape Control, Chester, England, April 1976, pp. 11-18.

9. T. Sheppard and J.M. Roberts, "Shape Control and Correction in Strip and Shape", International Metallurgical Reviews, Vol. 18, 1973, pp. 1-18.

10. O.G. Sivilotti, et al, "ASEA-ALCAN AFC System for Cold Rolling Flat Strip", AISE Year Book, 1973, pp. 263-270.

11. W.K.J. Pearson, "Shape Measurement and Control", Journal of the Institute of Metals, Vol. 93, 1964-1965, pp. 169-178.

12. "Shapemeter", Vollmer America Technical Bulletin, No. 100.

13. J. Mignon, et al, "New Development of the SIGMA-RO System for Computer Control in Wide and Narrow Hot Strip Mills", Proceedings of the International Conference on Steel Rolling, Tokyo, Japan, Sept. 29-Oct. 4, 1980, pp. 399-409.

14. "ROMETER Hot Strip Flatness Gauge", IRM: Industry Research and Metallurgy, P107 E-001/June 1984.

15. A. Iwawaki, et al, "Operating Experience With the Shapemeter for Hot Rolling Mill", Proceedings of the Metals Society Conference on Shape Control, Chester, England, April 1976, pp. 76-81.

16. V.B. Ginzburg, "Strip Rolling Mill Apparatus", U.S. Patent No. 4,771,622, Sept. 20, 1988.

17. Y. Uno, et al, "Dimension and Shape Measurement Using Water Column Distance Meter", Kawasaki Steel Report, 1984, pp. 132-138.

18. "Mitsubishi Shape Meter for Cold Strip Mills", Mitsubishi Publication, Feb. 1979.

19. T. Asamura, et al, "Development of a Shape Control for Cold Rolling Process and Practical Application of High Reduction Rolling", Nippon Steel Technical Report, No. 18, Dec. 1981, pp. 22-36.

20. J.P. Barreto and M.J. Hillier, "Shape Control in Cold Strip Rolling", Sheet Metal Industries, Vol. 45, Oct. 1986, pp. 707-709.

21. W. Fabian, et al, "On-Line Flatness Measurement and Control of Hot Wide Strip", MPT: Metallurgical Plant and Technology, Vol. 8, No. 4, 1985, pp. 68-75.

22. T. Kajiwara, U.S. Patent No. 3,475, 935, Nov. 4, 1969.

23. N. Ishimoto, et al, U.S. Patent No. 4,188,809, Feb.19, 1980.

24. V.B. Ginzburg, G.B. Jones, "Rolling Mill Strip Tension Monitoring and Shapemeter Assembly", U.S. Patent No. 4,680,978, July 21, 1987.

25. M. Borghesi and G. Chiozzi, "Shape Control Through Tension Distribution Control in Cold Strip Rolling", Proceedings of the International Conference on Steel Rolling, Tokyo, Japan, Sept. 29-Oct. 4, 1980, pp. 760-771.

26. T. Okabe, Japanese Patent No. 59-1013, Jan. 1984.

27. M. Okado, et al, "A New Shape Control Technique for Cold Strip Mills", Iron and Steel Engineer, June 1982, pp. 25-29.

28. V.B. Ginzburg, "Strip Tension Profile Apparatus and Associated Method", U.S. Patent No. 4,674,310, June 23, 1987.

29. R. Pirlet, et al, "A Noncontact System for Measuring Hot Strip Flatness", AISE Year Book, 1983, pp. 284-289.

30. "Laser Shapemeter, Shape Measurement at the Hot Strip Mill", Spie-Trindel, Uckange (France).

31. M. Kenyon, "Australian Developments in Steel Product Measurement and Inspection", Iron and Steel Engineer, July 1985, pp. 32-37.

32. H.F. van Mall and E. Meier-Engelen, "Improved Optical Scanning System for Gauging and Control", Wire World International, Vol. 16, May/June 1974.

33. J. Ichihara, "System for Detecting Camber of Rolled Material", U.S. Patent No. 4,528,756, July 16 (1985).

Part XI

Gauge and Width Control

33

Principle of Gauge Control

33.1 CAUSES OF GAUGE VARIATION

Causes of gauge variation in flat rolling may conveniently be analyzed from the following well-known expression which is often referred to as **the gaugemeter equation**

$$h_2 = C_o + \frac{P}{K_S} \tag{33-1}$$

where h_2 = exit thickness of rolled product

C_o = no-load roll gap

P = roll force

K_S = mill structural stiffness.

The gaugemeter equation is graphically shown in Fig. 33.1a. In this figure, slope of line A represents mill structural stiffness which is equal to

$$K_S = \frac{\Delta P}{\Delta C} = \tan\alpha \tag{33-2}$$

where ΔP = increment in roll force

ΔC = increment in no-load roll gap

α = slope of line A.

Slope of line B represents rolled material stiffness which is given by

$$K_M = \frac{\Delta P}{\Delta h} = \tan\beta \tag{33-3}$$

where Δh = change in rolled material thickness

β = slope of line B.

Intersection of the lines A and B is denoted by point C, which determines the values of the roll force F and exit thickness h_2.

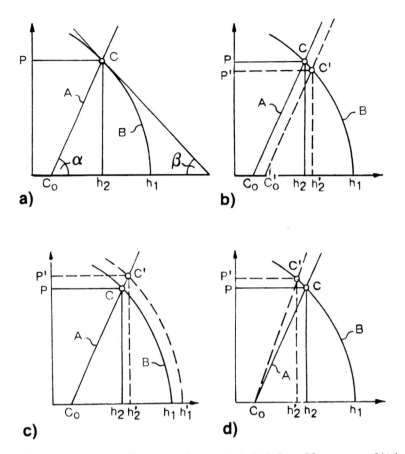

Fig. 33.1 Gaugemeter diagrams for: a) initial roll gap conditions; b) change in roll gap setting; c) change in entry thickness of rolled material; and d) change in mill stiffness.

Equation (33-1) and Fig. 33.1 are convenient to use for analysis of causes of gauge variation.

Effect of roll gap setting - Opening the roll gap shifts line A to the right (Fig. 33.1b). New equilibrium in the mill will be achieved at lower roll force P' with thicker exit thickness h_2' being rolled.

Variations of the no-load roll gap can be produced by:

a) roll eccentricity and ovality

b) roll thermal expansion or contraction

c) roll wear

d) roll bending

e) roll crossing

f) variation in oil film thickness of roll bearings

g) variation in lubricant film in roll bite.

Effect of entry thickness - Increase of entry thickness shifts line B to the right (Fig. 33.1c), so the equilibrium in the mill is achieved at higher roll force P'. This results in rolling of thicker exit thickness h_2'. Variation of entry thickness is usually due to imperfections of the gauge control in preceding rolling passes.

Effect of mill stiffness - Change in mill stiffness is equivalent to change in slope of the line A (Fig. 33.1d). An increase in the slope will produce equilibrium in the mill at higher force P' and thinner exit thickness h_2' will be rolled.

Main causes of variation of mill stiffness are:

a) variation in width of rolled product

b) change of the roll diameters

c) change of the roll crowns.

33.2 ACTUATORS FOR ROLL GAP CONTROL

Actuators for roll gap control may be divided into two major groups, mechanical and hydraulic.

Mechanical actuators are usually designed as the electrically driven mechanical screws with the nuts installed stationary in the housing posts. The electrically or hydraulically driven wedge type mechanism is another version of the mechanical actuators. To decrease coefficient of friction and increase load carrying capabilities of the wedge type actuators, the teflon cloth is installed at the interface between the wedge and the ram [1], as shown in Fig. 33.2.

Hydraulic actuators, known as the roll force cylinders, are usually installed either above top backup roll chocks or under bottom backup roll chocks [2], shown in Figs. 33.3 and 33.4. In some mill stands (Fig.33.5) the actuators are installed between top and bottom backup rolls [3].

Figure 33.3 illustrates a typical combination of a mechanical screwdown mechanism with a hydraulic actuator. In this arrangement, the screwdown mechanism is used for a rough adjustment of the roll gap whereas the fine roll gap control is achieved with the hydraulic cylinder. In some cases, the screwdown mechanism is eliminated and the roll gap control is solely provided by the hydraulic actuator. Figure 33.6

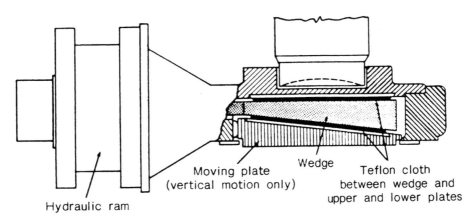

Fig. 33.2 Wedge type roll gap actuator. (From Lefoley, Hydraulic & Pneumatics, 1967. Copyright Penton Publishing. Reprinted with permission.)

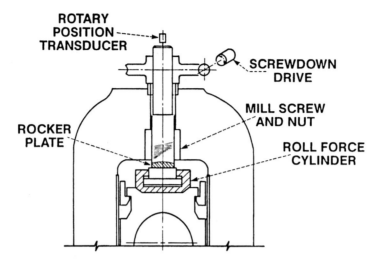

Fig. 33.3 A combination of a screwdown mechanism with a hydraulic roll force cylinder.

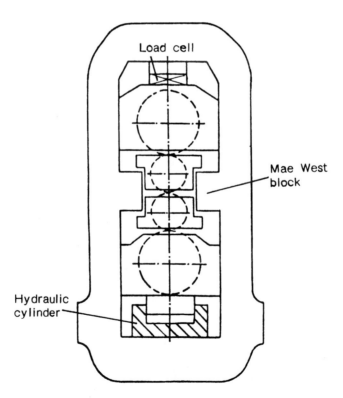

Fig. 33.4 Installation of the hydraulic actuator under bottom backup roll. Adapted from IHI Publication (1976).

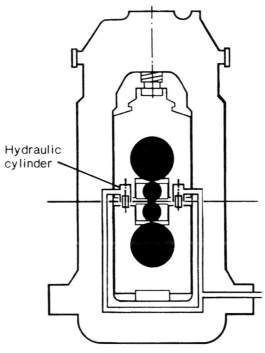

Fig. **33.5** A schematic presentation of the roll gap cylinders installed between backup roll chocks. Adapted from Troebs (1972).

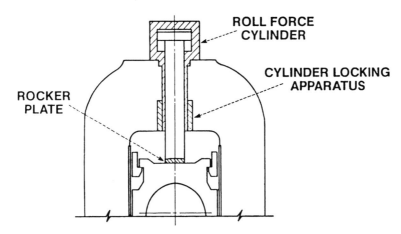

Fig. **33.6** A schematic illustration of a hydraulic force cylinder installed as a replacement for a screwdown mechanism.

shows schematically a mill stand arrangement in which a conventional screwdown mechanism is replaced with a hydraulic cylinders. To provide an alignment of the hydraulic cylinders with mill stack, the rocker plates or joints are installed either outside (Fig. 33.6) or inside (Fig. 33.7) of the cylinder [4].

One of the modifications of the hydraulic actuators (Fig. 33.8) is known as the Sermes beam [5]. The actuator is fabricated from a thick steel slab and is located between the main screwdown and the

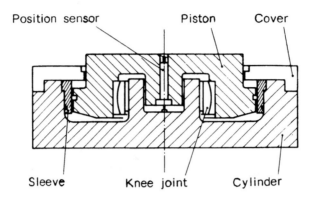

Fig. 33.7 CLECIM hydraulic cylinder used in HYDROGAGE system.
(From Morel, et al, Iron and Steel Engineer, 1984. Copyright
AISE, Pittsburgh, Pennsylvania. Reprinted with permission.)

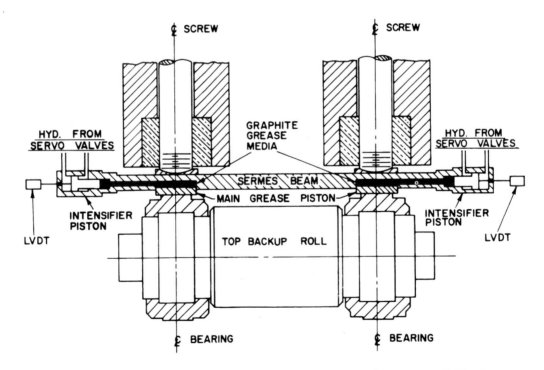

Fig. 33.8 The Sermes beam. (From Clarke, et al, Iron and Steel
Engineer, 1983. Copyright AISE, Pittsburgh, Pennsylvania.
Reprinted with permission.)

top backup roll bearing chocks. Two hydraulic cylinders are accommodated at both ends of the beam. They work as hydraulic intensifiers for the main cylinders in which the graphite grease is used as a hydraulic medium.

33.3 METHODS OF MEASUREMENT OF ROLL GAP

Direct measurement of the roll gap with the workpiece present in the roll bite is a problem yet to be resolved. Practically, only indirect methods of roll gap measurement have been developed.

One of the most common methods involves measurement of actual thickness of rolled material at the exit of a mill stand. A major deficiency of this method is a time delay due to the fact that the gauge sensor can not easily be installed close to the roll bite. This deficiency is eliminated by locating the roll gap sensors between work rolls [6], as shown in Fig. 33.9. In applying this method, however, another error, caused by roll bending, is being introduced.

In some roll gap control systems [2, 7] the sensors are installed either between backup roll chocks (Fig. 33.10) or between backup roll chocks and a part of the housing post called 'Mae West block' (Fig. 33.4).

The roll gap during rolling is often measured by using a method known as **gaugemeter principle**. In this method under loading conditions, the roll gap C is computed from the following equation

$$C = C_o + \frac{P}{K_S} \tag{33-4}$$

This equation is obtained from Eq. (33-1), assuming that under loading conditions the exit thickness of the rolled product is equal to the roll gap. So the roll gap during rolling can be measured if the no-load roll gap C_o, roll force P, and mill stiffness are known. The no-load gap measurement is usually provided with position transducers and measurement of roll force is made with either load cells or pressure transducers.

Upper part of transducer

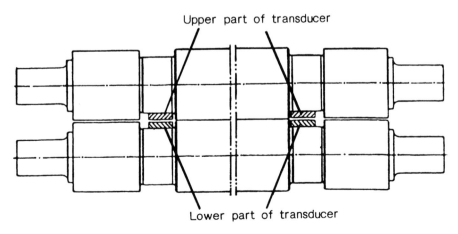

Lower part of transducer

Fig. 33.9 Roll gap sensors installed between work rolls. (From Liedtke and Fellenberg, Metallurgical Plant and Technology, 1983. Copyright Verlag Stahleisen mbH. Reprinted with permission.)

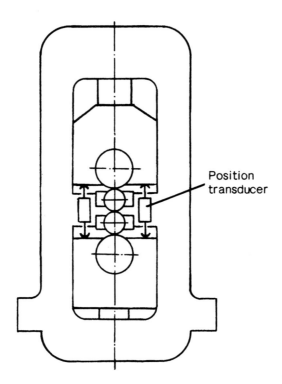

Fig. 33.10 Position sensors installed between backup roll chocks.

33.4 POSITION SENSORS

Although a number of different types of position sensors have been introduced, three basic types have mostly been utilized in the automatic gauge control systems. They include:

1. Analog induction type
2. Digital induction type
3. Magnetostrictive type.

Principle of operation and main characteristics of these transducers are briefly discussed below.

Analog induction type sensors - One of the modifications of this type of transducers was developed by Schaewitz Engineering [8] and is known as Linear Variable Displacement Transducers (LVDT). The LVDT is an electromechanical device that produces an electrical output proportional to the displacement of a separate movable core. It consists of a primary coil and two secondary coils symmetrically spaced in respect to the primary coil. A free-moving rod-shaped magnetic core inside the coil assembly provides a path for the magnetic flux linking the coils.

A cross-section of the LVDT and a plot of its operating characteristics are shown in Fig. 33.11. When the primary coil is energized through an oscillator, voltages are induced in the two secondary coils. These are connected series opposing so the two voltages are of opposite polarity. Therefore, the net output of the transducer, provided through a demodulator, is the difference between these voltages, which is zero when the core is at the center or null position.

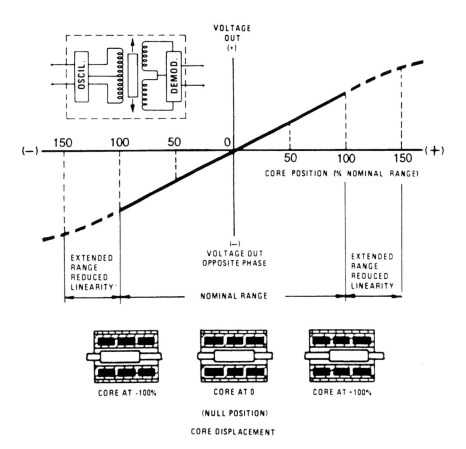

Fig. 33.11 Principle of operation of LVDT. (From Schaevitz Engineering Technical Bulletin 1002D. Copyright Schaevitz Engineering. Reprinted with permission.)

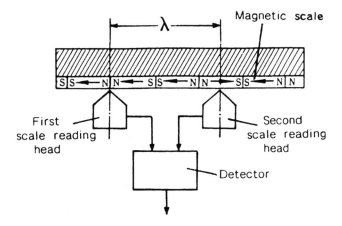

Fig. 33.12 Principle of operation of Sony Magnescale. Adapted from Sony Magnescale Publication (1972).

When the core is moved from the null position, the induced voltage in the coil toward which the core is moved increases, while the induced voltage in the opposite coil decreases. This action produces a differential voltage output that varies linearly with changes in core position. The phase of this output voltage changes abruptly by 180°, as the core is moved from one side of the null to the other.

Digital induction type sensors - One of these type of sensors was developed by Sony Magnescale, Inc. [9]. The sensor consists of three main parts: a magnetic scale, a scale-reading head, and a detector.

The magnetic scale is composed of a scale made of magnetic material on which a scale signal is magnetically recorded in a certain pitch (Fig. 33.12). The scale is made of a number of single-gap heads lined up with the same intervals as the magnetic pattern and connected to each other in series. When this kind of head is used, only the scale signal is sensed, offsetting signals of different wavelengths or the DC magnetic field.

When a carrier sine wave ($\sin\frac{\omega t}{2}$) is fed into a multi-gap head and the head is moved along the scale, it generates a pair of second harmonic signals:

$$e_1 = E\sin\frac{2\pi x}{\lambda}\sin\omega t \tag{33-5}$$

$$e_2 = E\cos\frac{2\pi x}{\lambda}\sin\omega t \tag{33-6}$$

where x = head displacement in relation to scale

ω = circular frequency

λ = distance between heads

t = time.

Detector is the electronic circuit which rectifies the signals expressed by Eqs. (33-5) and (33-6), and makes them applicable to the various uses. The detector generates pulses at the points where the sine and cosine waves of the scale-signal cross the zero level. It is used for the digital indication. More refined treatment of the sine and cosine waves allows for interpolation of the measurement between the zero level pulses and also to detect the direction of displacement.

Inductosyn is another modification of the digital induction type sensors. Its principle is similar to that of Sony Magnescale. The main difference is that the electromagnets rather than the permanent magnets are used in Inductosyn.

Magnetostrictive type sensors - This type of sensor has been developed by MTS Systems Corporations and is known by a trade name Temposonics [10]. The principle of operation of Temposonics (Fig. 33.13) is based on measurement of the time interval between an interrogating pulse and a return pulse. The interrogating pulse is transmitted through the transducer waveguide and the return pulse is generated by a permanent magnet representing the displacement to be measured. The sensor includes a Linear Displacement Transducer (LDT), a magnet, and the electronics necessary to generate the interrogating pulse, sense the return pulse and develop an analog output signal.

The electronics assembly sends the interrogating pulse to the LDT assembly and starts the leading edge of the pulse-width modulation signal. The pulse is conducted in a wire threaded through a magnetostrictive tube called a waveguide. The waveguide is mounted under tension in a non-magnetic

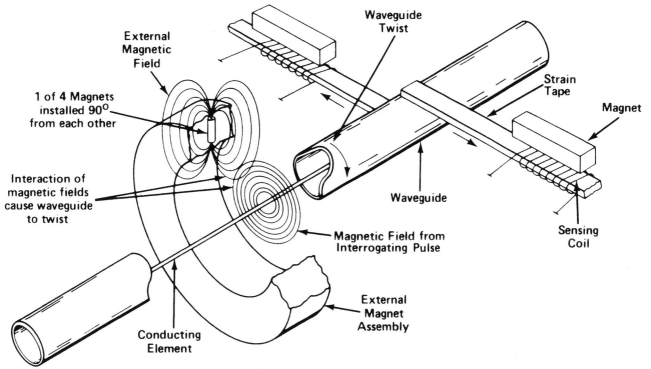

Fig. 33.13 Principle of operation of Temposonics. (From MTS Systems Corp. Publication, 1987. Copyright MTS Systems Corporation. Reprinted with permission.)

stainless steel rod. A permanent magnet is positioned along the stainless steel tube and mounted to the device from which displacement is to be measured.

The interaction of the magnetic field of the interrogating pulse and the magnetic field of the external permanent magnet causes a twist in the waveguide. The twist (or torsional strain pulse) is transmitted along the waveguide. The torsional strain pulse is dampened at the end of the waveguide and sensed in the transducer head. Two magnetic strain sensitive tapes are attached to the waveguide and coupled to sensing coils in the transducer head. The torsional strain pulse from the waveguide causes a small vibration of the tapes relative to the magnetic field of the sensing coils. This induces a voltage in the coils which is amplified and conditioned in the transducer head assembly; then it is sent back to the electronics box as the return pulse.

The performance characteristics of the position sensors vary with type of the sensor. For analog sensors they also vary with a measuring range. The best claimed characteristics are:

Accuracy	0.0015 - 0.003 mm
Resolution	0.001 - 0.0015 mm
Nonlinearity	0.025% of full scale
Hysteresis	0.001 - 0.004 mm
Repeatability	0.004 mm.

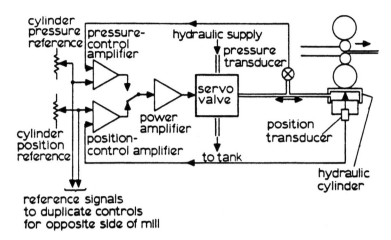

Fig. 33.14 Davy-Loewy position and roll force control of a hydraulic actuator. (From Dendle, Metals Society Conference Proceedings, Cardiff, Wales, 1978. Copyright Institute of Metals. Reprinted with permission.)

33.5 CLOSED LOOP CONTROL OF A HYDRAULIC ACTUATOR

A typical closed loop control system for one hydraulic actuator [11] is shown in Fig. 33.14. Two modes of the cylinder control are most frequently used:

  a) position control mode
  b) roll force control mode.

When position control mode is selected, the cylinder position reference signal is compared with a feedback signal provided by a cylinder position transducer. The error signal is then amplified and is fed into electrohydraulic servovalve. The servovalve translates this analog electrical signal into fluid flow, either into or out of the hydraulic cylinder, depending on the required direction of movement.

Identical closed-loop control systems are provided for both operating and drive side hydraulic cylinders. A common position reference is used for both systems to ensure that the cylinders move simultaneously. To achieve control of tracking of the strip throughout the mill, the additional reference of opposite polarities is provided to each cylinder control system.

When roll force control mode is selected, the roll force reference is compared with a feedback signal provided by either a load cell or a pressure transducer as shown in Fig. 33.14.

33.6 DYNAMIC CHARACTERISTICS OF ROLL GAP CONTROL

The following two types of characteristics are conveniently used to evaluate the dynamic response of the roll gap control system:

  a) step-function response
  b) frequency response
  c) phase shift.

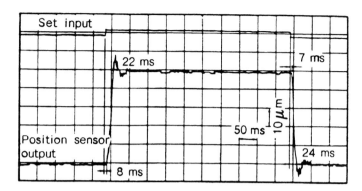

Fig. 33.15 Step response of the Mitsubishi Heavy Industries hydraulic roll gap control system. Adapted from Hayama, et al, Mitsubishi Heavy Industries Technical Review (1983).

These characteristics are briefly described below [12].

Step-function response - The step-function response is 'the characteristic curve or output plotted against time resulting from the input application of a step function'. It is also called **rise time.**

A step response of the hydraulic roll gap control system developed by Mitsubishi Heavy Industries [13] is shown in Fig. 33.15.

Frequency response - Generally, the frequency response is 'the portion of the frequency spectrum which can be passed by a device as it produces an output within specified limits of amplitude error'.

Decibels (db) are widely used to specify power gain (or attenuation). The decibel-power gain is equal to

$$dB = 10\log_{10} \frac{P_2}{P_1} \tag{33-7}$$

where P_1, P_2 = input and output power in a system respectively.

When power is delivered across active loads, we obtain

$$\frac{P_1}{P_2} = \frac{V_1^2 R_1}{V_2^2 R_2} \tag{33-8}$$

where V_1, V_2 = input and output voltage respectively

R_1, R_2 = input and output resistors respectively.

From Eqs. (33-7) and (33-8) the decibel-power gain is equal to

$$dB = 20\log_{10}\left(\frac{V_2}{V_1}\right) + 10\log_{10} \frac{R_1}{R_2} \tag{33-9}$$

if $R_1 = R_2$, then

$$dB = 20\log_{10}\left(\frac{V_2}{V_1}\right) \tag{33-10}$$

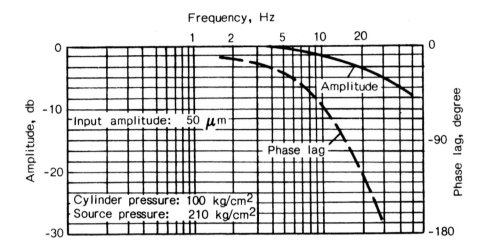

Fig. 33.16 Frequency response characteristics of the IHI hydraulic roll gap control system. Adapted from Saotome and Yamazawa (1974).

Phase shift - The phase shift is a change in the phase relationship between the input and output periodic functions.

A frequency response characteristics of the hydraulic roll gap control system developed by Ishikawajima-Harima Heavy Industries [14] is shown in Fig. 33.16.

33.7 GAUGEMETER CONTROL

Gaugemeter control may be used with both hydraulic and mechanical type of actuators. The gaugemeter control with hydraulic actuator, developed jointly by BISRA and Davy-United [11], is briefly described below.

The control implements the algorithm described by Eq. (33-1) and rewritten below in the form

$$C_R = H_{2R} - \frac{P}{K_S} \tag{33-11}$$

where C_R = no-load roll gap position reference

H_{2R} = exit gauge reference.

As shown in Fig. 33.17, a measurement of the roll force P is used by an electronic shaper to calculate mill deflection, or mill stretch P/K_S. This value is then subtracted from the exit gauge reference H_{2R}, so the no-load roll gap position reference C_R is obtained. The no-load roll gap position reference signal is then compared with measured value provided by a position transducer as previously described.

Gaugemeter system increases the mill stiffness beyond its natural value. By a complete (100%)

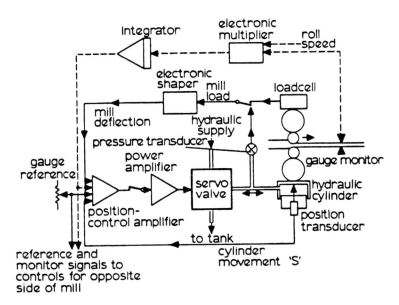

Fig. 33.17 BISRA-Davy-Loewy gaugemeter control system. (From Dendle, Metals Society Conference Proceedings, Cardiff, Wales, 1978. Copyright Institute of Metals. Reprinted with permission.)

compensating for the mill stretch, the mill stand can be made infinitely stiff so the roll gap will not vary with the variation in thickness or hardness of incoming product. It is found that in order to achieve optimum gauge and strip flatness characteristics, the gaugemeter control has to provide high mill stiffness for earlier mill stands and increasingly softer subsequent mill stands.

33.8 DIFFERENTIAL GAUGE CONTROL

The Differential Gauge Control (DGC), developed by Achenbach Buschhutten GmbH [3], utilizes a mill prestress concept to eliminate the gauge variation due to changes in roll force. The mill prestress force F_R is provided by the hydraulic cylinders installed between backup roll chocks (Fig. 33.18), so the screwdown force F_A is equal to

$$F_A = F_W + F_R, \tag{33-12}$$

where F_W = roll force

F_R = prestress force.

Thus, one part of the screwdown force F_A is transmitted through the roll bite producing the roll force F_W whereas another part F_R is transmitted between the top and bottom backup roll chocks. The value of the prestress force F_R is selected to be slightly greater than anticipated variation of the roll force during rolling. The control is designed in a such way that any increase in roll force dF_W will produce an equal decrease in the prestress force dF_R, so the screwdown force F_A remains constant. But constant screwdown force means constant mill stretch and thus constant roll gap.

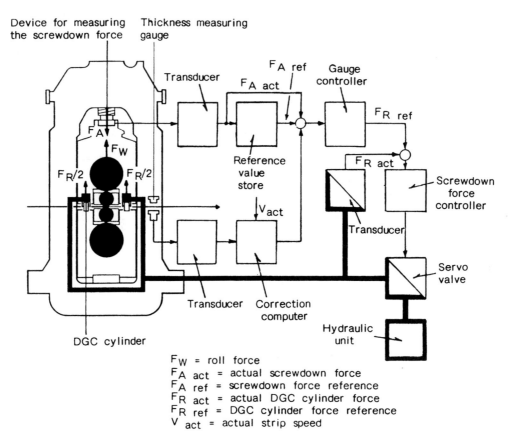

Fig. 33.18 Achenbach Buschhutten GmbH Differential Gauge Control (DGC) for a single mill stand. (From G. Troebs "Gauge Control Systems for Cold Rolling Mills", BBC Brown Boveri Nachrichten, Vol. 55, 1972. Copyright Brown, Boveri & Cie. Reprinted with permission.)

33.9 SPACER GAUGE CONTROL

The Spacer Gauge Control (SGC) method was developed by Achenbach BBC [15]. According to this method (Fig. 33.19), the distance between the back-up roll chocks is measured by a double arrangement to determine the mean value of tilting. The control deviation between the desired and actual values acts on a hydraulic adjusting SGC cylinder via a distance controller and pilot valve on each side of the stand.

The distance control circuit compensates for load-dependent changes in the resilience of the outer part of the stand. However, the roll gap and, thus, the strip thickness would still vary as a result of changes in the deformation of the roll. The compensation for the roll deformation is achieved by superimposition of a disturbance value into distance controller. This disturbance value is calculated from the pressure in the SGC cylinders taking into account calibration and roll bending forces.

The method resembles the gauge meter principle, but is more accurate because only the resilience of the rolls is required for the calculation and not the non-constant stiffness of the entire stand. The friction between chocks and columns also cannot take effect.

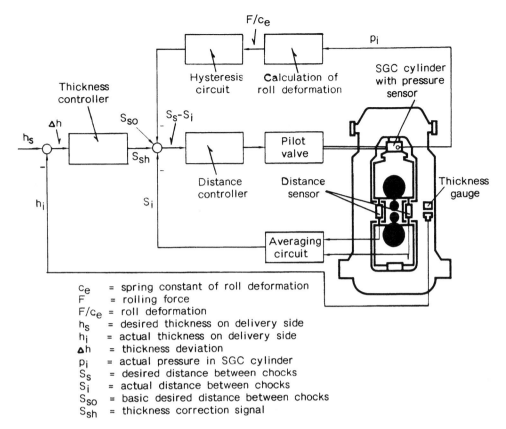

c$_e$ = spring constant of roll deformation
F = rolling force
F/c$_e$ = roll deformation
h$_s$ = desired thickness on delivery side
h$_i$ = actual thickness on delivery side
Δh = thickness deviation
p$_i$ = actual pressure in SGC cylinder
S$_s$ = desired distance between chocks
S$_i$ = actual distance between chocks
S$_{so}$ = basic desired distance between chocks
S$_{sh}$ = thickness correction signal

Fig. 33.19 Block diagram showing the principle of the Spacer Gauge Control (SGC). Adapted from Ludenscheild, et al (1976).

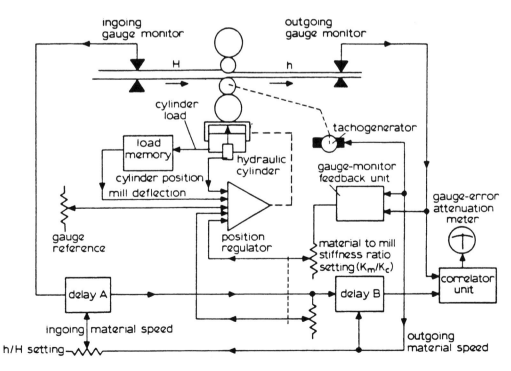

Fig. 33.20 Davy-Loewy feedforward control of a single mill stand. (From Dendle, Metals Society Conference Proceedings, Cardiff, Wales, 1978. Copyright Institute of Metals. Reprinted with permission.)

33.10 GAUGE DEVIATION CONTROL

Two methods are used for the gauge deviation control of a single mill stand: feedback method and feedforward method.

In the feedback method, the measured outgoing gauge error serves as a feedback signal in the gauge control system. When gaugemeter control is utilized (Fig. 33.19), the feedback signal from the outgoing gauge monitor provides a fine correction signal to compensate for some imperfections of the gaugemeter control.

The feedforward method of control [11] requires a thickness gauge to measure the thickness of the material entering the mill (Fig. 33.20). Since the incoming gauge monitor is installed at a distance from the roll bite, the measured incoming gauge error is fed into a block that computes a delay time taken by the strip to transverse the distance between the gauge monitor and the roll gap. The gauge error is then used to change the position of the cylinders.

33.11 STRIP TENSION CONTROL SYSTEM

The strip tension control system consists of a roll gap control system previously described with a tension closed loop control established around it [11]. The roll gap disturbances caused by incoming thickness and hardness variation, roll eccentricity, oil film thickness fluctuations, etc., may produce a change in the strip tension.

The tension meter, or tensiometer, detects actual strip tension (Fig. 33.21). The tensiometer output signal is compared with the tension reference signal and the error signal is fed into either the roll gap position regulator or the speed regulator of one of the adjacent mill stands. In former case, the control is usually referred to as the **tension control by gap,** while in latter case the control is known as the **tension control by speed.**

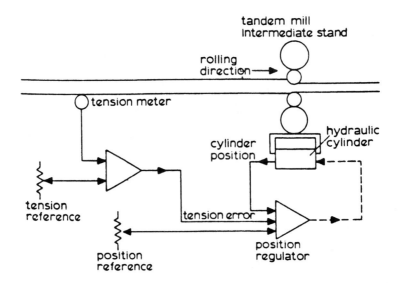

Fig. 33.21 Davy-Loewy strip tension control system. (From Dendle, Metals Society Conference Proceedings, Cardiff, Wales, 1978. Copyright Institute of Metals. Reprinted with permission.)

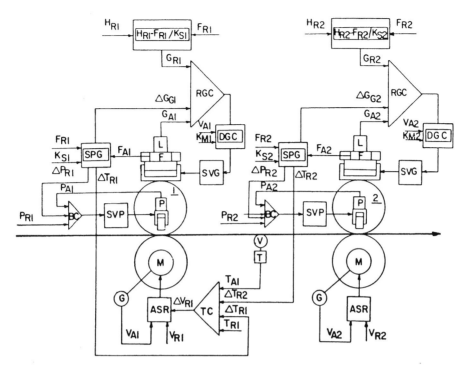

RGC = roll gap controller
DGC = dynamic gain controller
SVG = servovalve of the roll gap control system
SPG = gaugemeter controller
SVP = servovalve of the roll bending system
ASR = automatic speed regulator of the main drive
 BC = roll bending pressure controller
 TC = interstand strip tension regulator
  L = linear position transducer of the roll gap cylinder
  F = load cell
  P = pressure transducer of the roll bending
  M = main drive motor
  G = tachogenerator of the main drive

Fig. 33.22 Schematic presentation of Tippins AGC incorporating a gaugemeter control combined with a strip tension control and roll bending control. Adapted from Ginzburg and Snitkin, U.S. Patent No. 4,513,594 (1985).

Strip tension control can be combined with a gaugemeter control as proposed by Tippins Machinery Company [16]. In this control system (Fig. 33.22), the variations in mill stretch are compensated by adjusting the strip tension as long as the desired tension is within allowable range. Once there is a demand for the strip tension to be beyond this range, any further compensations for mill stretch are made by changing the roll gap as in the regular gaugemeter control.

33.12 THREE-STAGE AGC FOR TANDEM COLD MILL

The automatic gauge control system for the tandem cold mill, developed by Davy-Loewy [11], is a typical example of AGC for the tandem cold mill which consists of three distinct stages: (1) Entry AGC, (2) Interstand AGC, and (3) Exit AGC.

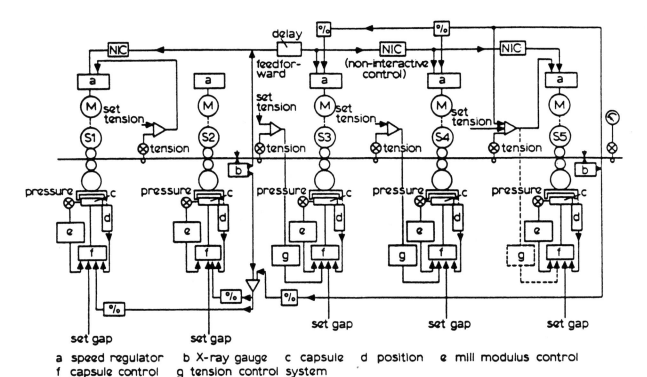

a speed regulator   b X-ray gauge   c capsule   d position   e mill modulus control
f capsule control   g tension control system

Fig. 33.23 Davy-Loewy three-stage AGC for tandem cold mill. (From Dendle, Metals Society Conference Proceedings, Cardiff, Wales, 1978. Copyright Institute of Metals. Reprinted with permission.)

The entry AGC (Fig. 33.23) incorporates the gaugemeter and gauge deviation control and, as an option, roll eccentricity compensation on stands S1 and S2. The interstand AGC provides constant mass flow on stand S3 and S4. The exit AGC regulates constant roll gap and also includes the gauge deviation feedback control on stand S5.

The gaugemeter at stands S1 and S2 along with gauge deviation feedback control assures constant gauge material that exits stand S2. Since the speed of stand S2 is also constant, the material will be fed into stand S3 with a constant mass flow.

Strip tension between stands S1 and S2 is maintained by adjusting the roll speed of stand S1. Strip tension between stands S2 and S3 as well as between stands S3 and S4 is controlled by changing the roll gap of a respective down-stream stand. Both types of strip tension control, as described above, are available between S4 and S5.

Gauge deviation feedback control provides adjustments of speeds of the stands S1, S2, S3, S4, and S5 whenever a correction of the exit gauge is required.

33.13 FEED-FORWARD AGC FOR TANDEM COLD MILL

In the most modern automatic gauge control (AGC) systems for tandem cold mills, the feedforward control is incorporated at the mill stands 1 and 2 [17]. The main objective of this control is to eliminate

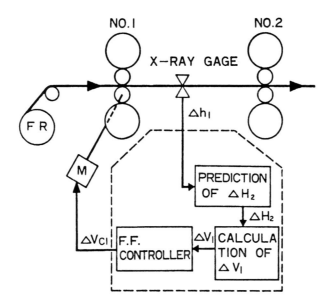

Fig. 33.24 Schematic arrangement of Sumitomo Metal Industries feedforward
AGC for tandem cold mill. (From Tajima, et al, Iron and Steel
Engineer, 1981. Copyright AISE, Pittsburgh, Pennsylvania. Reprinted
with permission.)

the deviation of the incoming gauge at stand 2. To achieve this objective, the tension between stands 1
and 2 is regulated by changing the roll speed of stand 1.

A schematic arrangement of the feedforward AGC developed by Sumitomo Metal Industries is shown
in Fig. 33.24. The thickness deviation is monitored by the thickness gauge installed after stand 1. The

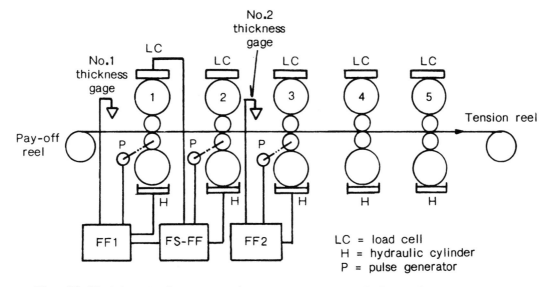

Fig. 33.25 Kobe Steel Corporation Flow-stress Feed-forward AGC for tandem
cold mill. Adapted from Nakada, et al (1982).

average thickness deviation ΔH_2 for a selected segment of the strip is calculated by a computer and the position of each segment is tracked as it passes through the mill stands.

The computer also predicts the time when each segment reaches stand 2 and the amount of speed change ΔV_1 of stand 1.

33.14 FLOW-STRESS FEED-FORWARD AGC

Kobe Steel Corporation has developed a modified version of feedforward AGC for tandem cold mill [18]. The system includes two conventional feedforward control systems (Fig. 33.25), which are applied for stands No. 1 and No. 2. Each of these systems measures the strip thickness in front of a rolling stand. Then an optimum roll gap is calculated and applied to control the roll gap of the stand as described earlier.

In addition to these conventional feedforward control systems, a so-called Flow Stress Feed Forward control system (FS-FF-AGC) has been utilized for Stand No. 2. Based on the information on the strip thickness measured with No. 1 thickness gauge, fluctuation in the resistance to deformation of the

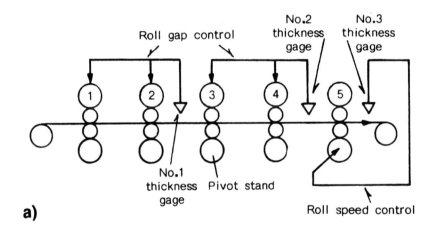

a)

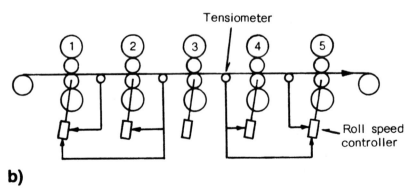

b)

Fig. 33.26 Sumitomo Metal Industries non-interactive AGC for tandem cold mill. Adapted from Okamoto (1976).

rolled material is calculated based on roll force measurement at the stand No. 1. The calculation of the optimum roll gap for stand No. 2 is made taking into account the time delay in the control system and the strip transfer time.

33.15 NON-INTERACTIVE AGC

In the non-interactive AGC, the tension control system and gauge control system are independent. One of these systems has been implemented by Sumitomo Metal Industries [19] in application for 5 stand tandem cold mill.

The roll gap setup errors and the variations in thickness and hardness of an incoming strip are corrected by automatic gauge control system (Fig. 33.26a) that incorporates three feedback control loops. Feedback for each loop is provided by the thickness gauges installed after stands 2, 4, and 5. The first control loop adjusts the roll gaps of stands 1 and 2. The second loop controls the roll gaps of stands 3 and 4. The third loop controls the gauge by adjusting the speed of stand No. 5 and, thus, affecting the exit strip tension.

The interstand strip tension is maintained constant by roll speed controller. Stand 3 is selected to be a pivot stand (key stand), as shown in Fig. 33.26b.

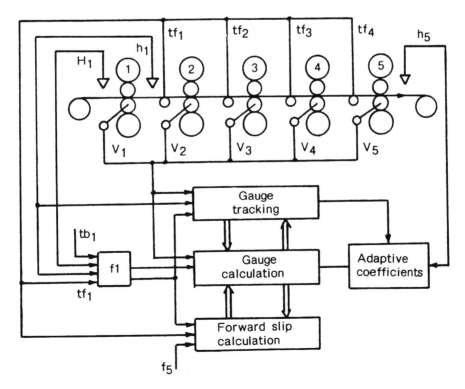

Fig. 33.27 Mass flow gauge calculating system of Hitachi AGC for tandem cold mill. Adapted from Eto, et al (1980).

33.16 AUTOMATIC TENSION AND GAUGE CONTROL SYSTEM

The objective of the automatic tension and gauge control system is to control the gauge during the entire rolling process from head end to tail end of the rolled strip. One of these systems has been developed by Hitachi [20].

The system incorporates the following three subsystems:

1. Mass flow gauge calculating system
2. Gauge and tension control system
3. Dynamic change control system.

Mass flow gauge calculating system - This system (Fig. 33.27) calculates the mass flow gauge of i-th stand by the equation

$$h_i = \frac{V_{i-1}}{V_i} \frac{1 + f_{i-1}}{1 + f_i} H_i \tag{33-13}$$

where H_i = mass-flow entry gauge

h_i = mass-flow exit gauge

f_i = forward slip.

The steady-state forward slip is calculated from the expression

$$f_i = \left[\frac{\sqrt{r_i}}{2} - \frac{1}{4\mu_i} \sqrt{\frac{h_i}{R_i'}} \left(\frac{2r_i}{2 - r_i} - \frac{tf_i - tb_i}{K_i} \phi \right) \right]^2 \tag{33-14}$$

where r_i = reduction

μ_i = coefficient of friction

R_i = deformed work roll radius

tf_i = forward tension

tb_i = backward tension

K_i = mean resistance to deformation

ϕ = modification coefficient.

Gauge and tension control system - Two modes of operation are used, low-speed mode and high-speed mode.

At low rolling speed, the control continuously maintains the desired interstand tension and quickly minimizes the gauge error at each stand by regulating the roll gaps and tension references. The interstand tensions are adjusted by changing the roll speeds. The required changes in the roll gap and the tension references are calculated by utilizing a multivariable optimum control concept. This concept is based on minimizing the integration of the square sum of the thickness deviation.

At high rolling speeds, the roll speeds of the mill stands are regulated by utilizing the following control loops:

a) feedforward control for entry gauge deviation

b) feedback control for delivery gauge deviation

c) gauge monitor, or gauge integration control.

The control compensates for lack of front tension during threading by an appropriate adjustment of the roll gaps. Also, the motor droop rate compensation is provided to avoid the roll slippage during threading.

Dynamic gauge control system - By using dynamic gauge control system, the strip is rolled to one gauge until the desired length of the rolled product is obtained. Then the gauge may be quickly changed to another scheduled value. The transition is made by smooth regulation of the roll gap and roll speed of each stand to a secondary set-up value while the mill is running.

33.17 INTERSTAND TENSION CONTROL IN HOT STRIP MILLS

In hot strip mills the interstand tension is usually maintained with so-called **loopers**. A looper is a mechanism that includes a free rotating roll which is elevated above pass line after threading the strip through the mill. The elevation of the looper and strip tension is continuously monitored. The aim of the control is to achieve the desired interstand tension when the looper roll is elevated to a predetermined position. If the desired tension is obtained at any different looper position then either the roll gap or roll speed of one of the adjacent stands is adjusted.

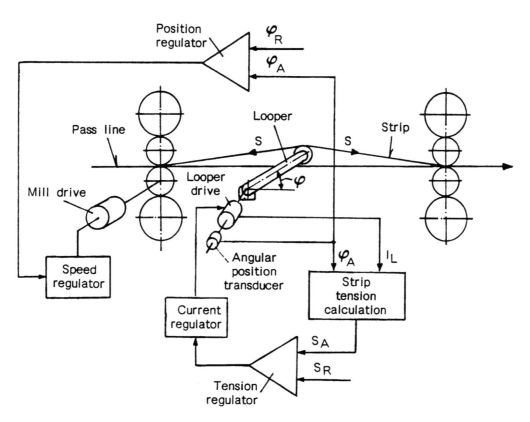

Fig. 33.28. Interstand tension control with an electric looper.

Three following types of loopers are known [21]:

1. Electric looper driven by an electric motor

2. Pneumatic looper driven by an air cylinder

3. Hydraulic looper driven by a hydraulic cylinder.

Figure 33.28 illustrates schematically the interstand tension control with an electric looper. The control has incorporated two control loops. The first control loop maintains a constant strip tension S. The actual strip tension S_A is calculated based on the measured values of the looper motor current I_L and of the looper angular position ϕ_A. The error signal due to a difference between the actual strip tension S_A and the strip tension reference S_R is fed into the current regulator of the looper motor. The current regulator adjusts the torque of the looper motor so the desired strip tension is achieved. The second control loop provides constant angular position of the looper. The error signal, due to a difference between the actual looper angular position ϕ_A and the position reference ϕ_R, is fed into the speed regulator of the mill motor of one of the adjacent mill stands. The speed regulator adjusts the speed of the mill stand so the desired looper position is provided.

More advanced design of a hydraulic looper developed by IHI [21] incorporates the load cells and an accelerometer to measure both the steady-state and dynamic strip tension. The hydraulic system utilizes a servovalve to enhance time response characteristics of the looper.

To objectively evaluate the dynamic characteristics of electric and hydraulic loop, Nagai, et al [22] have used a computer simulation model. The model is capable of simulating the strip tension variations

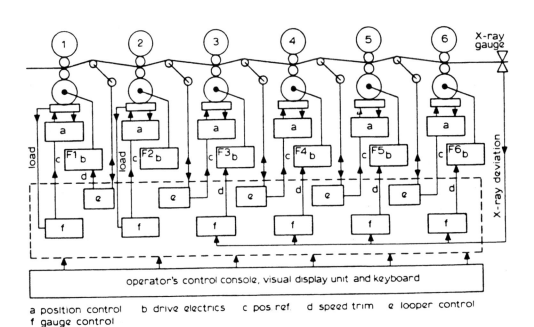

a position control   b drive electrics   c pos ref.   d speed trim   e looper control
f gauge control

Fig. 33.29 Davy-Loewy three-stage AGC for tandem hot strip mill. (From Dendle, Metals Society Conference Proceedings, Cardiff, Wales, 1978. Copyright Institute of Metals. Reprinted with permission.)

after the disturbances due to changes in roll gap, in looper angle, and in interstand tension occur. It was shown that the hydraulic looper insures more stable rolling conditions.

One of the latest developments in hot strip mills is an application of the **looperless control** to maintain the interstand tension [23]. In this control the interstand tension is estimated by measuring the roll force and roll torque. Once the tension is estimated, tension regulating can easily be realized by adjusting the roll speed of one of the adjacent mill stands.

33.18 THREE-STAGE AGC FOR TANDEM HOT STRIP MILL

A typical three-stage AGC for the hot strip mill can be exemplified by the gauge control system developed by Davy-Loewy [11].

The first stage of the AGC (Fig. 33.29) incorporates the gaugemeter control at stand 1 and 2. The second stage provides constant mass flow by maintaining constant interstand strip tension. The third stage includes the exit gauge error feedback system which controls the absolute delivery gauge by regulating the speed of the latter stands.

The interstand tension between stands 1 and 2 is maintained by regulating the roll speed of stand 1 whereas the interstand tension between the remaining stands is controlled by adjusting the roll gaps of the downstream stands.

33.19 FEED-FORWARD AGC FOR THE HOT TANDEM MILL

In the hot strip mills, the basic cause of gauge variations is the fluctuation of the strip temperature along the length of the strip [24]. The following two patterns of the strip temperature variations are usually recognized (Fig.33.30):

1. The overall temperature rundown (occasionally runup) from the head end to the tail end of the strip.

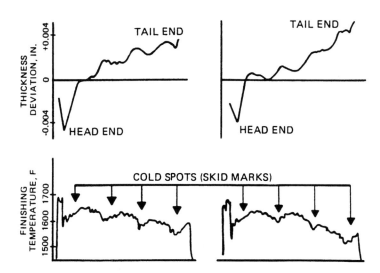

Fig. 33.30 Strip temperature and thickness variation in hot strip mill without gauge control. (From Pullen and List, AISE Year Book, 1969. Copyright AISE, Pittsburgh, Pennsylvania. Reprinted with permission.)

2. Local cold spots on the strip known as **skid marks** generated in the slab reheat furnaces at the contact surfaces of the slab with the furnace skids.

The Feed-forward AGC, developed by the Bethlehem Steel Corporation [25], is designed to reduce the effect of the temperature variation on both the strip gauge and the strip flatness performance.

The control block diagram is shown in Fig. 33.31. The first mill stand is the sampling stand, so no roll gap adjustment is made on this stand during rolling. The roll gap control of the next three stands is designed to overcorrect (negativity control) in-bar thickness variations caused by both overall and local temperature variations along the strip. The roll gap control of the last three stands allows the overcorrected or negative gauge to grow back, so that constant roll force is maintained at these stands.

The temperature readings from a sensor, installed after stand 1, are sampled 10 times per second and used to predict the strip temperature at the roll bite of each downstream stand. The growback predictor and negativity determinator, using these temperature data along with other strip characteristics, develop the gauge references for each sample of steel, so the necessary negativity and growback criteria are met.

33.20 COMPENSATION FOR IMPERFECTION OF MILL EQUIPMENT

Advanced AGC system usually incorporates the control functions which compensate for the gauge disturbances introduced by imperfections of the rolling mill equipment. Among these functions, the most important ones are:

1. Compensation for roll eccentricity
2. Compensation for variation of oil film thickness in oil-film type backup roll bearings

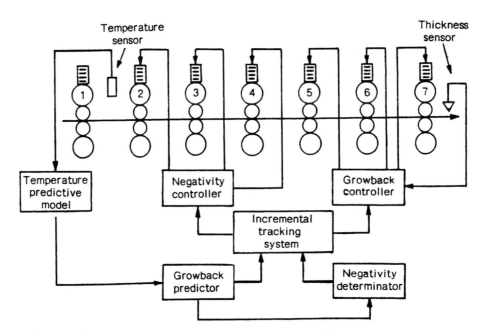

Fig. 33.31 Control block diagram of feedforward AGC for hot strip mill. Adapted from Wagner (1984).

3. Compensation for friction between moving parts of AGC actuators and between roll chocks and housing posts.

The term 'roll eccentricity' is usually applied to describe a more complicated imperfection of the mill equipment, that may include [26]:

a) out-of-roundness of work rolls

b) out-of-roundness of backup rolls

c) eccentricity between the center of the roll bearings and the center of the roll body.

Typical patterns of the roll force variation due to roll eccentricity are shown in Fig. 33.32. This pattern was recorded on a 4-high cold mill stand with oil-film type backup roll bearings. The roll force was 500 tons (1.1 mlb) with no strip in the roll bite. The amplitude of the roll force fluctuations depends greatly on mutual angular positions of the rolls and reaches its peak value when the eccentricities of the top and bottom backup rolls are in opposite directions [27].

There are basically three methods to reduce negative effect of roll eccentricity.

The first method utilizes the roll force mode to control the roll gap. The second method incorporates a dead-band circuit, that makes the roll gap control insensitive to the variation of the roll force within an established range. The third method utilizes a technique for compensation of the roll eccentricity.

Figure 33.33 shows a block diagram of the roll eccentricity control developed by IHI [26]. The control utilizes the Fourier Analyzer of Roll Eccentricity (FARE). The variation in rolling load due to backup roll eccentricity is synchronized with rotating angle of backup roll, which is detected by a pulse generator mounted on backup roll end. The synchronized component of the rolling load, which represents the roll eccentricity, is then extracted from the total roll force signal by using Fourier analysis

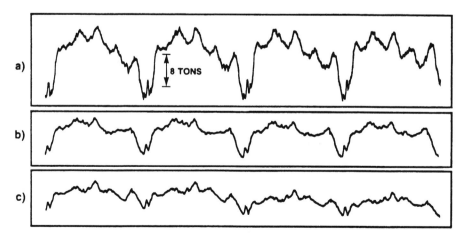

Fig. 33.32 Typical pattern of roll variation due to roll eccentricity when mutual orientations of eccentricities of top and bottom backup rolls are: a) opposite, b) approximately 90°, c) coincide. (From Ginzburg, Iron and Steel Engineer, 1984. Copyright AISE, Pittsburgh, Pennsylvania. Reprinted with permission.)

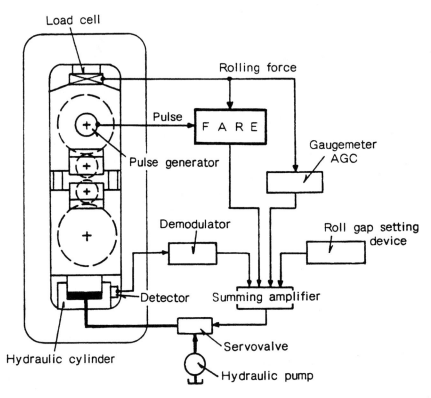

Fig. 33.33 Control block diagram of IHI roll eccentricity eliminator with FARE. Adapted from Imai (1978).

programmed in a computer. The generated signal is used to provide an additional roll gap reference that reduces roll force variations due to roll eccentricity.

Compensation for variation of the oil film thickness in oil-film type backup roll bearing is usually based on utilization of the experimental curves representing the relationship between the roll force and roll speed.

Compensation for friction between moving parts of AGC actuators and between roll chocks and housing posts may be achieved by introducing small low-amplitude oscillations into the system. Some measures, however, have to be taken to reduce effect of these oscillations on thickness of the rolled material. One of the solutions has been proposed by United Engineering, Inc. and International Rolling Mill Consultants, Inc. [28]. As illustrated in Fig. 33.34, the expansion of two telescopically arranged hydraulic cylinders are controlled with two independent position regulators. The position references H_R' and H_R'' provide a desired roll gap. In addition to these principal references, two periodic references H_O' and H_O'' of the same amplitude and opposite phase are introduced so the oscillations of the cylinders occur while the total expansion of two cylinders ($H_A' + H_A''$) remains independent of these oscillations.

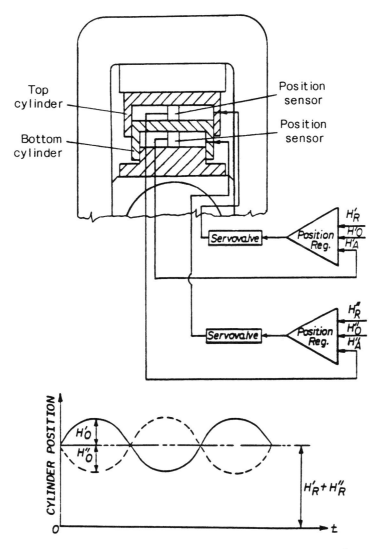

Fig. 33.34 Schematic presentation of the United Engineering/IRMC method of compensation for friction in the AGC actuators. Adapted from Ginzburg, U.S. Patent No. 4,757,746 (1988).

REFERENCES

1. G.M. Lefoley, "Servo Controlled Wedge Actuator Rolls Steel Sheet", Hydraulic & Pneumatics, October 1967, pp. 108–114.

2. "IHI" Hydraulic Mill", Ishikawajima-Harima Heavy Industries Co. Ltd., 1976.

3. G. Troebs, "Gauge Control Systems for Cold Rolling Mills", BBC Brown Boveri-Nachrichten, Vol. 55, 1972.

4. M. Morel, et al, "Quality Control and Production Optimization in Plate Mills Using HYDROPLATE System", Iron and Steel Engineer, May 1984, pp. 48–53.

5. J.F. Clarke, et al, "Automatic Gage Control of a 110-in. Plate Mill", Iron and Steel Engineer, Dec. 1983, pp. 17-20.

6. J. Liedtke and M. Fellenberg, "Modernization of a Universal Cold Rolling Mill Line", Metallurgical Plant and Technology, May 1983, pp. 82-87.

7. H.H. Hegenscheid, et al, "The SGC Method for Strip Thickness Control", BBC Brown Boveri Publication No. DIA7011E, 1976.

8. "LVDT and RVDT Linear and Angular Displacement Transducers", Schaevitz Engineering, Technical Bulletin 1002D, 1986.

9. "Sony Magnescale", Sony Magnescale Inc., Tokyo, 1972.

10. "Temposonics^{TM} Brand Absolute Linear Displacement Transducer Product Ordering Guide", MTS Systems Corporation, 1988.

11. D.W. Dendle, "Hydraulic Position-Controlled Mill and Automatic Gauge Control", Published in: Flat Rolling: A Comparison of Rolling Mill Types, Proceedings of International Conference held at University College, Cardiff, Sept. 26-29, 1978, pp. 103-111.

12. "Dictionary of Instrumentation Terminology", Machine design, November 23, 1967.

13. Y. Hayama, et al, "Development of High Response Type Hydraulic Roll Gap Control System ", Mitsubishi Heavy Industries Technical Review, 19:2 (reprint), 1983, pp. 1-7.

14. S. Saotome and K. Yamazawa", IHI New Position Sensing Type Hydraulic Mill", IHI Engineering Review, Vol. 7, No. 3, Sept. 1974.

15. H.H. Ludenschield, et al, "The SGC^R Method for Strip Thickness Control", Achenbach BBC Publication DIA70111E, 1976.

16. V.B. Ginzburg and S.R. Snitkin, "Method and Apparatus for Combining Automatic Gauge Control and Strip Profile Control", U.S. Patent No. 4,513,594, Apr. 30, 1985.

17. S. Tajima, et al, "Development of a New Type AGC System for a Tandem Cold Mill", Iron and Steel Engineer, June 1981, pp. 43-48.

18. T. Nakada, et al, "Feed-Forward AGC System for Cold Rolling Tandem Mills", Published in: Application of Mathematical and Physical Models in the Iron and Steel Industry, 3-rd Process Technology Conference, Pittsburgh, March 28-31, 1982, Proceedings, Vol.3 (Warrendale: AIME), pp. 117-121.

19. T. Okamoto, et al, "Advanced Gage and Tension Control of Tandem Cold Mill with Hydraulic Screwdown System", Transaction of Iron and Steel Institute of Japan, 161, 1976, pp. 614-622.

20. T. Eto, et al, "The Automatic Tension and Gauge Control at Tandem Cold Mill", Published in: The International Conference on Steel Rolling, Science and Technology of Flat Rolled Products, Tokyo, Sept. 29-Oct. 4, 1980, Proceedings, Vol. 1 (Tokyo: Iron and Steel Institute of Japan), pp. 439-450.

21. I. Imai, et al, "IHI New Type Hydraulic Looper", IHI Engineering Review, Vol. 11, No. 1, January 1978, pp. 29-37.

22. T. Nagai, et al, "Adoption of the Hydraulically Driven Push-up Device and Loopers in the Hot Strip Mill", Published in: The International Conference on Steel Rolling: Science and Technology of Flat

Rolled Products, Sept. 29-Oct. 4, 1980, Proceedings, Vol. 1 (Tokyo: Iron and Steel Institute of Japan), pp. 485-496.

23. S. Tanimoto, et al, "New Tension Measurement and Control System in Hot Strip Finishing Mill", Statistical Process Control in the Steel Industry, Iron and Steel Society, 1985.

24. C.C. Pullen and H.A. List, "Sparrows Point 56-in. Hot Strip Mill Gage Control System", Iron and Steel Engineer Year Book, 1969, pp. 339-348.

25. F. Wagner, "Feed Forward Control System for a Hot Strip Mill", Iron and Steel Engineer, October 1984, pp. 44-48.

26. I. Imai and T. Sukuzi, "FARE (Fourier Analyzer of Roll Eccentricity) Detector and Control System for Elimination of Roll Eccentricity", Ishikawajima-Harima Heavy Industries (IHI) Company, September 1973, pp. 1-12.

27. V.B. Ginzburg, "Dynamic Characteristics of Automatic Gage Control System with Hydraulic Actuators", Iron and Steel Engineer, January 1984, pp. 57-65.

28. V.B. Ginzburg, "Method and Apparatus for Control of a Force Applied to or the Position Assumed by a Work Effecting Element', U.S. Patent No. 4,757,746, July 19, 1988.

34

Modeling of Dynamic Characteristics of HAGC

34.1 BLOCK DIAGRAM OF A SINGLE STAND HAGC

Modeling of dynamic characteristics of HAGC allows one to evaluate stability and accuracy of the system as a function of engineering parameters of the mill and hydraulic actuator, type of servovalve and location, material stiffness, roll separating force, etc. The results of the modeling can be used for developing the algorithms for on-line computer models [1-8].

Development of the model involves the following three steps:

1. Development of the system's **block diagram**, each block representing a distinct elementary function. Figure 31.1 shows a block diagram of a single-stand HAGC based on a gaugemeter principle.

2. Derivation of **transfer function** for each block. The transfer function of each block represents functional relationship between output and input parameters of the block. Laplace transforms [9, 10] are commonly used for mathematical presentation of the transfer functions.

3. Derivation of the transfer function for a complete system which then allows one to obtain the **response characteristics** of the system, such as:

a) frequency response characteristics

b) time-domain response characteristics

Transfer functions for the individual blocks of the block diagram are reviewed below.

The block diagram shown in Fig. 34.1 is based on the gaugemeter equation given in the form

$$\frac{\Delta \dot{U}_C}{U_M} = \frac{\Delta \dot{U}_H}{U_M} + \frac{\Delta \dot{U}_C''}{U_M} \tag{34-1}$$

where $\Delta \dot{U}_C$ = increment in actuator position reference signal, volts

$\Delta \dot{U}_H$ = increment in material thickness reference signal, volts

$\Delta \dot{U}_C$ = increment in mill stretch compensation signal, volts

U_M = linear displacement transducer output corresponding to displacement H_M, volts.

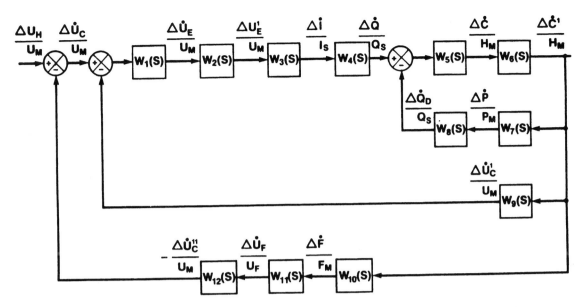

Fig. 34.1 Block diagram of a single-stand HAGC based on gaugemeter principle. (From Ginzburg, Iron and Steel Engineer, 1984. Copyright AISE, Pittsburgh, Pennsylvania. Reprinted with permission.)

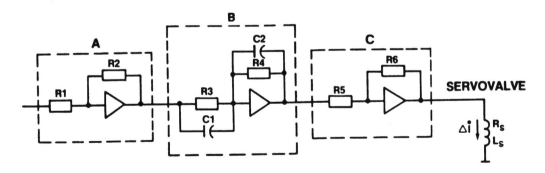

Fig. 34.2 Simplified electrical diagram of the position regulator including the position error amplifier (A), lead-lag network (B), and servovalve current amplifier (C). (From Ginzburg, Iron and Steel Engineer, 1984. Copyright AISE, Pittsburgh, Pennsylvania. Reprinted with permission.)

34.2 POSITION ERROR AMPLIFIER

Position error amplifier is represented by the first block in the block diagram (Fig. 34.1). Its transfer function is given by

$$W_1(s) = \frac{\dfrac{\Delta \dot{U}_E}{U_M}}{\dfrac{\Delta \dot{U}_C}{U_M}} = K_1 \qquad (34\text{-}2)$$

where $\Delta \dot{U}_E$ = position error amplifier output signal, volts

K_1 = amplification factor of the first block.

For the circuit shown in Fig. 34.2 (block A):

$$K_1 = K_{V1} = \frac{R_2}{R_1} \qquad (34\text{-}3)$$

where R_1, R_2 = resistors, ohms

K_{V1} = voltage amplification factor of the first block.

34.3 LEAD-LAG NETWORK

The second block in the block diagram (Fig. 34.1) represents the lead-lag network. Its transfer function is expressed in the form:

$$W_2(s) = \frac{\dfrac{\Delta \dot{U}_E'}{U_M}}{\dfrac{\Delta \dot{U}_E}{U_M}} = K_2 \frac{1 + T_{12}s}{1 + T_{22}s} \qquad (34\text{-}4)$$

where T_{12}, T_{22} = time constants, s

$\Delta \dot{U}_E'$ = lead-lag network output signal, volts

K_2 = amplification factor of the second block

$s = j\omega$ = Laplace operator

ω = circular frequency, 1/s.

And for the circuit shown in Fig. 34.2 (block B):

$$K_2 = K_{V2} = \frac{R_4}{R_3} \qquad (34\text{-}5)$$

$$T_{12} = R_3 C_1 \qquad (34\text{-}6)$$

$$T_{22} = R_4 C_2 \qquad (34\text{-}7)$$

where R_3, R_4 = resistors, ohms

K_{V2} = voltage amplification factor of the second block

C_1, C_2 = capacitors, farads.

34.4 CURRENT CONTROLLER

The third block in the block diagram (Fig. 34.1) is the current controller with transfer function described by the equation:

$$W_3(s) = \frac{\dfrac{\Delta\dot{I}}{I_s}}{\dfrac{\Delta\dot{U}_E'}{U_M}} = \frac{K_3}{1 + T_{23}s} \tag{34-8}$$

where $\Delta\dot{I}$ = increment in electrical current to servovalve, amps

I_s = servovalve rated current, amps

K_3 = amplification factor of the third block.

And for the circuit shown in Fig. 34.2 (block C):

$$K_3 = K_{V3}\frac{U_M}{I_s R_s} = \frac{R_6}{R_5}\frac{U_M}{I_s R_s} \tag{34-9}$$

$$T_{23} = \frac{L_s}{R_s} \tag{34-10}$$

where R_5, R_5 = resistors, ohms

R_s = servovalve coil circuit resistance, ohms

L_s = servovalve coil inductance, henries

T_{23} = time constant, s

K_{V3} = voltage amplification of the third block.

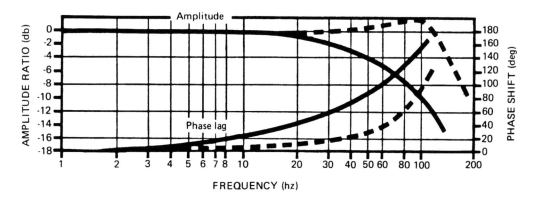

Fig. 34.3 Typical frequency response characteristics of two types of servovalve: type A – solid lines, type B – dotted lines. (From Ginzburg, Iron and Steel Engineer, 1984. Copyright AISE, Pittsburgh, Pennsylvania. Reprinted with permission.)

34.5 SERVOVALVE

The servovalve is shown as the fourth block in the block diagram (Fig. 34.1).
Its transfer function has the form:

$$W_4(s) = \frac{\frac{\Delta \dot{Q}}{Q_s}}{\frac{\Delta \dot{I}}{I_s}} \qquad (34\text{-}11)$$

where $\Delta \dot{Q}$ = increment in oil flow through servovalve control ports to the load due to change in current, in^3/s

$\quad Q_s$ = rated flow through servovalve control ports, in^3/s.

There is a difficulty in assigning a simplified (first, second or even third-order) transfer function for the servovalve. This is because of its complexity which exhibits high-order nonlinear responses. However, the frequency response characteristics of the servovalve are normally provided by their manufacturers. Typical frequency response characteristics are shown in Fig. 34.3; they are functions of both the servovalve current and pressure drop across the valve.

34.6 HYDRAULIC ACTUATOR

The fifth block in the block diagram (Fig. 34.1) is a hydraulic actuator with transfer function given by:

$$W_5(s) = \frac{\frac{\Delta \dot{C}}{H_M}}{\frac{\Delta \dot{Q}}{Q_s}} = \frac{1}{T_{25}s} \qquad (34\text{-}12)$$

where $\Delta \dot{C}$ = static increment in actuator position, in.

$\quad T_{25}$ = time constant, s.

Time constant T_{25} is equal to:

$$T_{25} = \frac{AH_M}{Q_s} \qquad (34\text{-}13)$$

where A = actuator working area, sq in.

$\quad H_M$ = position transducer nominal range, in.

34.7 MILL SPRING

Mill spring is represented by the sixth block in the block diagram (Fig. 34.1). Transfer function of the mill spring has the form:

$$W_6(s) = \frac{\frac{\Delta \dot{C}'}{H_M}}{\frac{\Delta \dot{C}}{H_M}} = \frac{1}{1 + T_{26}s + (T_{36}s)^2} \qquad (34\text{-}14)$$

where T_{26}, T_{36} = time constants, s.

$\quad \Delta \dot{C}$ = dynamic increment in actuator position, in.

The time constants may be expressed as:

$$T_{26} = 2\zeta T_{36}$$

$$(34\text{-}15)$$

$$T_{36} = \frac{1}{2\pi f_R}$$

$$(34\text{-}16)$$

where ζ = damping factor

f_R = mill resonant frequency, Hz.

The resonant frequency of the mill spring system is equal to:

$$f_R = \frac{1}{2\pi} \sqrt{\frac{Kg}{G}}$$

$$(34\text{-}17)$$

where K = equivalent mill stiffness, lb/in.

g = acceleration constant of free body, in./s/s

G = equivalent weight of mill structure per one side, lb.

34.8 MILL STIFFNESS AND MILL STRUCTURE WEIGHT

The equivalent mill stiffness K and equivalent mill structure weight G can be calculated from the following equations:

$$\frac{1}{K} = \frac{1}{K_H} + \frac{1}{K_S} + \frac{1}{K_M}$$

$$(34\text{-}18)$$

$$G = \sum_{j=1}^{n} \alpha_j G_j$$

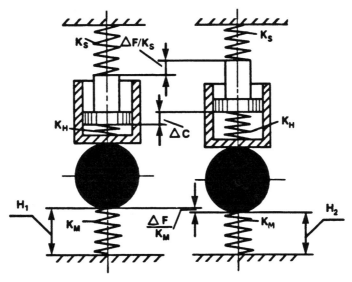

Fig. 34.4 Physical interpretation of mill structural stiffness K_S, hydraulic stiffness K_H, and material stiffness K_M. (From Ginzburg, Iron and Steel Engineer, 1984. Copyright AISE, Pittsburgh, Pennsylvania. Reprinted with permission.)

where K_H = hydraulic stiffness, lb/in.

  K_S = structural stiffness, lb/in.

  K_M = rolling material stiffness, lb/in.

  α_j = weight adjustment factor for j-th part of mill structure.

  G_j = weight of j-th part of mill structure per one side, lb

  n = number of mill parts.

Equations (34-16)-(34-18) present an approximate solution for a sophisticated mass-spring system with distributed parameters. Therefore, the range of validity of these equations has to be carefully evaluated for each particular case.

Figure 34.4 illustrates a physical interpretation of the mill structural stiffness K_S, hydraulic stiffness K_H and material stiffness K_M assumed in this study.

Hydraulic stiffness may be calculated from the equation:

$$K_H = \frac{A\delta}{h + L\left(\dfrac{d}{D}\right)^2} \tag{34-19}$$

where δ = oil bulk modulus, psi

  h = oil height inside of actuator, in.

  L = length of pipe between servovalve and actuator, in.

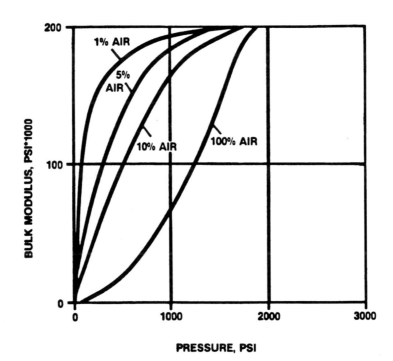

Fig. 34.5 Bulk modulus of hydraulic fluid as a function of system pressure and percentage of entrained air. (From Ginzburg, Iron and Steel Engineer, 1984. Copyright AISE, Pittsburgh, Pennsylvania. Reprinted with permission.)

d = ID of pipe between servovalve and actuator, in.

D = working diameter of actuator, in.

Under actual operating conditions, the hydraulic fluid will absorb air bubbles. This results in a decrease in the bulk modulus of the hydraulic fluid according to the volumetric fraction of air contained. However, the effect of absorbed air on the bulk modulus will be decreased with an increase in the fluid pressure [11], as shown in Fig. 34.5.

Structural stiffness K_S represents a slope of the mill spring curve and is equal to:

$$K_S = \frac{\Delta F}{\Delta C} \tag{34-20}$$

where ΔF = increment in roll separating force per actuator, lb.

Material stiffness may be determined from:

$$K_M = \frac{\Delta F}{\Delta h} = \frac{F}{2\Delta} \tag{34-21}$$

where Δh = increment in material thickness, in.

F = roll separating force per actuator, lb

Δ = draft, in.

34.9 SERVOVALVE DROOP

The eighth block of the block diagram (Fig. 34.1) represents the servovalve droop with transfer function given by:

$$W_8(s) = \frac{\dfrac{\Delta \dot{Q}_D}{Q_S}}{\dfrac{\Delta \dot{P}}{P_M}} = K_8 = \frac{\gamma H_M}{A Q_S} \frac{K_S K_M}{K_S + K_M} \tag{34-22}$$

where $\Delta \dot{Q}_D$ = increment in oil flow through servovalve control ports due to change in back pressure, cu in./s

P_M = actuator pressure corresponding to roll separating force F_m, psi

K_8 = amplification factor of the eighth block

γ = servovalve droop factor, in^3/s/psi

The servovalve droop factor is equal to

$$\gamma = \frac{\Delta Q}{\Delta P} = \frac{Q_S}{\sqrt{2P_S \left(2P_H - \dfrac{F}{A}\right)}} \tag{34-23}$$

where ΔP = increment in actuator pressure, psi

P_S = servovalve pressure drop corresponding to rated flow Q_S, psi

P_H = hydraulic pump pressure, psi.

34.10 TRANSDUCERS

The actuator position transducer is represented by the ninth block in the block diagram (Fig. 34.1). Its transfer function is expressed as:

$$W_9(s) = \frac{\dfrac{\Delta \dot{U}_C'}{U_M}}{\dfrac{\Delta \dot{C}'}{H_M}} = \frac{1}{1 + T_{29}s} \tag{34-24}$$

where $\Delta \dot{U}_C$ = increment in position transducer output signal, volts

T_{29} = time constant, s.

Similarly, the roll force transducer's transfer function (block 11) is given by:

$$W_{11}(s) = \frac{\dfrac{\Delta \dot{U}_F}{U_F}}{\dfrac{\Delta \dot{F}}{F_M}} = 1 \tag{34-25}$$

where $\Delta \dot{U}_F$ = increment in roll force transducer output signal, volts

U_F = roll force transducer output signal corresponding to roll separating force F_M, volts

F_M = roll separating force per actuator corresponding to actuator displacement H_M, lb

T_{210} = time constant, s.

34.11 ACTUATOR PRESSURE AND FORCE

The actuator pressure and force are represented by the seventh and tenth block respectively (Fig. 34.1). Their transfer functions have the forms:

$$W_7(s) = \frac{\Delta \dot{P}}{P_M} \bigg/ \frac{\Delta \dot{C}'}{H_M} = 1 \tag{34-26}$$

$$W_{10}(s) = \frac{\Delta \dot{F}}{F_M} \bigg/ \frac{\Delta \dot{C}'}{H_M} = 1 \tag{34-27}$$

34.12 MILL STIFFNESS MULTIPLIER

The mill stiffness multiplier is represented by the twelfth block (Fig. 34.1). Its transfer function is expressed by the equation:

$$W_{12}(s) = -\frac{\dfrac{\Delta \dot{U}_C''}{U_M}}{\dfrac{\Delta \dot{U}_F}{U_F}} = K_{12} = -\frac{mK_M}{K_S + K_M} \tag{34-28}$$

where K_{12} = amplification factor of the twelfth block

m = roll force feedback factor.

The roll force feedback factor $m = 0$, when the roll force feedback is disconnected.

The roll force transducer output signal is normally calibrated without strip in the roll bite and its sensitivity is set to be equal to:

$$S_F = \frac{S_H}{K_S} = \frac{U_M}{H_M} \frac{1}{K_S} \tag{34-29}$$

where S_H = linear displacement transducer sensitivity, volts/lb

S_F = roll force transducer sensitivity, volts/lb.

34.13 TRANSFER FUNCTIONS OF SYNTHESIZED BLOCKS

Once the transfer functions of individual elements of the control system are known, the block diagram illustrated in Fig. 34.1 can be reduced to the form shown in Fig. 34.6.

In the reduced block diagram, $\Delta U_H'$ is the increment in the material thickness feedback signal. The rest of the notations for control signals are the same as for the block diagram shown in Fig. 34.1.

Block 13 replaces the position error amplifier (block 1), lead-lag network (block 2), and current controller (block 3).

$$W_{13}(s) = W_1(s)W_2(s)W_3(s) = K_{13} \frac{1 + T_{113}(s)}{1 + T_{213}(s)} \tag{34-30}$$

where

$$K_{13} = K_1 K_2 K_3 = \frac{K_V U_M}{I_S R_S} \tag{34-31}$$

$$T_{113} = T_{12} \tag{34-32}$$

$$T_{213} = T_{22} \tag{34-33}$$

Block 14 replaces the actuator (block 5), mill spring (block 6), pressure transformer (block 7), and servovalve droop (block 8).

$$W_{14}(s) = \frac{W_5(s) \cdot W_6(s)}{1 + W_5(s)W_6(s)W_7(s)W_8(s)} \tag{34-34}$$

$$= \frac{1}{b_{14} + T_{214}s + (T_{314}s)^2 + (T_{414}s)^3}$$

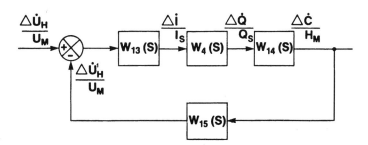

Fig. 34.6 Reduced block diagram of HAGC. (From Ginzburg, Iron and Steel Engineer, 1984. Copyright AISE, Pittsburgh, Pennsylvania. Reprinted with permission.)

where

$$b_{14} = K_8 \qquad\qquad\qquad (34\text{-}35)$$

$$T_{214} = T_{25} \qquad\qquad\qquad (34\text{-}36)$$

$$T_{314} = (T_{25}T_{26})^{1/2} \qquad\qquad\qquad (34\text{-}37)$$

$$T_{414} = (T_{25}T_{36}^2)^{1/3} \qquad\qquad\qquad (34\text{-}38)$$

Block 15 replaces the position transducer (block 9), feedback compensator (block 10), roll force transducer (block 11), and mill stiffness multiplier (block 12).

$$W_{15}(s) = W_9(s) + W_{10}(s)W_{11}(s)W_{12}(s) = \frac{a_{15} + T_{115}s}{1 + T_{215}s} \qquad (34\text{-}39)$$

where

$$a_{15} = 1 + K_{12} \qquad\qquad\qquad (34\text{-}40)$$

$$T_{115} = K_{12}T_{29} \qquad\qquad\qquad (34\text{-}41)$$

$$T_{215} = T_{29} \qquad\qquad\qquad (34\text{-}42)$$

34.14 AMPLITUDE RATIOS AND PHASE SHIFTS OF INDIVIDUAL BLOCKS

It follows from equations (34-30), (34-34) and (34-39), that each transfer function in the block diagram shown in Fig. 34.6 can be expressed in the following general form:

$$W_i(s) = \frac{K_i(a_i + T_{1i}s)}{b_i + T_{2i}s + (T_{3i}s)^2 + (T_{4i}s)^3} \qquad (34\text{-}43)$$

where K_i = amplification factor of i-th block

a_i = real component of numerator in transfer function of i-th block

b_i = real component of denominator in transfer function of i-th block

T_{ji} = time constants in transfer function of i-th block, s.

Therefore, the amplitude ratio A_i and phase shift ϕ_i for each transfer function can be calculated from following formula:

$$A_i = K_i\sqrt{\frac{a_i^2 + (T_{1i}\omega)^2}{[b_i - (T_{3i}\omega)^2]^2 + [T_{2i}\omega - (T_{4i}\omega)^3]^2}} \qquad (34\text{-}44)$$

$$\phi_i = \arctan\frac{T_{1i}\omega}{a_i} - \arctan\frac{T_{2i}\omega - (T_{4i}\omega)^3}{b_i - (T_{3i}\omega)^2} \qquad (34\text{-}45)$$

The amplitude ratio is commonly expressed in the units called **decibels** (db), according to the formula [4]:

$$A_{iD} = 20\log_{10}A_i \tag{34-46}$$

where A_{iD} = amplitude ratio for i-th block, db.

34.15 FREQUENCY RESPONSE CHARACTERISTICS OF CONTROL SYSTEM

The transfer function of the open-loop control control system, shown in the block diagram (Fig. 34.6), has the following form:

$$W(s) = W_4(s)W_{13}(s)W_{14}(s)W_{15}(s) \tag{34-47}$$

The amplitude ratio A_0 and the phase shift ϕ_0 for the open-loop control system are correspondingly equal to:

$$A_0 = A_4 A_{13} A_{14} A_{15} \tag{34-48}$$

$$\phi_0 = \phi_4 + \phi_{13} + \phi_{14} + \phi_{15} \tag{34-49}$$

The transfer function $W_C(s)$ of the closed-loop control system, shown in Fig. 34.6, describes the relationship between the increment in the actuator position ΔC and the reference signal ΔU_H:

$$W_C(s) = \frac{\left(\dfrac{W_0(s)}{W_{15}(s)}\right)}{1 + W_0(s)} \tag{34-50}$$

The amplitude ratio A_C and the phase shift ϕ_C for the closed loop control system, are correspondingly equal to:

$$A_C = \frac{\dfrac{A_0}{A_{15}}}{(1 + A_0\cos\phi)^2 + (A_0\sin\phi)^2} \tag{34-51}$$

$$\phi_C = \phi_0 - \phi_{15} - \arctan\frac{A_0\sin\phi_0}{1 + A_0\cos\phi_0} \tag{34-52}$$

In the steady state, i.e., when $s \to 0$, the amplitude ratio A_C becomes equal to:

$$A_{C0} = \frac{1 + \dfrac{K_M}{K_S}}{1 + \dfrac{K_M}{K_C} + \dfrac{K_M}{K_S}(1 - m)} \tag{34-53}$$

where

$$K_C = \frac{K_V A Q_S}{I_S R_S\gamma} S_H \tag{34-54}$$

and

$$K_V = K_{V1}K_{V2}K_{V3}$$ (34-55)

where A_{CO} = amplitude ratio for closed loop at zero frequency

K_C = apparent control system stiffness, lb/in.

K_V = total voltage amplification factor (gain) of position regulator

K_{Vi} = voltage amplification factor of i-th block of position regulator.

In the ideal case when $K_C \gg K_M$, A_{CO} reduces to

$$A_{CI} = \frac{1 + \dfrac{K_M}{K_S}}{1 + \dfrac{K_M}{K_S}(1 - m)}$$ (34-56)

where A_{CI} = amplitude ratio for closed loop at zero frequency and zero servovalve droop.

The frequency response characteristics of the closed-loop control system (Fig. 34.7) are normalized to provide a zero amplitude ratio (in decibels), when the exciting frequency is equal to zero:

$$A_{CD} = 20 \, \log_{10}\left(\frac{A_C}{A_{CI}}\right) = \psi(f)$$ (34-57)

where A_{CD} = amplitude ratio for closed-loop control system, Hz

f   = exciting frequency, Hz.

The exciting frequency is equal to

$$f = \frac{\omega}{2\pi}$$ (34-58)

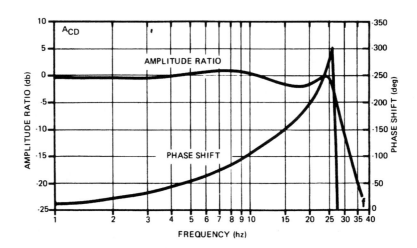

Fig. 34.7 Typical frequency response characteristics of closed-loop control system. (From Ginzburg, Iron and Steel Engineer, 1984. Copyright AISE, Pittsburgh, Pennsylvania. Reprinted with permission.)

34.16 POSITION ERRORS AND CONTROL MARGINS

Performance of the control system may be evaluated by the following four parameters:

1. Steady-state position error

2. Dynamic position error

3. Gain margin

4. Phase shift margin.

The steady-state position error can be determined from Eqs. (34-53) and (34-56) as follows:

$$\epsilon = 1 - \frac{A_{C0}}{A_{CI}} = \frac{1}{1 + \dfrac{K_C}{K_M} + \dfrac{K_C}{K_S}(1 - m)} \tag{34-59}$$

When the roll force feedback is connected (m = 1):

$$\epsilon = \frac{1}{1 + \dfrac{K_C}{K_M}} \tag{34-60}$$

Thus, the steady-state position error decreases with increase of apparent control system stiffness K_C in relation to both material stiffness K_M and structural stiffness K_S.

The dynamic position error is determined from the closed-loop response characteristics by attenuation and phase shift at the selected frequency.

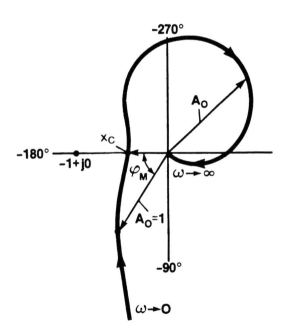

Fig. 34.8 Determination of gain margin and phase margin from polar plot of the open-loop transfer function. (From Ginzburg, Iron and Steel Engineer, 1984. Copyright AISE, Pittsburgh, Pennsylvania. Reprinted with permission.)

The gain margin K_{VM} is determined as the additional gain that, when added to the open-loop system, will cause the gain at the frequency corresponding to -180° phase shift to equal unity.

The phase margin ϕ_M is determined as the additional phase shift that, when added to the phase shift of the open-loop phaser where its amplitude equals 1, will result in total phase shift of -180°.

The gain margin, as well as the phase margin, may be obtained from the polar plot of the open-loop transfer function shown in Fig. 34.8. The gain margin is:

$$K_{VM} = \frac{1}{x_c} \tag{34-61}$$

where x_c = crossover point on the polar plot.

34.17 TIME DOMAIN RESPONSE CHARACTERISTICS

The transfer function for the closed-loop control system, described by Eq. (34-50), is a complex function of parameter s. It is difficult, or even impossible, to apply an inverse transformation process for finding its time domain function. Therefore, the approximation methods are usually used.

In one of the methods [12], the real component of the normalized amplitude ratio for the closed loop is calculated. A typical plot of A_{CR}, as a function of exciting frequency, is shown in Fig. 34.9. This

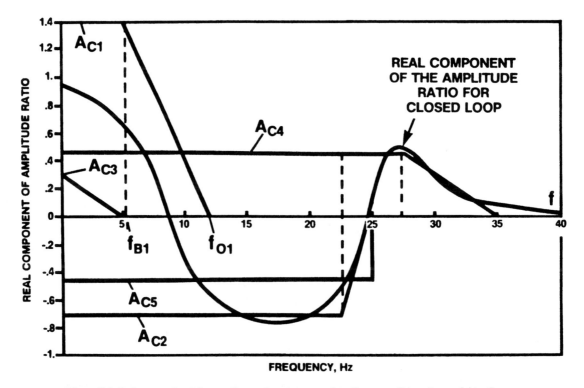

Fig. 34.9 Approximation of real component of normalized amplitude ratio for closed loop with trapezoids. (From Ginzburg, Iron and Steel Engineer, 1984. Copyright AISE, Pittsburgh, Pennsylvania. Reprinted with permission.)

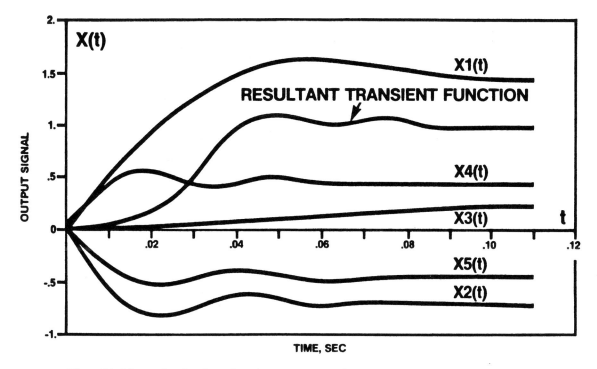

Fig. 34.10 Typical time domain response characteristics. (From Ginzburg, Iron and Steel Engineer, 1984. Copyright AISE, Pittsburgh, Pennsylvania. Reprinted with permission.)

function is then approximated with five trapezoids. For each trapezoid, the following parameters are determined: break frequency f_B; crossover frequency f_{oi}; and amplitude ratio A_{Ci}. Using a special table, the transient function $x_i(t)$ for each trapezoid is calculated. The resultant transient function is equal to:

$$X(t) = \sum_{i=1}^{5} x_i(t) \tag{34-62}$$

Results of the calculations are presented in Fig. 34.10.

34.18 COMPENSATION OF CONTROL SYSTEM

Compensation of the control system usually includes whatever procedure is required to improve its response characteristics and eliminate instability. Although various techniques are used to achieve these goals, one of the most popular approaches to the compensation problem has been used: lag compensation placed in the forward loop of the control system. This approach is used in combination with the proper selection of the gain of the position regulator.

In spite of the simplicity of this technique, the compensation of AGC presents a significant problem due to the destabilizing factors varying significantly in magnitude during the rolling process. The most important of these variable destabilizing factors are: material stiffness, oil height inside of the actuator, structural stiffness and roll separating force.

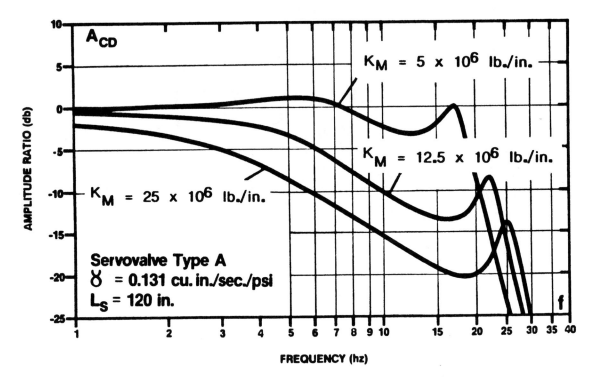

Fig. 34.11 Effect of material stiffness on frequency response characteristics of adaptive closed-loop control system optimized for only one value of material stiffness. (From Ginzburg, Iron and Steel Engineer, 1984. Copyright AISE, Pittsburgh, Pennsylvania. Reprinted with permission.)

34.19 COMPENSATION FOR MATERIAL STIFFNESS

Variation of the material stiffness becomes more explicit when the material is being rolled on a single-stand reversing hot strip mill. In this case, material stiffness changes not only from pass to pass but also within each pass due to temperature variation of the rolled product along its length.

The effect of material stiffness on the frequency response characteristics of the closed-loop control system is illustrated in Fig. 34.11. The graphs represent the case when the system is compensated for a minimum material stiffness ($K_M = 5 \times 10^6$ lb/in.) only. When stiffer material is rolled, the attenuation of the frequency response increases. Figure 34.12 illustrates the case when adaptive compensation of control system is done for each material stiffness using the lag compensation technique described. When the material stiffness increases, the gain K_V has to be increased and the lag time constant T_{22} has to be decreased. It follows from a comparison of the optimized frequency response characteristics that the resonant frequency decreases and the attenuation at 10 Hz increases when rolling softer materials.

34.20 COMPENSATION FOR OIL HEIGHT IN ACTUATOR

Variation of the oil height inside of the actuator may be limited by shortening the actuator working stroke. In this case, any additional roll gap opening will be achieved with mechanical systems like the

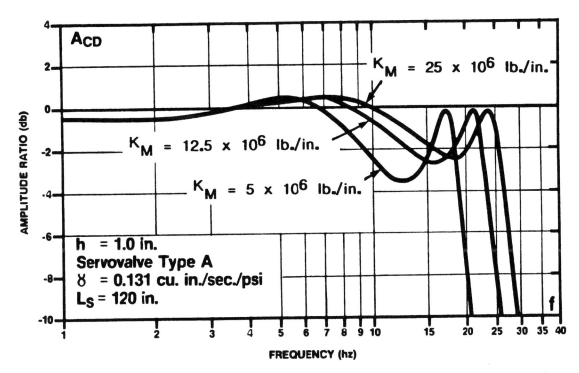

Fig. 34.12 Effect of material stiffness on frequency response characteristics of adaptive closed-loop control system. (From Ginzburg, Iron and Steel Engineer, 1984. Copyright AISE, Pittsburgh, Pennsylvania. Reprinted with permission.)

screw-down or wedge-type mechanism. However, besides calling for additional hardware, this arrangement requires a more sophisticated roll-gap control system. Therefore, it appears to be attractive to utilize long-stroke actuator as long as the deterioration of the dynamic characteristics of the AGC system is within acceptable limits.

The effect of oil height inside the actuator on the frequency response characteristics of the closed-loop control system is shown in Fig. 34.13. As expected from Eqs. (34-18) and (34-19), an increase in oil height will have the same effect as a decrease in the material stiffness. Therefore, the optimization of the system can be achieved in the same manner.

34.21 COMPENSATION FOR ROLL FORCE

Variations in roll separating force will cause the servovalve droop factor to change in accordance with Eq. (34-23). Assuming that the pump pressure stays constant, an increase in roll separating force will result in an increase in the servovalve droop factor γ. This, in turn, will decrease the apparent control system stiffness K_C as predicted from Eq. (34-54). The final result is that the steady-state error decreases as described by Eqs. (34-59) and (34-60).

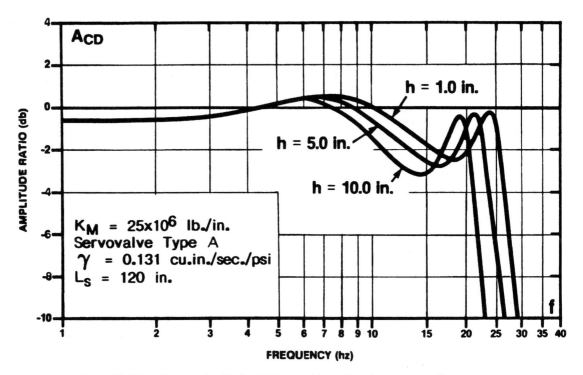

Fig. 34.13 Effect of oil height inside of actuator on frequency response characteristics of adaptive closed-loop control system. (From Ginzburg, Iron and Steel Engineer, 1984. Copyright AISE, Pittsburgh, Pennsylvania. Reprinted with permission.)

The effect of the servovalve droop factor on the frequency response characteristics is shown in Fig. 34.14. The less the servovalve droop the less attenuation of the amplitude ratio. Adaptive adjustment of the gain and lead-lag parameters may allow to reduce deterioration of the response characteristics of the system.

34.22 PERFORMANCE OF THE SYSTEM WITHOUT STRIP

Knowledge of the correlation between performance of the system with and without strip in the roll bite may be of great help during the tuning of dynamic characteristics of HAGC.

Let us assume that when rolled material is present in the roll bite, the control system has to perform with 100% roll force feedback signal, i.e., m = 1. Then from Eq. (34-56) we obtain:

$$A'_{CI} = 1 + \frac{K_M}{K_S} \tag{34-63}$$

To obtain an amplitude ratio for the case without strip in the mill, let us first modify Eq. (34-56) to the form:

$$A''_{CI} = \frac{1 + \dfrac{K_S}{K_M}}{1 + \dfrac{K_S}{K_M} - m_o} \tag{34-64}$$

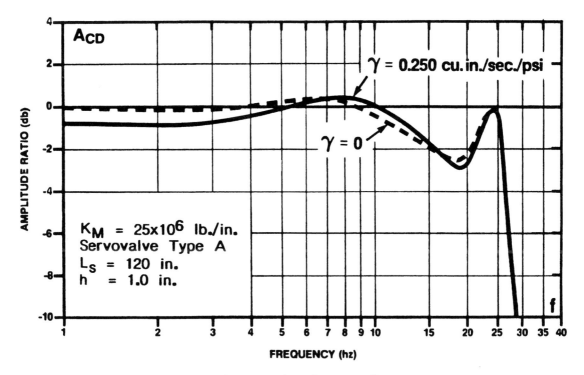

Fig. 34.14 Effect of servovalve droop on frequency response characteristics. (From Ginzburg, Iron and Steel Engineer, 1984. Copyright AISE, Pittsburgh, Pennsylvania. Reprinted with permission.)

where m_0 = roll force feedback factor in the mill without strip.

When there is no strip in the mill ($K_M \to \infty$), we obtain from Eq. (34-64) that:

$$A_{CI}^{''} = \frac{1}{1 - m_0} \tag{34-65}$$

By equating A_{CI} and A_{CI} in Eqs. (34-63) and (34-65), we obtain the value of the roll force feedback factor m_0 in the mill without strip

$$m_0 = \frac{1}{1 + \dfrac{K_S}{K_M}}, \tag{34-66}$$

Equation (34-66) gives the value for the roll force feedback factor m_0 that provides the same dynamic characteristics of HAGC without strip in the mill as those when the strip, having material stiffness K_M, is present in the roll bite and the roll force feedback factor $m = 1$.

CHAPTER 34

REFERENCES

1. V.B. Ginzburg, "Dynamic Characteristics of Automatic Gauge Control System with Hydraulic Actuators", Iron and Steel Engineer, Jan. 1984, pp. 57-65.

2. K. Wiedmer, "Significance of Mill Stand Stiffness in Relation to Rolling Accuracy", Metals Technology, April 1974, pp. 181-185.

3. A. Yamashita, et al, "Development of Feedforward AGC for 70-in. Hot Strip Mill at Kashima Steel Works", The Sumitomo Search, No. 16, Nov. 1976, pp. 34-39.

4. J. Davies, R.W. Jackson, and J.A. Tracy, "Design Criteria, Development and Test Activities for 6-stand Tandem Mill Hydraulic Screwdown System", Journal of Iron and Steel Institute, July 1972, pp. 489-500.

5. T. Okamoto, et al, "Advanced Gauge and Tension Control of Tandem Cold Mill with Hydraulic Screwdown System", Transactions ISIJ, Vol. 16, 1976, pp. 614-622.

6. R. Jackman, R.W. Gronbech, and G.A. Forster, "The Position-Controlled Hydraulic Mill", reprint of a paper presented at a conference on "Hydraulic Control of Rolling Mills and Forging Plants", The Iron and Steel Institute and The West of Scotland Iron and Steel Institute, May 1971.

7. T. Okamoto, Y. Misaka, and H. Takeuchi, "Advanced Gauge and Tension Control of Computerized Tandem Cold Mill", International Federation of Automatic Control, Symposium on Automatic Control in Mining, Mineral and Metallurgical Processes, 2nd Sydney, Australia, Aug. 13-17, 1973, Published by Inst. of Eng., Sydney, Australia, 1973.

8. K. Yoshida, et al, "Development of a New Type of AGC System in the Tandem Cold Mill", Sumitomo Metal Industries, Ltd.

9. T.F. Bogart, Jr., "Control Systems Theory, Laplace Transforms and Control Systems Theory for Technology", 1st Edition, John Wiley and Sons, New York, 1982, pp. 1-211.

10. R.N. Clark, Introduction to Automatic Control Systems, John Wiley and Sons, New York, London, pp. 300-308 (1962).

11. J.M. Nightingale, "Hydraulic Servos-2, Practical Methods for Analyzing Typical Systems", Machine Design, March 7, 1957, pp. 100-105.

12. A.C. Kluev, Automation Control, Energy, Moscow, USSR, pp. 265-268, 378, 379 (1973).

35

Principles of Width and Plan View Control

35.1 LATERAL SPREAD

Lateral spread is the most important factor which affects width control of the rolled product.

Lateral spread may be defined as an increase in width of a workpiece which is being reduced in thickness and is given by:

$$\Delta w_s = w_2 - w_1, \tag{35-1}$$

where w_1, w_2 = workpiece width before and after thickness reduction respectively.

Lateral spread is often expressed as a difference between maximum values of widths before and after rolling. This definition may not be precise since the edge of the rolled products is usually not straight. Figure 35.1 illustrates two most common edge profiles: a convex profile (a) and a convex-concave or double-barreled profile (b). In these cases, use of a mean cross-sectional width may be more appropriate for calculation of lateral spread. For a convex edge profile, the mean cross-sectional width is equal to [1]:

$$w_{2m} = w_2'' - (w_2'' - w_2')/3, \tag{35-2}$$

where w_2', w_2'' = minimum and maximum values for width respectively, as shown in Fig. 35.1a.

The lateral spread is often presented by the **width spread coefficient** which is equal to

$$S_w = [\ln(w_2/w_1)]/[\ln(h_1/h_2)], \tag{35-3}$$

where h_1, h_2 = workpiece thickness before and after thickness reduction respectively.

Lateral spread is found to be a function of various geometrical factors. The dominant absolute geometrical factors are:

　　a) initial workpiece thickness w_1

　　b) initial workpiece thickness h_1

c) draft in thickness reduction Δ

d) work roll radius R.

The most important relative geometrical factors are:

a) ratio of initial width to roll contact length w_1/L

b) ratio of initial width to initial thickness w_1/h_1

c) ratio of initial thickness to work roll radius h_1/R

d) relative reduction in thickness r.

Friction in the roll bite is found to be one of the principal nongeometrical factors affecting the lateral spread.

35.2 WUSATOWSKI'S FORMULA FOR SPREAD

Wusatowski's formula for spread [2] has been obtained by collecting different results from several sources and analyzing them. The formula is:

$$(w_2/w_1) = (h_1/h_2)^W, \tag{35-4}$$

$$\log_{10}W = -1.269(w_1/h_1)(h_1/D)^{0.56}, \tag{35-5}$$

where D = roll diameter.

Equations (35-4) and (35-5) can be rearranged to obtain the width spread coefficient in the form:

$$S_w = \exp[-1.982(w_1/h_1)(h_1/R)^{0.56}] \tag{35-6}$$

As follows from Eqs. (35-4)-(35-6), lateral spread increases with decrease in both ratios w_1/h_1 and h_1/R.

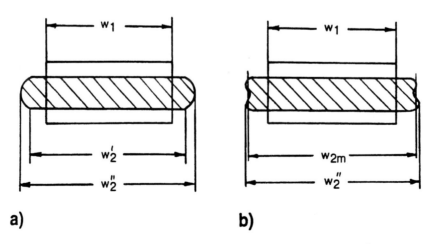

Fig. 35.1 Slab edge profiles: a) convex and b) convex/concave.

35.3 HILL'S FORMULA FOR SPREAD

Hill has derived the simplest formula for the width spread coefficient which has a theoretical basis [3]. The formula is given by:

$$S_w = 0.5\exp(-0.707cw_1/L), \qquad (35\text{-}7)$$

where c = constant.

Original value for c, suggested by Hill, is 0.5. McCrum [4] showed that the value of 0.525 gave the best agreement with his experimental results. Equation (35-7) shows that lateral spread increases with decrease in ratio w_1/L.

35.4 EL-KALAY AND SPARLING'S FORMULA FOR SPREAD

El-Kalay and Sparling [5] have proposed an empirical formula for calculation of the width spread coefficient:

$$S_w = a\exp[-b(w_1/h_1)^c(h_1/R)^d r^e], \qquad (35\text{-}8)$$

where a, b, c, d, e = constants

R = radius of horizontal roll

r = relative reduction in thickness.

The constants a through e depend on the work roll surface finish and scale conditions as shown in Table 35.1.

It was found that lateral spread is not very sensitive to temperature. Also, that the effect of scale on spread seems to be independent on geometrical variables. The effect of roll finish on spread, however, does depend upon geometrical parameters. The spread is highest for high reductions, smooth rolls and lower w_1/h_1 ratios.

Table 35.1 Coefficients for El-Kalay and Sparling formula for spread. Data from El-Kalay and Sparling (1968).

Condition	a	b	c	d	e
Smooth rolls:					
light scale	0.851	1.766	0.643	0.386	−0.104
heavy scale	0.955	1.844	0.643	0.386	−0.104
Rough rolls:					
light scale	0.993	2.186	0.569	0.402	−0.123
heavy scale	0.980	2.105	0.569	0.402	−0.123
Previous formula (Sparling):					
smooth rolls light scale	0.981	1.615	0.900	0.550	−0.250

35.5 HELMI AND ALEXANDER'S FORMULA FOR SPREAD

Formula for lateral spread derived by Helmi and Alexander [3] has the following form:

$$S_w = 0.95(h_1/w_1)^{1.1}\exp[-0.707(w_1/L)(h_1/w_1)^{0.971}] \tag{35-9}$$

This formula was experimentally verified for the range of ratios w_1/h_1 between 1 and 13. Percentage error in predicting spread was calculated from an expression:

$$\text{Error} = (w_{2a} - w_{2p})/(w_{2p} - w_1), \tag{35-10}$$

where w_{2a}, w_{2p} = actual and predicted exit widths respectively.

Table 35.2 shows a distribution of error calculated by Eq. (35-9) as well as by Eqs. (35-6), (35-7) and (35-8).

35.6 BEESE'S FORMULA FOR SPREAD

The formula proposed by Beese [6] is given by:

$$S_w = 0.6(h_1/w_1)\exp[-0.32(h_1/L)] \tag{35-11}$$

According to Beese, the predictions of Eq. (35-11) have been verified during a series of production rolling tests and the results have been found to be in good agreement.

35.7 EKELUND'S FORMULA FOR SPREAD

Formula for lateral spread derived by Ekelund [7], introduces coefficient of friction in the roll bite as an additional variable. The formula has the form:

$$w_2^2 - w_1^2 = 1.6L[\Delta - h_a\ln(w_2/w_1)](4\mu L - 3\Delta)/h_a, \tag{35-12}$$

Table 35.2 Distribution of error in predicting of lateral spread by different formula. Data from Helmi and Alexander (1968).

Eqs.	Authors	Number of tests	Number of tests with calculated results in error by less than			
			5%	10%	15%	20%
35-6	Wusatowski	217	15	33	25	37
35-7	Hill	217	34	20	20	12
35-8	El-Kalay, Sparling	149	52	41	31	15
35-9	Helmi, Alexander	217	63	112	19	8

where μ = coefficient of friction

h_a = average workpiece thickness.

Thus, Eq. (35-12) indicates that lateral spread increases with increase in coefficient of friction. For steel rolls, the coefficient of friction is defined by the equation:

$$\mu = 0.55 - [0.0005(t - 1000)], \qquad (35\text{-}13)$$

where t = rolling temperature, ^oC.

The effect of the coefficient of friction on lateral spread is also analytically expressed by the formulae derived by Bakhtinov and Tselikov [8] as described in Chapter 22.

35.8 PRINCIPLE OF EDGING

The process of reducing the workpiece width is known as **edging**. The edging can be accomplished by either rolling or pressing. In case of rolling, the following three types of edging rolls are usually used:

1. Flat cylindrical rolls (Fig. 35.2a)
2. Slightly tapered rolls with a single collar (Fig. 35.2b)
3. Double-collar, or grooved, rolls (Fig. 35.2c).

There is a substantial difference between the edging process and the thickness reduction process, mainly in respect to plastic deformation [8]. During thickness reduction, almost one hundred percent of the reduction is transferred into elongation of the workpiece. This elongation is rather equally distributed across the width of the workpiece. During width reduction, just a small part (6 to 7 percent) of the reduction is transferred into elongation of the workpiece. The major part of the width reduction produces an increase in the thickness of the rolled material. This increase of the thickness is greater near the edges of the workpiece where the distinct bulges are usually formed.

As will be shown below, this nonuniform increase in thickness of the workpiece significantly impairs the efficiency of the edging process and also creates difficulties in achieving tight width tolerances and high yield. Thickening of the workpiece near its edges produced by an edging process is usually referred to as **bulging** and the resulting shape of the workpiece near the edges is known as a **dog-bone shape.**

The cross profile of bulging depends on type of rolls used for edging. The bulging effect is maximum when edging with flat cylindrical rolls. The maximum height of the dog bone can be described quantitatively as (Fig. 35.3):

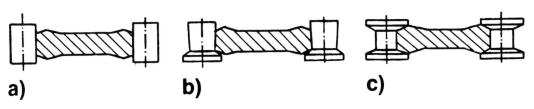

a)   b)   c)

Fig. 35.2 Three types of edging rolls.

$$B = h_b - h_o,$$ (35-14)

where h_b = workpiece thickness at the peak bulging

h_o = thickness of the workpiece prior to edging.

35.9 EDGING WITH FLAT CYLINDRICAL ROLLS

When edging with flat cylindrical rolls, the workpiece thickness at the peak bulging can be calculated by using the empirical equation derived by Tazoe, et al [9]

$$h_b = h_o[1 + a(d_e/w_e)^b(R_e/w_e)^c(h_o/w_o)^d],$$ (35-15)

where a, b, c, d = constants

d_e = draft in width

w_o, w_e = workpiece widths before and after edging respectively (Fig. 35.4)

R_e = radius of edging roll.

The draft in width is equal to:

$$d_e = w_o - w_e$$ (35-16)

Figure 35.3 illustrates the measured cross profile of bulging as published by Shibahara, et al [10]. Equation (35-15) becomes compatible with the cross profile of bulging illustrated in Fig. 35.3, when the values for the constants are: a = 0.34, b = 0.60, c = -0.35, and d = -0.29.

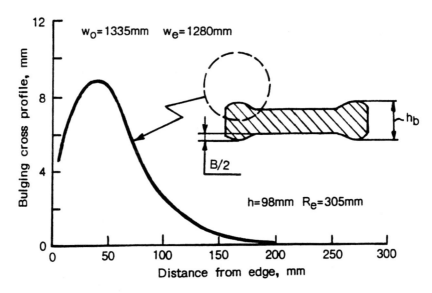

Fig. 35.3 Cross-sectional profile of bulging produced by edging between flat cylindrical rolls. Adapted from Shibahara, et al (1981).

Okado, et al [11], have defined a typical dog-bone shape by the following four parameters indicated in Fig. 35.5:

1. Maximum height of dog bone B

$$B = 0.098h_0^{0.56}d_e^{0.70} \text{ [mm]} \tag{35-17}$$

2. Increase in roll contact $(h_r - h_0)$

$$h_r - h_0 = 0.028h_0^{0.72}d_e^{0.73} \text{ [mm]} \tag{35-18}$$

3. Peak position of dog bone A

$$A = 4.35h_0^{0.35}d_e^{0.07} \text{ [mm]} \tag{35-19}$$

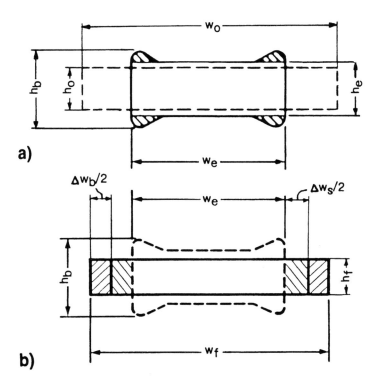

Fig. 35.4 Formation of the bar cross-section: a) after edging, b) after horizontal pass. Adapted from Shibahara, et al (1981).

4. Sphere of influence of dog bone C

$$C = 1.65h_o^{0.77}d_e^{0.20} \quad [mm] \tag{35-20}$$

Effect of edging reduction on parameters of the dog-bone shape is shown in Fig. 35.5. The graphs are based on experimental data obtained by Huismann on a laboratory mill by rolling with plasticine [12].

35.10 EDGING FOLLOWED BY REDUCTION IN THICKNESS

When edging is followed by reduction in thickness, the bar width after the thickness reduction will be equal to (Fig. 35.4):

$$w_f = w_e + \Delta w_s + \Delta w_b, \tag{35-21}$$

where w_e = bar width after edging

Δw_s = bar width spread after reduction in thickness excluding the width increase caused by spread of bulges

Δw_b = bar width spread after reduction in thickness due to bulging.

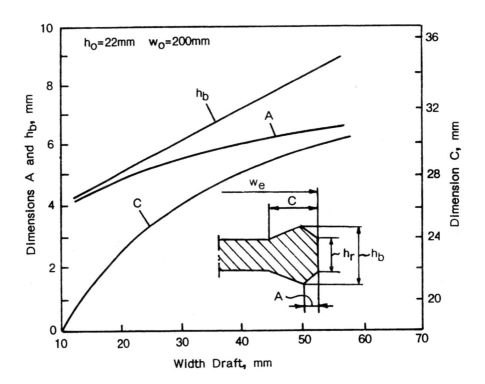

Fig. 35.5 Effect of width draft on the bulge geometry. (From Huismann, Commission of the European Communities Report, 1983. Copyright Commission of the European Communities. Reprinted with permission.)

According to Shibahara, et al [10], the bar width spread due to thickness reduction is given by:

$$\Delta w_s = w_e [(h_o/h_f)^a - 1], \tag{35-22}$$

where h_o, h_f = bar thickness prior to edging and after reduction in thickness respectively

a = geometrical parameter.

The additional bar width spread caused by bulging is expressed by the formula:

$$\Delta w_b = bd_e(1 + \Delta w_s/w_e), \tag{35-23}$$

where b = geometrical parameter.

The geometrical parameters a and b are presented in the following forms:

$$a = \exp[-1.64m^{0.376}(w_e/L)^{0.016m}(h_o/R)^{0.015m}], \tag{35-24}$$

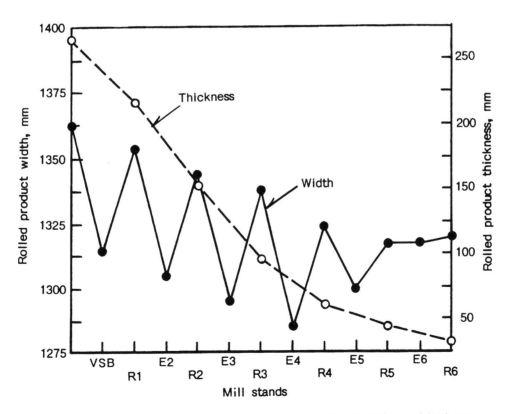

Fig. 35.6 Width change in roughing stands. Adapted from Shibahara, et al (1981).

$$b = \exp[-1.877(d_e/w_0)^{0.063}(h_0/R_e)^{0.441}(R_e/w_0)^{0.989}(w_0/w_e)^{7.591}, \qquad (35\text{-}25)$$

where R = work roll radius used in thickness reduction

R_e = work roll radius used in edging

L = roll contact length in thickness reduction.

Parameter m is equal to

$$m = w_e/h_0 \qquad (35\text{-}26)$$

The length of the bar after edging l_e is defined by the equation

$$l_e = l_0(w_e + d_e)/(w_e + bd_e), \qquad (35\text{-}27)$$

where l_0 = bar length prior to edging.

Equations (35-22)-(35-27) were verified during trials at continuous hot strip mill. A typical variation of the strip width after edging and thickness reduction in a roughing train of the hot strip mill is shown in Fig. 35.6.

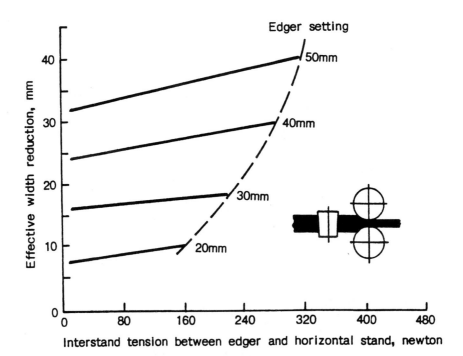

Fig. 35.7 Effect of interstand tension between edger and horizontal mill on effective width reduction. (From Huismann, Commission of the European Communities Report, 1983. Copyright Commission of European Communities. Reprinted with permission.)

35.11 EFFECTIVE WIDTH REDUCTION

Effective width reduction is defined as a difference between the workpiece width prior to edging w_0 and the width after edging followed by reduction in thickness w_f (Fig. 35.4), i.e.

$$d_{eff} = w_0 - w_f \qquad\qquad (35\text{-}28)$$

From Eqs. (35-16), (35-21) and (35-28) we obtain that

$$d_{eff} = d_e - \Delta w_s - \Delta w_b \qquad\qquad (35\text{-}29)$$

The edger efficiency can also be expressed in relative terms by using the following edger efficiency factor [13]:

$$\eta = (w_0 - w_f)/(w_0 - w_e) = d_{eff}/d_e \qquad\qquad (35\text{-}30)$$

Effective width reduction increases with increase of interstand tension between edger and horizontal stand as was proven in rolling experiments conducted by Huismann [12] with plasticine on a laboratory mill (Fig. 35.7).

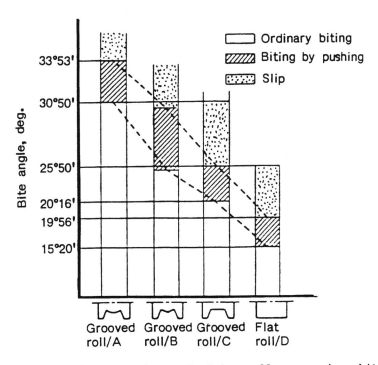

Fig. 35.8 Effect of groove shape of edging rolls on maximum bite angle. Adapted from Takeuchi, et al (1983).

35.12 EDGING WITH GROOVED ROLLS

One of the principal questions in edging is whether to use flat or grooved rolls. Flat rolls are usually used when the width reduction of the slabs of different thicknesses has to be made by the same edger. Application of flat rolls reduces both rolling load and torque as well as increases roll life. The stability of edging with flat rolls, however, may become inadequate when heavier edging reduction has to be taken.

Application of the grooved rolls allows one to achieve the following three goals:

1. Improvement in stability of edging process
2. Increase in reduction per one pass
3. Increase in effective width reduction.

Biting properties of the grooved rolls with different groove shape have been investigated by Takeuchi, et al [13] in model experiment using lead. As shown in Fig. 35.8, the bite angle of the grooved rolls is substantially greater than that of flat rolls.

Increase in effective width reduction with grooved rolls is mainly due to the fact that these rolls shift the peak of dog bone closer to the center of the slab and so, when the dog bone is flattened by horizontal rolls, the amount of the spread due to the dog bone is minimized. This is confirmed by actual rolling data [14]. Figure 35.9 illustrates the edger efficiency factor as a function of the width reduction and the slab width. The edger efficiency is greater with narrow slabs and also increases with an increase in width reduction. As shown in Fig. 35.10 for narrow slabs, the edger efficiency factor with grooved rolls is almost two times greater than that with the flat rolls.

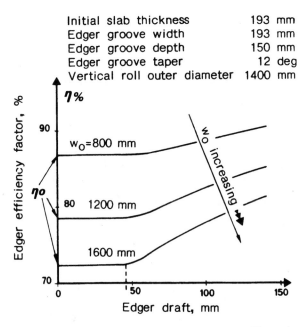

Fig. 35.9 Edger efficiency factor for grooved rolls. (From Vathaire, et al, Proceedings of 4th International Steel Rolling Conference, Vol. 1, 1987. Copyright IRSID, France. Reprinted with permission.)

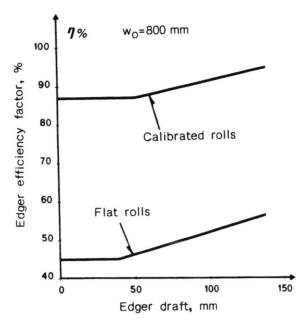

Fig. 35.10 Comparison of the edger efficiency for grooved and flat rolls. (From Vathaire, et al, Proceedings of 4th International Steel Rolling Conference, Vol. 1, 1987. Copyright IRSID, France. Reprinted with permission.)

Experiments conducted by Huismann [12] have shown (Fig. 35.11) that the spread after horizontal pass decreases and, hence, the edging efficiency increases, when the groove body height h_g is made less than the slab thickness h_o prior to edging.

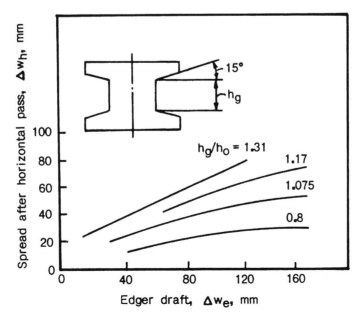

Fig. 35.11 Influence of the groove height (h_g) to slab thickness (h_o) ratio on spreading after horizontal pass. (From Huismann, Commission of the European Communities Report, 1983. Copyright Commission of European Communities. Reprinted with permission.)

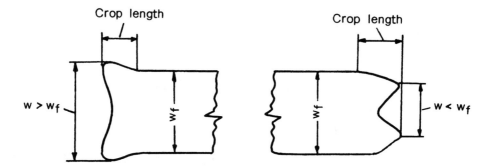

Fig. 35.12 Two types of fish tail.

35.13 FISH TAIL

One of the consequences of the edging process is a formation of the fish tails at the ends of a workpiece [13-17] as shown in Fig. 35.12. This is caused by elongation of bulges during flat pass. The fish tails are usually cut prior to rolling at finishing mill. As illustrated by experiments conducted by Takeuchi, et al [13], the larger the slab width, the larger the crop length. The crop length decreases with an increase in the edging roll diameter.

Width variation is somewhat related to formation of the fish tails (Fig. 35.12). The metal at the head and tail ends of a slab has tendency to flow more freely in rolling direction rather than to spread. The variation in width is further increased after leveling the dog bone by horizontal rolls. The rolling

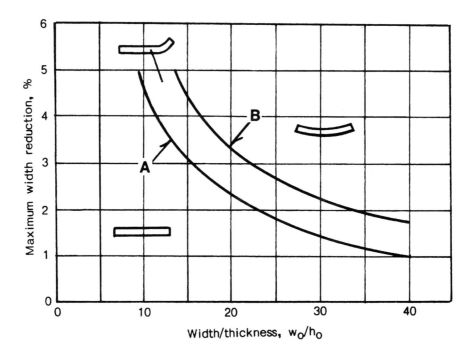

Fig. 35.13 Limits of buckling. Adapted from Okado, et al (1981).

experiments show that the width variation increases with an increase in both the width reduction and slab width. This phenomena was analytically described by Nagai, et al [18].

35.14 BUCKLING

Buckling is known as a loss of stability of the rolled piece during edging. Figure 35.13 shows the results of experiments conducted by Okado [11] with pure lead samples of 5 to 7 mm thick and 100 to 160 mm wide. The edging roll diameter was 90 mm. Curve A identifies maximum width reductions that are possible without buckling. The width reductions corresponding to partial buckling ('build up in slab') are defined by area between the curves A and B. A complete buckling occurs when the width reduction exceeds the values determined by the curve B.

The results of the experiments are reportedly confirmed by actual rolling data from a production mill.

REFERENCES

1. F. Sassani and N. Serehri, "Prediction of Spread in Hot Flat Rolling Under Variable Geometry Conditions", Journal Materials Shaping Technology, Vol. 5, 1987, pp. 117-123.

2. Z. Wusatowski, "Hot Rolling: A Study of Draught, Spread and Elongation", Iron and Steel, Vol. 28, Feb. and March 1955, pp. 49-54, 89-94.

3. A. Helmi and J.M. Alexander, "Geometric Factors Affecting Spread in Hot Flat Rolling of Steel", Journal of the Iron and Steel Institute, No. 206, Nov. 1968, pp. 1110-1117.

4. A.W. McCrum, "Progress Report on the Experimental Investigation of Spread, Load, and Torque in Hot Flat Rolling", BISRA Report MW/AL, 10/56.

5. A.K.E.H.A. El-Kalay and L.G.M. Sparling, "Factors Affecting Friction and Their Effect Upon Load, Torque, and Spread in Hot Flat Rolling", Journal of the Iron and Steel Institute, No. 206, Feb. 1968, pp. 152-163.

6. J.G. Beese, "Some Problem Areas in the Rolling of Hot Steel Slabs", AISE Yearbook, 1980, pp. 360-363.

7. L.G.M. Sparling, "Formula for Spread in Hot Flat Rolling", Proceedings of the Institution of Mechanical Engineers, Vol. 175, No. 11, 1961, pp. 604-611.

8. Z. Wusatowski, Fundamentals of Rolling, Pergamon Press, Oxford, pp. 82-95 (1969).

9. N. Tazoe, et al, "New Forms of Hot Strip Mill Width Rolling Installations", Paper Presented at 1984 AISE Spring Conference, Dearborn, Mich., April 30-May 2, 1984.

10. T. Shibahara, et al, "Edger Set-Up Model at Roughing Train in Hot Strip Mill", Tetsu-to-Hagane, Vol. 67, No. 15, 1981. pp. 2509-2515.

11. M. Okado, et al, "New Light on Behavior of Width of Edge of Head and Tail of Slabs in Hot Strip Rolling Mills", Tetsu-to-Hagane, Vol. 67, No. 15, 1981, pp. 2516-2525.

12. R.L. Huismann, "Large Width Reductions in Hot Strip Mill", Commission of the European Communities Report, 1983.

13. M. Takeuchi, et al, "Heavy Width Reduction in Rolling of Slabs", Nippon Steel Technical Report, No. 21, June 1983, pp. 235-246.

14. M. de Vathaire, et al, "Automatic Operation of SOLLAC Reversing Roughing Mill", Proceedings of the 4th International Steel Rolling Conference: The Science and Technology of Flat Rolling, Vol. 1, Deauville, France, June 1-3, 1987, pp. A.9.1-A.9.7.

15. A. Faessel, et al, "Reduction of Crop Losses by Optimal Setting of the Roughing Stands in a Hot Strip Mill", Proceedings of the International Conference on Steel Rolling: Science and Technology of Flat Rolled Products, Vol. 1, Tokyo, Sept. 29-Oct. 4, 1980, pp. 252-262.

16. N. Tazoe, "Prevention of Fishtail During Intensive Edging in Hot Roughing Mill Line", IHI Engineering Review, Vol. 14, No. 3, July 1981, pp. 42-47.

17. O. Pawelski and V. Piber, "Possibilities and Limits of Deformation in Width Direction in Hot Flat Rolling', Stahl und Eisen, Vol. 100, No. 17, Aug. 25, 1980, pp. 937-949.

18. T. Nagai, et al, "Improving Strip Width, Profile and Shape Control in Sumitomo Hot Strip Mill", Restructuring Steelplants for the Nineties, Institute of Metals, London, 1986, pp. 238-255.

36

Width Change and Control in Rolling Mills

36.1 MAIN OBJECTIVES AND METHODS OF WIDTH CHANGE

The principle objective of the width change is to obtain the desired width of the rolled product. Since the width adjustment by plastic deformation is practically impossible when the rolled material is thin and cold, the term 'width control' is usually applicable to hot rolling process.

When a flat product is rolled from an ingot, the desired slab width is readily obtained by rolling the slab on the edge in the slabbing mill as shown in Fig. 36.1. With an advent of continuous casting process, the problem of the width change has become more complicated. Indeed, in devising the width control strategy with continuously cast slabs one has to take into consideration the following four major factors.

1. Production rate of a continuous casting machine increases with an increase in slab width.

2. When the slab width is changed during casting, a tapered portion of the slab is produced. This may reduce yield and complicate the width control during rolling.

3. In order to decrease the slab inventory, the desired slab width should preferably be obtained either during or shortly before hot rolling of the required product.

4. If some special measures are not taken, high width reductions would lead to formation of fish tails and, consequently, to excessive crop losses.

The width change is usually performed at three principal locations:

a) on-line location, after continuous casting machine (Fig. 36.2a)

b) on-line location, in hot rolling mill

c) off-line location (Fig. 36.2b).

The most common methods of the width change are:

1. Width reduction by rolling

2. Width reduction by pressing

3. Width spreading by rolling.

These methods are briefly discussed below.

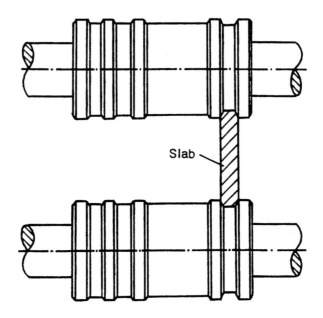

Fig. 36.1 Width reduction with a slabbing mill.

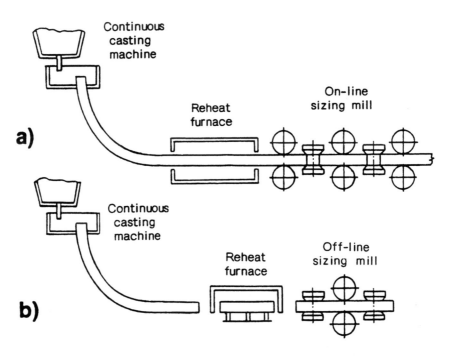

Fig. 36.2 Width reduction with: a) on-line sizing mill and
b) off-line sizing mill. Adapted from IHI Publication (1988).

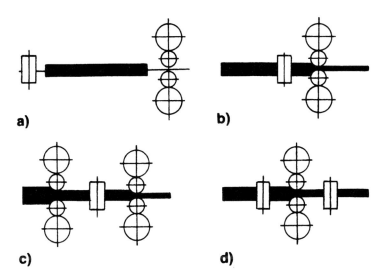

Fig. 36.3 Types of rolling mill arrangements for edging in combination with flat rolling.

36.2 WIDTH REDUCTION BY ROLLING

The vertical mill stands used for reduction of width of a workpiece are conventionally called **vertical edgers.** The vertical edgers installed right after the reheating furnaces are known as **vertical scalebreakers** because of their ability to enhance breaking of scale during squeezing of a slab.

The edging process conducted on vertical mill stands, is usually combined with flat rolling process which is implemented in horizontal mill stands. The following main arrangements of vertical (V) and horizontal (H) mill stands are known:

   a) an open mill stand arrangement which allows for a free bar to be rolled by the edger (Fig. 36.3a)

   b) VH-arrangement, in which a vertical edger is close-coupled with a horizontal mill stand (Fig. 36.3b)

   c) HVH-arrangement, in which a vertical edger is installed between two close-coupled horizontal mill stands (Fig. 36.3c)

   d) VHV-arrangement, in which a horizontal mill stand is installed between two close-coupled vertical edgers (Figs. 36.2b and 36.3d)

   e) HVHVH-arrangement, in which two vertical edgers are installed between three close-coupled horizontal mill stands (Fig. 36.2a).

   A rolling system that incorporates two vertical edgers close-coupled with one or more horizontal mills is known as a **sizing mill.**

36.3 SLAB SIZING MILLS

The main purpose of the slab sizing mills is to reduce a number of the required slab widths produced by a continuous casting machine. Two major arrangements were practically implemented, HVHVH and VHV.

The HVHVH-arrangement (Fig. 36.2a) was used in the slab sizing mill installed on-line with a continuous caster at U.S. Steel Gary Works [1]. The horizontal stands of the sizing mill were used to produce high interstand tension which enhanced the edging process. Reductions in both width and thickness were provided.

The capability of the slab sizing machine was as follows:

1. Number of the cast slab widths   2
2. Maximum size of the continuously cast slab:
   thickness   254 mm (10 in.)
   width   1930 mm (76 in.)
3. Minimum size of the slab after the sizing mill:
   thickness   152 mm (6 in.)
   width   813 mm (32 in.)

The U.S. Steel Gary Works sizing mill was the first serious attempt to minimize a number of required concast slab widths. Its operation was abandoned after an introduction of a caster with an adjustable mold width.

The VHV-arrangement (Fig. 36.2b) is used in the slab sizing mill developed by IHI for Nippon Steel Oita Works Hot Strip Mill [2]. This is an off-line reversing mill which provides reduction in both thickness and width. The edging mill uses grooved rolls. The principal specification of the slab sizing mill is as following:

1. Number of the cast slab widths   1
2. Continuously cast slab sizes:
   thickness   280 mm (11 in.)
   width   1900 mm (74.8 in.)
   length   6-29 m (19.7-95.1 ft)
3. Slab sizes after the sizing mill:
   thickness   160-280 mm (6.3-11 in.)
   width   750-1900 mm (29.5-74.8 in.)
4. Average production rate   840 t/h.

36.4 WIDTH REDUCTION BY PRESSING

Width reduction by pressing is produced by the machines known as **squeezing** or **sizing presses.** In these presses, the press tools are reciprocated in the horizontal plane in the direction perpendicular to the slab length. The press tools are driven either by a crank-shaft mechanism [3] or by a hydraulic cylinder.

Figure 36.4 illustrates a hydraulic sizing press developed by United Engineering and Foundry Company [4]. In this press, the initial gap between the press tools is set with the screw-nut mechanism, this gap being slightly greater than initial slab width. The squeezing action is produced by a hydraulic cylinder located at one side of the press. Prior to squeezing, the slab is lifted against a hold-down mechanism to prevent buckling of the slab. The slab is advanced with the table rolls.

By the use of the grooved press tools in some designs of the sizing presses, the peak of the bulge is moved closer to the center of the slab allowing for better edging efficiency.

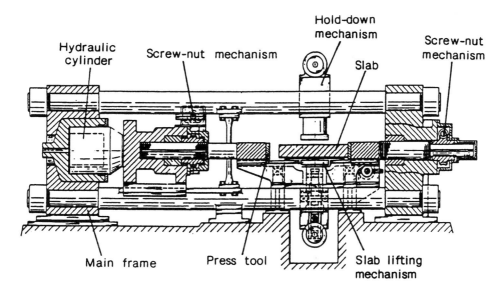

Fig. 36.4 Sizing press with hydraulic actuator developed by United Engineering and Foundry Company. Adapted from Stone and Talbot, U.S. Patent No. 3,580,032 (1971).

The sizing presses can be divided into two principal groups:

1. Long-tools sizing presses
2. Short-tools sizing presses.

A brief description of these presses is given below.

36.5 LONG-TOOL SIZING PRESSES

Main purpose of the long-tool sizing presses is twofold. Firstly, it is to provide a true rectangular shape of a slab and, secondly, to reduce the slab width. To accomplish that, the length of the sizing tools is made slightly longer than the plate length.

These types of sizing presses were installed in a number of Generation I hot strip mills. The sizing had usually taken place after broadsiding of a slab. The main technical characteristics of a sizing press developed by United Engineering and Foundry Company is as follows [3]:

1.	Press tool length	20 ft
2.	Maximum slab thickness	6.25 in.
3.	Maximum single stroke width draft for a 6.25 inch-thick slab	3.0 in.
4.	Type of drive	Motor-driven crank-shaft mechanism
5.	Motor power	700 HP
6.	Motor speed	500 rpm
7.	Squeezing speed	5 strokes/min.

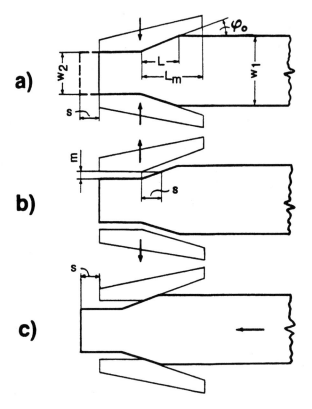

Fig. 36.5 Principle of operation of the short-tool sizing press.

36.6 SHORT-TOOL SIZING PRESSES

The length of the press tools of the short-tool sizing presses is substantially less than the length of a workpiece. As a rule, the press tools of this type of presses have both parallel and tapered surfaces. This allows to reduce the slab width by reciprocated displacements of the press tools with a stroke being much shorter than the width draft.

The principle of operation of the short-tool sizing press is shown in Fig. 36.5. Prior to squeezing a slab, the press tools are separated and the slab is advanced forward, so the head end of the slab is located within the area between the parallel surfaces of the press tools. The press tools then squeeze the slab until the desired width w_2 of the head end of the slab is obtained (Fig. 36.5a). Squeezing of the remaining portion of the slab is provided by oscillating movement of the press tools.

During each oscillating cycle the press tools retract to provide a gap m between parallel portions of the press tools and the slab (Fig. 36.5b). This would allow the slab to be advanced a distance s that is equal to (Fig. 36.5c):

$$s = m/\tan\phi_0, \tag{36-1}$$

where ϕ_0 = slope angle of the tapered portion of the press tool.

The average slab feed speed is given by

$$V = mn/\tan\phi_0, \tag{36-2}$$

where n = frequency of oscillation.

Two types of the short-tool sizing presses have recently been developed: start-stop type and flying type. Their main features are described below.

36.7 START-STOP TYPE SIZING PRESS

Figure 36.6 illustrates the main parts and principle of operation of the start-stop type sizing press developed by Hitachi, Ltd. with Kawasaki Steel Corporation [5, 6].

The slab squeezing is provided by the press tools which have both parallel and tapered surfaces. Prior to squeezing of a new slab, the press tools are separated and the slab is advanced forward with the entry pinch rolls so that the head end of the slab is located within the area between the parallel surfaces of the press tools. Then the press tools squeeze the slab with a screw-nut mechanism

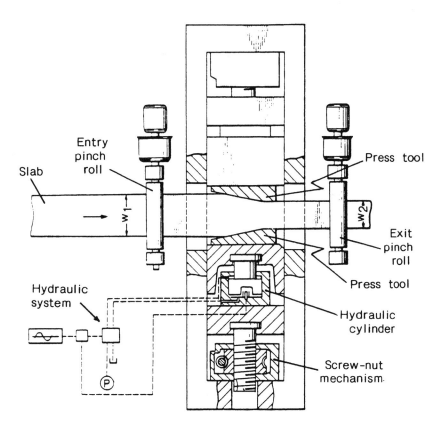

Fig. 36.6 Schematic presentation of the start-stop type sizing press developed by Hitachi, Ltd. with Kawasaki Steel Corporation. Adapted from Kimura, U.S. Patent No. 4,578,983 (1986).

until the desired width w_2 of the head end is obtained. Squeezing of the remaining portion of the slab is provided by oscillating the hydraulic cylinders with an appropriate advancement of the slab. As soon as the tail end of the slab leaves the entry pinch roll area, the slab continues to be fed into the press with the exit pinch roll. Technical specification of the sizing press is as follows:

1. Maximum press force   2,000 tons
2. Slab width range   1000-2200 mm (39.4-86.6 in.)
3. Maximum width draft per pass   350 mm (13.8 in.)
4. Width reduction efficiency   90%.

The slab feeding speed varies with the width draft. When the draft is equal to 300 mm (11.8 in.), the average slab feed speed is equal to 200 mm/s (7.9 in./s).

36.8 FLYING TYPE SIZING PRESS

In the flying type sizing press developed by IHI [7], a continuous slab sizing operation without start-stop motion of the slab is carried out by a crank-shaft mechanism. The mechanism realizes the motions necessary for simultaneous width reduction and feeding of the slab. Reportedly, this feature allows one to eliminate surface defects during sizing and transferring operations.

The flying type sizing press utilizes press tools similar to those used in the start-stop type sizing press. Technical data of this sizing press is as follows:

1. Maximum press force   2,700 tons
2. Maximum width draft per pass   350 mm (13.8 in.)
3. Width reduction efficiency   96-98%
4. Crop loss due to fish tail   0.1-0.2%
5. Slab feeding speed   330 mm/s (13.0 in./s).

One of the main features of the flying type sizing press is that the slab feeding speed is constant and independent of the width draft.

36.9 IMPROVING THE EDGING EFFICIENCY

As was discussed in the previous chapter, the edging efficiency can be substantially improved with use of grooved rolls. The application of this method, however, is limited to the cases when the slab thickness is compatible with the groove size. Indeed, once the slab thickness becomes thinner than the groove height, the bulge formation will no longer be affected by the collars of the edging rolls and the resulting slab cross-section will be similar to that obtained with flat edging rolls. Two solutions to this problem are discussed below.

Adjustable groove size rolls - One of the designs of the adjustable groove size roll have been proposed by United Engineering and Foundry Company [4]. In this design (Fig. 36.7), a composite roll consists of two main parts, one part being a roll body with non-adjustable collar and the other part being an adjustable collar. Each part has threads to permit the adjustment of the groove size. The adjustment is made by the mechanism which consists of two motors, installed on the top of the adjustable collar, and two sets of worm and cylindrical gears.

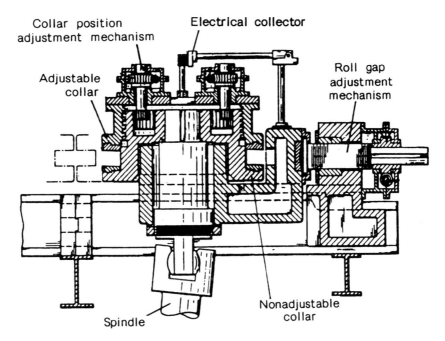

Fig. 36.7 Edging roll with adjustable groove size. Adapted from Stone and Talbot, U.S. Patent No. 3,580,032 (1971).

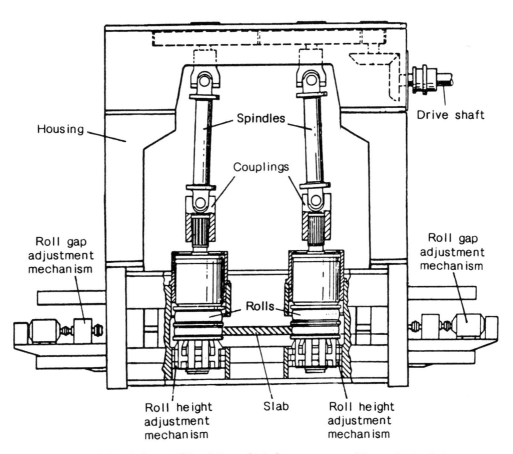

Fig. 36.8 Edging mill with multiple groove rolls. Adapted from Lemper, U.S. Patent No. 3,670,587 (1972).

Multiple groove rolls - The edging capability can also be improved by using two or more grooves of different sizes. In the design proposed by Mesta Machine Company [8], the vertical edging rolls are connected with spindles through splined type couplings (Fig. 36.8). The roll height adjustment mechanism, located underneath each edging roll, provides positioning of a selected groove at the mill pass line.

36.10 DECREASING THE SLAB DISTORTION

There are three major types of the slab distortion which occur during edging process:

1. Out-of-square slab cross-section
2. Slab buckling
3. Slab edge overlap.

Out-of-square slab cross-section - The main cause of the out-of-square slab cross-section is the ascension of one side of the slab. This is generally prevented by using either grooved rolls or tapered rolls with bottom collar as shown in Fig. 35.2b.

Another method was recently proposed in Japan [9]. This method is illustrated schematically in Fig. 36.9. When the axes of both edging rolls are perpendicular to the advancing direction of a slab, the latter will tend to ascend at one side. To prevent this ascension, the edging roll at this side is tilted

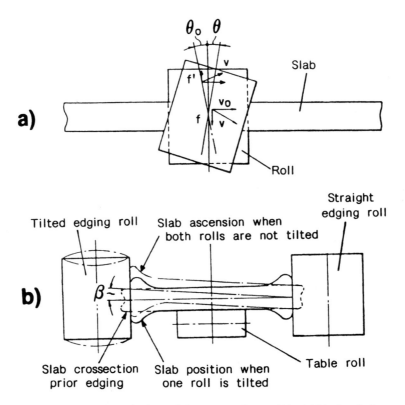

Fig. 36.9 Prevention of the slab ascension with tilted edging rolls. Adapted from Kokubo, et al, U.S. Patent No. 4,712,414 (1987).

toward the same direction as the advancing direction of the stock material. This would create a component of the rolling force f (Fig. 36.9a) that will push the slab down against the table rolls as shown in Fig. 36.9b.

Slab buckling - As was mentioned in the previous chapter, an excessive width reduction may lead to buckling of the slab. The slab buckling is usually avoided by limiting the maximum allowable value of the edge draft. This value, however, can be increased with use of support rolls and skids. Three buckling prevention systems are shown in Fig. 36.10, with center support, with both ends support, and with three points support. A system with three points was developed by Kawasaki Steel Corporation and IHI [10] in application to plate rolling. In this system the support rolls at the end are also capable of suppressing the formation of a dog bone.

In the method proposed by Hitachi, Ltd. [11], the buckling of a thin slab is prevented by conveying the slab over a turning roll, so the slab is curved at a certain contact angle as shown in Fig. 36.11. Both back and front tension are applied to the thin slab forcing the slab against the turning roll. This assures stability of the slab while it is being squeezed with the press tools located above the turning roll.

The slab edge overlap may be prevented by application of the V-rolls as proposed by Kawasaki Steel Corporation and IHI [10]. Each V-roll (Fig. 36.12) has a flat portion and a caliber portion, each portion being selected by simultaneous shifting both V-rolls in vertical direction according to the rolling condition. The caliber portion is used for chamfer rolling, thereby preventing edge overlap.

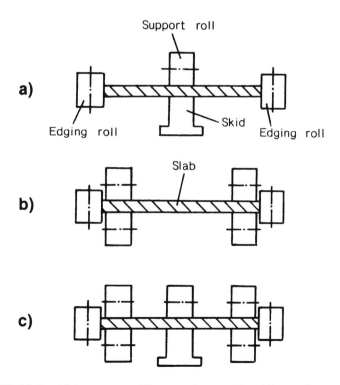

Fig. 36.10 Buckling prevention systems: a) with center support, b) with both ends support, and c) with three points support. Adapted from Inoue, et al (1988).

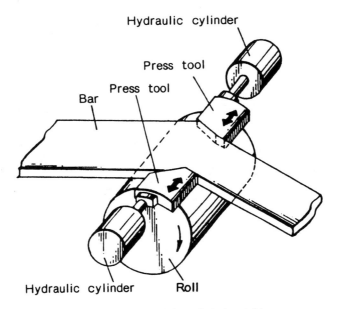

Fig. 36.11 Prevention of the thin slab buckling as proposed by Hitachi, Ltd. Adapted from Nihei and Kimura, U.S. Patent No. 4,651,550 (1987).

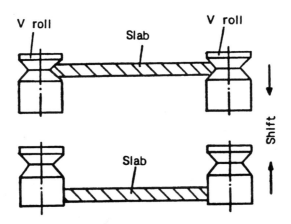

Fig. 36.12 Prevention of the edge overlap with V-roll. Adapted from Inoue, et al (1988).

36.11 METHODS OF PREVENTING A FISH TAIL

As was described in the previous chapter, intensive edging increases the crop losses caused by formation of fish tails. A number of methods have been proposed to alleviate this problem [12]. Some of these methods are briefly discussed below.

1. **Utilization of convex crown slabs** - According to this method, the convex crown slabs are produced with a continuous casting machine. When these slabs are rolled with horizontal flat rolls, the effect of slab center elongation is achieved. This reduces a fish tail.

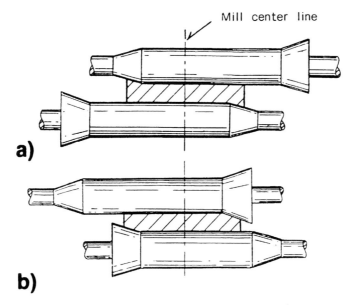

Fig. 36.13 Adjustable width crown rolling of slabs with:
a) convex crown rolls and b) concave crown rolls. Adapted from
Ginzburg, U.S. Patent No. 4,730,475 (1988).

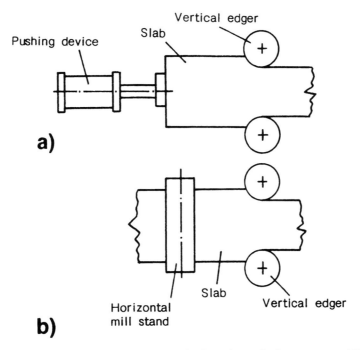

Fig. 36.14 Squeezed slab rolling method. Adapted from Tazoe (1981).

2. **Rolling with lubrication** - In this method, the lubrication is provided during edging with grooved rolls. Since the roll lubrication decreases the coefficient of friction in the roll bite, the growth of dog bone becomes smaller.

3. **Convex crown rolling** - This method involves rolling of part of a slab with convex crown rolls before edging. In order to accommodate various slab widths, the shifting of tapered rolls (Fig. 36.13) can be used as proposed jointly by United Engineering, Inc. and International Rolling Mill Consultants, Inc. [13].

4. **Squeezed slab rolling** - In this method, the edging is produced by pushing the slab into a vertical edger from backward. The pushing can be accomplished either by a pushing device (Fig. 36.14a) or by a horizontal mill (Fig. 36.14b).

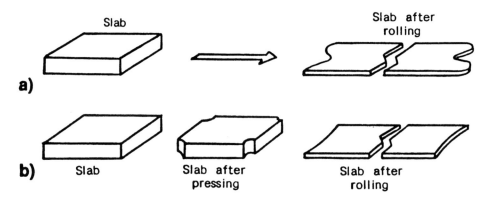

Fig. 36.15 Plan view of a plate produced by: a) conventional rolling and b) working slab corner method.

5. **Working slab corner** - This method incorporates squeezing the slab end corners prior to edging. The squeezing is achieved either by pressing or by rolling. Figure 36.15 illustrates schematically the effect of the working slab corner method, when the edging press developed by IHI [14] is being used. The edging press squeezes the slab head and tail corners prior to edge rolling of the slab. The slab yield improvement produced by this method is claimed to be as much as 30% in comparison with a conventional edge rolling.

Another modification of the working slab corner method is a so-called **'bite back rolling process'** (Fig. 36.16) developed by Kawasaki Steel Corporation [15]. In application to rolling from ingots, the process consists of the following five steps:

A. Initial reduction to remove primary scale and eliminate the ingot taper in direction of width and thickness

B. Primary bite and back rolling in direction of thickness which forms recesses at the head and tail ends of the ingot, with subsequent flow of metal to recesses

C. Bite and back rolling in direction of width which results in formation of recesses at the head

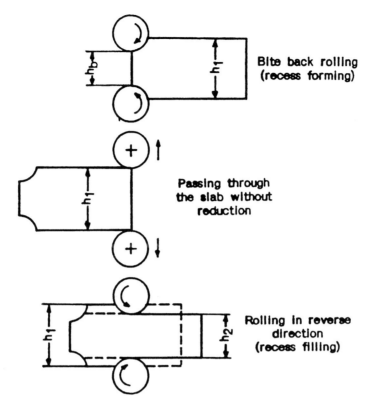

Fig. 36.16 Schematic presentation of the bite back rolling process. Adapted from Matsuzaki, et al (1981).

and tail ends of the ingot, followed by flow of metal to recesses

D. Secondary bite and back rolling in direction of thickness which produces recesses at the head and tail ends of the ingot and absorbs the dog bone formed at the stage C

E. Final light reduction while flowing metal to remaining recesses.

The bite back rolling process reportedly increases slab yield by 4% in comparison with a conventional rolling practice.

36.12 WIDTH ENLARGEMENT

Rolling with width enlargement means producing a bar wider than the initial slab width. The two basic methods of width enlargement can be classified as passive and active.

In the passive method [16], the width enlargement is provided by natural spread of the workpiece during horizontal flat passes. The active method utilizes a technique that enhances the spread. One of these techniques, known as 'HI-Spread', has been developed by Davy McKee, Ltd. [17]. This technique involves rolling a single or multiple number of longitudinal grooves in the slab to produce sideways spread with subsequent rolling in a conventional horizontal mill stand. The overall change in width with this method is claimed to be as high as 25%.

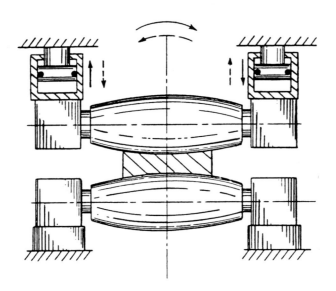

Fig. 36.17 United Engineering/IRMC spreading method utilizing rocking movements of high-crowned rolls. Adapted from Ginzburg, U.S. Patent No. 4,735,116 (1988).

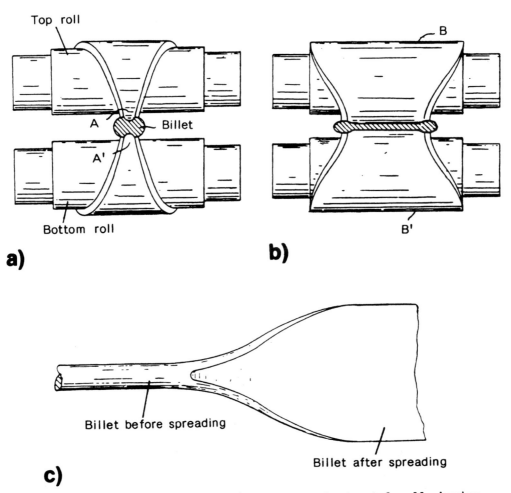

Top roll

B

A

Billet

A'

Bottom roll

B'

a)

b)

Billet before spreading

Billet after spreading

c)

Fig. 36.18 Spreading of a billet with two horizontal rolls having complementary, diverging work surfaces. Adapted from Ginzburg, U.S. Patent No. 4,793,169 (1988).

Two different techniques for the active width enlargement have been proposed jointly by United Engineering, Inc. and International Rolling Mill Consultants, Inc. One of these techniques promotes spreading of the rolled material in a horizontal mill by rocking movements of high-crowned work rolls in both horizontal and vertical planes [18] as shown in Fig. 36.17.

Another technique is particularly effective in application to spreading the billets having various cross-sections [19]. An advantage of using billets as an initial hot rolling stock is that radiation heat losses are minimized with that shape. Further, because the rolling of billets to strip involves both longitudinal and transverse elongation, the occurence of anisotropy in the finished strip is reduced.

As shown in Fig. 36.18a,b the work rolls have complementary, diverging work surfaces, each beginning with a narrow region A and A' at the midpoint of the roll and diverging to a wider region B and B', extending across the width of the roll. Initially, billet is brought into the roll bite where first contact is made by narrow regions A and A'. As the forward pass proceeds, the work surfaces in contact with the billet become progressively wider; the result is a flattening and spreading of the rolled

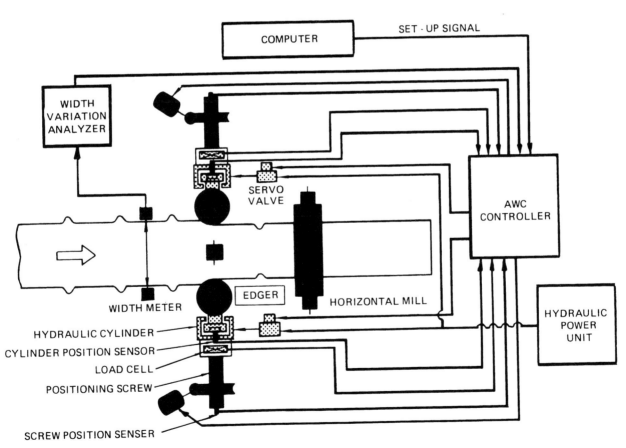

Fig. 36.19 Schematic diagram of IHI automatic width control system. From IHI Publication (1984).

material. As shown in Fig. 36.18c when wider regions B and B' come into contact with the material, the roll gap is relieved and the rolled material is partially retracted in a back pass. The roll gap is again closed and the narrow region again contacts the material to further the flattening and spreading, eventually producing the strip.

36.13 AUTOMATIC WIDTH CONTROL DURING EDGING

Automatic width control during edging is accomplished by installation of the hydraulic closed-loop control system such as a system developed by IHI [20]. The control diagram of the system is schematically shown in Fig. 36.19. The width variations are detected by an optical type width meter and compared with the width setup reference provided by on-line computer. The developed control signal is then used to drive the hydraulic servomechanism which adjusts a gap between edging rolls. Similar control system is developed by Mitsubishi Heavy Industries, Ltd. [21].

Adjustment of the roll gap during edging is aimed to reduce the fish tail as well as the width variation due to skid marks.

Width control for fish tails - As was mentioned before, a conventional edging results in formation of fish tails (Fig. 36.20a). To reduce the fish tails, the gap between edging rolls gradually opens to

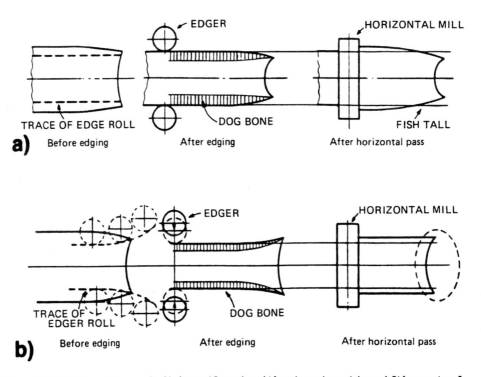

Fig. 36.20 Formation of fish tail: a) without automatic width control and b) with automatic width control. From Mitsubishi Heavy Industries Publication.

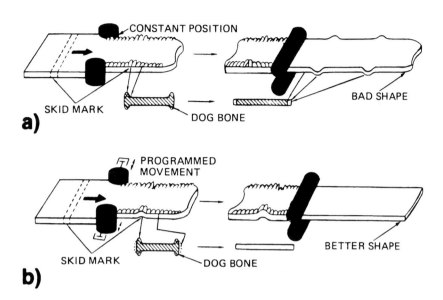

Fig. 36.21 Effect of skid marks on the plate plan view: a) without automatic width control and b) with automatic width control. From IHI Publication (1984).

produce wider slab at its head and tail ends as shown in Fig. 36.20b. This slab widening offsets the narrowing of the slab caused by formation of fish tail during edging and a subsequent horizontal pass.

Width control for skid marks - In conventional edging, when skid marks are present in a slab, the dog bone at the skid marks becomes thicker. This dog bone will produce an additional spread during following horizontal pass (Fig. 36.21a). To compensate for this spread, the width control system provides an additional squeezing of the slab near the skid marks during edging as shown in Fig. 36.21b.

36.14 AUTOMATIC WIDTH CONTROL IN HOT STRIP MILL

Contemporary automatic width control system in hot strip mill provides continuous monitoring and appropriate correction of the product width in both roughing and finishing mills [20-22].

The system developed by Nippon Steel Corporation [23] includes (Fig. 36.22):

1. Edger setup control (ESU)
2. Rougher automatic width control (RAWC)
3. Finisher automatic width control (FAWC).

Rougher automatic width control is accomplished by variable control of the gap between edging rolls. The RAWC system provides control of average width (bar-to-bar control). It also allows one to reduce width variations in each bar (in-bar control).

Finisher automatic width control is achieved by variable control of tension between finishing mill stands. Based on the width measured by the final width gauge on the roughing mill, the feedforward control signals are provided to modify the interstand tension references. Feedback width control signals are supplied by the width meters installed after the last finishing mill stand and near downcoiler.

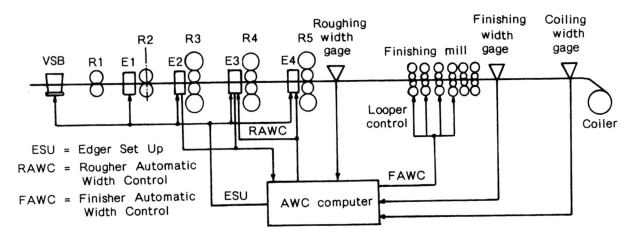

Fig. 36.22 Schematic presentation of Nippon Steel Corporation's automatic width control system for Hot Strip Mill. Adapted from Sasada, et al (1981).

36.15 PLAN VIEW CONTROL

The term 'plan view control' is customarily applied to a technique designed to produce a true rectangular shape of rolled plate.

Plate rolling is roughly divided into the following three stages:

1. **Sizing rolling** - In this stage the slab is rolled in longitudinal direction to produce required intermediate thickness.

2. **Broadside rolling** - To obtain the required plate width the slab is turned around 90 degrees and rolled in transverse direction.

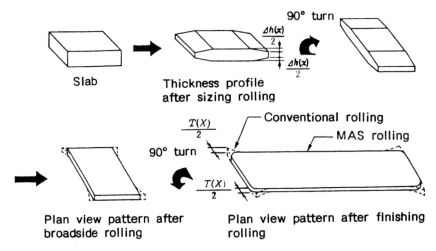

Fig. 36.23 Principle of sizing MAS to control plan view pattern. (From Yanazawa, et al, Kawasaki Steel Technical Report, No. 1, 1980. Copyright Kawasaki Steel Corporation. Reprinted with permission.)

3. Finishing rolling - The slab is turned around 90 degrees again and rolled to the final thickness.

In conventional plate rolling, the plan view of rolled plate shows various shapes which depend on the slab thickness, reduction, broading ratio, etc. A number of the plan view control methods have recently been developed [24-27]. One of the methods has been developed by Kawasaki Steel Corporation and is referred to as MAS rolling method [25]. Two types of the method are known, sizing MAS and broadside MAS. Both methods involve a calculation of the predicted plan view pattern.

In the sizing MAS (Fig. 36.23), the amount of inferior patterns is converted to a plate thickness difference at the final pass in sizing rolling.

The amount of plate thickness modification is obtained from the equation:

$$\Delta h(x) = T(X)\frac{H}{W}, \qquad (36-3)$$

where $\Delta h(x)$ = amount of plate thickness modification at distance x from the tail or head end of the slab in the longitudinal direction

$T(X)$ = side crops at distance X in the longitudinal direction after completion of rolling

H = final plate thickness

W = final plate width.

In the broadside MAS (Fig. 36.24), the amount of inferior patterns is converted to a plate thickness difference in the broadside rolling stage.

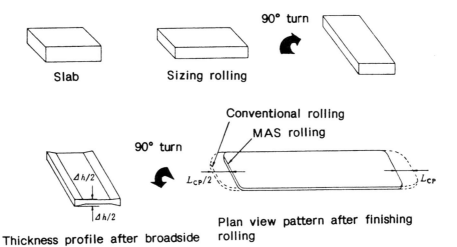

Fig. 36.24 Principle of broadsizing MAS to control the head and tail ends crop. (From Yanazawa, et al, Kawasaki Steel Technical Report, No. 1, 1980. Copyright Kawasaki Steel Corporation. Reprinted with permission.)

The amount of thickness modification is obtained from the equation:

$$\Delta h(y) = L_{cp}(Y)\frac{H}{L},$$ (36-4)

where $\Delta h(y)$ = amount of thickness modification at distance y in the longitudinal direction

$L_{cp}(Y)$ = head and tail crops at distance Y in the width direction after plate rolling is completed

L = plate width in broadside rolling, in.

United Engineering, Inc. and International Rolling Mill Consultants, Inc. have jointly proposed a method for plan view control that does not require broadside rolling to produce tapered thickness. In this method [13], an increase or decrease in reduction near the rolled plate edges is achieved by appropriate positioning of the tapered rolls as shown in Fig. 36.13.

REFERENCES

1. A.V. Wiebel, "United States Steel's Breakthrough in Continuous-Continuous Casting", Linear Casting of Steel Shapes, AISE, Pittsburgh, Pa., 1970, pp. 135-144.

2. M. Takeuchi, et al, "Heavy Width Reduction in Rolling of Slabs", Nippon Steel Technical Report, No. 21, June 1983, pp. 235-246.

3. J. Kelly, Jr., "New Hot Strip Mill at McDonald, Ohio", AISE Yearly Proceedings, 1935-36, pp. 337-345.

4. M.D. Stone and H.H. Talbot, U.S. Patent No. 3,580,032, May 25, 1971.

5. T. Kimura, "Press Type Method of and Apparatus for Reducing Slab Width", U.S. Patent No. 4,578,983, Apr. 1, 1986.

6. T. Naoi, et al, "Development of a Slab Sizing Press", Hitachi Review, Vol. 37, No. 4, 1988, pp. 189-184.

7. "Slab Sizing Press", Ishikawajima-Harima Heavy Industries Company, Ltd., Tokyo, 1988.

8. H. Lemper, "Vertical Mill", U.S. Patent No. 3,670,587, June 20, 1972.

9. I. Kokubo, et al, "Rolling Method of Plate-like Stock Material by Edger and Continuous Hot Rolling Mill", U.S. Patent No. 4,712,414, Dec. 15, 1987.

10. M. Inoue, et al, "Development of a Process for Manufacturing Trimming Free Plates", Transactions of the Iron and Steel Institute of Japan, Vol. 28, 1988, pp. 448-455.

11. M. Nihei and T. Kimura, "Method of Decreasing Width of Thin Slab and Apparatus Therefor", U.S. Patent No. 4,651,550, March 24, 1987.

12. N. Tazoe, "Prevention of Fish Tail During Intensive Edging in Hot Roughing Mill Line", IHI Engineering Review, Vol. 14, No. 3, July 1981, pp. 42-47.

13. V.B. Ginzburg, "Rolling Mill Method", U.S. Patent No. 4,730,475, March 15, 1988.

14. N. Tazoe, et al, "New Forms of Hot Strip Mill Width Rolling Installations", Paper presented at 1984 AISE Spring Conference, Dearborn, Mich., April 30-May 2, 1984.

15. M. Matsuzaki, et al, "A New Method of Slab Rolling for Prevention of Growth of Crops", Tetsu-to-Hagane, 1981 (15), pp. 2350-2355.

16. J.O. Pera, et al, "Optimal Width Reductions in Hot Strip Mills", Transactions of the Iron and Steel Institute of Japan, Vol. 26, 1986, pp. 206-211.

17. T. Hope, et al, "HI-Spread: A New Hot Rolling Process Which Can Affect Major Changes of Slab Width", Proceedings of 4th International Steel Rolling Conference: The Science and Technology of Flat Rolling, Vol. 1, Deauville, France, June 1-3, 1987, pp. A.13.1-A.13.5.

18. V.B. Ginzburg, "Spreading Rolling Mill and Associated Method", U.S. Patent No. 4,735, 116, April 5, 1988.

19. V.B. Ginzburg, "Continuous Backpass Rolling Mill", U.S. Patent No. 4,793,169, Dec. 27, 1988.

20. "IHI Hydraulic AWC Edger for Hot Rolling", Ishikawajima-Harima Heavy Industries Co. Publication, Tokyo, pp. 1984.

21. "RAWC Edger for Hot Strip Mill", Mitsubishi Heavy Industries, Ltd. Publication, Tokyo, pp. 1-10.

22. T. Hagai, et al, "Improving Strip Width, Profile and Shape Control in Sumitomo Hot Strip Mill", Restructuring Steelplants for the Nineties, Institute of Metals, London, 1986, pp. 238-255.

23. T. Sasada, et al, "Modernization Technology of Conventional Hot Strip Mills", Nippon Steel Technical Report, No. 18, Dec. 1981, pp. 1-21.

24. Y. Haga, et al, "Development of New Plan View Control Technique in Plate Rolling (NKK-DBR)", Nippon Kokan Technical Report, Overseas No. 39, 1983, pp. 21-30.

25. T. Yanazawa, et al, "Development of the New Plan View Pattern Control System in Plate Rolling", Kawasaki Steel Technical Report, No. 1, Sept. 1980, pp. 33-47.

26. M. Morel, et al, "Quality Control and Production Optimization in Plate Mills Using the HYDROPLATE System", Iron and Steel Engineer, May 1984, pp. 48-53.

27. M. Kenyon, et al, "Process Development Research - Slab and Plate Products", SEAISI Quarterly, July 1984, pp. 37-47.

Part XII

Strip Profile
and Flatness Control

37

Strip Profile and Flatness Actuators

37.1 STRATEGY OF PROFILE AND FLATNESS CONTROL

Basic strategy of profile and flatness control is to achieve a desired strip profile with strip flatness not exceeding prerequisite tolerances. In order to maintain good flatness, only certain changes in the strip crown to thickness ratio are possible in one rolling pass without introducing undesirable disturbances to the strip shape.

Shohet, Townsend and Sommers [1, 2] developed a simple empirical criterion for predicting the onset of bad shape in steel during hot rolling. This criterion proposes that good shape of the strip will result if, at each reduction taken at a mill stand, the change in the crown to thickness ratio is within the following limits:

$$-80\left(\frac{h_2}{w}\right)^{1.86} < \left(\frac{c_1}{h_1} - \frac{c_2}{h_2}\right) < 40\left(\frac{h_2}{w}\right)^{1.86} \quad , \tag{37-1}$$

where c_1, c_2 = strip crowns before and after reduction respectively

h_1, h_2 = strip thicknesses before and after reduction respectively

w = strip width.

Figure 37.1 illustrates a simplified block-diagram of the on-line strip profile and flatness control system. The desired strip profile reference is compared with actual strip profile measured by a profile sensor. The strip profile error is then input into strip profile regulator that generates control signal for strip profile and flatness actuator. At the same time the actual strip flatness, measured by a strip flatness sensor, is compared with strip flatness tolerances and the flatness error signal is then input into flatness limiter. Whenever the actual flatness exceeds the tolerances, the output signal from profile regulator into actuator is accordingly adjusted, so good flatness is maintained.

37.2 TYPES OF STRIP PROFILE AND FLATNESS ACTUATORS

The strip profile and flatness actuator is a complex system that incorporates the work rolls, supportive rolls as well as devices for adjusting the roll gap profile.

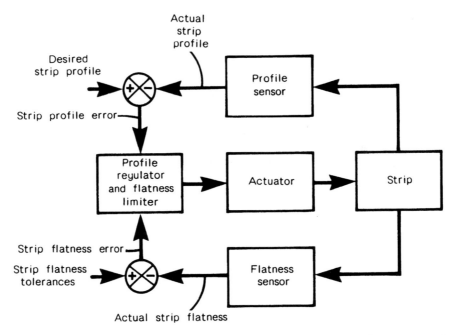

Fig. 37.1 Simplified block-diagram of strip profile and flatness control.

In the following review, known actuator systems have been divided into separate groups according to their specific method for adjusting the roll gap profile:

1. Vertical plane roll bending systems [3-6].
2. Horizontal plane roll bending systems [7-11].
3. Roll crossing systems [12-14].
4. Nonadjustable stepped backup rolls [15].
5. Adjustable stepped backup rolls [16-19].
6. Axially shifted cylindrical rolls [20-24].
7. Axially shifted noncylindrical rolls [25-28].
8. Axially shifted sleeved rolls [29-31].
9. Flexible body adjustable crown rolls [32-38].
10. Flexible edge adjustable crown rolls [39-43].
11. Flexible edge self-compensating rolls [44-46].

A brief description of these systems is given below.

36.3 MAIN CHARACTERISTICS OF ACTUATORS

Characteristics which can be used for quantitative evaluation of the capability and efficiency of different actuators include the strip crown control range, strip crown control rate, rolling force disturbance rate, backup roll diameter disturbance rate, strip thermal crown disturbance rate and roll wear disturbance rate [47].

Strip crown control range C_{max} is the maximum change in overall center crown $\Delta(chI)$ corresponding to the maximum value of the regulating parameter. When the regulating parameters are positive and negative roll bending forces F_{WW} and F_{WB} respectively, then the strip crown control range is equal to

$$C_{max} = \Delta(chI)_{WW} + \Delta(chI)_{WB}, \qquad (37\text{-}2)$$

where $\Delta(chI)_{WW}$ = strip crown control range provided by the roll bending cylinders installed between two work rolls at the maximum value of the roll bending force F_{WW}

$\Delta(chI)_{WB}$ = strip crown control range provided by the roll bending cylinders installed between work rolls and backup rolls at the maximum value of the roll bending force F_{WB}.

Strip crown control rate S_F is the variation of overall strip center crown with variation of the regulating parameter. If this parameter is the roll bending force F then:

$$S_F = \frac{\Delta c}{\Delta F} \qquad (37\text{-}3)$$

Rolling force disturbance rate S_p is the ratio of a variation of overall strip center crown to a corresponding variation of rolling force per unit of strip width w per unit of draft Δ:

$$S_p = \frac{\Delta c}{\Delta p} w\Delta \qquad (37\text{-}4)$$

Backup roll diameter disturbance rate S_D is the variation of overall strip center crown with variation of backup roll diameter D_B:

$$S_D = \frac{\Delta c}{\Delta D_B} \qquad (37\text{-}5)$$

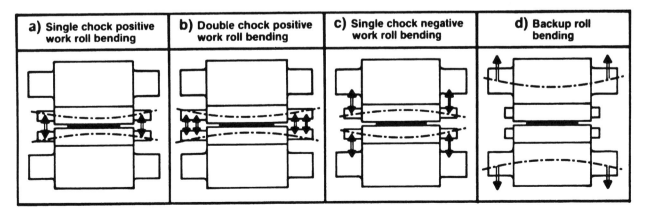

a) Single chock positive work roll bending	b) Double chock positive work roll bending	c) Single chock negative work roll bending	d) Backup roll bending

Fig. 37.2 Vertical plane roll bending systems. (From Ginzburg, Iron and Steel Engineer, 1987. Copyright AISE, Pittsburgh, Pennsylvania. Reprinted with permission.)

Strip thermal crown disturbance rate S_T is the variation of strip crown with variation of work roll thermal crown ΔC_T:

$$S_T = \frac{\Delta c}{\Delta C_T} \tag{37-6}$$

Roll wear disturbance rate S_W is a ratio of the ridge hs to a stepped roll wear C_r:

$$S_W = \frac{hs}{C_r} \tag{37-7}$$

37.4 VERTICAL PLANE ROLL BENDING SYSTEMS

The vertical plane roll bending systems include single-chock positive roll bending (Fig. 37.2a), double-chock positive roll bending (Fig. 37.2b), single-chock negative roll bending (Fig. 37.2c) and a backup roll bending system (Fig. 37.2d).

The advantage of the vertical plane roll bending system is in the possibility of providing a continuous control of the strip profile during rolling. It is also cost effective. However, the crown control range is limited, mainly by the maximum load that can be withstood by the work roll chock bearings.

The double-chock roll bending system implemented by Nippon Steel [4] alleviates the problem. However, other factors such as higher stresses between work rolls and backup rolls, as well as higher costs, may limit application of this system. The crown control range achieved by utilizing a double-chock work roll bending system is shown in Fig. 37.3.

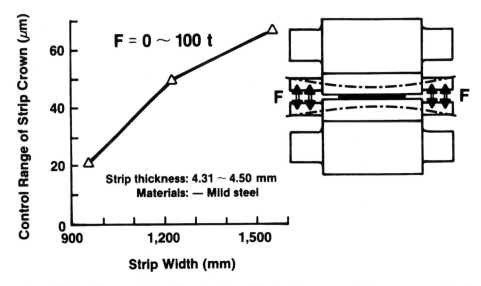

Fig. 37.3 Effect of double-chock roll bending on strip crown control range. Adapted from Takeshima, et al (1979).

The negative roll bending system [5] has found recognition in many rolling mills. It doubles the crown control range when used together with positive roll bending. Application of backup roll bending system [6] is relatively limited because of its complexity.

37.5 HORIZONTAL PLANE ROLL BENDING SYSTEMS

A number of horizontal plane roll bending systems have been proposed: single bending roll [7] (Fig. 37.4a), multiple bending rolls [8] (Fig. 37.4b) and a segmented roll acting on the work roll through an intermediate roll [9] (Fig. 37.4c). The last design has been implemented in a 5-h mill with the flexible flatness control system (FFC) developed by IHI [10]. FFC mills allow one to substantially reduce the edge drop.

A horizontal roll bending system with peripheral bending rolls (Fig. 37.5) has been proposed jointly by United Engineering, Inc. and International Rolling Mill Consultants Inc. [11]. An installation of bending rolls further from the strip center increases bending lever arm and, thus, allows one to substantially increase the crown control range.

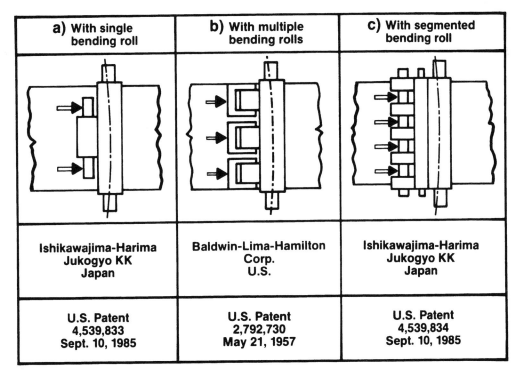

Fig. 37.4 Horizontal plane roll bending systems. (From Ginzburg, Iron and Steel Engineer, 1987. Copyright AISE, Pittsburgh, Pennsylvania. Reprinted with permission.)

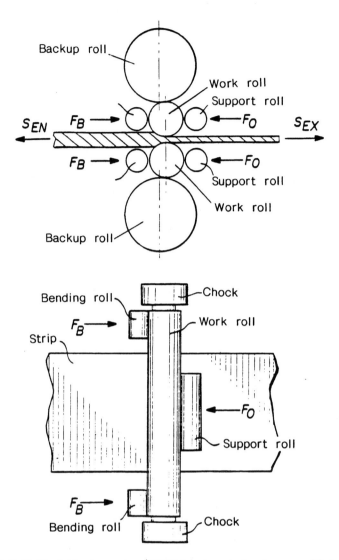

Fig. 37.5 United Engineering/IRMC horizontal plane roll bending system with peripheral application of bending rolls. Adapted from Ginzburg, U.S. Patent No. 4,724,698 (1988).

37.6 ROLL CROSSING SYSTEMS

Roll crossing is another approach to increase the crown control range. Three types of roll crossing systems are known. The early developed systems [12] have provided the crossing of either the backup roll axis (Fig. 37.6a) or the work roll axis (Fig. 37.6b). However, these arrangements cause relative slip in the axis direction between the work roll and the backup roll. It produces excessive thrust and power loss due to friction as well as excessive roll wear. This problem has been solved with the introduction of the pair roll crossing (PC) system (Fig. 37.6c) developed jointly by Mitsubishi Heavy Industries, Ltd. and Nippon Steel Corporation [13, 14].

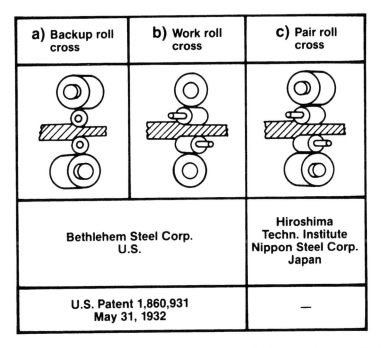

Fig. 37.6 Roll crossing systems. (From Ginzburg, Iron and Steel Engineer, 1987. Copyright AISE, Pittsburgh, Pennsylvania. Reprinted with permission.)

The roll gap between the upper and lower work rolls of the PC mill forms the parabola by crossing the upper and lower parts of the roll stack. This produces the same effect as a convex crown on the work roll. The equivalent roll crown C_e is expressed by the equation:

$$C_e = \frac{w^2 \tan^2 \theta}{2d} \cong \frac{w^2 \theta^2}{2d} \quad , \tag{37-8}$$

where w = strip width, mm

θ = cross angle, deg.

d = work roll diameter, mm.

When the cross angle is equal to 1°, this system can produce an equivalent mechanical crown of approximately $900 \, \mu$m (0.035 in.), as shown in Fig. 37.7.

It is presumed that the adjustments of the roll cross angle are preferably done between rolling the coils and, therefore, a conventional roll bending system will still be required to provide a continuous strip profile control function.

37.7 NONADJUSTABLE STEPPED BACKUP ROLLS

Another step in overall improvement of strip profile control was made as a result of an ingenious analysis of the limitations of a conventional 4-high rolling mill and a discovery of so-called **undesirable**

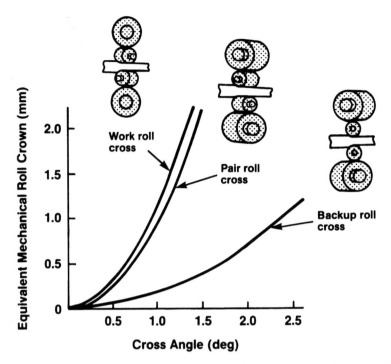

Fig. 37.7 Effect of the cross angle on equivalent mechanical roll crown. Adapted from Nakajima, et al (1985).

contact zones A between work rolls and backup rolls outside the material being rolled (Fig. 37.8a).

It was demonstrated [15] that by using stepped backup rolls (Fig. 37.8b) and, thus, eliminating the undesirable contact zones, the effectiveness of a conventional roll bending system can be significantly improved. With stepped backup rolls, the change Δc in the strip crown with the change ΔP in the rolling

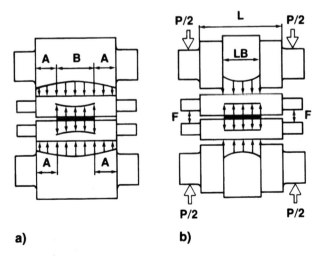

Fig. 37.8 Concept of stepped backup rolls: a) conventional, and b) stepped. (From Nakanishi, et al. Reproduced from HITACHI REVIEW Vol. 34, No. 4, 1985, by courtesy of Hitachi, Ltd., Japan.)

load becomes considerably smaller compared to that of a conventional 4-h mill (Fig. 37.9a). Simultaneously, with a decrease in the rolling force disturbance rate, the crown control rate $\Delta c/\Delta F$ is substantially increased as shown in Fig. 37.9b.

However, these improvements become less pronounced with an increase in strip width w and, eventually, the gain becomes negligible when the strip width is close to the barrel length L of a conventional backup roll. Another drawback is a necessity to replace the backup roll with every substantial change in the strip width.

37.8 ADJUSTABLE STEPPED BACKUP ROLLS

Adjustable stepped backup rolls permit changes to be made in the barrel length. One proposal (Fig. 37.10a) is to perform a gradual grinding of the backup rolls in the mill [16]. Another solution was to adjust the barrel length of the rolls by either supplementing the stepped roll with narrow width segments (Fig. 37.10b, c) [17, 18] or by adjusting the supporting length of a mandrel installed inside of a hollow roll as illustrated in Fig. 37.10d [19].

37.9 AXIALLY SHIFTED CYLINDRICAL ROLLS

A bi-directional shifting of the cylindrical backup rolls [20] is another way to simulate the stepped backup roll concept (Fig. 37.11a). A similar effect has been obtained in the high crown (HC) mill developed by Hitachi, Ltd. [21, 22] by shifting the intermediate rolls (Fig. 37.11b). Figure 37.11c illustrates a bi-directional shifting of the work rolls which accomplishes either a reduced edge drop or reduced roll wear [23]. In the former, the work rolls tapered at one side are shifted so that the tapered

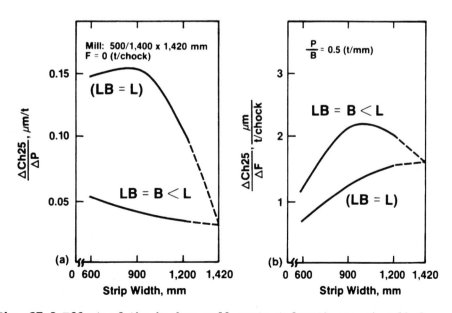

Fig. 37.9 Effect of the backup roll contact length on: a) roll force disturbance rate and b) strip crown control range. (From Furuya, et al, Metals Society Conference Proceedings, Cardiff, Wales, 1978. Copyright Institute of Metals. Reprinted with permission.)

ADJUSTMENT OF THE ROLL BARREL LENGTH		ROLL DESIGN	COMPANY	PATENT
a)	By in-mill grinding of the rolls		SMS Schloemann-Siemag AG W. Germany	U.S. Patent 4,479,374 Oct. 30, 1984
b)	By engaging the desired number of segmented sleeves		Hitachi KK Japan	Japanese Patent 51-103058 Sept. 11, 1976
c)	By spacing the outer segmented sleeves		Mitsubishi Heavy Ind KK Japan	Japanese Patent 55-10366 Jan. 24, 1980
d)	By spacing the inner supports of hollow roll		Davy-Loewy, Ltd. England	U.S. Patent 4,407,151 Oct. 4, 1983

Fig. 37.10 Adjustable stepped backup rolls. (From Ginzburg, Iron and Steel Engineer, 1987. Copyright AISE, Pittsburgh, Pennsylvania. Reprinted with permission.)

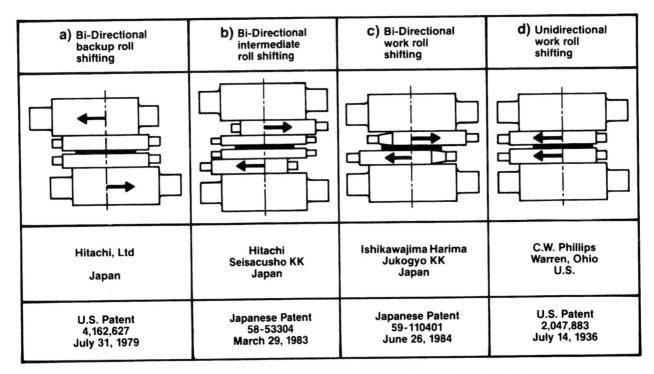

a) Bi-Directional backup roll shifting	b) Bi-Directional intermediate roll shifting	c) Bi-Directional work roll shifting	d) Unidirectional work roll shifting
Hitachi, Ltd Japan	Hitachi Seisacusho KK Japan	Ishikawajima Harima Jukogyo KK Japan	C.W. Phillips Warren, Ohio U.S.
U.S. Patent 4,162,627 July 31, 1979	Japanese Patent 58-53304 March 29, 1983	Japanese Patent 59-110401 June 26, 1984	U.S. Patent 2,047,883 July 14, 1936

Fig. 37.11 Axially shifted cylindrical rolls. (From Ginzburg, Iron and Steel Engineer, 1987. Copyright AISE, Pittsburgh, Pennsylvania. Reprinted with permission.)

portion of the rolls is positioned in the vicinity of the strip edge. In the latter case, the cylindrical work rolls are shifted cyclically to smooth the roll wear contour and thermal crown, so that uniform strip profile without ridges can be obtained.

Figure 37.12 shows the effect of amount of overlapping the strip by the tapered portion of the rolls on the strip crown. Figure 37.13 illustrates the case when bi-directional roll shifting is used to reduce roll wear. In the figure, S is the roll shift stroke and the roll wear groove depth ΔW is equal to

$$\Delta W = 2C_r = 2(C_e - C_m), \qquad (37\text{-}9)$$

where C_e = peak roll wear near the edge of the roll wear pattern, μm

C_m = roll wear at the roll center, μm

C_r = wear groove depth, μm.

Another way of smoothing the roll wear is by shifting the work rolls in the same direction (Fig. 37.11d) [24]. The advantage of the unidirectional roll shifting is its cost effectiveness.

37.10 AXIALLY SHIFTED NONCYLINDRICAL ROLLS

A novel technology for increasing the crown control range is the continuous variable crown (CVC) control system developed by SMS Schloemann-Siemag AG [25, 26]. In this system, the increased crown control range has been achieved by bi-directional shifting of the backup rolls (Fig. 37.14a), intermediate

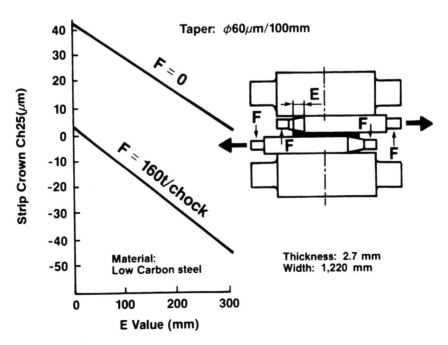

Fig. 37.12 Effect of amount of overlapping the strip by the tapered portion of the rolls at F6 stand on the strip crown. (From Nakanishi, et al. Reproduced from HITACHI REVIEW Vol. 34, No. 4, 1985, by courtesy of Hitachi, Ltd., Japan.)

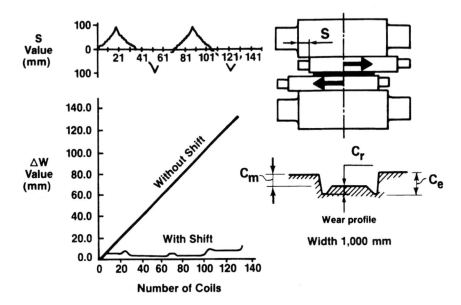

Fig. 37.13 Effect of roll shifting on reduction of roll wear.
(From Nakanishi, et al. Reproduced from HITACHI REVIEW Vol. 34,
No. 4,1985, by courtesy of Hitachi, Ltd., Japan.)

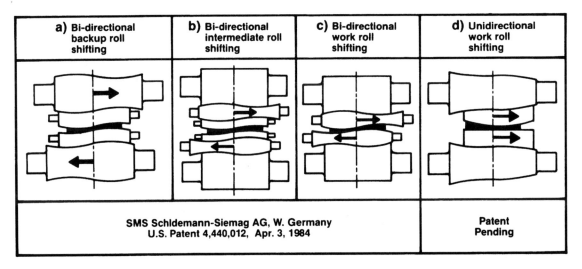

Fig. 37.14 Axially shifted noncylindrical rolls. (From Ginzburg,
Iron and Steel Engineer, 1987. Copyright AISE, Pittsburgh,
Pennsylvania. Reprinted with permission.)

rolls (Fig. 37.14b) and work rolls (Fig. 37.14c). The pair shifting rolls are ground to have a bottle-shaped contour with convex contours located on opposite sides of the adjacent rolls.

In the systems with shifting work rolls, the roll gap profile is usually adjusted during a gap time between the rolling a number of coils. An axial displacement of the rolls of ±100 mm (±4.0 in.) has the same effect as altering conventionally ground positive roll camber between 100 and 500 μm (0.004 and

0.020 in.) as shown in Fig. 37.15. To provide a continuous control function, these systems are supplemented with conventional work roll bending jacks.

An alternate system, known as Universal Profile System (UPS) has been proposed by Mannesmann Demag [27]. In this system the work rolls of 4-high mill stand have a cigar-shaped convex contour. An increased roll shifting stroke, up to ± 200 mm (± 8.0 in.) allows one to achieve not only an increased crown control range but also to utilize roll shifting to reduce roll wear.

In both CVC and UPS systems the rolls are shifted in opposite directions. In the system shown in Fig. 37.14d, the work rolls are shifted in the same direction. A change in roll gap crown is obtained by special shaping of both the shifted work rolls and non-shifted backup rolls [28].

37.11 AXIALLY SHIFTED SLEEVED ROLLS

The stepped backup roll concept can also be simulated by using assembled sleeved backup rolls. In the design proposed by Nippon Steel Corporation [29], the strip profile adjustment is made by either shifting the backup roll arbors (Fig. 37.16a) or the sleeves (Fig. 37.16b). In designs proposed by Kobe Steel Corporation [30, 31] strip profile adjustment is made by shifting either the inner sleeve (Fig. 37.16c) or the tapered outer sleeve residing on the tapered arbor (Fig. 37.16d).

37.12 FLEXIBLE BODY ADJUSTABLE CROWN ROLLS

A number of modifications of flexible body adjustable crown rolls have been proposed [32]. The most well known designs are the Sumitomo variable crown (VC) rolls [33] and the Blaw-Knox inflatable crowns (IC) backup rolls [34]. Their principle of operation is based on inflation of the central zone of the sleeve which is shrunk-fitted over a solid mandrel at the edges of the roll (Fig. 37.17a).

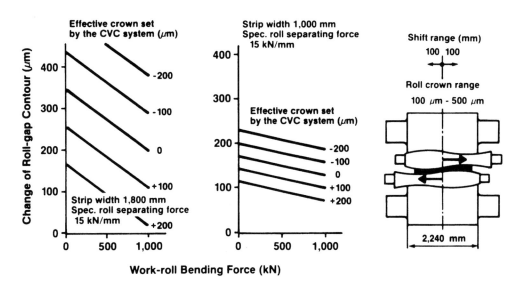

Fig. 37.15 Effect of roll shifting and bending force of CVC rolls on equivalent mechanical roll crown. (From Wilms, et al, Metallurgical Plant and Technology, Vol. 8, No. 6, 1985. Copyright Verlag Stahleisen mbH. Reprinted with permission.)

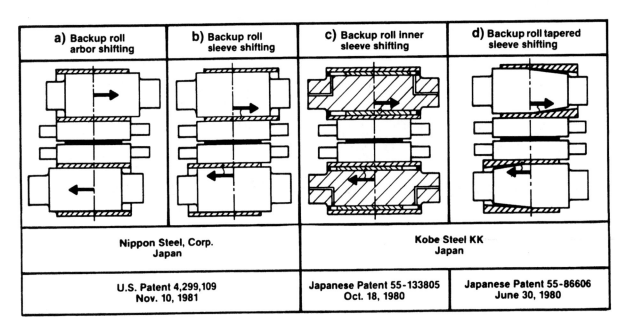

Fig. 37.16 Axially shifted sleeved rolls. (From Ginzburg, Iron and Steel Engineer, 1987. Copyright AISE, Pittsburgh, Pennsylvania. Reprinted with permission.)

	DESIGN FEATURES	ROLL DESIGN	COMPANY	PATENT
a)	With single cavity in a middle of the roll body for hydraulic fluid		BWG Bergwetk Und Walzwerk-Maschinenbau G.m.b.H. W. Germany	U.S. Patent 3,457,617 July 29, 1964
b)	With multiple cavities along the roll body for hydraulic fluid		T. Sendzimir, Inc. U.S.	U.S. Patent 3,355,924 Dec. 5, 1967
c)	With wedge type mechanism		White Consolidated Industries, Inc. U.S.	U.S. Patent 4,553,297 Nov. 19, 1985
d)	With segmented hydro-static bearings		Escher Wyss Aktiengesellschaft Switzerland	U.S. Patent 4,429,446 Feb. 7, 1984

Fig. 37.17 Flexible body adjustable crown rolls. (From Ginzburg, Iron and Steel Engineer, 1987. Copyright AISE, Pittsburgh, Pennsylvania. Reprinted with permission.)

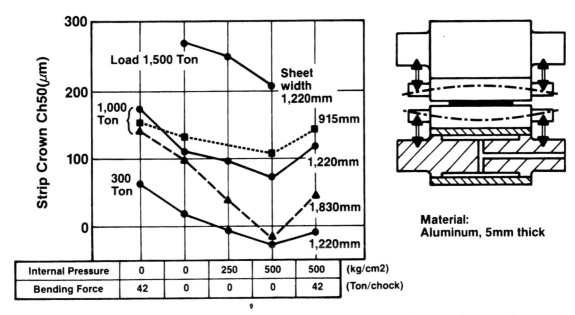

Fig. 37.18 Effect of internal pressure and bending force of VC roll on strip crown. (From Yamada, et al, Iron and Steelmaker, Vol. 9, 1982. Copyright The Metallurgical Society of AIME. Reprinted with permission.)

The effect of internal pressure inside of the center zone of VC rolls on the crown of aluminum plate is shown in Fig. 37.18. Internal pressure applied inside the cavity may be as high as 500 kg/cm² (7070 psi). This places some strict requirements on the design of the rotary hydraulic joint through which the oil is supplied to the roll. In IC rolls this problem has been alleviated by installing a hydraulic intensifier inside the roll, so the fluid will be passing through the rotary joint under low pressure.

In the sleeved roll proposed by Sendzimir [35] multiple cavities are provided along the roll body (Fig. 37.17b) allowing for individual adjustment of the pressure in each cavity. A hydraulically driven wedge-type mechanism is used in the roll shown in Fig. 37.17c to expand the outer shell of the roll. This design was implemented in the hydromechanical inflatable crown (ICHM) roll developed by Blaw-Knox Foundry and Mill Machinery Company [34, 36].

In the roll shown in Fig. 37.17d, the roll crown is adjusted with piston-like support elements installed between a stationary arbor and a sleeve rotating on the arbor [37]. This principle is utilized in Nipco rolls [38], which have been used in paper machines.

37.13 FLEXIBLE EDGE ADJUSTABLE CROWN ROLLS

Adjustment of the roll crown of the flexible edge rolls is usually achieved by adjusting the pressure inside the chambers located at the edges of the rolls. A roll with two pressurized chambers between an arbor and a sleeve is illustrated in Fig. 37.19a. In one application, it was proposed to use this roll as an intermediate roll in a 5-h mill [39]. A roll design with restricted sleeve deflection near the edges is depicted in Fig. 37.19b [40, 41].

DESIGN FEATURES		ROLL DESIGN	COMPANY	PATENT
a)	Non-restricted sleeve deflection near the edge	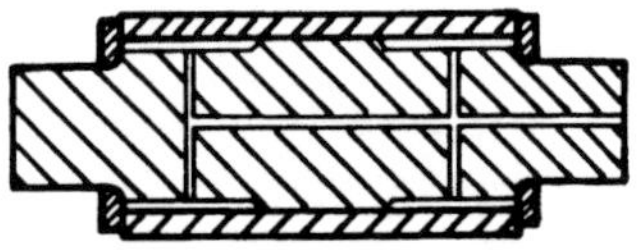	Ishikawajima Harima Jukogyo KK Japan	Japanese Patent 57-68206 Apr. 26, 1982
b)	Restricted sleeve deflection near the edge	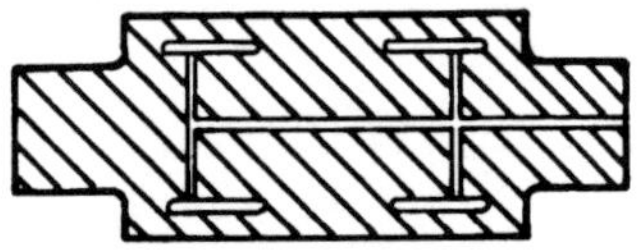	Same	Japanese Patent 59-54401 March 29, 1984
c)	Adjustment made by shifting the wedges between sleeve and arbor	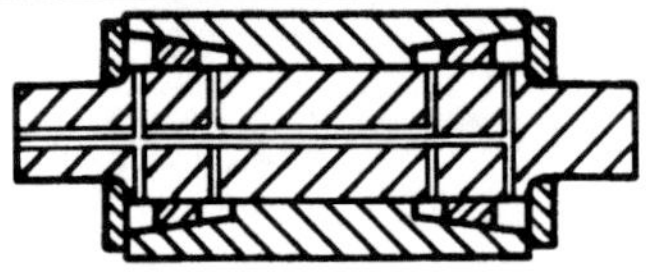	Same	U.S. Patent 4,599,770 July 15, 1986
d)	Adjustment made by shifting the segmented sleeves on tapered arbor	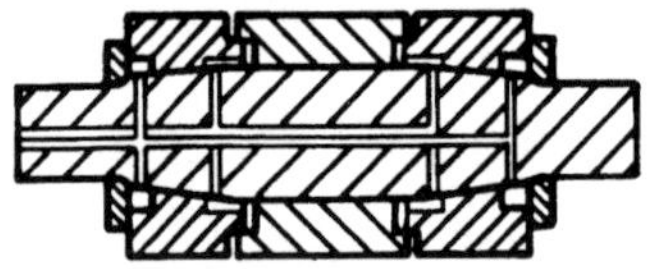	Same	Japanese Patent 58-196104 Nov. 15, 1983

Fig. 37.19 Flexible edge adjustable crown rolls. (From Ginzburg, Iron and Steel Engineer, 1987. Copyright AISE, Pittsburgh, Pennsylvania. Reprinted with permission.)

In the roll shown in Fig. 37.19c, the hydraulically driven tapered pistons are located near the roll edges between an arbor and the tapered portions of an outer sleeve [42]. The roll crown can be adjusted by lateral displacement of the tapered pistons.

An arrangement with shifting tapered segmented sleeves located near the roll edges is shown in Fig. 37.19d [43].

37.14 FLEXIBLE EDGE SELF-COMPENSATING ROLLS

A number of ideas have been proposed on how to design a roll with reduced rigidity near the edges. One idea was to insert a metal with lower modulus of elasticity between an arbor and a roll sleeve (Fig. 37.20a) [44]. The same effect may be achieved by drilling horizontal holes near the edges of a solid roll (Fig. 37.20b) or by making longitudinal slots at the arbor edges of a sleeved roll (Fig. 36.20c). A more sophisticated roll shape with a counter lever of the sleeve on the arbor (Fig. 37.20d) was proposed for paper machines [45].

The basic concept of the new, Self-Compensating (SC) roll is relatively simple. To compensate for deflection of the backup roll under the rolling load, the roll is provided with a sleeve which is shrunk-fitted to the arbor in its middle part and is flexible at its edges (Fig. 37.21). Under the action of a rolling load, the sleeve will deflect in the direction opposite to the deflection of the arbor, so the overall change of the backup roll contour contacting the work roll becomes smaller.

METHOD FOR RELIEVING THE ROLL EDGES		ROLL DESIGN	COMPANY AND PATENT
a)	By filling the gap between arbor and sleeve with material having lower modulus of elasticity	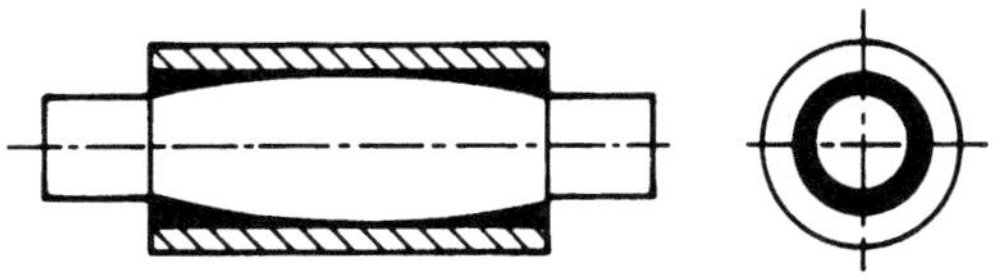	American Rolling Mill Company, U.S. U.S. Patent 2,187,250 Jan. 16, 1940
b)	By drilling the holes at the roll edges	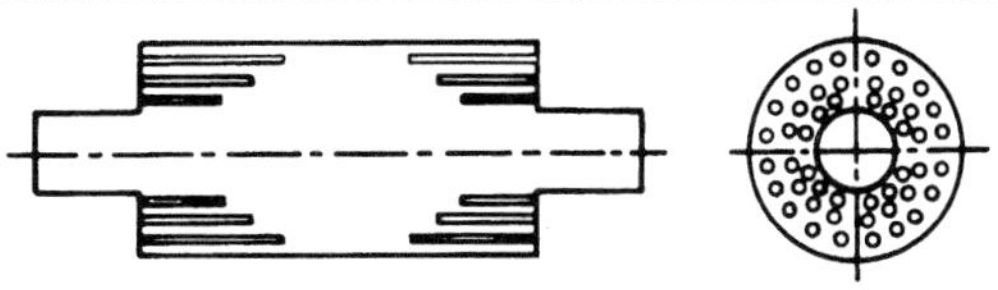	Same
c)	By milling the slots at the arbor edges	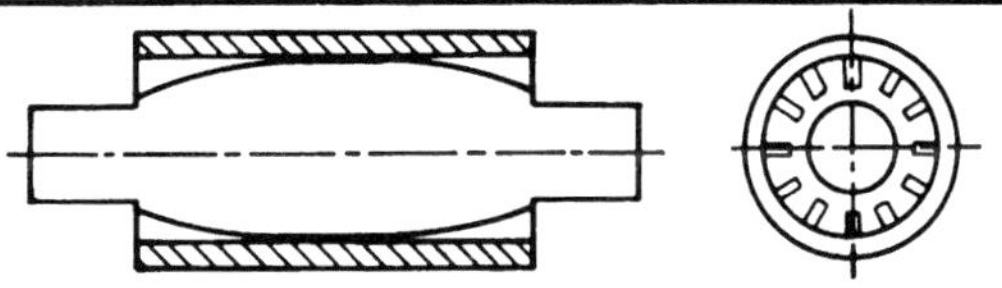	Same
d)	By counterlever support of the sleeve on the arbor	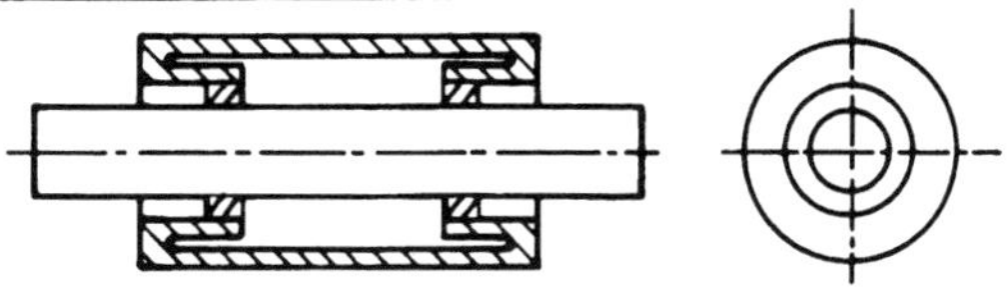	Beloit Iron Works U.S. U.S. Patent 3,097,590 July 16, 1963

Fig. 37.20 Flexible edge self-compensating rolls. (From Ginzburg, Iron and Steel Engineer, 1987. Copyright AISE, Pittsburgh, Pennsylvania. Reprinted with permission.)

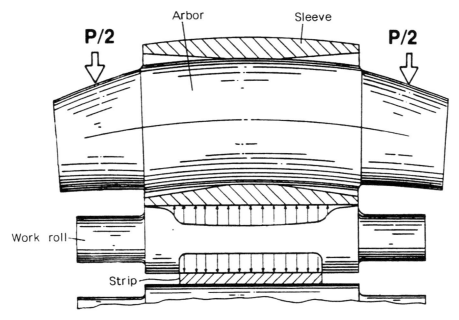

Fig. 37.21 Principle of operation of the United Engineering/IRMC Self-Compensating (SC) backup roll. Adapted from Ginzburg and Guo, U.S. Patent No. 4,722,212 (1988).

Two principal designs of the SC backup rolls with shrunk-fitted sleeves were developed jointly by United Engineering, Inc. and International Rolling Mill Consultants, Inc. In the first design, the cavity between flexible portion of the sleeve and the arbor remains open under rolling load. In the second design [46], the contact length between inner surface of the sleeve and the arbor varies with load. This feature allows one to provide self-compensating for wider strips and higher rolling loads.

REFERENCES

1. K.N. Shohet and N.A. Townsend, "Flatness Control in Plate Rolling", Journal of the Iron and Steel Institute, Oct. 1971, pp. 769-775.

2. R.R. Somers, et al, "Verification and Applications of a Model for Predicting Hot Strip Profile, Crown and Flatness", AISE Year Book, 1984, pp. 441-450.

3. E.A. Pauls, U.S. Patent No. 2,430,410, Nov. 4, 1947.

4. Y. Takeshima, et al, "Studies on the Strip Crown Control for Hot Strip Rolling - Double Chock Work Roll Bending System (DC-WRB)", IHI Engineering Review, Vol. 12, No. 3, Oct. 1979, pp. 28-34.

5. T.A. Fox, U.S. Patent No. 3,024,679, March 13, 1962.

6. M.D. Stone and R. Gray, "Theory and Practical Aspects in Crown Control", AISE Yearly Proceedings, 1965, pp. 657-667.

7. H. Kato, et al, U.S. Patent No. 4,539,833, Sept. 10, 1985.

8. G. Cozzo, U.S. Patent No. 2,792,730, May 21, 1957.

9. T. Iwanami, U.S. Patent No. 4,539,834, Sept. 10, 1985.

10. F. Fujita, et al, "Development of a New Type of Cold Rolling Mill for Sheet Products", AISE Year Book, 1985, pp. 264-271.

11. V.B. Ginzburg, U.S. Patent No. 4,724,698, Feb. 16, 1988.

12. A.T. Keller, U.S. Patent No. 1,860,931, May 31, 1932.

13. K. Nakajima, et al, "Basic Characteristics of Pair Cross Mill", Mitsubishi Heavy Industries Technical Review, Vol. 22, No. 2, June 1985, pp. 143-148.

14. H. Hino, et al, "Shape and Crown Control Mill - Pair Cross Mill", Innovative Hot Strip Mill Technology, AISI Symposium No. 16, Pittsburgh, Pa., Oct. 30-31, 1985, pp. 14-21.

15. T. Furuya, et al, "High Crown Control Mill: A Newly Developed 6-high Cold Mill to Solve Shape Problems", Flat Rolling: A Comparison of Rolling Mill Types, Metals Society Conference Proceedings, Cardiff, Wales, Sept. 1978, pp. 147-154.

16. H. Feldman, et al, U.S. Patent No. 4,479,374, Oct. 30, 1984.

17. Hitachi KK, Japanese Patent No. 51-103058, Sept. 11, 1976.

18. Mitsubishi Heavy Industries KK, Japanese Patent No. 55-10366, Jan. 24, 1980.

19. R.W. Gronbech, U.S. Patent No. 4,407,141, Oct. 4, 1983.

20. S. Shida and T. Kajiwara, U.S. Patent No. 4,162,627, July 31, 1979.

21. S. Onda, et al, "Profile Control of Hot Rolled Steel Strip by Shifting Work Roll with Tapered Crown", Advanced Technology of Plasticity, Vol. 2, 1984, pp. 1366-1371.

22. T. Kimura, Japanese Patent No. 58-53304, March 29, 1983.

23. T. Nakanishi, et al, "Application of Work Roll Shift Mill HCW-Mill to Hot Strip and Plate Rolling", Hitachi Review, Vol. 34, No. 4, 1985, pp. 153-160.

24. C.W. Phillips, U.S. Patent No. 2,047,883, July 14, 1936.

25. H. Feldmann, et al, U.S. Patent No. 4,440,012, April 3, 1984.

26. W. Wilms, et al, "Profile and Flatness Control in Hot Strip Mills", MPT: Metallurgical Plant and Technology, Vol. 8, No. 6, 1985, pp. 74-90.

27. E. Kersting and H. Teichert, "The UPC Technology: Modernization of Hot Strip Mills, Specifically in Regard to Profile Control", Proceedings of the 4th International Steel Rolling Conference: The Science and Technology of Flat Rolling, Vol. 1, Deauville, France, June 1-3, 1987, pp. A.19.1-A.19.7.

28. V.B. Ginzburg, U.S. Patent No. 4,656,859, April 14, 1987.

29. H. Matsumoto, et al, U.S. Patent No. 4,299,109, Nov. 10, 1981.

30. Kobe Steel KK, Japanese Patent No. 55-133805, Oct. 18, 1980.

31. Kobe Steel KK, Japanese Patent No. 55-86606, June 29, 1980.

32. O. Noe, et al, U.S. Patent No. 3,457,617, July 29, 1969.

33. J. Yamada, et al, "The Development of Sumitomo VC (Variable Crown) Roll System", Iron and Steelmaker, Vol. 9, June 1982, pp. 37-42.

34. W.W. Eibe, "Inflatable Crown Rolls - Characteristics, Design and Applications", AISE Year Book, 1984, pp. 426-432.

35. T. Sendzimir, U.S. Patent No. 3,355,924, Dec. 5, 1967.

36. W.W. Eibe, U.S. Patent No. 4,553,297, Nov. 19, 1985.

37. R. Lehmann, U.S. Patent No. 4,429,446, Feb. 7, 1984.

38. H.P. Wiendahl, "The Nipco Roll: From Idea to New Product", Escher Wyss News, Feb. 1978 - Jan. 1979, pp. 47-52.

39. I. Imai, japanese Patent No. 57-68206, April 26, 1982.

40. H. Honjiyou, Japanese Patent No. 59-54401, March 29, 1984.

41. V.B. Ginzburg and N.M. Kaplan, U.S. Patent No. 4,683,744, Aug. 4, 1987.

42. H. Kato and H. Shiozaki, U.S. Patent No. 4,599,770, July 15, 1986.

43. H. Shiozaki, Japanese Patent No. 58-196104, Nov. 15, 1983.

44. T. Sendzimir, U.S. Patent No. 2,187,250, Jan. 16, 1940.

45. E.J. Justus, U.S. Patent No. 3,097,590, July 16, 1983.

46. V.B. Ginzburg and Remn M. Guo, U.S. Patent No. 4,722,212 , Feb.2, 1988.

47. V.B. Ginzburg, "Strip Prifile Control with Flexible Edge Backup Rolls", Iron and Steel Engineer, July 1987, pp. 23-34.

38

Roll Deformation Models

38.1 TYPES OF ROLL DEFORMATION MODELS

A history of development of the roll deformation models dates back to 1958, when the first comprehensive study of a four-high mill was made by Saxl [1]. Since that time this field of research has been enriched by the introduction of a number of mathematical models which can be divided into three groups [2].

1. Simple beam models
2. Slit beam models
3. Finite element analysis models.

Below is a brief review of these models.

38.2 SIMPLE BEAM DEFLECTION MODELS

In the simple beam deflection models, both the backup rolls and work rolls are considered as straight elastically stressed beams. In deriving the formulas for the beam deflection, it is assumed that [3]:

1. The beam is of homogeneous material which has the same modulus of elasticity in tension and compression.
2. The beam cross-section is uniform.
3. The beam has at least one longitudinal plane of symmetry.
4. All loads and reactions are perpendicular to the axis of the beam.
5. The span-to-depth ratio of the beam is eight or greater for metal beams of compact section.

The strip profile is determined by super-position of the following components of the roll deflection:

a) roll deflection due to bending forces generated by rolling load

b) roll deflection due to shear forces generated by rolling load

c) roll deflection due to the forces generated by the roll bending mechanism.

Let us examine these three components in more detail.

Roll deflection due to bending forces - The deflection for this case is described by the differential equation [4]:

$$E_B I_B \frac{d^2 y_{B1}}{dx^2} = -\frac{P}{2}\left[x - \frac{(x-e)^2}{w}\right] \qquad (38\text{-}1)$$

where E_B = roll modulus of elasticity

    I_B = moment of inertia of roll section at distance x (Fig. 38.1)

    P = rolling load

    y_{B1} = roll deflection at distance x.

In the zone of roll contact with the strip, x varies within the range:

$$(L - w)/2 \leq x \leq L/2 \qquad (38\text{-}2)$$

The solution of Eq. (38-1) in application for a two-high mill, corresponding to this range, is given by Larke [5]:

$$y_{B1} = cP\left[x(\beta - 4x^2) + \frac{2(x-e)^4}{w} + 2u^3(\alpha - 4)\right] \qquad (38\text{-}3)$$

where

$$c = \frac{4}{3\pi E_B D_B^4}$$

$$\beta = 3L^2 - w^2$$

$$\alpha = 4\left(\frac{D_B}{d_B}\right)^4 \qquad (38\text{-}4)$$

$$e = \frac{L - w}{2}$$

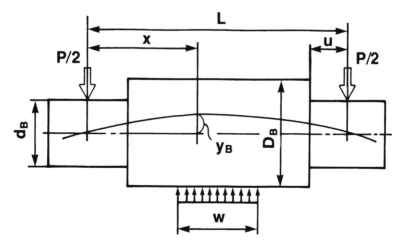

Fig. 38.1 Simple beam deflection model for a 2-high mill roll-strip system. From Ginzburg, et al (1987). Copyright AISE, Pittsburgh, Pennsylvania. Reprinted with permission.

Roll deflection due to shear forces - In deriving the deflection due to the action of shearing forces, it is assumed that uniform distribution of shear stress exists across the vertical sections of the rolls. Within the range of x, given by Eq. (38-2), deflection is expressed by the differential equation [5]:

$$G_B \frac{dy_{B2}}{dx} = \frac{2P(L - 2x)}{\pi w D_B^2} \tag{38-5}$$

where G_B = roll modulus of rigidity

y_{B2} = roll deflection at distance x.

The solution of Eq. (38-5) is given by Larke as:

$$y_{B2} = \frac{2P}{\pi w G_B D_B^2} \left[x(L - x) + \left(\frac{D_B^2}{d_B^2} - 1 \right) wu - e^2 \right] \tag{38-6}$$

The total roll deflection due to rolling load is equal to:

$$y_B = y_{B1} + y_{B2} \tag{38-7}$$

Roll deflection produced by the roll bending mechanism - The work roll deflection in a four-high mill is described by Stone [6] as the deflection of a simple beam on elastic foundation (Fig. 38.2). The differential equation, representing the deflection, is expressed by [6, 7]:

$$E_w I_w \frac{d^4 y_w}{dx^4} = -k y_w \tag{38-8}$$

where E_w = work roll modulus of elasticity

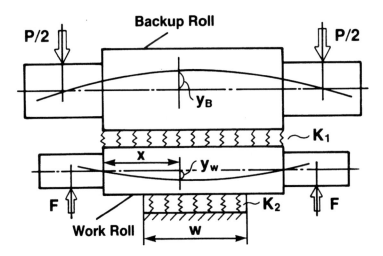

Fig. 38.2 Simple beam deflection model for a 4-high mill roll-strip system. From Ginzburg, et al (1987). Copyright AISE, Pittsburgh, Pennsylvania. Reprinted with permission.

I_W = work roll moment of inertia

k = beam-on-elastic foundation equivalent constant

y_W = work roll deflection at distance x.

General solution for the deflection curve is given by:

$$y_W = e^{\beta x}[A\cos\beta x + B\sin\beta x] + e^{-\beta x}[C\cos\beta x + D\sin\beta x], \qquad (38-9)$$

where A,B,C,D = integration constants determined by the type of loading and nature of the boundary conditions.

$$\beta = \sqrt[4]{\frac{k}{4EI_W}} \qquad (38-10)$$

The elastic beam conditions and loadings are shown in Fig. 38.2, where k_1 is the elastic beam constant representing the elastic contact conditions between the backup roll and adjacent work roll, and k_2 is the elastic beam constant for the two contacting work rolls, including the effect of metal being rolled. According to Stone, $k = k_1 + 2k_2$.

The values of k_1 and k_2 are determined from [6] as follows:

$$k_i = \frac{P}{l_i \delta_i}, \qquad i = 1, 2 \qquad (38-11)$$

where l_i = contact length in longitudinal direction between two contacting rolls

δ_i = compressive decrease in the distance between centers of the two contacting rolls.

Based on a study by Foppl [8], when $E_B = E_W = E$, the compressive decrease between the centers of a pair of rolls δ_F, such as a work roll and backup roll, is given by:

$$\delta_F = \frac{2P(1 - \nu^2)}{\pi Ew}\left(\frac{2}{3} + \ln\frac{2D_B}{b} + \ln\frac{2D_W}{b}\right) \qquad (38-12)$$

where D_W, D_B= diameters of work roll and backup roll under compression

b = width of the flattened contact area between work roll and backup roll that is given by:

$$b = \sqrt{\frac{16P(1 - \nu^2)}{\pi Ew}\frac{D_B D_W}{D_B + D_W}} \qquad (38-13)$$

where ν = Poisson's ratio of the roll material.

The simple beam deflection model allows the incorporation of important factors which affect the strip profile. However, this model has a number of limitations such as:

a) The span-to-depth ratios of the backup rolls used in rolling mills are usually much less than eight, which is a minimum value required in order to produce accurate results from the equations introduced above.

b) Equations (38-3) through (38-7), derived for a two-high mill, are applied to a four-high mill, assuming that in the latter case the load between backup and work rolls is applied only along the roll contact length that is equal to the strip width. This contradicts the actual loading conditions in the mill stands in which the load is transmitted throughout an entire contact zone between backup and work rolls.

c) The model does not allow the simulation of uneven transverse load distribution between the work roll and the workpiece, as well as between the work roll and the backup roll. As a consequence, it does not take into account the effects of such important factors as the roll crown, incoming strip profile, material hardness distribution along the strip width, roll wear, etc.

These main reasons have justified the development of more sophisticated models for calculation of the strip profile. One of the solutions for simulating the effects of the roll and strip crowns, as well as uneven load distribution along the roll length, was introduced by Poplawski [9] and McDermott [10]. In theirs models, the interfaces between the backup roll and work roll as well as the interfaces between the strip and work roll were simulated with a series of springs, whereas the rolls were still conceived to be simple elastic beams.

38.3 SLIT BEAM MODELS

The slit beam deflection model was first proposed by Shohet and Townsend [11]. In their model the plate profile and gauge are found by determining the following three unknown quantities:

    a) the transverse load distribution between the work roll and the plate

    b) the transverse load distribution between the work roll and backup roll

    c) the rigid body movement of the work rolls.

Since the mill is symmetrical about the mid-span of the rolls, the calculating process involves only one half of a roll span. The numerical method consists of splitting this half of the roll span into m elements (Fig. 38.3) and replacing the load distribution on the roll by a concentrated load applied at the

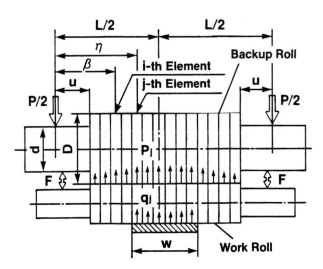

Fig. 38.3 Slit beam deflection model for a 4-high mill roll-strip system. From Ginzburg, et al (1987). Copyright AISE, Pittsburgh, Pennsylvania. Reprinted with permission.

middle of each element. Since the workpiece width is less than the roll barrel, the number of roll elements **n** in contact with the workpiece, is less than **m**.

Highlights of the slit beam model are briefly described below.

Deformation of the backup roll - The vertical displacement of the roll surface at location i is given by:

$$y_B(i) = -\sum_{j=1}^{m} p_j \alpha_{ijb} \Delta x - Z_b(i) - K_b \tag{38-14}$$

where p_j = load per unit width between the work and backup rolls at the middle of the j-th element

α_{ijb} = influence coefficient for the backup roll

Δx = length of the element

$Z_b(i)$ = local contact deformation of backup roll surface at location i

K_b = rigid body movement of the backup roll.

The local contact deformation $Z_b(i)$ is due to roll flattening and can be found by linear approximation of Eq. (38-12). The rigid body movement K_b is defined by the mill stiffness characteristics and by loads applied to the rolls.

Deformation of the work rolls - The vertical displacement of the work roll surface at location i is expressed as:

$$y_W(i) = \sum_{j=1}^{m} p_j \alpha_{ijw} \Delta x - \phi_i \sum_{j=1}^{m} q_j \alpha_{ijw} \Delta x + Z_w(i) - K_w \tag{38-15}$$

where q_j = load per unit width between the work roll and the workpiece at the middle of the i-th element

α_{ijw} = influence coefficient for the work roll

ϕ_i = multiplying factor for i-th location (ϕ_i = 1 at the locations where the workpiece is present and ϕ_i = 0 at the locations where there is no workpiece).

$Z_w(i)$ = local contact deformation of the work roll surface at location i

K_w = rigid body movement of the work roll.

Calculation of the influence coefficients - The influence coefficient α_{ij} is defined as the deflection at the middle of the i-th element due to a unit load at the middle of the j-th element (Fig. 38.3). For $j \geq i$:

$$\alpha_{ij} = \frac{32}{3\pi E} \left[(1 + \nu)\left(\frac{u}{d^2} + \frac{\beta - u}{D^2}\right) + \frac{2u^3}{d^4} + \frac{1}{D^4}(3\eta\beta L - 2u^3 - 3\beta\eta^2 - \beta^3) \right] \tag{38-16a}$$

For $j < i$:

$$\alpha_{ij} = \frac{32}{3\pi E} \left[(1 + \nu)\left(\frac{u}{d^2} + \frac{\eta - u}{D^2}\right) + \frac{2u^3}{d^4} + \frac{1}{D^4}(3\beta\eta L - 2u^3 - 3\eta\beta^2 - \eta^3) \right] \tag{38-16b}$$

Coefficients α_{ijw} and α_{ijb} can be obtained when the corresponding values for work roll and backup roll parameters are used in Eqs. (38-16a,b).

Compatibility for contact of work roll and backup roll - When there is no load, the cambered work roll and backup roll are only in point contact with each other and are separated beyond this point by a gap $\gamma(i)$. Under rolling load this gap will be closed. Thus, the compatibility equation takes the form:

$$\gamma(i) = y_B(i) - y_W(i) \qquad\qquad (38\text{-}17)$$

Compatibility for contact of work roll and workpiece - The loaded roll gap height at any point, and hence the exit thickness of the workpiece at that point, varies with the sum of roll flattening and the deformation of the work roll axis. Thus, the compatibility equation takes the form:

$$h(i) - C(i) = -y_W(i) + Z_w(i), \qquad\qquad (38\text{-}18)$$

where $h(i)$ = semi-thickness of the workpiece exiting at location i

$C(i)$ = semi-height of the no-load roll gap at location i.

The exit thickness of the workpiece is calculated by using the linearized equation of the rolling theory describing $h(i)$ as a function of the entry thickness of the workpiece, rolling load, strip tension and resistance to deformation.

Static equilibrium of the work rolls - Static equilibrium is obtained by summing up the loads acting between the rolls, between the workpiece and the work roll, and the load applied to the work rolls by the roll bending mechanism. The resulting equation is:

$$\sum_{j=1}^{m} p_j \Delta x - \phi_i \sum_{j=1}^{m} q_j \Delta x = F \qquad\qquad (38\text{-}19)$$

where F = total roll bending force (Fig. 38.3).

Solution of equations - The problem is confined to finding $(m + n + 1)$ unknowns, i.e., m values defining the load distribution between the work roll and backup roll $(p_1, p_2, ...p_m)$, n values defining the load distribution between the work roll and workpiece $(q_1, q_2,...q_n)$, and the rigid body movement K_w of the work roll. With the three basic equations (38-17) through (38-19), there are $(m + n + 1)$ equations available to obtain p, q, and K_w. These equations are solved using matrix algebra.

The slit beam deflection model allows one to take into consideration the strip tension distribution across its width, which produces n additional unknowns. This difficulty can be overcome by iterative procedures.

In the model developed by Guo [12], the rolls are presented with spring elements (Fig. 38.4). Interrelation between deformations of the spring is calculated using the influence-line technique that takes into account a beam deflection due to both bending moment and shearing force. The model replaces the strip with a series of springs. The stiffness of these springs is then assumed to be a function of the strip rolling characteristics. Another version of the slit beam model was described by Hollander [13].

The slit beam deflection model was a substantial step in improving the capability to simulate the strip profile during rolling. However, this model has its own limitations such as:

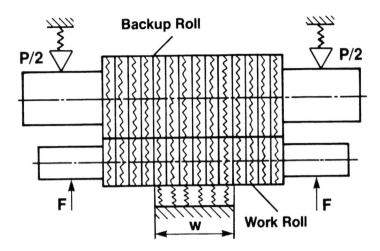

Fig. 38.4 Modification of the slit beam deflection model for a 4-high mill roll-strip system. From Ginzburg, et al (1987). Copyright AISE, Pittsburgh, Pennsylvania. Reprinted with permission.

1. The model is based on the assumption that under rolling load there is a complete contact between the backup roll and work roll. This may not be true when rolls with special configurations such as CVC rolls [14], UPC rolls [15], or tapered rolls [16] are utilized, In these cases, the model should allow for a possible interface mismatch (Fig. 38.5).

2. The model calculates the influence coefficients based on the equations for deflection of a simple beam. However, as mentioned before, the validity of these equations for the rolls with small length-to-diameter ratios is questionable.

3. The model provides a 2-dimensional presentation of a 3-dimensional system and, therefore, the accuracy of simulation may be found inadequate for some cases.

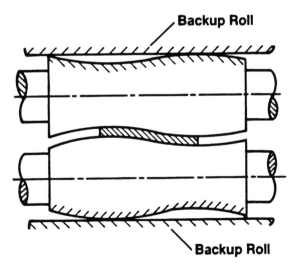

Fig. 38.5 Interface mismatch between the work rolls and backup rolls of a 4-high mill. From Ginzburg, et al (1987). Copyright AISE, Pittsburgh, Pennsylvania. Reprinted with permission.

38.4 PRINCIPLES OF FINITE ELEMENT ANALYSIS

The basis of the finite element method is the representation of a body or a structure by an assemblage of subdivisions called **finite elements** which is often referred to as a **mesh** [17, 18] (Fig. 38.6). These elements are considered to be interconnected at joints which are called **nodes** or **nodal points.** The finite elements may be triangles, a group of triangles, or quadrilaterals for a 2-dimensional continuum. For 3-dimensional analysis, the finite elements may be in the shape of a tetrahedron, rectangular prism, or hexahedron.

The following six steps summarize the finite element analysis procedure:

1. Selection of the displacement model
2. Derivation of the strain-displacement function
3. Derivation of the stress-strain function
4. Derivation of the element stiffness matrix
5. Assembly of the algebraic equations
6. Solution for the unknown parameters.

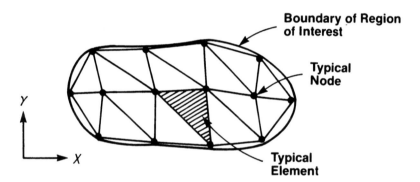

Fig. 38.6 Two-dimensional region represented as an assemblage of triangular elements. From Ginzburg, et al (1987). Copyright AISE, Pittsburgh, Pennsylvania. Reprinted with permission.

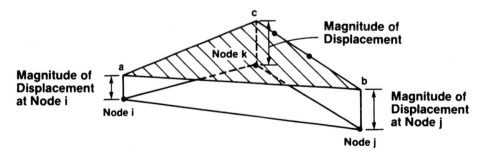

Fig. 38.7 Isometric view of triangular element with linear displacement model plotted in the third dimension. From Ginzburg, et al (1987). Copyright AISE, Pittsburgh, Pennsylvania. Reprinted with permission.

Selection of the displacement model - The displacement model establishes the displacements over each finite element as a function of the nodal displacements (Fig. 38.7) and can be expressed in general terms as:

$$\{u\} = [N]\{d\}, \tag{38-20}$$

where $\{u\}$ = element general displacement matrix

$[N]$ = element shape matrix

$\{d\}$ = element nodal displacement matrix.

In derivation of the displacement models, the distribution of the actual displacements along a given direction is usually presented with a simple function such as a polynomial of the type (Fig. 38.8):

$$u = a_1 + a_2x + a_3x^2 + \ldots + a_{n+1}x^n, \tag{38-21}$$

where u = displacement along x-direction.

The coefficients of the polynomial, represented as the a's, are known as **generalized displacement amplitudes** which determine the shape of the displacement model. These coefficients are derived from the boundary conditions. The number of equations used to define finite-elements depends on the geometry of the structure to be modeled and the number of directions that the model is permitted to move, which is called **degrees of freedom** [19]. Simple spring elements can undergo tension and have two degrees of freedom because each end of the beam is permitted to move independently in the direction of the major axis of the spring. However, elements with more degrees of freedom may be required to allow other movements such as bending.

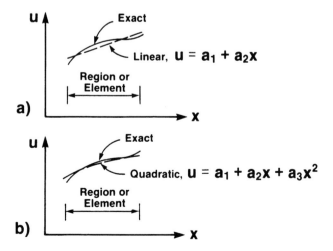

Fig. 38.8 Polynomial approximation of the displacement function in one dimension: a) linear, b) quadratic. From Ginzburg et al (1987). Copyright AISE, Pittsburgh, Pennsylvania. Reprinted with permission.

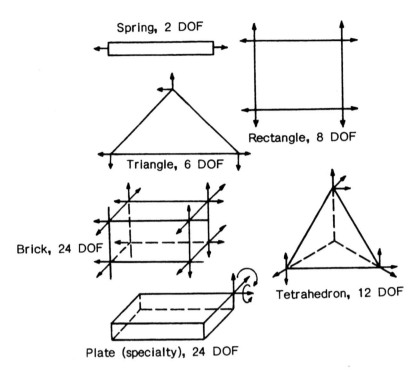

Fig. 38.9 Degrees of freedom (DOF) for different types of elements.
(From Jeffrey M. Steele. Reprinted from COMPUTER-AIDED ENGINEERING,
May/June 1984. Copyright 1984, by Penton Publishing, Inc., Cleveland,
Ohio. Reprinted with permission.)

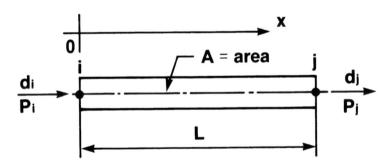

Fig. 38.10 Spar element. From Ginzburg, et al (1987). Copyright AISE,
Pittsburgh, Pennsylvania. Reprinted with permission.

The selection of element type also determines the degrees of freedom in a model. The number of degrees of freedom next to each element type shown in Fig. 38.9 is for simple linear formulations. With a quadratic displacement function applied to a brick, there may be 60 degrees of freedom for a single element. As an example, let us consider a spar element subjected to external loads P_i and P_j (Fig. 38.10). For this case, there are two displacement degrees of freedom, so a linear displacement model is chosen:

$$u = a_1 + a_2x = [1 \quad x] \begin{Bmatrix} a_1 \\ a_2 \end{Bmatrix} \tag{38-22}$$

From the boundary conditions we obtain:

$$\begin{aligned}
\text{at} \quad x &= 0 \quad u = d_i \\
\text{at} \quad x &= L \quad u = d_j
\end{aligned} \tag{38-23}$$

where d_i and d_j = displacements at nodes i and j respectively.

Solving equations (38-22) and (38-23) gives $a_1 = d_i$ and $a_2 = (d_j - d_i)/L$, or in matrix form:

$$\begin{Bmatrix} a_1 \\ a_2 \end{Bmatrix} = \begin{bmatrix} 1 & 0 \\ -\dfrac{1}{L} & \dfrac{1}{L} \end{bmatrix} \begin{Bmatrix} d_i \\ d_j \end{Bmatrix} \tag{38-24}$$

Substituting a's from Eq. (38-24) into Eq. (38-22) gives the displacement function in terms of nodal displacements:

$$u = \begin{bmatrix} 1 - \dfrac{x}{L} & \dfrac{x}{L} \end{bmatrix} \begin{Bmatrix} d_i \\ d_j \end{Bmatrix} = [N]\{d\} \tag{38-25}$$

Derivation of the strain-displacement function - The strain-displacement function can be presented in general terms as:

$$\{\epsilon\} = [B]\{d\}, \tag{38-26}$$

where $\{\epsilon\}$ = element strain matrix

[B] = element strain-displacement matrix.

The strain-displacement function is then obtained by taking a partial derivative of the displacement function. For a spar element (Fig. 38.10), we obtain:

$$\epsilon = \frac{du}{dx} = \begin{bmatrix} -\dfrac{1}{L} & \dfrac{1}{L} \end{bmatrix} \begin{Bmatrix} d_i \\ d_j \end{Bmatrix} = [B]\{d\} \tag{38-27}$$

Derivation of the stress-strain function - The stress-strain relationship may be expressed in general terms as:

$$\{\sigma\} = [C]\{\epsilon\} \tag{38-28}$$

where $\{\sigma\}$ = element stress matrix

[C] = element material stress-strain matrix.

In case of elastic deformation the material stress-strain matrix [C] is usually derived from the generalized Hooke's Law, which establishes the relationship between linear strains ϵ, shear strains γ, normal stresses σ, and shear stresses τ:

$$\begin{Bmatrix} \sigma_x \\ \sigma_y \\ \sigma_z \\ \tau_{xy} \\ \tau_{yz} \\ \tau_{zx} \end{Bmatrix} \begin{bmatrix} c_{11} & c_{12} & c_{13} & c_{14} & c_{15} & c_{16} \\ c_{21} & c_{22} & c_{23} & c_{24} & c_{25} & c_{26} \\ c_{31} & c_{32} & c_{33} & c_{34} & c_{35} & c_{36} \\ c_{41} & c_{42} & c_{43} & c_{44} & c_{45} & c_{46} \\ c_{51} & c_{52} & c_{53} & c_{54} & c_{55} & c_{56} \\ c_{61} & c_{62} & c_{63} & c_{64} & c_{65} & c_{66} \end{bmatrix} \begin{Bmatrix} \epsilon_x \\ \epsilon_y \\ \epsilon_z \\ \gamma_{xy} \\ \gamma_{yz} \\ \gamma_{zx} \end{Bmatrix} \qquad (38\text{-}29)$$

The coefficients in the matrix [C] are usually expressed in terms of Young's modulus E and Poisson's ratio ν. In case of elastic deformation of a spar element (Fig. 38.10) [c] = E. Therefore:

$$\{\sigma\} = E\{\epsilon\} = \begin{bmatrix} -\dfrac{E}{L} & \dfrac{E}{L} \end{bmatrix} \begin{Bmatrix} d_i \\ d_j \end{Bmatrix} \qquad (38\text{-}30)$$

Derivation of the element stiffness matrix - The element stiffness relates the nodal displacement to the applied nodal forces, i.e.:

$$[k]\{d\} = \{P\}, \qquad (38\text{-}31)$$

where [k] = element stiffness matrix

$\{P\}$ = element nodal force vector.

The stiffness matrix is derived by using the principle of minimum potential energy and can be expressed in general form as:

$$[k] = \int_V [B]^T [C][B] dV, \qquad (38\text{-}32)$$

where $[B]^T$ = transpose of matrix [B]

V = volume of an element.

For a spar element (Fig. 38.10) V = AL, therefore:

$$[k] = A \int_0^L \begin{Bmatrix} -1/L \\ 1/L \end{Bmatrix} [E] \begin{bmatrix} -\dfrac{1}{L} & \dfrac{1}{L} \end{bmatrix} dx = \dfrac{AE}{L} \begin{bmatrix} 1 & -1 \\ -1 & 1 \end{bmatrix} \qquad (38\text{-}33)$$

Assembly of the algebraic equations - This process includes the assembly of the overall stiffness matrix [K] from the individual element stiffness matrices [k] and the overall force vector $\{R\}$ from the element nodal force vectors $\{P\}$.

The basis for an assembly method is that the nodal interconnections require the displacements at a node to be the same for all elements adjacent to the node. The equilibrium relations between the overall stiffness matrix [K], the overall force vector $\{R\}$, and the overall displacement vector $\{r\}$ will be expressed as a set of simultaneous equations:

$$[K]\{r\} = \{R\} \qquad (38\text{-}34)$$

As an example, let us consider the assembly procedure for a system that consists of two spars (Fig. 38.11). Let the stiffnesses for these elements be equal to $k_1 = A_1E_1/L_1$ and $k_2 = A_2E_2/L_2$ respectively. Then the element matrices can be written in the form similar to Eq. (38-33):

$$\begin{array}{cc} d_1 & d_2 \\ \begin{bmatrix} k_1 & -k_1 \\ -k_1 & k_1 \end{bmatrix} \end{array} \text{ and } \begin{array}{cc} d_2 & d_3 \\ \begin{bmatrix} k_2 & -k_2 \\ -k_2 & k_2 \end{bmatrix} \end{array} \tag{38-35}$$

If the element matrices are expanded to the total number of degrees of freedom in the system, they become:

$$\begin{array}{ccc} d_1 & d_2 & d_3 \\ \begin{bmatrix} k_1 & -k_1 & 0 \\ -k_1 & k_1 & 0 \\ 0 & 0 & 0 \end{bmatrix} \end{array} \text{ and } \begin{array}{ccc} d_1 & d_2 & d_3 \\ \begin{bmatrix} 0 & 0 & 0 \\ 0 & k_2 & -k_2 \\ 0 & -k_2 & k_2 \end{bmatrix} \end{array} \tag{38-36}$$

The orders of matrices are identical so the like terms may be added by the use of superposition. The resulting assembled stiffness matrix is:

$$[k] = \begin{array}{ccc} d_1 & d_2 & d_3 \\ \begin{bmatrix} k_1 & -k_1 & 0 \\ -k_1 & k_1 + k_2 & -k_2 \\ 0 & -k_2 & k_2 \end{bmatrix} \end{array} \tag{38-37}$$

The equation for the spar system becomes:

$$\begin{bmatrix} k_1 & -k_1 & 0 \\ -k_1 & k_1 + k_2 & -k_2 \\ 0 & -k_2 & k_2 \end{bmatrix} \begin{Bmatrix} d_1 \\ d_2 \\ d_3 \end{Bmatrix} = \begin{Bmatrix} P_1 \\ P_2 \\ P_3 \end{Bmatrix} \tag{38-38}$$

Solution for the unknown parameters - The algebraic equations (38-34) are solved for the unknown displacements by considering both the geometric and force boundaries of the body. In linear equilibrium problems this becomes a relatively straightforward application of the matrix algebra technique:

$$\{r\} = [K]^{-1}\{R\} \tag{38-39}$$

Since the boundary conditions of the spar system shown in Fig. 38.11 are: $P_3 = P$ and $d_1 = 0$, therefore, we obtain the following two matrices from Eq. (38-38) by using a matrix partitioning:

$$\begin{bmatrix} k_1 + k_2 & -k_2 \\ -k_2 & k_2 \end{bmatrix} \begin{Bmatrix} d_2 \\ d_3 \end{Bmatrix} = \begin{Bmatrix} 0 \\ P \end{Bmatrix} \tag{38-40}$$

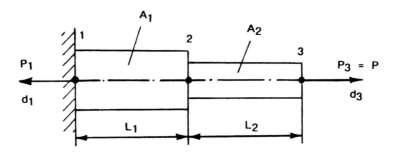

Fig. 38.11 Assembly of two spar elements.

and

$$[-k_1 \quad 0] \begin{Bmatrix} d_2 \\ d_3 \end{Bmatrix} = P_1 \tag{38-41}$$

Solving the system of the Eqs. (38-40) and (38-41) gives:

$$d_2 = \frac{P}{k_1} \tag{38-42}$$

$$d_3 = \frac{k_1 + k_2}{k_1 k_2} P \tag{38-43}$$

$$P_1 = -P \tag{38-44}$$

38.5 ACCURACY OF FINITE ELEMENT ANALYSIS MODELS

Application of finite element analysis allows to substantially improve accuracy of calculations of strip profile. Both 2-dimensional [20] and 3-dimensional [2, 21] computer models have been developed.

In application to strip profile calculations, the finite element analysis usually involves development of a finite element mesh for a complete system (Fig. 38.12) that includes work rolls, backup rolls, backup roll bearings, rolled strip, and the interface between them [2]. When developing this mesh, the number of elements or mesh size can be determined by the required accuracy of the strip profile calculations. However, this number must also be evaluated with regard to calculation time and, therefore, cost. The following two parameters may be utilized to evaluate the accuracy of the calculations:

    a) relative center crown error

    b) relative roll flattening effect error.

The relative center crown error e_c is expressed as:

$$e_c = \frac{c_n - c_\infty}{h_2} \times 100\% \tag{38-45}$$

where c_n, c_∞ = center crown corresponding to n and infinite number of elements in the finite element
          mesh respectively

      h_2 = exit strip gauge.

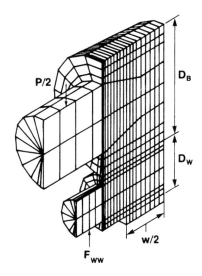

Fig. 38.12 Three-dimensional finite element analysis mesh for a 4-high mill roll-strip system. From Ginzburg et al (1987).

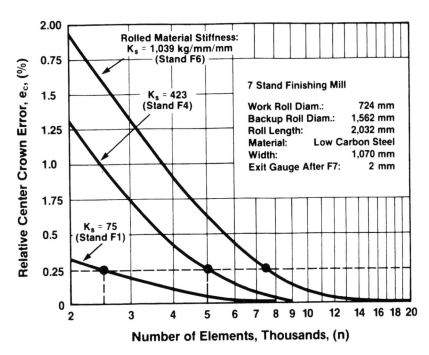

Fig. 38.13 Effect of the number of elements in the finite element analysis mesh representing the 4-high mill stand on accuracy of the strip crown calculations. From Ginzburg et al (1987).

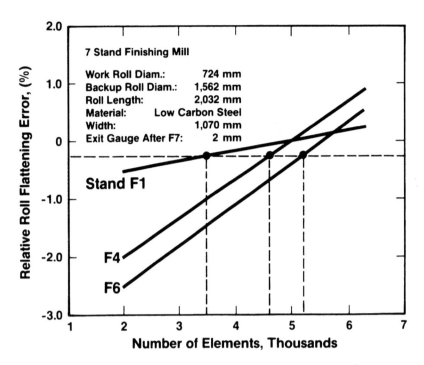

Fig. 38.14 Effect of the number of elements in the finite element analysis mesh representing the 4-high mill stand on accuracy of the calculations of roll flattening. From Ginzburg et al (1987).

The values of c_∞ are found by assuming the exponential law for diminishing the relative center crown error with an increase in the number of elements.

As can be seen from Fig. 38.13, the number of elements required to obtain the same accuracy grows with the increase in stiffness of rolled material. In order to reduce the error to 0.25%, the mesh size representing finishing stands F1, F4, and F6 of hot strip mill must be as high as 2500, 5000, and 7500 elements respectively.

The relative roll flattening effect error e_F is calculated from the following equation:

$$e_F = \frac{\delta_n - \delta_F}{h_2} \times 100\% \tag{38-46}$$

where δ_n = compressive decrease between the centers of work roll and backup roll as obtained from finite element analysis with **n** elements

δ_F = compressive decrease between the centers of work roll and backup roll as obtained from Eqs. (38-12) and (38-13) for roll flattening effect.

The number of elements required to reduce the relative roll flattening effect error to 0.25% for stands F1, F4, and F6 is equal to 3500, 4600, and 5200 respectively (Fig. 38.14).

Major drawback of increasing the number of elements is an increase in time for computation. Savings in the required number of elements may be obtained by converting an asymmetrical model into a symmetrical one as shown in Fig. 38.15.

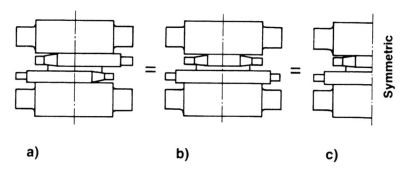

Fig. 38.15 Converting an asymmetrical model (a) into a symmetrical model (b, c). From Ginzburg et al (1987).

REFERENCES

1. K. Saxl, "Transverse Gauge Variation in Strip and Sheet Rolling", Proceedings of the Institution of Mechanical Engineers, Vol. 172, 1958, pp. 727-742.

2. V.B. Ginzburg, F. Bakhtar and C. Tabone, "Application of the Computer Model ROLLFLEX™ for Analysis of the Strip Profile and Shape in Rolling Mills", Paper presented at the AISE Annual Convention, Pittsburgh, PA, Sept. 21-24, 1987.

3. R.J. Roark, Formulas for Stress and Strain, Fourth Edition, McGraw-Hill, Inc., New York, pp. 96-98, 155-157 (1965).

4. S. Timoshenko, Strength of Materials, Part 1, Elementary Theory and Problems, Third Edition, D. Van Nostrand Company, Inc., Princeton, New Jersey, pp. 137-139 (1958).

5. E.C. Larke, The Rolling of Strip, Sheet and Plate, Science Paperbacks and Chapman and Hall, Ltd., London, pp. 71-126 (1967).

6. M.D. Stone and R. Gray, "Theory and Practical Aspects in Crown Control", AISE Yearly Proceedings, 1965, pp. 657-667.

7. S. Timoshenko, Strength of Materials, Part 2, Advanced Theory and Problems, Third Edition, D. Van Nostrand Company, Inc., Princeton, New Jersey, pp. 1-4 (1958).

8. A. Foppl, Technische Mechanik, Fourth Edition, Vol. 5, p. 350.

9. J.V. Poplawski, et al, "Mathematical Modeling System for Cold Tandem Mills: Influence of Rolling and Roll Bending Forces on Strip Crown", Proceedings of 1981 AIME Mechanical Working and Steel Processing Conference, Pittsburgh, Pa., Oct. 27-30, 1981.

10. J.F. McDermott, "Computer Roll Deflection Program for Predicting Strip Crown", AISE Year Book, 1984, pp. 301-305.

11. K.N. Shohet and N.A. Townsend, "Roll Bending Methods of Crown Control in Four-High Plate Mills", Journal of the Iron and Steel Institute, Nov. 1968, pp. 1088-1098.

12. R.M. Guo, "Computer Model Simulation of Strip Crown and Shape Control", Iron and Steel Engineer, Nov. 1986, pp. 35-42.

13. F. Hollander and A.G. Reinen, "Automatic Shape Control-Hoogovens' 88-in. Hot Strip Mill", AISE Year Book, 1976, pp. 135-143.

14. W. Wilms, et al, "Profile and Flatness Control in Hot Strip Mills", Metallurgical Plant and Technology, Vol. 8, No. 6, 1985, pp. 74-90.

15. E. Kersting and H. Teichert, "The UPC Technology: Modernization of Hot Strip Mills, Specifically in Regard to Profile Control", Proceedings of the 4th International Steel Rolling Conference: The Science and Technology of Flat Rolling, Vol. 1, Deauville, France, June 1-3, 1987, pp. A.19.1-A.19.7.

16. T. Nakanishi, et al, "Application of Work Roll Shift Mill HCW Mill to Hot Strip and Plate Rolling", Hitachi Review, Vol. 34, No. 4, 1985, pp. 153-160.

17. C.S. Desai and J.F. Abel, Introduction to the Finite Element Method, Van Nostrand Reinhold Company, New York, pp. 3-88 (1972).

18. R.H. Gallagher, Finite Element Analysis Fundamentals, Prentice-Hall, Inc., Englewood Cliffs, New Jersey, pp. 1-104 (1975).

19. J.M. Steele, How to Select an Element, Computer-Aided Engineering, May/June 1984, pp. 92,93.

20. C. Xianlin and Z. Jiaxiang, "A Specialized Finite Element Model for Investigating Controlling Factors Affecting Behavior of Rolls and Strip Flatness", Proceedings of the 4th International Steel Rolling Conference: The Science and Technology of Flat Rolling, Vol. 2, Deauville, France, June 1-3, 1987, pp. E.4.1 - E.4.7.

21. J. Kihara, et al, "Application of BEM to Calculation of the Roll Profile in Flat Rolling", Proceedings of the 4th International Steel Rolling Conference: The Science and Technology of Flat Rolling, Vol. 2, Deauville, France, June 1-3, 1987, pp. E.1.1 - E.1.12.

39

Roll Contour and Strip Flatness Models

39.1 ROLL THERMAL CONTOUR MODELS

Roll thermal contour is the result of its thermal expansion during rolling. This is a complex heat transfer problem which must take into account the following factors [1]:

1. Heat content of the strip prior to rolling.
2. Heat generated in the strip in the arc of contact due to work of deformation and friction.
3. Heat conducted into the roll in the arc of contact.
4. Heat removed from the roll surface by the coolant.
5. Heat carried away from the deformation zone by the strip.

The roll temperature consists of basic and periodic components [2]. The basic component is equal to the average roll temperature. The periodic component has a cycle time equal to the time required for one revolution of the work roll. Since the periodic component is localized at the roll surface (Fig. 39.1), the roll thermal crown is usually determined taking into account the basic component only.

Both two-dimensional and three-dimensional roll thermal contour models were developed. Some of them are briefly described below.

39.2 TWO-DIMENSIONAL ROLL THERMAL CONTOUR MODEL

In the model described by Sumi [3], the axial and radial temperature distribution in the work roll is obtained from the two-dimensional heat conduction equation:

$$c\rho \left(\frac{\delta T}{\delta t} \right) = \frac{\lambda}{r} \frac{\delta}{\delta r} \left(r \frac{\delta T}{\delta r} \right) + \lambda \frac{\delta^2 T}{\delta z^2} \tag{39-1}$$

The boundary condition at the roll surface is:

$$- \lambda \frac{\delta T}{\delta r} = h_w (T - T_w) - q \tag{39-2}$$

and at the roll edge:

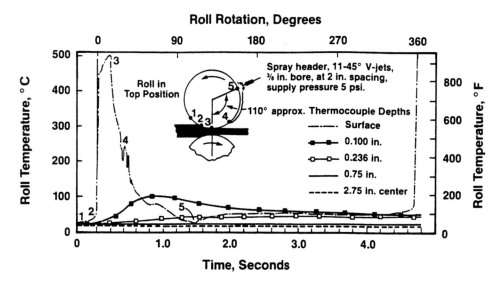

Fig. 39.1 Variation of roll temperature during first working revolution. (From Parker and Baker, AISE Year Book, 1972. Copyright AISE, Pittsburgh, Pennsylvania. Reprinted with permission.)

$$- \lambda \frac{\delta T}{\delta z} = h_a (T - T_a) \qquad\qquad\qquad\qquad (39\text{-}3)$$

where c, ρ, λ = specific heat, density, and thermal conductivity of the roll respectively

T = roll temperature at the point with axial coordinate, z, and radial coordinate, r

t = time

h_w, h_a = heat transfer coefficient of roll coolant and atmosphere respectively

T_w, T_a = temperature of coolant and atmosphere respectively

q = heat flux from the workpiece to the roll.

Equations (39-1) through (39-3) allow the calculation of the temperature build-up at any point on the roll and thus determine the roll thermal crown. According to the calculations performed by Cerni [1], the roll temperature at the different distances r from the roll center increases exponentially with time (Fig. 39.2).

39.3 THREE-DIMENSIONAL ROLL THERMAL CONTOUR MODEL

The three-dimensional roll thermal contour model developed by Research Laboratory of U.S. Steel Corporation [4-7] is based on the following assumptions:

1. Mean roll temperature is considered to play a major role in diameter expansion of the roll.

2. Axial conduction is considered so the variation of the roll expansion along the roll barrel is taken into account.

3. Both heat input from hot slab and heat removal by water sprays are assumed to act uniformly around the roll circumference at any given circular cross-section of the roll.

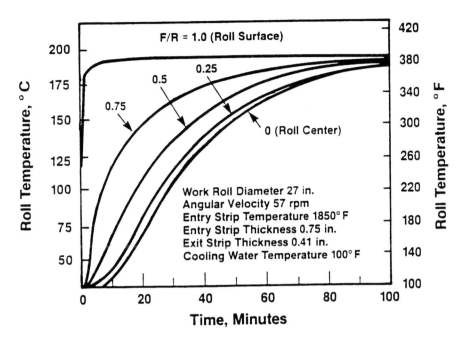

Fig. 39.2 Roll temperature build-up during hot rolling. (From Cerni, et al, AISE Year Book, 1963. Copyright AISE, Pittsburgh, Pennsylvania. Reprinted with permission.)

4. Constant values of the heat transfer coefficients are adopted between the slab and the roll throughout the slab width as well as between the water spray and the roll.

5. The heat transfer from the slab to the roll is assumed to be by conduction with the contact conductance coefficient selected as a function of a mean roll temperature.

In the model, the roll is divided into two separate regions (Fig. 39.3). In the Region 1, both the slab and water sprays participate in the heat transfer, whereas only the water sprays act in Region 2.

Based on the law of conservation of energy, the following differential equations for axially symmetric time-varying temperature T(z, t) have been derived in cylindrical coordinates.

For Region 1:

$$\frac{\delta T_1}{\delta t} = \alpha \frac{\delta^2 T_1}{\delta z^2} + \frac{2h_t \phi}{\rho c \pi d} (T_s - T_1) - \frac{2h_c \psi}{\rho c \pi d} (T_1 - T_w) \qquad (39\text{-}4)$$

For Region 2:

$$\frac{\delta T_2}{\delta t} = \alpha \frac{\delta^2 T_2}{\delta z^2} - \frac{2h_c \psi}{\rho c \pi d} (T_2 - T_w) \qquad (39\text{-}5)$$

where T_1, T_2 = work roll temperatures in Region 1 and 2 respectively, $^\circ$F

T_w = cooling water temperature, $^\circ$F

T_s = slab temperature, $^\circ$F

h_t = coefficient of thermal contact conductance, Btu/hr/ft^2/$^\circ$F

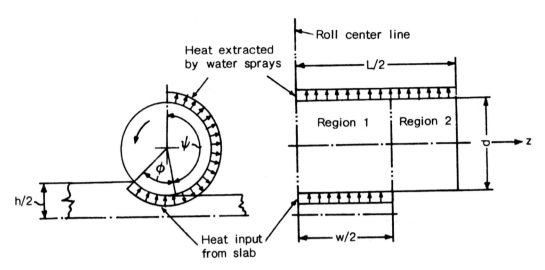

Fig. 39.3 Heat transfer in the slab-roll-water sprays system. Adapted from Pallone (1983).

ϕ = angle of roll contact with slab, rad

ψ = angle of convective cooling, rad

h_c = coefficient of forced convection, Btu/hr/ft^2/^{o}F

α = work roll thermal diffusivity, ft^2/hr

c = work roll specific heat, Btu/lb/^{o}F

ρ = work roll density, lb/ft^3

d = work roll diameter, ft

t = time, hr

z = axial distance from the roll center line, ft.

The common boundary conditions at the interface, where Regions 1 and 2 meet, are

$$\frac{\delta T_1(w/2,t)}{\delta z} = \frac{\delta T_2(w/2,t)}{\delta z} \tag{39-6}$$

and

$$T_1(w/2,\ t) = T_2(w/2,\ t), \tag{39-7}$$

where w = slab width, ft.

Since the temperature gradient is symmetrical about the roll center line, we obtain

$$\frac{\delta T_1(0,t)}{\delta z} = 0 \tag{39-8}$$

Before rolling begins, the roll temperature is assumed to be equal to the spray water temperature T_w. It is also assumed that the ends of the roll are always maintained at temperature T_w. These conditions are described by

$$T_1(z, 0) = T_2(z, 0) = T(L/2, t) = T_w, \tag{39-9}$$

where L = work roll barrel length, ft.

Solution of the Eqs. (39-4)-(39-9) allows one to define the mean temperature distribution along the barrel length of the work roll. This temperature is then converted to diametral expansion by the formula

$$\Delta d = \alpha d_0(T_{1,2} - T_w), \tag{39-10}$$

where d_0 = initial work roll diameter, ft

$\quad \Delta d$ = work roll thermal expansion.

$\quad \alpha$ = thermal expansion coefficient, $1/^\circ$ F.

39.4 EFFECT OF ROLL THERMAL CONTOUR ON STRIP CROWN

Two major components of the variation of the roll thermal contour have to be considered:

    a) thermal build-up during rolling of each coil

    b) changes in the roll thermal contour from coil to coil.

The on-line measurements conducted by Knox [8] have shown that the roll thermal contour changes significantly during rolling of each roll. The shape of this contour is a function of the strip width; narrower strip produces an 'inverted bell' shape, whereas the wide strip produces a shape that approaches a semi-circular form as shown in Fig. 39.4. It was also shown that the roll thermal crown reduces progressively downstream of the hot strip mill as illustrated in Fig. 39.5. This can be attributed

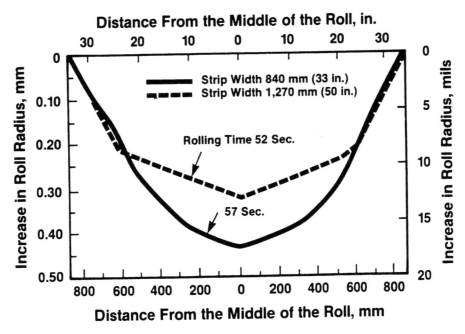

Fig. 39.4 Increase in the roll radius at stand F2 of the hot strip mill after rolling of one coil. Data from Knox (1987).

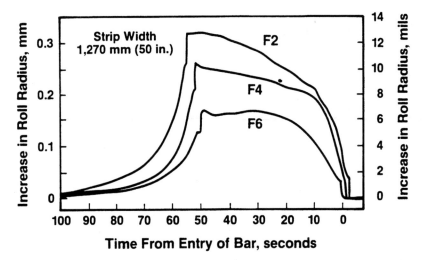

Fig. 39.5 Increase in the roll radius at stands F2, F4, and F6 of the 68-inch hot strip mill during rolling of one coil. Data from Knox (1987).

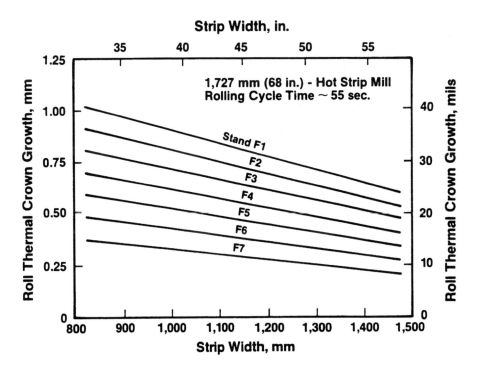

Fig. 39.6 Maximum increase in the roll thermal crown during hot rolling of one coil. From Ginzburg, et al (1987). Copyright AISE, Pittsburgh, Pennsylvania. Reprinted with permission.

to both the lower strip temperature and the lower energy of deformation in the last mill stands. The graphs shown in Fig. 39.6 allow one to determine maximum roll thermal crown as a function of mill stand and strip width. These graphs were derived by extrapolating the test data provided by Knox.

The effect of changes in roll thermal contour from coil to coil on the strip crown was investigated by Somers [9]. His study shows that during typical rolling cycles at the hot strip mill, the strip crown decreased rapidly and almost linearly during rolling the first 20 to 40 coils with newly redressed work rolls in the finishing mill stands (Fig. 39.7). After the initial loss of strip center crown takes place, smaller but rapidly varying changes in strip center crown occur. These changes are attributed to a variety of causes such as mill delays, changes in rolling rate, variations in strip and cooling water temperature, changes in mill drafting, etc., but all are related to the thermal aspects of rolling.

39.5 ROLL WEAR MODELS

Both the work roll wear (Fig. 39.8) and the backup roll wear may affect the strip profile. According to Oike [10], the work roll wear on diameter C_m can be calculated from the following equation:

$$C_m = \alpha \sum_{i=1}^{n} \left(\frac{P_i}{w_i l_i} \right)^a (r_i l_i)^b \frac{L_i}{\pi D} \delta_i(z) \tag{39-11}$$

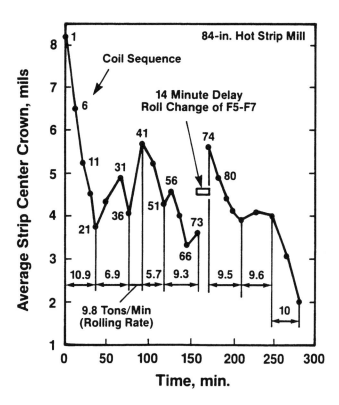

Fig. 39.7 Variation of strip center crown during hot rolling of a tin-mill product cycle. Adapted from Somers (1980).

where $\delta_i(z) = 1$ when $0 < z \le w_i/2$

$\quad\quad \delta_i(z) = 0$ when $z > w_i/2$

$\quad\quad\quad\quad$ i = rolling pass number

$\quad\quad\quad\quad$ n = total number of rolling passes

$\quad\quad\quad\quad$ P = roll separating force

$\quad\quad\quad\quad$ w = strip width

$\quad\quad\quad\quad$ l = roll contact length

$\quad\quad\quad\quad$ r = reduction

$\quad\quad\quad\quad$ L = exit strip length

$\quad\quad\quad\quad$ D = work roll diameter

α, a, b = empirical coefficients which depend on the roll material, strip temperature, roll bite
$\quad\quad\quad\quad\quad$ lubrication, roll coolant, etc.

Somers [11] has found, on a simplified basis, that the work roll wear rate varied inversely with the work roll diameter. He also states that the wear rate for the work rolls of the first two stands of a hot strip mill differ from the remaining stands. The average rate of radial wear for grain iron rolls during a typical tin mill cycle is expressed by:

$$W_{1,2} = 99.25 - 3.33D \quad\quad \text{for F1 and F2} \tag{39-12}$$

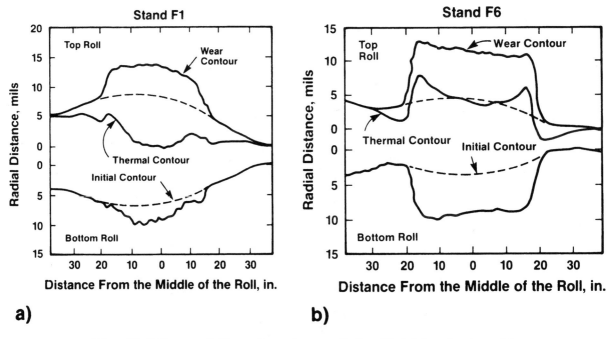

Fig. 39.8 Wear and thermal contours of the finishing train rolls for a tin mill: a) stand F1; b) stand F6. (From Somers, et al, AISE Year Book, 1984. Copyright AISE, Pittsburgh, Pennsylvania. Reprinted with permission.

$$W_{3-7} = 122.47 - 4.26D \quad \text{for F3 to F7} \tag{39-13}$$

where W = average rate of radial wear, mils/10^5 ft. of strip rolled

   D = work roll diameter, in.

Studies conducted by Sibakin [12] in hot strip mills have shown that no direct relationship between the strip crown and the roll wear could be established. The distribution of the wear may be responsible for the absence of such a relationship. However, the latest studies have shown that the wear greatly affects the strip edge profile and the edge drop [13].

Another outcome from roll wear is a deterioration of surface quality of the rolled strip. Nakanishi [14] shows that the amount of roll wear C_e at the edge of the strip is (Fig. 39.9):

$$C_e = kC_m, \tag{39-14}$$

where k = roll wear increase coefficient at the strip edges against the roll wear at the middle of the strip, C_m.

When a = 10 mm and b = 50 mm, k = 1.3.

The local roll wear, that affects the strip surface, is defined as:

$$C_r = C_e - C_m \tag{39-15}$$

It was established that when the local roll wear C_r exceeds 10μm, it will leave marks on the strip surface. The imprint that the roll wear leaves on a strip varies with the type of roll and also depends on the stand number. The local roll wear imprinted on the strip C_s may be expressed as:

$$C_s = k_w(2C_r), \tag{39-16}$$

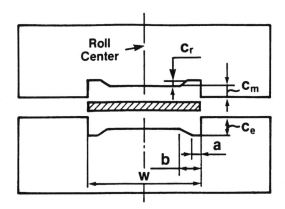

Fig. 39.9 Modeling of the wear profile of the work rolls.
(From Nakanishi, et al. Reproduced from HITACHI REVIEW,
Vol. 34, No. 4, 1985, by courtesy of Hitachi, Ltd., Japan.)

where k_w = imprinting coefficient of roll wear to strip

$2C_r$ = local roll wear per roll diameter.

The experimental values for C_s, k_w, and $2C_r$ are shown in Fig. 39.10 for a 7-stand finishing train. The imprinting coefficient k_w is close to zero at stands F1 and F2 and increases almost linearly at stands F3 through F7. The local roll wear C_r is very small at stands F1 and F2, but increases substantially and attains its maximum value at stands F5. The resulting local roll wear imprinted on the strip C_s, is small at the stands F1 through F3 and then increases sharply, reaching its maximum value at stand F6.

Although the backup roll does not contact the strip directly, it influences the strip crown through its effect on the roll deflection. Sibakin [12] had derived the following empirical formula based on statistical analysis of measurement of the strip crown on 367 coils:

$$c = 0.37 - 0.0663t_s + 0.0459w + 0.108n - 0.0034n^2, \tag{39-17}$$

where c = strip crown, mils

t_s = strip contact time during 30 minutes preceding measurement, min

w = strip width, in.

n = number of turns rolled on the backup rolls.

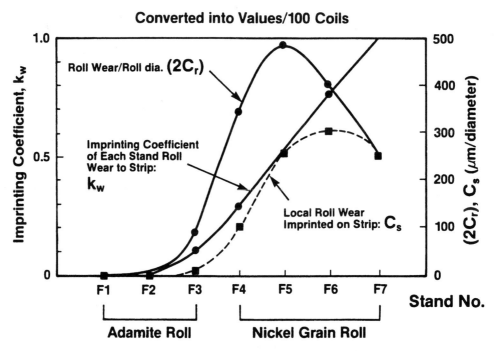

Fig. 39.10 Imprint of the roll wear on the strip in the hot strip finishing train. (From Nakanishi, et al. Reproduced from HITACHI REVIEW, Vol. 34, No. 4, 1985, by courtesy of Hitachi, Ltd., Japan.)

39.6 STRIP FLATNESS MODELS

Bad shape of rolled strip is caused by differential elongation across its width. This elongation is directly related to the strip crown change during the pass reduction:

$$\Delta \left(\frac{c}{h} \right) = \frac{c_1}{h_1} - \frac{c_2}{h_2} \tag{39-18}$$

where $\Delta (\frac{C}{h})$ = change in relative strip crown

   c_1, c_2 = entry and exit strip crowns respectively

   h_1, h_2 = entry and exit strip thicknesses respectively.

In the strip flatness model developed by Shohet and Townsend [13], and further studied by Somers [11], good flatness during hot rolling will result, if the change in relative strip crown is within the following limits (Fig. 39.11):

$$-80(\frac{h_2}{w})^a < \Delta(\frac{c}{h}) < 40(\frac{h_2}{w})^b \tag{39-19}$$

where a, b = constants.

   For low carbon steel, a = b = 1.86.

   In the strip flatness model that is described by Takashima, et al [15], the strip shape is expressed

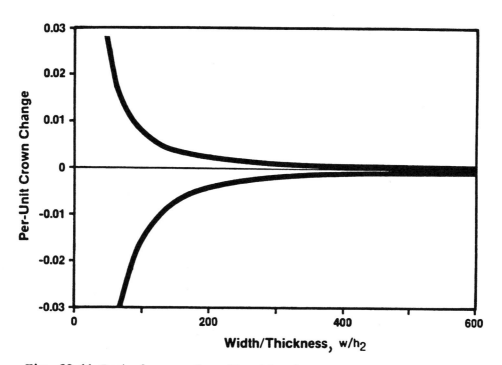

Fig. 39.11 Typical range for allowable change in crown-to-thickness ratio for hot rolling of low carbon steel.

as follows:

$$\lambda_o = \pm\, a\, \sqrt{\frac{c_2}{h_2} \times 100 - b} \qquad\qquad (39\text{-}20)$$

where a = variation flatness coefficient

b = flatness dead point.

In Eq. (39-20), if $b = b_1$ denotes the flatness dead point in the edge wave zone, and $b = b_2$ denotes that in the center buckle zone, then the region expressed by $B = b_1 - b_2$, is defined as the

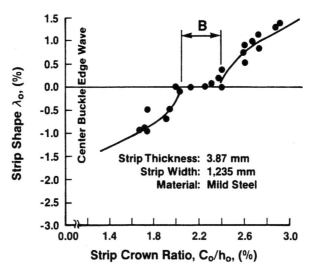

Fig. 39.12 Relationship between strip flatness and relative exit center crown in hot rolling of mild steel. Adapted from Takashima, et al (1979).

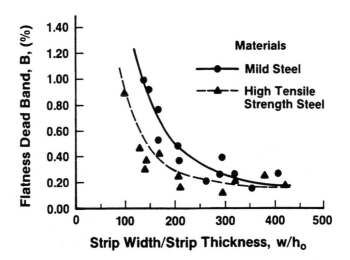

Fig. 39.13 Relationship between flatness dead band and strip width-to-thickness ratio in hot rolling of different types of steel. Adapted from Takashima, et al (1979).

flatness dead band (Fig. 39.12). The flatness dead band represents a range of the relative exit strip crown within which the strip shape does not change. This range is decreased with the increase in the width-to-thickness ratio and resistance to deformation of rolled material (Fig. 39.13).

REFERENCES

1. S. Cerni, et al, "Temperatures and Thermal Stresses in the Rolling of Metal Strip", AISE Yearly Proceedings, 1963, pp. 717-725.

2. P.G. Stevens, et al, "Increasing Work-Roll Life by Improved Roll-Cooling Practice", Journal of Iron and Steel Institute, Jan 1971, pp. 1-11.

3. H. Sumi, et al, "A Numerical Model and Control of Plate Crown in the Hot Strip or Plate Rolling", Advanced Technology of Plasticity, Vol. 2, 1984, pp. 1360-1365.

4. G.T. Pallone, "Transient Temperature Distribution in Work Rolls During Hot Rolling of Sheet and Strip", AISE Year Book, 1983, pp. 496-501.

5. E.J. Patula, "Steady-State Temperature Distribution in a Rotating Roll Subject to Surface Heat Fluxes and Convective Cooling", ASME Journal of Heat Transfer, Feb. 1981, p. 36.

6. H.S. Carslaw and J.C. Jaeger, Conduction of Heat in Solids, Oxford University Press (1959).

7. V.S. Arpaci, Conduction Heat Transfer, Addison-Wesley Publishing Co. (1966).

8. T.J. Knox and J.M. Moore, "Improving Dimensional Control in the Hot Mill", Proceedings of the 4th International Steel Rolling Conference: The Science and Technology of Flat Rolling, Vol.1, Deauville, France, June 1-3, 1987, pp. A.23.1-A.23.13.

9. R.R. Somers, "Effect of Hot-and Cold-Rolling Operations on Strip Crown and Feather Edge", Proceedings of the International Conference on Steel Rolling: Science and Technology of Flat Rolled Products, Vol. 1, Tokyo, Japan, Sept. 29-Oct. 4, 1980, pp. 701-712.

10. Y. Oike, et al, Tetsu-to-Hagane, Vol. 63, No. 4, 1977, p. S222.

11. R.R. Somers, et al, "Verification and Applications of a Model for Predicting Hot Strip Profile, Crown and Flatness", AISE Year Book, 1984, pp. 441-450.

12. J.G. Sibakin, et al, "Factors Affecting Strip Profile in Cold and Hot Strip Mill", Flat Rolled Products: Rolling and Treatment, Vol. 1, 1959, pp. 3-45.

13. K.N. Shohet and N.A. Townsend, "Flatness Control in Plate Rolling", Journal of the Iron and Steel Institute, Oct. 1971, pp. 769-775.

14. T. Nakanishi, et al, "Application of Work Roll Shift Mill HCW Mill to Hot Strip and Plate Rolling", Hitachi Review, Vol. 34, No. 4, 1985, pp. 153-160.

15. Y. Takashima, et al, "Studies on Strip Control for Hot Strip Rolling-Double Chock Work Roll Bending System (DC-WRB)", IHI Engineering Review, Vol. 12, No. 3, Oct. 1979, pp. 28-34.

40

Selection of Strip Profile and Flatness Actuators

40.1 STRATEGY IN SELECTION OF ACTUATORS

Various approaches have been proposed to select appropriate strip profile and flatness actuators. One of the approaches [1, 2] includes following steps:

1. Derivation of the original strip crown range
2. Derivation of the flatness cones
3. Improving the compatibility of the original strip crown range with the flatness cones
4. Selection of the actuators.

This strategy involves analysis of the existing conditions in the mill in respect to the strip profile and flatness control and determining its deficiencies. Modifications of the existing operating practice are then derived to improve performance of the existing strip profile and flatness control system.

If after these modifications some further improvements are still required, then the appropriate actuators are selected to achieve this goal. The process of selection of the actuators involves evaluation of the contribution by each disturbance parameter to the original strip crown range. These parameters can be divided into two categories: dynamic and static.

The **dynamic disturbance parameters,** such as the variations of roll thermal crown, roll wear, and rolling load, may change during rolling the same coil. The **static disturbance parameters,** such as roll diameters, strip width and material grade, may change only from coil to coil.

40.2 ORIGINAL STRIP CROWN RANGE

An examination of the original strip crown range at a 4-high mill stand F1 of a typical 7-stand finishing train illustrates the concept of the original crown control range. The term 'original strip crown range' is applied to the case when all rolls are flat and also when no corrective adjustments of the strip crown are introduced in the mill.

The mill stand data used in the 3-dimensional finite element analysis (Fig. 40.1) are shown in Table 40.1. Two materials have been selected as examples: low carbon steel (LCS) and high-strength low-alloy (HSLA) steel, which is assumed to have a resistance to deformation 50% higher than low carbon steel.

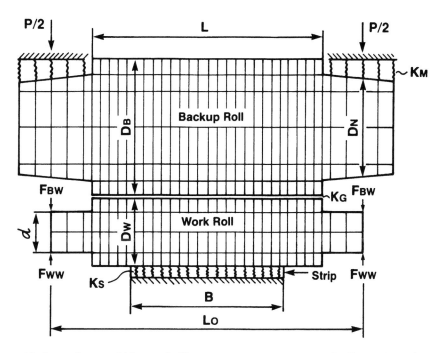

Fig. 40.1 Strip profile and flatness actuator as a 3-dimensional spring-beam system. (From Ginzburg, Iron and Steel Engineer, 1987. Copyright AISE, Pittsburgh, Pennsylvania. Reprinted with permission.)

Table 40.1 Mill stand data used in finite element analysis.

Parameter	Symbol	Value
Backup roll outer diameter (min/max, mm)	D_B	1486/1562
Roll force length, mm	L	2032
Backup roll neck diameter, mm	D_N	940
Work roll diameter, mm	D_W	724
Work roll neck diameter, mm	d	483
Distance between bearings, mm	L_O	1565
Rolls modulus of elasticity, tons/mm²	E	21.14
Mill housing spring, tons/mm	K_M	430
Roll bending force per chock, tons:		
between work and backup roll	F_{WB}	150
between work rolls	F_{WW}	150

For each of these materials, the strip crown range has been calculated for both head and tail end of the strip. The pass schedule, strip stiffness, desired strip crown and the effective thermal crown are summarized in Table 40.2.

The calculated strip crown range, when both the work rolls and the backup rolls do not have mechanical crown, is illustrated in Fig. 40.2. The upper limit of the strip crown range A corresponds to

Table 40.2 Workpiece data used in finite element analysis.

Parameter	Units	Stand number						
		F1	F2	F3	F4	F5	F6	F7
Entry thickness,	mm	33.0						
Exit thickness,	mm	18.16	10.54	6.35	4.06	2.85	2.29	2.00
Strip springs,	kg/mm/mm							
LCS: Head end		75.0	138	245	423	694	1039	1380
Tail end		87.5	156	270	454	726	1061	1380
HSLA: Head end		112.5	207	368	635	1041	1559	2070
Tail end		131.2	234	405	681	1089	1592	2070
Exit crown (min/max),	μm							0/40
Effective thermal crown								
at B = 1830 mm,	μm	150	150	150	100	100	90	75

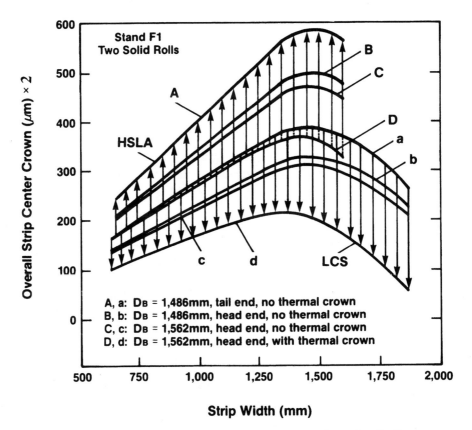

Fig. 40.2 Typical original strip crown range for stand F1.
(From Ginzburg, Iron and Steel Engineer, 1987. Copyright AISE,
Pittsburgh, Pennsylvania. Reprinted with permission.)

the rolling of the tail end of high-strength low-alloy steel with minimum backup roll diameter and with no thermal crown. The lower limit of the strip crown range d corresponds to the rolling of the head end of low carbon steel with maximum backup roll diameter and with thermal crown.

40.3 STRIP FLATNESS CONES

The shape cones are derived for both minimum and maximum exit strip crowns. The shape cones determine the upper C_U and lower C_L entry crown limits within which no disturbance in strip flatness occurs. These limits can be derived from the strip flatness criterion expressed by Eqs. (39-18) and (39-19):

$$C_{U1} = h_1 \left[\frac{c_2}{h_2} + 40 \left(\frac{h_2}{w} \right)^b \right] \qquad (40\text{-}1)$$

$$C_{L1} = h_1 \left[\frac{c_2}{h_2} - 80 \left(\frac{h_2}{w} \right)^a \right] \qquad (40\text{-}2)$$

where a, b = constants; for steel, a = b = 1.86.

Knowing the desired crown after the last stand, the flatness cones for stand F1 can be calculated from Eqs. (40-1) and (40-2), using the reduction schedule shown in Table 40.2.

40.4 COMPATIBILITY OF ORIGINAL STRIP CROWN RANGE

Figure 40.3 depicts in broken lines the shape cone for stand F1, which corresponds to a value of 40 -m for the strip crown after stand F7 (i.e., 2%). Superimposed on this shape cone is the previously calculated original strip crown range for stand F1 (Fig. 40.2).

When the desired relative exit crown is equal to 2%, a high degree of compatibility is provided between the original strip crown range and the shape cone (Fig. 40.3a). This may require only a small or no adjustment of the strip crown at stand F1 to roll the desired crown. However, this is not the case when the desired exit strip crown after stand F7 is equal to zero, because most of the original strip crown range lies beyond the shape cone (Fig. 40.3b).

The compatibility of the original strip crown range with the flatness cones may be improved without an introduction of additional strip profile and shape actuators. This can be achieved by optimizing both the reduction schedules and roll grinding practice. If these measures are not sufficient to meet the desired strip profile and shape tolerances, then the application of specific actuators must be considered.

40.5 FUNCTIONAL CLASSIFICATION OF THE ACTUATORS

The strip profile and flatness actuators may be classified by function into three groups:

1. **Dynamic modifiers of the strip profile** - These actuators allow one to adjust the strip profile and shape in the process of rolling the same coil. The examples of these actuators are the roll bending systems, flexible body crown rolls, and flexible edge adjustable crown rolls.

2. **Static modifiers of the crown control range** - These actuators allow one to make adjustment during the gap time between rolling the coils. The examples of these actuators are the axially shifted noncylindrical rolls and roll crossing systems.

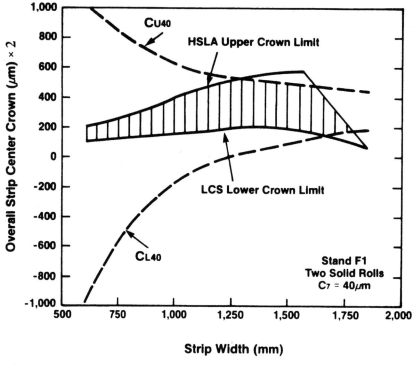

a)

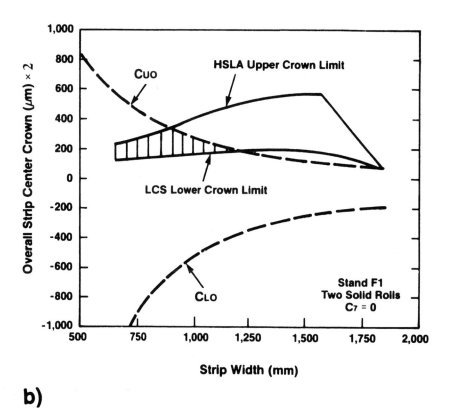

b)

Fig. 40.3 Compatibility of original strip crown range for stand
F1 with the flatness cones for different exit crowns after stand
F7 equal to: a) $c_7 = 40 \mu m$, b) $c_7 = 0$. (From Ginzburg, Iron and Steel
Engineer, 1987. Copyright AISE, Pittsburgh, Pennsylvania. Reprinted
with permission.)

3. Optimizers of the controllability of the strip profile - These actuators have a dual function. They reduce the adverse effect of the disturbance parameters, such as variation of the rolling load, and at the same time enhance the crown control range of the roll bending system. The examples of these actuators are the flexible edge backup rolls and the axially shifted intermediate rolls.

40.6 SELECTING THE TYPE OF ACTUATOR

The process of selecting the type of actuator may begin with examining the dynamic modifiers, and particularly, roll bending systems. In some cases the selected system may compensate for the variation of the strip profile caused by both dynamic and static disturbance parameters.

If this is not possible, the next step would be to consider the use of static modifiers of the crown control range. Unfortunately, the application of static modifiers, utilizing the roll shifting of noncylindrical rolls or roll crossing systems, is not always acceptable, especially when roll shifting is used to reduce the roll wear or edge drop. In the latter case, it is appropriate to consider the application of actuators which optimize the controllability of the strip profile, such as flexible edge backup rolls, or the intermediate roll shifting system. This arrangement makes the axially shifted cylindrical rolls available for reduction of either the roll wear or edge drop.

One of the main criteria in selecting the strip profile and flatness actuators is to provide a sufficient crown control range to compensate for possible variations in strip profile. This can be achieved by one or a combination of the following three methods:

a) increasing the crown control range
b) increasing the crown control rate
c) reducing the disturbance rate of the parameters affecting the strip profile.

40.7 SELECTION OF SELF-COMPENSATING BACKUP ROLLS

The concept of application of Self-Compensating (SC) backup rolls (Fig. 37.21) is illustrated in Fig. 40.4a by comparing two strip crown ranges. The first range relates to stand F1 with two conventional solid backup rolls. The second range relates to stand F1 with two self-compensating SC rolls.

The crossover point A corresponds to the material width that produces a complete self-compensation of the backup roll deflection. That means, if both backup and work rolls are flat, the strip of this specific width will also be flat under any rolling load. When the material width is less than indicated by cross-over point A, the backup roll deflection will be undercompensated. In the opposite case, when the material width is greater than indicated by point A, the backup roll deflection will be overcompensated. When no special measures are taken, this crossover point lies within the strip width range from 75 to 80% of the roll body length which is equal to 2032 mm (80 in.).

One of the simplest ways to avoid overcompensation is to utilize only one SC roll in combination with a regular solid backup roll (Fig. 40.4b). This arrangement still allows a substantial reduction in the strip crown variation due to rolling force. Another way of reducing the overcompensation is to utilize the SC rolls with variable contact length between inner surface of the sleeve and the arbor [3].

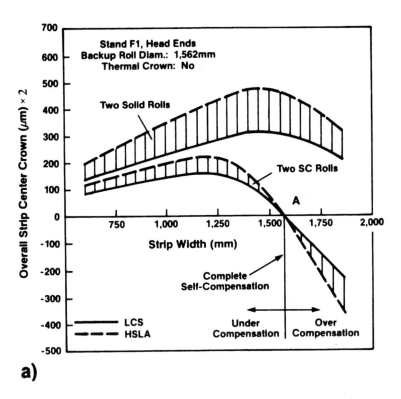

a)

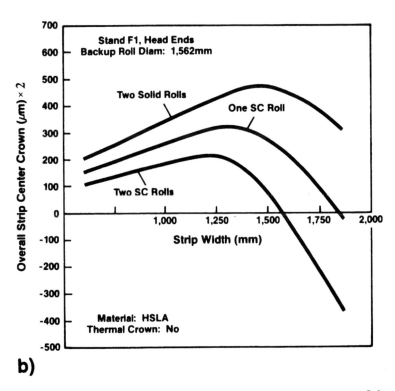

b)

Fig. 40.4 Comparison of strip crowns with different types of backup rolls. (From Ginzburg, Iron and Steel Engineer, 1987. Copyright AISE, Pittsburgh, Pennsylvania. Reprinted with permission.)

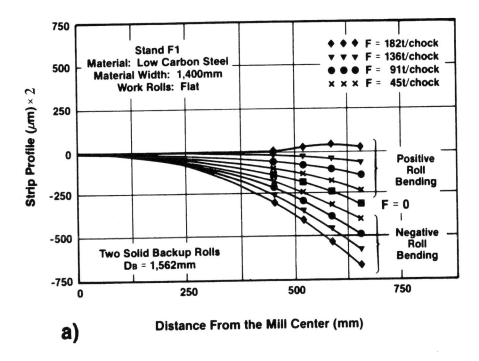

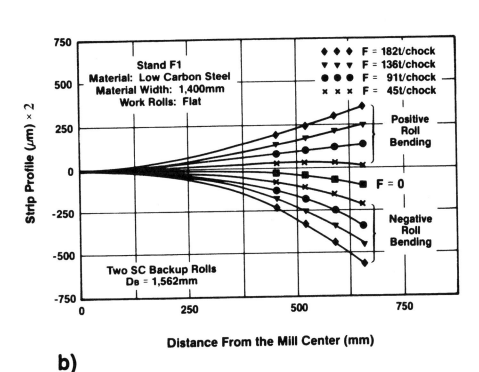

Fig. 40.5 Strip profile variation with roll bending: a) two solid backup rolls and b) two SC backup rolls. (From Ginzburg, Iron and Steel Engineer, 1987. Copyright AISE, Pittsburgh, Pennsylvania. Reprinted with permission.)

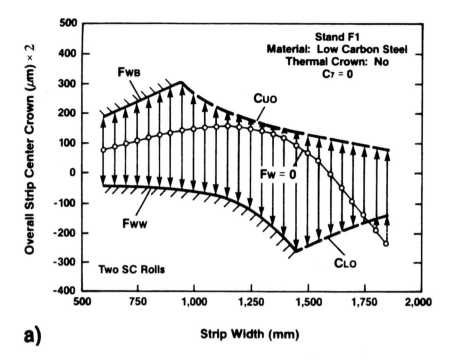

a)

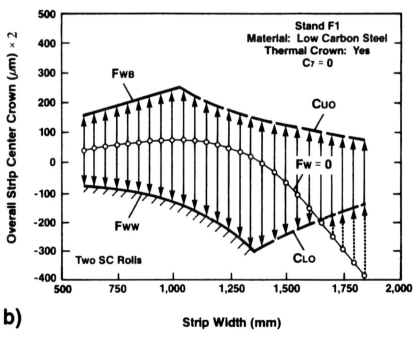

b)

Fig. 40.6 Available and required strip crown control range with two flexible edge self-compensating (SC) backup rolls and with positive and negative roll bending systems. (From Ginzburg, Iron and Steel Engineer, 1987. Copyright AISE, Pittsburgh, Pennsylvania. Reprinted with permission.)

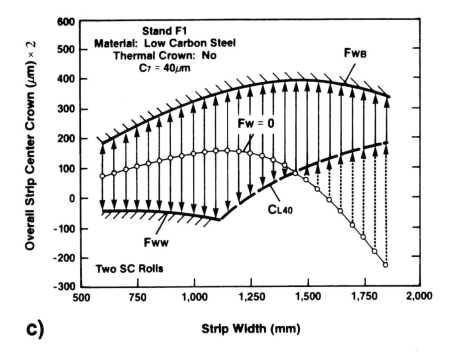

c)

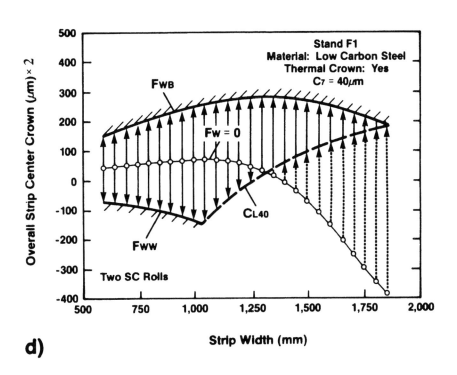

d)

Typical strip profiles corresponding to various roll bending forces are shown for two cases in Fig. 40.5. The first case (Fig. 40.5a) relates to a mill with two regular solid backup rolls. The second case (Fig. 40.5b) relates to a mill with two SC backup rolls and provides up to a 33% increase in the crown control range compared to the first case. This increase is a function of the strip width.

The calculated strip crown control range at stand F1 for low carbon steel is illustrated in Fig. 40.6. The calculations are made for the head end of the strip and for a width range between 610 to 1830 mm with a backup roll outer diameter of 1562 mm. Both work and backup rolls are assumed to have no mechanical crown. The lines with the circles indicate the strip crown when the roll bending force is not being applied. The broken lines represent the upper C_U and the lower C_L, crown limits of the shape cones. The solid lines show the roll bending force limits for the positive roll bending cylinder F_{WW}, installed between the work roll chocks and also for the negative roll bending cylinders F_{WB}, installed between the backup and work roll chocks.

The area between these four limits represents the allowable crown control range of the actuator. Figures 40.6a and 40.6b show that limited adjustment of the strip crown (broken lines) will be necessary when rolling 2.0 mm (0.079 in.)-thick strip with zero crown at the exit of stand F7 ($c_7 = 0$), whereas almost full available crown control range will be used when rolling the same gage with the exit crown $c_7 = 40\,\mu m$ (2%) (Figs. 40.6c and 40.6d). This is due to overcompensation of the backup roll deflection at a strip width greater than 1575 mm (approximately 62 in.).

40.8 EFFECT OF DISTURBANCE FACTORS

The values for the rolling force disturbance rate S_p and backup roll diameter disturbance rate S_D may be calculated from Eqs. (37-4) and (37-5) for three following types of actuators: two solid backup rolls; one SC and one solid backup roll; and two SC backup rolls.

The results illustrated in Fig. 40.7 show that the rolling force disturbance rate can be substantially reduced by using either two or one SC backup roll. The preference is for two SC rolls when the product width does not exceed 80% of the roll barrel length (Fig. 40.7a). In respect to the backup roll diameter disturbance rate, it may be more advantageous to use only one SC roll rather than two (Fig. 40.7b), because the backup roll diameter disturbance rates S_D of the solid rolls and SC rolls have opposite signs.

Although a number of parameters can be used for evaluating different types of actuators, the following three factors appear to integrate most of them: strip crown compatibility, strip crown stability, and crown control range margin.

40.9 STRIP CROWN COMPATIBILITY FACTOR

Strip crown compatibility factor α evaluates the capability of the strip profile actuator to provide compatibility of the original strip crown with the shape cone. It is expressed as:

$$\alpha = 1 - \frac{c_o - c_d}{\Delta c_{max}} \qquad (40\text{-}3)$$

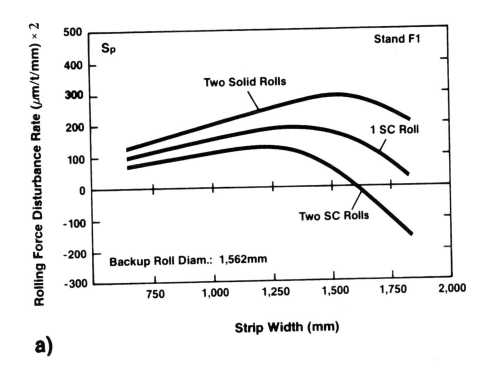

a)

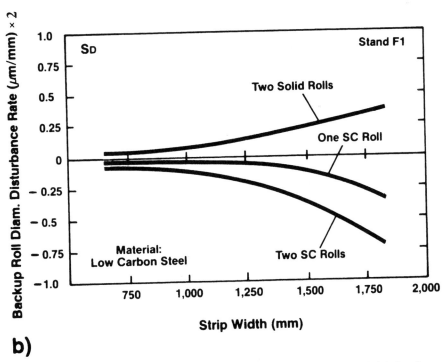

b)

Fig. 40.7 Effect of disturbance factors: a) roll force and b) backup roll diameter. (From Ginzburg, Iron and Steel Engineer, 1987. Copyright AISE, Pittsburgh, Pennsylvania. Reprinted with permission.)

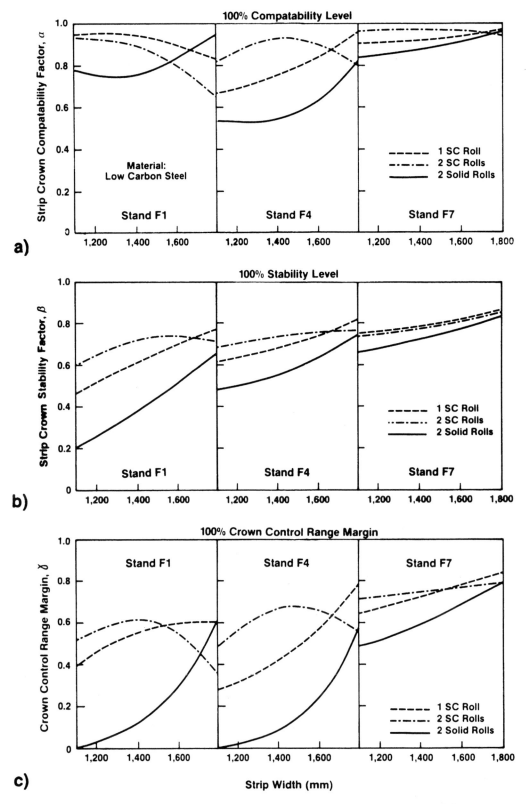

Fig. 40.8 Performance characteristics of roll bending system with different types of backup rolls: a) strip crown compatibility factor, b) strip crown stability factor, and c) crown control range margin. (From Ginzburg, Iron and Steel Engineer, 1987. Copyright AISE, Pittsburgh, Pennsylvania. Reprinted with permission.)

where c_o = original strip crown

c_d = average desired strip crown based on constant crown to thickness ratio maintained throughout the mill

Δc_{max} = maximum available crown control range.

The strip crown compatibility factor indicates the crown control range margin which would be available after the compatibility of the original strip crown with the average desired strip crown has been provided. When $\alpha = 1$, the compatibility and, therefore, the available crown control range margin will also be equal to 100%.

In the case of the 7-stand finishing mill example previously discussed, the maximum improvement (up to 70%) in strip crown compatibility is obtained for stand F4 when two SC backup rolls are used. With one SC roll, the improvement is as high as 40%. Application of two SC rolls at the stands F1 and F7 would result in up to 27% and 15% improvement respectively (Fig. 40.8a).

40.10 STRIP CROWN STABILITY FACTOR

Strip crown stability β factor evaluates the capability of the strip profile actuator to maintain the desired strip crown in the presence of disturbance factors. It is expressed as

$$\beta = 1 - \frac{\delta c_{max}}{\Delta c_{max}} \qquad (40\text{-}4)$$

where δc_{max} = total maximum variation of strip crown due to variation of rolling force, backup roll diameter and work roll thermal crown.

The strip crown stability factor indicates the crown control range margin which would be available after compensating for the effects of disturbance factors on strip crown. When $\beta = 1$, the stability level is equal to 100%. It means that no adjustment of the strip profile is needed to compensate for the effects of disturbance factors and, therefore, the available crown control range margin will also be equal to 100%.

In the case of the 7-stand finishing mill example, the maximum improvement (up to three times) in strip crown stability is obtained for stand F1 when two SC backup rolls are used. With one SC roll, the improvement is as high as 120%. Application of two SC rolls at stands F4 and F7 would produce up to 40% and 14% improvement respectively (Fig. 40.8b).

40.11 CROWN CONTROL RANGE MARGIN

Crown control range margin γ indicates the crown control range margin which will be available after the strip crown compatibility has been achieved and also after compensating for the effects of the disturbance factors on the strip crown. It is equal to:

$$\gamma = \alpha + \beta - 1 \qquad (40\text{-}5)$$

The effect of actuators at stands F1, F4 and F7 on crown control is illustrated, for the 7-stand finishing mill example, in Fig. 40.8c. Actuators with two solid rolls provide a very limited crown control range margin for stands F1 and F4, especially for narrow strip widths. The application of one or two SC rolls at these stands would produce the crown control range margin for narrow strip widths as as large as 50% of the available crown control range. An improvement of up to 40% will also be made with two SC rolls at stand F7.

REFERENCES

1. V.B. Ginzburg, "Strip Profile Control with Flexible Edge Backup Rolls", Iron and Steel Engineer, July 1987, pp. 23-34.
2. V.B. Ginzburg, et al, "Application of the Computer Model ROLLFLEX^{TM} for Analysis of the Strip Profile and Shape in Rolling Mills", Paper presented at the AISE Annual Convention, Pittsburgh, Pa, Sept. 21-24, 1987.
3. V.B. Ginzburg and R.M. Guo, U.S. Patent No. 4,722,212, Feb. 2, 1988.

Index

A

Abrasion, resistance to 45
Accuracy, 589
Actuator(s):
  hydraulic, 652
  roll gap, 617-620, 626
  strip profile and flatness, 762,
    763
    classification of, 765
    selection of, 762
Affinity for oxygen, 146
Aggregate, 10, 78, 84
Aging, 102
  quench, 133
Aerospace Material Specification
  (AMS), 43
Air combustion, 486, 487
AISI designation system, 43, 46
AISI-SAE designation system, 43, 44
  for alloy steels, 45, 52
  for carbon H-steels, 50
  for carbon steels, 49
Alloying elements, 88-106, 464
  effect on
    corrosion resistance, 95
    critical temperature, 91
    electrical resistance, 96
    eutectoid composition, 94
    eutectoid temperature, 94
    grain growth, 95
    grain refinement, 464
    hardenability, 94
    hardness, 94
    microstructure, 91
    precipitation strengthening, 464
    tensile strength, 94

    transformation rate, 94
    yield strength, 464
Allotropic changes, 4
Alloys:
  aluminum, 20
    annealed, 20
    cold-worked, 28
  cobalt-base
    solid-solution, 66
    precipitation-hardening, 67
  heat-resistant casting
    cobalt-base, 100-103
    nickel-base, 100-103
  iron-base
    heat-resisting, 47
    permanent-magnet, 102
    precipitation-hardening, 66
    solid-solution, 66
  iron-carbon, 74
  molten, 74, 75
  nickel-base
    solid-solution, 66
    precipitation-hardening, 67
Aluminum, 10, 31, 47, 102, 489
Amonton's law, 369
Analysis of plastic deformation:
  upper-bound, 247
  lower-bound, 247
Annealing:
  cycle, 124
  full, 124-128
  furnaces, 527
  grain restoration by, 204
  intercritical continuous, 132
  isothermal, 124, 126, 127
  of steel, 477
  process, 126
  recrystallization, 126

Angstrom unit, 3
Anisotropy:
  of magnetic permeability, 6, 7
  of magnetic properties, 6, 7
  of mechanical properties, 6, 7
  of physical properties, 7
Arc of roll contact, 233-235, 269
  arithmetic average, 280, 284, 289,
   293.
  parabolic mean, 280, 289
  geometric mean, 280, 296
Arc-reheating:
  gas-stirring, 160, 162
  induction-stirring, 160, 163
Arc-remelting, vacuum 163
Argon bubbling:
  non-vacuum
    capped, 156, 159
    composition adjustment by
     sealed, 156, 159
Argon stirring, 156, 158
Arsenic, 10
ASME specification, 43
Aspect ratio of roll bite zone:
  arithmetic average, 272, 276, 281,
   285, 309
  geometrical mean, 272, 288, 312
  parabolic mean, 272, 281, 309
ASTM classification, 53
ASTM grain size chart, 10, 11
ASTM specification, 43, 44, 50
Atmosphere:
  oxidizing, 486
  reducing, 488
  scale-free, 486
Atom:
  solute, 12
  solvent, 12
Atomic planes, 5, 6
Ausforming, 457
Austempering, 126, 131, 132
Austenite, 74, 75

B

Bainite, 83, 84
Basic oxygen processes:
  bottom-blown, 142, 143
  combination-blown, 142, 143
  top-blown, 141, 142
Bending, 121
Bite angle, maximum 371
Black body, 36
Bleed in slabs and ingots, 194
Blowholes in slabs and ingots, 195
Bootleg in ingots, 192
Boron, 104
Brinell hardness number, 25-28
Brittleness, 105
  temper, 129

Bruises in slabs and ingots, 195
Buckles in rolled products, 575,
  682-683, 695-696, 683
Bulging:
  in cast slabs, 195, 196
  in rolled products, 673-675
Burner(s), 409

C

Camber, 578
  of rolled product, 579
    measurement of, 611
Carbide, 11
  chromium, 90, 99
  columbium , 90
  iron, 98
  molybdenum, 90, 99
  silicon, 31
  solid solution of, 74
  titanium, 90
  tungsten, 90, 101
  vanadium, 90, 101
Carbon, 10, 44, 98, 489
  equivalent (CE) formula, 121
  free, 76
  graphitic, 76
Carbonitriding, 133
Carburizing, 133
  in cast steel, 193
Caster, thin, 523, 524, 525
Casting:
  close stream or shrouded, 170
  horizontal, 169-171
  inside-the-ring, 179
  machine, 172
    types of, 183-185
  methods of continuous, 169
  of thick slabs, 170
  open stream, 170
  process, near-net-shape 173
  roller, 173, 174
    belt, 174
    single, 178
    twin, 178, 179
  semi-horizontal or curved mold,
   170
  single-belt, 176, 179
  stationary-mold, 173-175
  traveling-mold, 173, 174
  traveling twin-belt, 173-177
  vertical or stick, 167, 183
  vertical plus bending, 170
  wheel-belt, 180
Cementite, 11, 74-76
  spheroidized, 76
Change, allotropic, 35
Chatter:
  fifth-octave-mode, 392
  third-octave-mode, 392

Chromium, 46, 89, 99, 489
Classification, of steels and
    alloys 43
Clinks in slabs and ingots, 192
Cobalt, 102
Coefficient:
    geometrical, 285, 288, 294
Coilbox, 521, 555, 561
    basic concept of, 561
    inverse arrangement of, 561
Coil car, 525
Coil conveyer, 525
Coiler, 519, 523
Coiling, 493
Cold mill:
    arrangements, 526
    continuous, 526
    double-reduction, 526
    high-reduction, 526
    single-stand, 526
    stand-alone, 526
    tandem, 525, 526
    twin-stand, 526
Columbium (See niobium)
Composition:
    of AISI-SAE standard carbon
        H-steels, 50
    of alloy steels, 52
    of heat-resisting alloys, 64, 65
    of HSLA steels, 53
    of stainless steels, 57, 58
    of superalloys, 66, 67
    of tool steels, 55
Compound, 11
    intermetallic, 10, 11
    layer-lattice, 354
Compression,
    plane-strain, 225-228, 249, 285
        upper-bound solution for, 250
    slip-line field analysis of, 259
Concavity in slabs, 195, 196
Conditions in stainless steels:
    austenitic, 33-35
    ferritic, 33-35
Conductivity, thermal, 32-34
Continuous-cooling transformation
        diagram, 86
Convection:
    forced, 422
    free, 422
Cooling:
    accelerated, 473-475
    curve, 7
    dynamic accelerated, 477
    facilities, 527
    interstand, 519
    multipurpose accelerated, 475
    on-line accelerated, 475
    rate, 9, 450, 451, 451
    zone, 523
Copper, 20, 31, 44, 45, 104, 489

Core loss, total, 41
Corrosion:
    definition of, 38
    electrochemical, 38
    galvanic, 38, 39
    rate, 39
    resistance, 46
Cracking:
    stress, 224
    stress-corrosion, 224
Cracks in slabs and ingots:
    external, 189-192
        basal, 190, 192
        corner, 192
        double scin, 190, 192
        fin, 190
        hanger, 190
        longitudinal facial, 190, 192
        transverse facial, 190, 192
    internal, 188, 189
        diagonal, 188
        halfway, 188
        star, 189
        withdrawal roll, 189
Cracking:
    hydrogen-induced cold, 121
    internal, 106
Crazing, 192, 364
Creep, 32
    rate, 32
Crown, 573
    center, 573
    partial center, 573
Crystal(s), 9
    boundary, 9
    cubic, 7
    equiaxed, 205
    growth, 8
    lattice(s)
        body-centered cubic (b.c.c.),
            3, 5 ,6
        close-packed hexagon (c.p.h.),
            3, 5
        constant, 3
        face-centered cubic (f.c.c.),
            3, 5
    of solid solutions, 12
Crystallization, 3, 7
    center of, 7
Crystallographic axes, 5
Cutting model, 351

D

Decarburization, 145, 151, 489
    argon-oxygen, 156
    vacuum-oxygen, 156, 157
Defects:
    in ingots, 187-196
    in slabs, 187-196

Deformation:
  axial, 211
  bands, 439, 441
  biaxial, 212, 214-216
  by slip, 200, 203
  by twinning, 202
  elastic, 199
  fields, 248, 249
  homogeneous, 232, 236
  ideal work of, 221
  plastic, 199, 225, 231
    lower-bound analysis, 247
    slab analysis of, 225-246
    slip-line field analysis,
      255-266
    upper-bound analysis, 247-254
  quasi-forging, 546
  quasi-rolling, 546
  restricted plastic, 232, 234
  triaxial, 214-216
  uniaxial, 209, 210, 216
  zone, 229, 230, 270, 311
    aspect ratio, 272
Degassing, 151
  induction-stirring ladle, 157
  ladle-to-mold, 151
  ladle-to-ladle, 152, 153
  recirculation, 151, 153
  tap, 153, 154
  vacuum, 155
    ladle, 153
    stream, 151
    with heating, 151, 160
Dendrite:
  metallic, 7, 8
  growth, 8
Density:
  definition of, 35
  of iron, 4
Deoxidation, 44, 98
  of steel, 147, 148
  practice:
    aluminum-killed, 44
    capped, 44
    fully killed, 44
    rimmed, 44
    semi-killed, 44
Deoxidizer, 103
Dephosphorization, 145
Depression in slabs:
  longitudinal, 193
  transverse, 193
Descaling, 419, 526
  box, 568
  system, 519
Desiliconization, 145
Desulfurization, 98, 145, 147, 151
  ladle steel, 161
Deviation:
  out-of-square, 578, 579
  standard, 579-581

Diagram:
  constitutional, 73
  continuous-cooling transformation,
    86, 463
  equilibrium, 73
  iron-carbon phase, 74
  iron-cementite phase, 74
  isothermal transformation, 85
  phase, 73
Diffusivity, thermal 35
Dirt in steel, 97
Dislocation in crystals:
  definition of, 201
  edge, 201, 203
  screw, 202, 203
Draft, 229
Drawability, 118
Ductility, 20, 24, 25, 30

E

Edge:
  drop, 574
  thicknesses, 573
  waves, 575
Edge heater, 522, 523,
Edge preheating, 429
Edger:
  attached, 568
  back, 568
  front, 568
Edging, 673-674, 676, 679
Elements:
  alloying, 88-106, 488
  pure, 96
  residual, 96, 488
Elongation, 32
  definition of, 230
  limit of uniform, 16
  percent, 20, 21
Emissivity:
  coefficient of thermal, 37
  of black body, 37
  thermal, 36, 37
Energy:
  shelf, 31
  specific, 16
Energy consumption:
  electrical, 552
  fuel, 552
Energy savings:
  electrical, 552
  fuel, 552
Error, combined, 589
Eutectic, 74, 78
Eutectoid, 74, 78
  composition, 92
  point, 78
  reaction, 78
  structure, 78

temperature, 92
Expansion:
   coefficient of linear thermal, 38
   thermal, 37

F

Factor, geometrical 281, 283
Feather in workpiece profile, 574
Ferrite, 74, 75
Films:
   metal, 353
   oxide, 353
   polymer, 354
Fin in slabs and ingots, 195
Finishing train, 559
Finite element analysis:
   accuracy of, 744
   principles of, 738
   three-dimensional, 745
Firecracking, 364
Fish tail in rolled products, 682, 696
Flash in ingots, 19
Flatness:
   actuators, 711
   control, 711-712
   definition of, 577
   measurement of, 594-597
   models, 759
   of rolled product, 575
Flat product, geometrical characteristics of 571
Flow-tempering, 457
Fluid:
   Newtonian, 357
   Non-Newtonian, 357
Force:
   coercive, 41
   compressive, 13
   formulae for, 299, 307, 309, 312, 313
   frictional, 366
   in heavy draft rolling, 319
   method for calculation, 324, 327, 331, 334, 337
   roll separating, 269, 270, 298
   shear, 13, 14
   tensile, 13, 209
Formers, phase 88-90
   austenite, 88
   carbide, 88, 90
   ferrite, 88, 90
Fourer's general law, 36
Fracture, 30, 204
   shattering, 24
Freezing point, 7
Friction:
   coefficient of, 279, 282,

344, 368
   compensation for, 645
   definition of, 343
   dry, 347
   dry slipping, 232-234, 240
   entry coefficient of, 368, 370, 372
   hill, 227, 231, 240
   in compression test, 22, 24
   instabilities, 363, 391
   measurement of coefficient of, 377
   static, 364
   steady-state coefficient of, 368, 386
   sticking, 232, 241, 250, 251, 345
   transient coefficient of, 368
   viscous slipping, 232
Furnace:
   coil holding, 523
   direct-arc electric
      alternating current (A.C.), 143
      direct current (D.C.), 143, 144
      induction, 143, 144
   open-hearth, 140
   reheating, 517
      batch-type
      continuous-type, 405-410
      electrical resistance, 404, 405
      fuel-fired, 403, 405
      induction slab, 404, 405
      multiple zone, 409, 410
      pusher, 405, 406,
      reverse-fired, 409, 410
      roller-hearth, 405, 408
      roof-fired, 409, 410
      rotary hearth, 405, 406
      single zone, 409, 410
      walking beam, 405, 407
      walking hearth, 405, 407
   soaking-pit, 399

G

Galvanic series, 39
Gases:
   oxidizing, 485
   reducing, 485
Gauge:
   center, 573
Gauge control:
   differential, 629
   feedforward, 631, 634, 635, 641, 642
   flow-stress feedforward, 635, 636

non-interactive, 636, 637
  spacer, 630, 631
  three-stage, 633, 640, 641
Gaugemeter:
  control, 628, 629
  equation, 615
Geometry:
  plan view, 579
  workpiece, 579
Gouges, 493, 506, 507
  contraction, 493, 505, 506
  pickle line, 493
Grain(s), 9
  ASTM number of, 10
  boundary, 9, 205
  chill, 10
  coarse, 206
  coarsening, 440
  columnar, 10
  equi-axed, 10
  growth, 204-206, 398, 449,
          450, 451
  polyhedral, 75
  refinement, 109
  restoration, 204
  size, 9-11, 207, 444, 447
          distribution, 440
  shape, 9
  structure, 10, 46
Graphite, 75, 76
  crystals, 355
Graphitization, 102

H

Hardenability:
  definition of, 28
  multiplying factors, 94, 93
  test, 28, 29
Hardness, 45
Hardening:
  case, 132
  decremental, 132
  definition of, 25
  effect, 93
    coefficient, 278
  flame, 132
  induction 132
  strain or work, 204
  surface, 132
  tests, 25-27
  versus strength, 26
Hastelloy X, 48
Heat:
  balance in soaking pits, 402
  conduction, 36
  insulator, 522
  radiation, 36
  retention box, 523
  specific, 33, 35

transfer, 395-430
  in coilbox, 433
  in ingots, 399
  in reradiating heat covers, 430
  in rolling mills, 395
  in soaking pits, 401
treatment, 124-132
  dual-phase, 126, 132
  types of, 124
Heating:
  practice, 403-415
    in ingots, 403
    in slabs, 411-415
  process, 397-411
    analysis of, 411
    purpose of, 397
    requirements for, 397
Hematite, 483, 484
Histogram, 580
Homogenizing, 125, 129
Hook's law, generalized, 215
Hot-cold working, 457
Hot metal, 137, 139
Hot shortness, 192
Hot strip mill:
  compact, 521
  continuous, 559
  continuous tandem, 559
  fully continuous, 520
  generation I, 520
  generation II, 520
  generation III, 520
  high-production, 517, 519
  layouts, 518
  modernization of, 550
  optimization of, 550
  semi-continuous, 519, 560
  single-stand reversing, 521
  temperature conditions in, 553
  three-quarter continuous, 520
  twin reversing, 521
Hot working, 125
Housing, 511
Hydrogen, 106
Hysteresis, 589

I

Impurities, 96, 145
  gaseous, 97
  insoluble, 207
Inclusion(s), 96
  microscopic, 182
  morphology, 151
  non-metallic
  oxide, 181, 182
  shape control, 122
Inconel, 31, 48
Induction, magnetic:
  residual, 41

saturation, 41
Indentation, 25
  slip-line field analysis of, 258
Ingot, 9, 10, 165
  big-end-down, 168
  big-end-up, 168
  capped, 167
  casting of, 165
  fully-killed, 167
  hot-topped, 167
  non hot-topped, 167
  rimmed, 167
  rolling from, 166
  semi-killed, 167
  structures, 168
  types of, 167
Injection method:
  lance powder, 162
  REM canister, 162
Instruments, precision of, 589
Interface:
  macroscopic viewpoint of, 346
  microscopic viewpoint of, 346
  shear factor, 345
  tool-workpiece, 346
Iron:
  allotropic forms of, 4, 5
  alpha, 4, 5, 74, 75
  cast, 76
  gamma, 4, 5, 74, 75
  delta, 4, 5, 74

L

Ladle
  injection, 160
  metallurgy, 151
Lamellar tearing, 121
Lap in slabs and ingots, 194
Latent heat, 7
Lead, 104
  segregation, 104
Ledeburite, 74, 79
Lever arm, 241
  coefficient, 308-313
    formulae for, 314
Limit:
  elastic, 15
  proportional, 15
Loopers, 639
Lubricants, rolling 384, 389
Lubrication:
  boundary, 355, 35
  by dispersion, 389
  elastohydrodynamic (EHD), 357
  extreme pressure, 355
  fluid thick-film, 357
  full-fluid-film, 385
  hydrodynamic, 356, 357, 381
  mechanism, 360, 381, 385

  mixed-film, 359, 381, 385
  plastohydrodynamic (PHD), 357
  solid, 382
    film, 353
  with emulsions, 389
  with separate water, 389
Luder lines, 479

M

Machinability:
  definition of, 41
  index, 41
Magnesium, 31
Magnetic hysteresis loop, 40
Magnetic properties, 6
  anisotropy of 7, 46
Magnetite, 483, 484
Main drive:
  motors, 516
  train, 511
Malleability, 41
Manganese, 44, 45, 89, 489
Marks in cast slabs:
  guide, 193
  heavy reciprocation, 193
Marstraining, 457
Martempering, 126, 130, 131
Martensite, 28, 83, 84
  tempered, 84
Materials:
  brittle, 15, 16
  ductile, 15, 16, 24
  elastic, linearly strain-
    hardening, 18, 19
  engineering, 17
  perfectly elastic, 17-19
  perfectly plastic, 17, 18
  rigid
    perfectly plastic, 17-19
    linearly strain-hardening,
      18, 19
Mechanical properties:
  of alloy steels, 54
  of carbon steels, 51
  of electric steels, 63
  of stainless steels, 59-62
  of superalloys, 68
  of tool steels, 56
Metal:
  molten, 9
  pure, 10, 11
  ceramics, 31
Metallic bond, 4
Metalloids, 10
Metallurgical length, 170
Method(s), rolling parameters
    calculation 276-347
  aspect ratio, 296
  specific power, 296

Microalloying, 446
Microcleanliness, 151
Micropitting, 364
Microstructure:
  control of, 107
  of bainite, 84
  of cementite, 77
  of ferrite, 76
  of graphite, 76
  of ledeburite, 79
  of martensite, 85
  of pearlite, 77, 78
Mill:
  finisning, 519
  planetary, 524
  sizing, 686, 687
  skin-pass, 526
  spring, 652
  Steckel, 525
  stiffness, 615, 617, 653
    hydraulic, 654
    structural, 654
  reversing finishing, 521
  roughing, 519
  temper, 526
Mill drive, efficiency, 552
Mill stand(s):
  arrangement,
    close-coupled, 517
    open, 517
  classification of, 511
  cluster type, 514
  five-high, 514
  four-high , 511
  horizontal, 512
  main components of, 511
  roll arrangements, 513
  six-high, 514
  special design of, 516
  three-high, 511
  types of, 514
  two-high, 511
  vertical, 513
Modulus:
  of elasticity, 16, 32
  of resilience, 16
  of rigidity, 25
  shear, 25
Mohr's circle:
  for axial deformation, 211, 216
  for biaxial deformation, 215, 216
  for triaxial deformation, 216
Mold(s), 9, 170, 171
  adjustable width, 173
  car, 175
  divided, 173
  ingot, 165
    big-end-down, 165, 166
    big-end-up, 165, 166
    straight, 167
  stationary, 175

traveling belt, 177
  traveling caterpillar, 177
Molybdenum, 31, 99
Mould (see Mold)

N

Necking of specimen, 22, 24
Nickel, 100, 489
Niobium, 104
Nitriding, 133
Nitrogen, 106
Noise in control systems, 589
Nonlinearity, 589
Normal distribution curve, 581
Normalizing, 125-129
Nuclei, 9
Nucleus, 7, 205

O

Oil-film thickness, 234
Operating parameters, 556
Oxidation, 47, 145
  rate, 488
Oxide(s), 96, 145
  acid, 146
  aluminum, 103
  basic, 146
  distribution, 181
  inclusions, 181, 182
  iron, 97, 483
  phases, 181, 184

P

Pearlite, 74, 78, 83
Peening effect, 311, 312
Phase:
  diagram, 73
  transformation , 80-83
    hysteresis, 82
    in steel, 80-82
    temperature, 83
Phases:
  as constituents in steel, 75
  in eutectoid steel, 81
  in hypereutectoid steel, 81
  in hypoeutectoid steel, 80
  solid state, 11
Phase shift, 628
Phosphide, 105
Phosphorus, 10, 45, 105
Pinholes in slabs and ingots, 195
Pipe:
  in cast steel, 167, 187
  secondary, 187
Pit(s):

bottom-two-way fired, 400
circular, 400
continuous-fired, 400
electric soaking, 400
one-way fired, 400
regenerative, 399
top-two-way fired, 400
vertically fired, 400
Plan view control, 704
Plane(s):
    neutral, 234
    shear, 250
Plasticity:
    coefficient of, 279
    microscopic nature of, 209
Plate, definition of, 582
Plating in cast steel, 194
Point:
    critical, 79
    freezing, 7, 9
    neutral, 231
Population, 579
    mean, 579
    variance, 579
Power:
    curve, 292, 293
    in rolling, 318
    rolling horse, 291
    specific, 291
Precision, 589
Press:
    sizing, 688-692
    squeezing, 688
Pressure, rolling 288, 290
Profile of rolled product, 571
    actuators, 711
    control, 711, 712
    definition of, 573
    measurement of, 592, 593
Properties of steel product:
    cost, 107
    functional, 107
    surface quality, 107

Q

Quality, requirements, 553
Quenching, 126, 130, 131

R

Rate:
    cooling, 82, 86
        critical, 86
Ratings, of machinability 41
Rare earths, 103
Ratio:
    air-gas, 486
    aspect, 22

gas, 486, 487
Poisson's, 16
Recovery, 19, 204, 205, 448
    dynamic, 443
    elastic, 19, 242, 243
    partial, 448
    static, 443
Recrystallization, 204, 205
    complete, 448
    dynamic, 443
    metadynamic, 443
    static, 441, 443
Reduction:
    critical rolling, 448
    effect of amount of, 447
    in area, 20, 21, 32
    percent, 20, 21
    relative, 229
Refining, 145
    of stainless steel, 158
Reheat facilities, intermediate, 428
Remelting, electro-slag 160, 161
Reoxidation, 147
Repeatability, 590
Residuals, 146
Resistance:
    to abrasion, 41, 45
    to corrosion, 46
    to deformation, 269
        for alloy steels, 289
        isothermal, 295
        normalized, 293
    to oxidation, 47
    to softening, 46
    to wear, 41
Resistivity, electrical, 39, 40
Resolution, 591
Response:
    characteristics, 648
    frequency, 627, 659
    step-function, 627
Restoration process:
    dynamic, 442
    grain, 204, 442
    metadynamic, 442
    static, 442, 443
Rhomboidity in slabs, 195, 196
Ridge, 574, 575
Rigidity of metal, 16
Rimming action, 167
Rippled surface in ingots, 192
Roll:
    arrangement, 511
        shifted, 719-724
    backup, 511
        center, 514
        self-compensating (SC), 727-728,
            767
        shifted, 719-722
        stepped, 717-720
        side, 514

bending system, 714-715
bite, 237
bite angle, 230, 234, 269
contact length, 230
contact, 271
  projected area of, 271
coolant system, 519
crossing system, 716-717
deformation model, 730
driven, 532
eccentricity, 642-644
flat, 679, 680, 681
flattening, 238, 746
flexible body, 723-724
flexible edge, 725-727
gap, 511
grooved, 679, 680, 681
intermediate, 514, 532
temperature biuld-up, 751
tension measuring, 525
thermal contour model, 749, 750
thermal crown, 764
variable crown (VC), 725
wear, 573, 722
wear model, 755
work, 511, 514
Rollability, limit of 244
Roll crown, thermal 571
Roll gap:
  control, 626
  setting, 616
Rolling-drawing process:
  double, 536
  PV, 534
  PV-E, 535
Rolling method:
  bite back, 698, 699
  convex crown, 698
  high-speed entry, 557
  squeezed slab, 697, 698
  with lubrication, 698
Rolling mill(s):
  C-B-S, 538, 539
  classification of, 516
  cluster, 531
  cycloidal, 542-545
  flexible flatness control (FFC),
    531
  high-reduction, 529
  IPV, 536
  Krause reciprocating, 540
  MKW cold, 529, 530
  pendulum, 539, 540
  planetary, 547
  Platzer planetary, 547, 548
  Platzer reciprocating, 541, 542
  rolling-drawing (PV), 534
    double, 537
  Sendzimir, 532, 545, 546
    cluster, 532
    planetary, 545, 546

Rolling process:
  contact-bend-stretch, 537
  continuum, 459
  controlled, 459
  conventional hot, 459
  hot direct charging, 522
  low finishing temperature, 459
  temper, 479, 527
Rubbing model, 351
Rupturing, 41

S

SAE designation system, 43
  for HSLA steels, 53
Scabs:
  in ingots, 193
  in rolled products, 493, 507
Scale:
  classification of, 493
  factor, 591
  furnace, 493, 497, 498
  heat pattern, 493, 504
  heavy primary, 493, 497
  losses, 488
  plugged nozzle, 493
  primary, 490, 493, 499
  rebound, 493, 503
  red oxide, 493, 500
    blotchy, 501
    line type, 501
    tail end, 502
    teardrop, 500
  refractory, 493, 498
  roll wear, 493, 504, 505
  salt and pepper, 493, 504
  scale streaks and plugged nozzle,
    502
  secondary, 490, 493, 499
  sensitivity, 591
  streaks, 493
  structure of, 484
Scaling, 483
  effect of atmosphere on, 485
  rate, 484
  time, 552
Scalebreaker, 568
Schaeffler diagram, modified 91
Scrap, 137, 139
Scratches in rolled products, 493,
  506, 507
Scum, entrapped in slabs and ingots,
  195
Segregation in ingots, 188
  inverse-V, 188
  major, 188
  minor, 188
Sendzimir Z-high insert, 533
Sensor:
  built-in roll, 595, 602

contact, 597
  external, 603
  non-contact, 598, 602
  position, 622, 623
  roll gap, 621
Servovalve, 652
Shape actimeter, 605, 607
Shapemeter, 595-598, 601, 602, 603
  classification of, 599, 600
  stressometer, 596, 600
Shape of rolled product, 575
  affected by tension, 576
  dog-bone, 673
  dual, 576, 577, 605
  ideal, 576
  latent, 576, 605
  manifest, 576, 577
Shear, 519
  modulus, 25
  strain at fracture, 25, 26
Sheet:
  deep-drawing, 479
  definition of, 582
  electrical, 479
  high-strength cold-rolled, 479
Shelf energy, 31
Silicon, 10, 44, 45, 47, 103, 489
Skid:
  coolant temperature, 417
  insulated, 417
  marks, 398, 416, 417, 641
  non-insulated, 417
  system, 415
Slab(s):
  cast
    thick, 165, 166
    thin, 165, 166
  charging temperature, 413
  defects, 187-196
  shape defects, 195
  slitting, 173
Slab charging, hot 415
Slab distortion, 694
Slag, 145
Slip:
  bands, 200
  direction, 200
  forward, 375
  line field(s), 255-266
  lines, 256, 262
  mechanism of, 201
  plane, 200, 205
Slivers, 493, 507
Softening, 205
  resistance to 46
Solidification, 3, 8
Solution:
  interstitial solid, 12, 84
  liquid, 73, 78
  solid, 10, 11, 73, 78
  substitutional solid, 12

Space lattice (see Crystal lattice)
Spalling, 364
Specification, definition of 43
Speed effect coefficient, 278, 282,
  290
Spheroidizing, 126, 128, 129
Splash in slab and ingots, 195
Spongy top in ingots, 192
Spot-weld peeling, 122
Spray deposition process, 174, 180
Spread in rolled products:
  factor, 375
  lateral, 376
State of metals:
  amorphous, 3
  crystalline, 3
  liquid, 7
  solid, 7
Steckel mill, intermediate 562
Steel:
  alloy, 43, 45, 54
  annealed, 18, 20, 28
  arctic grade, 471
  boron, 104
  capped, 147
  carbon, 28, 30, 44
  chromium, 100
  chromium-vanadium, 101
  cold-rolled, 18, 20
  cold-worked, 28, 47
  continuously cast, 44
  dual-phase, 116, 468-471
  electrical sheet, 39, 46, 63
    non-oriented, 46
    oriented, 46
  eutectoid, 81
  full-hard, 47
  high-alloy, 88
  high-carbon, 44, 76
  high-speed, 46, 98
  high-strength low-alloy (HSLA),
    45, 47, 103
  hot-rolled, 18
  hypereutectoid, 44, 81
  hypoeutectoid, 44, 80
  killed, 34, 35, 38, 40, 45, 147
  leaded, 104
  low-alloy, 88
  low-carbon, 18, 20, 44, 45
  manganese, 99
  maraging, 102
  medium-carbon, 44
  mild, 37
  molybdenum, 99
  nickel, 100
  oil-hardening, 46
  pearlitic, 95
  plain carbon, 74, 98, 99
  produced by teeming into ingots, 44
  rimmed, 147
  semi-killed, 147

shock-resisting, 46
silicon-killed, 489
stainless, 18, 20, 34, 35, 38, 40
  annealed, 47
  austenitic, 31, 33, 37, 46, 47, 57
    62, 99-101
  cold-worked, 47
  definition of, 46
  ferritic, 46, 47, 57, 59, 99,
    100, 103
  martensitic, 46, 47, 57, 59,
    99-101
  precipitation-hardening, 46,
    47, 57, 59, 100, 101, 103
  quenched and tempered, 47
tool, 41, 45, 46, 55
  air-hardening medium-alloy
    cold work, 55
  cold-work, 98-102
  chromium hot work, 55
  high-carbon high-chromium cold
    work, 53
  high-speed, 99-102
  hot-work, 98-102
  low-alloy special purpose, 55
  molybdenum high-speed, 55
  molybdenum hot work, 55
  oil-hardening cold work, 55
  shock-resisting, 55, 102
  tungsten high-speed, 55
  tungsten hot work, 55
Steelmaking process(es):
  acid, 146
  acid open-hearth, 44
  acid electric-furnace, 143
  basic, 146
  basic electric-furnace, 44, 143
  basic open-hearth, 44
  basic oxygen, 44
  comparison of, 138
  electric-furnace, 137
  open-hearth, 137
  oxygen, 137, 141
  secondary
    capability of, 164
    comparison of, 163
    purpose of, 151
Steepness in rolled products, 578
Steffan-Boltzmann and Kirchhoff's
  laws, 37, 420
Stirring, induction 156
Strain:
  aging, 457
  effective, 220
  engineering, 14, 15
  hardening exponent, 17
  hardening material, 19
  lateral, 16
  longitudinal, 16
  plane, 218-220
  shear, 24

tempering, 457
types of, 14
true, 14, 15
Strain rate:
  definition of, 21
  engineering, 21
  in flat rolling, 273
  sensitivity exponent, 22
  true, 21
Strength, 30
  coefficient, 17
  creep, 32
  rupture, 32
  tensile, 32
  ultimate, 16
    tensile, 28
  versus hardness, 26
  yield, 15, 458
Strengthening:
  dislocation, 115
  effect of grain size on, 109
  interstitial solid solution, 110
  precipitation, 111, 112, 465, 466
  second-phase, 116
  subgrain, 461
  substitutional-interstitial solid
    interaction, 110
  substitutional solid-solution, 109
  with columbium, 111
  with titanium, 115
  with vanadium and nitrogen, 114
Stress(es):
  axial, 19, 25, 26
  bearing, 20
  bending, 19, 20
  compressive, 13, 22, 24, 25, 28
  contact, 19, 20
  effective, 220
  engineering, 14
  hydrostatic, 256
  internal, 205
  kinds of, 13
  normal, 20, 30, 209
  plane, 218, 219
  principal, 213, 215
  relieving, 126, 129
  residual, 205
  shear, 13, 14, 24, 30, 210
  tensile, 13, 25
  torsional, 19, 20
  true, 14
  types of, 19
  uniaxial, 28
  yield, 28, 326, 327
    constrained, 283-285, 300, 301
    shear, 285-287
Stress-strain diagram:
  engineering, 15, 16
  idealized, 18, 19
  true, 17
Stretch forming, 120

Stretcher-strain markings, 479
Strip:
  cast, 165, 166
  cleaning, 527
  cold-rolled, 166
  crown, 762, 765
    compatibility factor, 772, 774
    control range margin, 774, 775
    stability factor, 774, 775
  definition of, 582
  flatness cone, 765
  tension, 269
  waviness, 579
Strip shape, parameters, 575
Structure, grain 46
Structural changes:
  during controlled rolling, 461
  during continuum rolling, 462
  during cooling, 450
    controlled, 464
  during reheating, 439
Sulfide, 105
  inclusion-shape control, 121
Sulfur, 10, 45, 105
  control, 148
Superalloys:
  cobalt-base, 47
    solid-solution, 66
    precipitation-hardening, 66
  iron-base, 47
    solid-solution, 66
    precipitation-hardening, 66
  nickel-base, 47
    precipitation-hardening, 67
    solid-solution, 66
Supercooling of austenite, 82, 83
Superplasticity, 207
Surface:
  contact area, 24
  finish, 23
Surface roughness:
  characteristics, 360
  effect of tool, 362
  effect of workpiece, 362
Surge in slabs and ingots, 194

T

Table:
  runout, 519
  transfer, 519
Tapping system:
  bottom, 147, 149, 150
  slag free, 149
Temperature:
  annealing, 206
  coiling, 471
  control in casting, 181
  critical, 79
    breakdown, 356

curves, 293
  distribution, 557
  effect, 591
  effect coefficient, 276, 294
  equilibrium, 82, 486
  finishing rolling, 450, 451, 458,
    469, 470, 472
  gradient, 33, 223
  grain-coarsing, 439
  holding, 85, 467
  impact-transition, 30, 31, 108,
    116
  intermediate, 470
  mean differential, 553
  operating, 589
  recrystallization, 206, 445, 446
  rolling, 516
  rundown, 432, 550, 551, 555, 556,
    564
  slab dropout, 552
  slab reheating, 472
  static recrystallization, 452
  subcritical, 85
  teeming, 151
  transformation, 79
  upper critical, 28
  workpiece, 419
Temperature loss:
  due to conduction to work rolls,
    424
  due to convection, 422
  due to radiation, 420
  due to water cooling
Temperature rise due to mechanical
  work, 426
Tempering, 126, 130, 131
Temp-forming, 457
Tensile strength:
  effect of composition on, 117
  effect of structure on, 117
Tension, strip 317
  control, 632, 638, 639
  distribution, 606
    controller, 607
  interstand, 678
  reel, 525
  regulator, 607
Test:
  bending, 30
  combined deformation, 25
  compression, 22, 28, 30
  dynamic impact, 29, 30
    Charpy, 29, 30
    Izod, 29, 30
  end-quench, 29
  hardenability, 28, 29
    Jominy, 29
  hardness, 25, 28
    Brinell, 25
    diamond pyramid, 26,
    Knoop, 26

Mohr's, 26
    Rockwell, 25
    Scleroscope, 26
    Vickers, 26
  tension, 15, 17, 20, 30
  torsion, 24, 30
    compression, 25
    tension, 25
  wear, 41
Texture, 46
  control, 122
  parameter, 463
Theory of friction:
  adhesion, 347
  asperity interaction, 347, 349
  junction-growth, 347, 348
  molecular, 347, 350
Thermal covers, 420, 555
  insulating, 429
  reflecting, 429
  reradiating, 429, 563
Thermomechanical treatment:
  classification of, 454
  during rolling, 457
  facilities, 476
  of steel, 454
Thickness of rolled product:
  average, 229
  mean, 229
  measurement of, 591
Threading method:
  conventional, 558
  zigzag, 558
Time:
  air cooling, 399
  constant, 591
  delay, 450
  heating, 402, 403
  holding, 449, 467
  mold cooling, 399
  transit, 402
Time-Temperature-Transformation
    (T.T.T.):
  curves, 85-87
  diagram, 455, 456
Titanium, 10
Tolerances of rolled product:
  absolute, 579
  camber, 585
  flatness, 586, 587
  geometrical, 581
  relative, 579
  thickness, 582, 583
  width, 584
Torque, 24
  driving, 318
  formulae for, 299, 307, 309, 312,
    313
  geometrical factor, 299
  in heavy draft rolling, 319
  method of calculation of, 334,

336, 337
    roll, 298
Toughness:
  definition of, 30
  of metal, 17, 46
  effect of metallurgical factors
    on, 118
Tramps, 146
Transducer:
  position, 607
  pressure, 607
Transfer, function 648
Tresca yield criterion, 217
Tribology, 343
Tundish, 170-172
Tungsten, 31, 101

U

Uncoiler, 523, 525
Unified Numbering System (UNS), 43
UNS primary series, 48
UNS secondary division, 49

V

Valley, 574, 575
Vanadium, 101
Velocity:
  discontinuity, 248, 250
  hodograph, 247
  in slip-line field, 258
Von Mises yield criterion, 217

W

Warping, 223
Wear:
  abrasive, 364, 391
  adhesive, 364, 391
  chemical, 364
  classification of, 364
  corrosive, 391
  fatique, 364
  model, 351
  localized, 383
  roll, 382, 391
  uniform, 383
Weldability, 121
Wedge, 573
Width:
  arithmetic average, 271
  change, 685
  control, 173, 701-703
  enlargement, 699
      by spreading, 700
  geometric mean, 271

parabolic mean, 271
reduction, 678, 686, 687
Width of rolled product:
measurement of, 609-611
Width spread:
coefficient, 669
formula for, 670-672
lateral, 669
Work, in rolling 318
Work-hardening:
curves, 325
Wustite, 483, 484

Y

Yield:
criteria, 217
criterion:
distortion energy, 217-219
maximum-shear-stress, 217,
218
point, 15, 16
requirements, 552
strength, 15, 237, 450
at elevated temperature,
45-48
effect of columbium on,
113
effect of manganese on,
113
effect of nitrogen and
vanadium on, 114
effect of titanium on, 115
effect of vanadium on, 113
offset, 32

Z

Zirconium, 103